SAP PRESS e-books

Print or e-book, Kindle or iPad, workplace or airplane: Choose where and how to read your SAP PRESS books! You can now get all our titles as e-books, too:

- By download and online access
- For all popular devices
- And, of course, DRM-free

Convinced? Then go to www.sap-press.com and get your e-book today.

SAP® on AWS®

SAP PRESS

Bardhan, Baumgartl, Choi, Dudgeon, Górecki, Lahiri, Meijerink, Worsley-Tonks
SAP S/4HANA: An Introduction (4th Edition)
2021, 648 pages, hardcover and e-book
www.sap-press.com/5232

Densborn, Finkbohner, Höft, Rubarth, Klöß, Mathäß
Migrating to SAP S/4HANA: Operating Models, Migration Scenarios,
Tools, and Implementation (3rd Edition)
2024, 633 pages, hardcover and e-book
www.sap-press.com/5816

Finkbohner, Höft, Roth, Kinold, Kuchelmeister, Widera
Data Migration for SAP: SAP S/4HANA and Cloud Solutions
2024, 568 pages, hardcover and e-book
www.sap-press.com/5817

Saueressig, Stein, Boeder, Kleis
SAP S/4HANA Architecture (2nd Edition)
2023, 544 pages, hardcover and e-book
www.sap-press.com/5675

Banda, Chandra, Gooi
SAP Business Technology Platform: An Introduction
2022, 570 pages, hardcover and e-book
www.sap-press.com/5440

Ravi Kashyap, Rajendra Narikimelli, Rozal Singh

SAP® on AWS®

Architecture, Migration, and Operation

Rheinwerk
Publishing

Editor Rachel Gibson
Acquisitions Editor Hareem Shafi
Copyeditor Yvette Chin
Cover Design Graham Geary
Photo Credit Shutterstock: 2089684921/© FrentaN
Layout Design Vera Brauner
Production Graham Geary
Typesetting III-Satz, Germany
Printed and bound in the United States, on paper from sustainable sources

ISBN 978-1-4932-2490-6
© 2024 by Rheinwerk Publishing, Inc., Boston (MA)
1st edition 2024

Library of Congress Cataloging-in-Publication Control Number: 2024004033

Contents at a Glance

Contents

3   Framework for Designing Solutions on AWS   103

4   AWS Landing Zones   127

8 Automation Tools and Technologies — 393

9 Multicloud Considerations — 429

10 Transformation

447

11 What's Next?

485

Preface

"I just got involved with my company's SAP migration to AWS initiative; what documents can I refer to learn more?" This question came from a coworker, and after some intense googling, we compiled a list of links that we thought would be helpful. Now, if you've ever tried finding information online, you know that there is a lot of it and often times, it's missing the key ingredient of a learning process—the sequence in which information is presented. For example, you can't comprehend high availability (HA) for SAP unless you know what availability zones (AZs) are and, perhaps more importantly, why HA is important in the first place.

This book addresses these problems in the following ways:

- By discussing how to start with business focused view and use technology as an enabler.

- By providing context into why and how and not just what, for example, why organizations need HA and how to achieve HA in Amazon Web Services (AWS) according to your organization's uptime requirements rather than just saying, "Here's how you do HA for SAP."

- By presenting the information in the sequence that best enables learning the topics covered. We also use anecdotes and metaphors from daily life and on-premise deployment terms to help relate to your current understanding.

Objective of This Book

The key objective of this book is to answer the question we discussed earlier with "Start with this book." Rather than jumping straight into technical configuration steps, we first want to discuss why organizations are moving to the cloud and how AWS specifically enables the flexibility, availability, scalability, and other business needs required for SAP.

Throughout this book, our goal is to relate AWS concepts back to familiar on-premise terminology as a bridge in understanding. Through anecdotes and metaphors, we hope not just to explain which options to choose, but why cloud capabilities matter for business objectives. Whether you're an executive seeking context, an architect exploring offerings, or an admin tasked with cloud deployments, this book holds relevant insights for your organization's SAP deployments in AWS. We also think this book will be a handy reference guide after you learn the initial concepts and explore other learning methods such as online documentation or videos.

> **Note**
>
> The authors represent their own thoughts, points of view, and experience and not that of their employers.

Target Audience

This book is meticulously crafted for a specific group of professionals, primarily SAP architects. Designed to cater to the unique needs and challenges faced by SAP technology architects, this comprehensive guide goes beyond the conventional and delves into the intricacies of SAP on AWS including RISE with SAP.

Ideal for individuals who possess a deep understanding of SAP but who may not be well versed in the intricacies of Amazon Web Services (AWS), or conversely, those with a solid foundation in AWS but seek to enhance their proficiency in SAP, this book bridges the gap between these two powerful technologies.

Whether you're a seasoned SAP technology architect seeking to integrate AWS into your skill set or an experienced AWS professional looking to seamlessly incorporate SAP into your repertoire, this book is tailored to meet your requirements. Our intended readership encompasses not only SAP architects but also cloud architects and team members actively involved in the installation, operation, and maintenance of SAP systems on the AWS platform. By addressing the specific needs of this diverse audience, we hope this book serves as a valuable resource that empowers individuals to harness the full potential of SAP and AWS in tandem.

Structure of This Book

This book is arranged in 11 chapters. The following list provides an overview of the topics covered in each chapter:

- **Chapter 1**
  This chapter introduces the available cloud services offerings, highlighting their role beyond traditional data centers and focusing on economies of scale and automation. We'll guide you in converting your business requirements into cloud specifications, covering foundational aspects such as reliability, automation capabilities, compliance, security, cost, performance, innovation, integration, and sustainability. This chapter serves as a comprehensive overview for understanding and leveraging cloud services to meet your diverse business needs.

- **Chapter 2**
  This chapter details various cloud deployment models for SAP systems on AWS, including infrastructure as a service (IaaS), platform as a service (PaaS), software as a

service (SaaS), RISE with SAP, and the SAP Business Technology Platform (SAP BTP). We explore how AWS services integrate into SAP deployments and operations, focusing on the initial step of connecting corporate data centers to AWS. This chapter encompasses AWS Direct Connect, virtual private networks (VPNs), and AWS Transit Gateway. This chapter also examines AWS services pertinent to SAP, covering aspects like global infrastructures, regions, compute resources, storage, networking, key management, databases, identity, security, and management governance, thus providing a comprehensive guide to SAP on AWS.

- **Chapter 3**
This chapter focuses on architecting SAP applications in AWS, utilizing the AWS Well-Architected Framework. This framework comprises six pillars—reliability, security, performance efficiency, operational excellence, cost optimization, and sustainability—to guide cloud architects in creating optimal cloud infrastructures. Additionally, this chapter contrasts business service level agreements (SLAs) with AWS SLAs, offering insights into how these SLAs should be considered in your SAP solution architecture. We include a look at AWS SLAs, the concept of a composite SLA, and balancing business uptimes with AWS infrastructure SLAs.

- **Chapter 4**
This chapter shows you how to establish an initial baseline infrastructure on AWS for SAP deployments. We cover tools and strategies for setting up a landing zone architecture, utilizing AWS Service Catalog, AWS Control Tower, and AWS Organizations. Additionally, this chapter delves into common patterns for ensuring HA and effective disaster recovery (DR), including resilient architectural patterns that range from HA setups to DR strategies and considerations for single-region and multiregion deployments.

- **Chapter 5**
This chapter provides reference architectures for various operating system and database (OS/DB) combinations for SAP deployments on AWS. We cover support for SAP and related products, detailing distributed versus standard deployment options and the SAP application and technology stacks. This chapter also discusses application- and database-specific architectures for SAP HANA, Microsoft SQL Server, Oracle, SAP ASE, SAP MaxDB, and IBM Db2. Additionally, we explore architectural trade-offs based on cost, availability, and performance and apply the AWS Well-Architected Framework Lens specifically to SAP.

- **Chapter 6**
This chapter delves into the methodologies, approaches, and tools for migrating SAP applications to AWS. We begin with current state discovery, focusing on application discovery, sizing, dependencies, and priorities. This chapter then outlines migration frameworks, including the 6 Rs strategy and the AWS Cloud Adoption Framework (CAF), which encompasses people, processes, and technology. Various SAP migration approaches are discussed, such as standard, downtime-optimized, and large

database migrations. We also review migration tools and services from SAP, AWS, and partners. Additionally, this chapter covers project and portfolio management, highlighting the cloud center of expertise (CCoE) and cloud financial management.

- **Chapter 7**

This chapter details the operation of SAP systems after migrating to the cloud, emphasizing how cloud flexibility can optimize operations and reduce costs. We explore the trade-offs when optimizing SAP landscapes and underscores the importance of adopting cloud innovations. Key topics include observability, incident response and recovery, SAP-specific systems administration, change management, and restore and recovery processes. Furthermore, this chapter zeroes in on cost optimization by discussing cost visualization and continuous optimization opportunities.

- **Chapter 8**

This chapter focuses on automation methods, tools, and technologies for SAP deployment and operations on AWS. We cover AWS automation tools including the AWS Command Line Interface (CLI), AWS Launch Wizard, and AWS Migration Hub Orchestrator. The chapter also delves into open-source tools such as Terraform, Ansible, and various automation templates. Additionally, we discuss the AWS software development kit (SDK) for ABAP, provide SAP on AWS code samples, and explore the role of DevOps in SAP on AWS environments.

- **Chapter 9**

This chapter examines some use cases and the operational factors for organizations considering multicloud deployments with SAP. We address the common challenges and potential benefits of a multicloud strategy, highlighting issues such as vendor lock-in, skill requirements, and portability. Topics include how SAP customers might end up using multiple clouds, operational and management considerations, observability, data synchronization, hybrid and edge scenarios, and cost implications. This chapter also discusses the skills needed for multicloud environments, the use of open-source tools to avoid vendor lock-in, and situations where a multicloud approach may not be effective.

- **Chapter 10**

This chapter covers transforming SAP workloads using architectures, cloud-native services, and best practices, enhanced by our customer experiences and common challenges. Topics include event-driven architectures with event-based triggers, rule-based orchestration, pub-sub notifications, and loosely coupled processing. We explore serverless application design, emphasizing cloud application gateways, scaling consumption models, and building without infrastructure. This chapter also delves into application extensibility with cloud-native service integrations, modern data architectures for SAP workloads, and multicloud integration patterns as we discuss building microservices, extension strategies, container technologies, cloud data lakes, data integration patterns, cloud database models, SAP Integration Solution Advisory, and consistent operations across multiple clouds.

- **Chapter 11**
  This chapter presents a vision for SAP workloads on AWS, focusing on their integration with cloud-native services to enhance capabilities and achieve cloud maturity. Topics include SAP modernization, self-service IT with analytics and citizen developer initiatives, and artificial intelligence (AI)/machine learning-powered SAP value chains featuring process automation and business process extensions. We also cover autonomous operations and security, including automated incident response, self-healing architecture, and AIOps. Additionally, this chapter discusses model-driven architectures and the role of low-code/no-code development in this context.

Throughout the book, we've also provided several elements that will help you access useful information:

Tips and Tricks [+]

Boxes with this symbol provide you with recommendations as to how you can simplify your work.

Notes [«]

Boxes marked with this symbol contain additional information or important content that you should keep in mind.

Examples [Ex]

Boxes marked with this symbol provide practical scenarios and explain, in detail, how particular functions can be applied.

Warnings [!]

Boxes with this symbol contain details worth considering. Moreover, it warns you of common errors or problems that might occur.

Acknowledgments

The saying "it takes a village" is as true for writing a book as it is for raising kids!

To start with, I would like to thank my co-authors, Raj and Rozal, without whom the project would never have taken off, much less be completed. Thank you for your collaboration, insightful ideas, and contributions throughout the writing and editing process. Your expertise and dedication have been invaluable, and I am honored to have had the opportunity to work with you.

Once we decided to proceed with the project and reached out to Hareem, our acquisitions editor from SAP PRESS, she shared our enthusiasm and helped us think through all the details. Thank you for exploring our idea and your guidance in shaping the book. Our editor Rachel provided us thoughtful feedback and helped my metaphors make sense; without her help, only people deep in SAP and AWS ecosystem would be able to make sense of it, which beats the whole purpose of the book. Our copyeditor, production team, and other SAP PRESS folks that we didn't directly interact with have been of great help as well throughout the editing and publishing process!

Finally, I couldn't do any of this without the support of my lovely wife Katie and my kids Ellie and Neil, who asked why I am writing a book and whether they are in it. (And now they are!)

—Ravi Kashyap

I am profoundly grateful to my parents who, despite their own lack of formal education, instilled in me the value of learning. Their steadfast support provided me with the opportunities to pursue my academic and professional aspirations.

To my dear wife, your constant belief in my abilities and your selfless support have been essential in making this journey possible.

My sincere thanks go to my co-authors, Rozal and Ravi. Their expertise, camaraderie, and shared passion for our subject have greatly enhanced the journey of writing this book.

Lastly, I express my deep gratitude to our editors, Hareem Shafi and Rachel Gibson. Their editorial expertise, guidance, and support have been crucial in shaping this book.

—Rajendra Narikimelli

Writing this book has been a journey fueled by a profound passion for spreading knowledge and a commitment to bridging the gap between AWS and SAP. The idea was born out of a genuine desire to contribute to the collective understanding of these transformative technologies.

I owe a debt of gratitude to two individuals without whom this endeavor would never have come to fruition. Rajendra and Ravi, your unwavering support, guidance, and collaboration have been invaluable. Ravi, drawing from your rich experience in authoring *SAP on Microsoft Azure*, you brought a wealth of insights that enriched the content of this book. Special thanks to you for connecting us with the exceptional team at SAP PRESS—Hareem and Rachel. Hareem, your efficiency in navigating the approval process with precision and timeliness ensured our seamless collaboration.

Rachel, your editorial prowess, marked by an impeccable attention to detail, elevated the quality of the manuscript. The SAP PRESS team has indeed been exceptional, turning what could have been a challenging journey into a remarkably smooth ride.

Lastly, I want to dedicate this book to my pillars of strength—my daughter Inaayat and my wife Shivali. Your motivation has been the driving force behind this endeavor. This book is a testament to our collective journey, and I am grateful for the inspiration you provide every day. Thank you all for being part of this incredible experience.

—*Rozal Singh*

Chapter 1
Cloud Providers and Business Requirements

In the last fifteen years or so, cloud service has gone from an obscure notion to supporting some of the largest businesses in the world! SAP and cloud providers have been working together to bring business-critical applications to the cloud, which has been a great step forward in enabling organizations to focus on business models and customers rather than worrying about the reliability of IT infrastructures. Even though moving to the cloud is often thought of as a technological decision, when used wisely, cloud solutions can help your organization innovate and create new business models as well.

Surveys conducted by the Americas' SAP Users' Group (ASUG) consistently show organizations evaluating and moving SAP systems to cloud providers. If this describes you and you are either evaluating a cloud provider or trying to learn about how the cloud works for SAP systems, you're at the right place.

In this chapter, we'll discuss what makes public clouds versatile and, at the same time, robust enough to handle mission-critical workloads such as SAP. We'll also highlight a top-down approach to identify your business requirements and solve them in the cloud rather than learning about cloud features first and then mapping these features to your organization's requirements.

1.1 What Are Cloud Providers?

Just as electricity is delivered to your home via an energy grid, cloud computing provides access to computing resources and services via the internet. With the flip of a switch, electricity is available on demand to power your appliances and devices. Similarly, the cloud allows you to access computing power whenever needed through various services; several configurations are available for resiliency, performance, and scalability, that you can control and architect for applications such as SAP.

> **Public Cloud and SAP Data**
>
> The basic premise of a public cloud is that computing resources are available over internet; however, this availability doesn't mean that SAP data is available over internet. As you'll learn in Chapter 2 through Chapter 4, several ways exist to restrict how, where, and who can access the data.
>
> Other common features of a public cloud are its on-demand availability of resources, its pay-as-you-go pricing model, and its abstraction of physical infrastructures. We'll explore those topics in this book and discuss how these theme fits into the larger "SAP-on-the-cloud" story.

Several cloud providers support SAP, the largest among those are: Amazon Web Services (AWS), Google Cloud (GCP), and Microsoft Azure. We won't go into much detail about how to choose among these cloud providers since the focus of this book is AWS. Thus, we highlight AWS recommendations for SAP deployments. In this section, we'll discuss a few salient features of the cloud, such as its global scale and automation capabilities, which provide insights and context into why the model has been so successful.

1.1.1 Beyond Data Centers

We often hear the classification of a cloud as "someone else's data center," which, in our experience, is the same as comparing a home office setup to an office building—it's not a fair comparison.

We like to think of the cloud as globally connected services that work together to run applications, such as SAP, and to manage data at scale. Understanding how these services are designed, their interdependencies, and fault isolation are crucial aspects of designing a solution. Let's start with some fundamental concepts.

Control and Data Planes

Cloud operations in general can be divided in terms of a control plane and a data plane, in the following ways:

- **Control plane**
  This plane is used to manage cloud resources—such as create, read, update, and delete (CRUD) operations—through application programming interfaces (APIs). An example of an activity on the control plane is creating storage or creating a virtual machine (VM), for instance, Amazon Elastic Compute Cloud (EC2).

- **Data plane**
  This plane provisions the main capabilities of a service, for example, read or write operations to storage, interacting with VMs, and database queries.

Control Plane and Service Availability

Think of the control plane as an orchestration engine that lets you start with the services that you'd like to use for any application. Also, since the control and data planes are independent of each other (although working together), service availability is improved. If you follow service interruption notifications, at times you might find creating new resources are impacted, but using existing resources are not. If so, a problem is probably impacting the control plane and not the data plane, thus illustrating the independence of these planes and the benefit to availability.

Service Boundaries and Fault Isolation

Fault tolerance is a large part of how cloud services are designed. In terms of boundaries, AWS services fall under following three categories:

- **Zonal**
  This kind of service exists and operates in a particular availability zone (AZ), which is a collection of data centers with independent power and cooling. We'll cover AZs in detail in Chapter 2, but for now, you should consider that an Amazon EC2 instance is a zonal service that operates independently. This independence in operation guards against failure. If an Amazon EC2 instance in an AZ fails, other zones will not be affected, unless there is a regional outage. We hope you see how this isolation is useful when designing applications to make a system fault tolerant. Zonal services have *zonal data planes.*

- **Regional**
  This type of service is deployed across AZs in a particular region, such as AWS us-east-1 (N. Virginia) and uses regional endpoint for interaction; it also uses AZ independence for data protection and availability. Amazon Simple Queue Service (SQS) and Amazon DynamoDB are examples of regional services. Regional control planes, in addition to providing regional services, also aggregate zonal control planes.

- **Global**
  This kind service doesn't have a region-specific resource. In this case, the data plane and the control plane for global service don't exist in each region. However, the control plane is hosted in a region while the data plane is globally distributed. An example would be AWS Identity and Access Management (IAM).

Service Boundaries

You can often deduce whether a service is regional versus zonal from the available creation options. When creating an Amazon EC2 instance, you'll be asked for an AZ (making Amazon EC2 a zonal service) while Amazon SQS has no such option to choose an AZ.

1.1.2   Economies of Scale

If you've considered digging a well and compared this cost and effort with just using the city water service, you'll probably conclude that city water service is the way to go. You'll have a similar *aha moment* with cloud services. Cloud providers often have both internal economies of scale (e.g., developing proprietary software or hardware) and external (e.g., taking advantage of cheaper hardware or extending its useful life) economies of scale and pass the cost savings on to their customers.

Let's consider a few other sources of scale that are less talked about, such as the following:

- **Technology**
  Scale from commodity hardware is only part of the story and often not a differentiator by definition. What makes a cloud provider unique is the type of technology they invest in and how they apply that on top of commodity hardware. For example, AWS has invested in AWS Nitro systems and AWS Graviton processors. Also, the amount of investment made by cloud providers in security is often far beyond other organizations' abilities, perhaps even your own.

- **Organization culture**
  Culture and scale is little difficult to correlate, but Amazon is known for long-term thinking and believing in big ideas. Some of these ideas scale well and contribute to better technology platform for cloud.

- **Personnel**
  Cloud companies hire the most skilled folks and provide them independence to do their best work. We often hear of companies coming up with great ideas, but at the micro level, it's the employees that make the difference. Smaller ideas that increase productivity, at scale, might greatly impact the overall reliability of a cloud platform, which in turn benefits customers.

1.1.3   Automation

Our vision of the cloud would not be complete without automation. Less than a decade ago, the lead time to provision the infrastructure and install SAP was *weeks*; automation has fast tracked deployment to *hours*.

Infrastructure as code (IaC), combined with tools from SAP and cloud providers, has made deploying and managing SAP landscapes easier and faster, which translates into greater cost savings and employee effectiveness. We'll cover cloud automation in detail in Chapter 8, but a few examples of automation use cases in SAP lifecycle management include the following:

- Infrastructure provisioning and SAP systems installation
- Automated provisioning or deprovisioning based on schedule or by utilization
- Steps for an SAP migration that supports an unattended path

- SAP maintenance and infrastructure patching
- SAP system copies

Automation has made it possible to create *immutable architectures*, where servers are never modified or patched after their initial deployments, but are instead replaced by another server, as well.

We have seen folks doing total cost of ownership (TCO) calculations for each infrastructure line item of on-premise list mapped to cloud service; while an exercise like that is very useful, those calculations often don't take automation aspects into account. Cloud TCO exercise is not comprehensive unless you account for efficiencies provided by automation and translate that into some dollar figure. Efficiency from automation can be tangible (e.g., saving few hours' worth of effort during deployment) or intangible (e.g., preventing a mistake or misconfiguration that would have caused application downtime and impacted a business-critical process).

The Cloud in Two Words

If asked to define cloud in just two words, we would say *scale* and *automation*.

1.2   Translating Business Use Cases into Cloud Solutions

Given that SAP migration projects can span anywhere between 6 months and 2 years, you must get the business context and requirements right, or you risk pushing the timeline. Even worse is the chance of suffering higher business downtime. Start with identifying the right stakeholders and gather more details about their most common activities and problems, as we discuss in this section. The idea is to start with the right thing to do for the business and then explore solutions based on that goal rather than get lost in the technical advancements that the cloud has made over the years!

Let's walk through some example conversations illustrating what we mean by *understanding requirements*:

- **Trying to design solution without requirement**

  Person A: What kind of high availability (HA) does cloud offer?

  Person B: What uptime requirement does the business have for the SAP system?

  Person A: We are not sure, but asking business folks, we will probably hear 100% uptime.

  Person B: Trying to design an HA solution in cloud without uptime information will cost more and still not provide 100% availability.

- **Exploring a solution when you know what you are looking for**

  Person A: We are a global business and perform a disaster recovery (DR) test every year. What would it look like in a cloud?

Person B: DR test and drills are recommended and compared to on-premise. They're fairly easy to execute in the cloud and can be done in isolation from your primary environment; before we get into details about execution, what are the recovery time objective (RTO) and recovery point objective (RPO) for the systems?

Person A: There are different tiers depending on what the system is used for, but it's 15 mins RPO and 1 hour RTO for the SAP S/4HANA system.

Person B: Yes, we can work with that; let's explore a design for your systems based on multi-region DR architectural pattern.

Note how requirements should drive the solution—not the other way around! We'll explore some common requirement areas in the following sections.

1.2.1   Reliability

One of the top three triggers for an organization to look into moving to the cloud, reliability is most commonly a combination of availability, performance, and resiliency. Before digging into cloud service provider reliability metrics and their recommendations, you should review your organization's targets.

Table 1.1 identifies some commonly used metrics that you should consider in your evaluation. Ask yourself the following questions:

- What are the uptime requirements for your SAP systems?
- Do you have different uptime requirements for different systems or tiers of systems?
- What are current performance metrics being measured, and how do you measure them?
- How does your current architecture meet your uptime requirements?
- What are your organization's business continuity plans?
- What are the RTO and RPO targets for critical systems?
- When a failure happens, what is the mean time to recover (MTTR)?
- What are your backup and restore requirements?
- Does the backup need to be saved offsite and how often?

Event/Configuration	Metric
HA	Uptime
DR	RTO and RPO
Availability incident	MTTR

Table 1.1 Metrics to Look at the Different Reliability Events

HA versus DR Metrics

HA and DR often get mixed up when talking about reliability and business continuity. When you think HA, the uptime requirement is the key metric, while with DR, you only need to pay attention to RTO and RPO. We often hear folks saying they want the system to be highly available and then spell out the exact RTO/RPO metrics to meet. Be aware of this trap!

For SAP end users, a reliable application can be accessed when they need it and operates with acceptable performance. What happens behind the scenes is where the architecture comes into play; a resilient architecture requires a holistic view of the application along with options from the cloud service provider.

Figure 1.1 shows the building blocks to create a more resilient, and thus reliable, SAP architecture on the cloud. Among these three aspects of resiliency, application design and resiliency configuration activities fall under your responsibility, and platform availability falls under your cloud provider's responsibility.

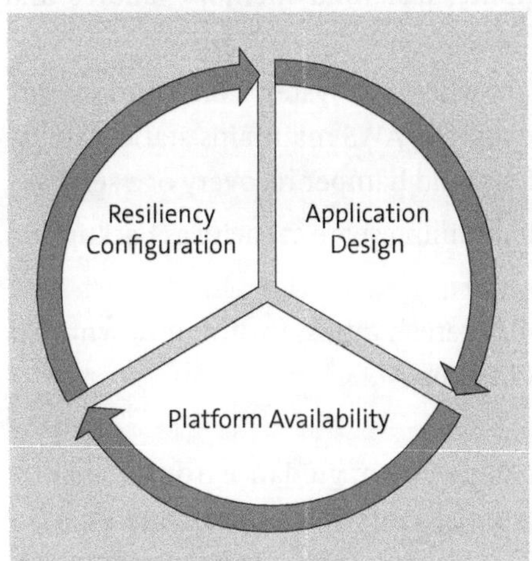

Figure 1.1 Building Blocks for Reliability in Cloud

Let's look at these building blocks in a little more detail:

- **Application design**

  Application design refers to how an application responds to failure and the resiliency options provided by the application itself. In the context of ABAP stack applications, ABAP SAP Central Services failover using the new standalone enqueue replicator 2 is an example of how design of an application plays an important role in reliability.

  In general, mitigating all single point of failures, such as ABAP SAP Central Services (ASCS), database, application servers, along with business continuity and disaster recovery (BCDR) planning are some things you should consider from an application perspective. Other factors that affect application reliability include:

- Other dependent applications or components such as SAP Business Warehouse (SAP BW) reporting not showing correct data without SAP ERP availability
- Operational procedures of your organization
- Sizing and scaling of the application

- **Platform availability**
  Cloud providers themselves are committed to platform availability (also called the "availability of the cloud"), and in this area, you have the least control. Availability is governed by the service level agreement (SLA) provided for corresponding services (more details in Chapter 3), and you can decide how to use a particular service or combination of services to match your application availability requirements. Some common methods used by cloud providers for enhanced availability include the following:

  - Geographical separation: Distributing services in different zones and regions
  - Predictive maintenance: Predicting compute, disk, and memory failures and replacing before they occur to prevent outages
  - Static stability: Maintaining a design pattern where the system continues to work even with an impaired dependency. For example, AWS maintains static stability by preventing circular dependencies, which could hamper recovery of a service
  - Hardware redundancy and replication: Maintaining spare capacity as backup and replicating data to maintain durability
  - Live migration: Migrating resources to different hardware without downtimes either during maintenance or as a result of failure

- **Resiliency configuration**
  Resiliency configuration relates to following deployment guidance from cloud providers about how to map application *design* components to the infrastructure.

  This view considers platform availability, service SLAs, and any application certification requirements. We'll discuss deployment architectures in Chapter 4 in more detail, but some common examples of resiliency configurations include the following:

  - Using an infrastructure that is certified to run SAP
  - Multi-AZ configuration for single point of failure components, such as SAP Central Services
  - Load balancers in front of web components
  - Multi-region deployments for DR
  - Backup replication to different zones or regions

1.2.2   Performance

Application performance has a broad definition. Performance could be defined as users being able to access the application efficiently with acceptable latency. Making data-driven decision is especially important when assessing and improving performance. In our experience, if an application performs poorly, some organizations consider this poor performance as an availability event. After all, often, the end-user impact is the similar for either a severely degraded or entirely unavailable application.

Before you start looking at options and configurations to optimize throughput, input/output operations per second (IOPS), and response times, you should collect data about the performance metrics your organization already uses and evaluate how your current system stacks against those metrics. Some data that could be helpful to gather include the following:

- Latency to end users
- Execution times of all business-critical processes, such as month-end close
- SAP dialog response times
- Background job runtime and resource utilization, specifically for long running jobs (e.g., archiving)
- Database response times
- Database backup and restore times
- Inter-system connectivity and data loads (such as SAP ERP to SAP BW)
- Interface and connectivity latency among perimeter applications
- SAP HANA load times to memory

Application efficiency and performance evaluation is an iterative process, and sustaining good performance in the long run requires periodic review and new choices to be made. Figure 1.2 shows the building blocks to design a high-performance SAP landscape in the cloud.

These building blocks include the following:

- **Observability of systems**
  Observability is the next step in the evolution of monitoring, which according to Information Technology Infrastructure Library (ITIL) is reactive rather than proactive since the incident has already occurred. Observability is about how well you understand the state of your whole system (not just its applications or just its infrastructure).

  How does this topic relate to designing a system? The better you understand the system, the more data you have for performance-related metrics, and the more opportunities you have to optimize your target landscape for performance and efficiency.

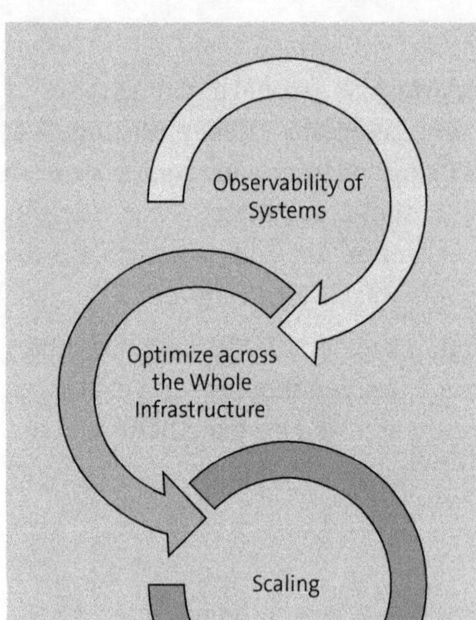

Figure 1.2 Building Blocks of a High-Performance Architecture

- **Optimize across the infrastructure**

  Quite often, SAP (application) folks try to optimize SAP parameters while infrastructure folks look at hardware-related metrics for performance issues; however, everyone is working in their own silos, often without safe ways to experiment.

  Looking at the bigger picture *across application, compute, storage, and network* and testing different configurations is the best way to observe the effect of a change across the board. Consider the following examples:

  - When evaluating SAP HANA performance, also look at the storage configuration that is certified by SAP. Often, certain throughputs and IOPS are required for the certification, and using that configuration is required to be supported by SAP.

  - When evaluating which AZs to pick for an HA SAP configuration, some level of latency testing would probably show that not all zones are created equally. This example illustrates how choosing the right infrastructure eventually benefits application performance.

  - Performance is not isolated from cost discussion; there is always a trade-off. As a result, you'll hear phrases such as *price-performance ratio*.

  Performance optimization is an iterative and ongoing process. Therefore, every time you do an application or infrastructure change, always review whether any performance-related metric has also changed. Some organizations have built this activity into their operational procedures as well.

- **Scaling**
  Scaling (either up or down) is an important aspect of performance management of an application. We already recommend not to size the SAP system for the next 3 or 5 years like it's done on-premise. Let the cloud's inherent elasticity help you as you grow!

> **Autoscaling**
>
> You may have heard of autoscaling in the cloud. Platform as a service (PaaS) solutions often have scaling built in while infrastructure as a service (IaaS) requires you to build automation solution. A large part of SAP on the cloud is IaaS, so an automation solution for scaling is required.

Use the framework we've described in this section to start gathering performance requirements. We'll discuss AWS recommendations for SAP landscapes in more detail in Chapter 4, Chapter 5, and Chapter 6.

1.2.3 Automation Capabilities

Automation is at the heart of cloud—without automation the cloud business model would not be successful. Cloud providers have all sorts of automation capabilities, from infrastructure and SAP provisioning to system copying. For use cases that lack existing tools, you may be able build a solution yourself or partner with a system integrator.

A good step in this direction is to look at all your current tasks and divide them into one-time activities versus recurring activities. Then, you can prioritize the recurring tasks in your automation decision. Table 1.2 shows some common SAP use cases that you can use as a starting point.

Activity	Frequency
SAP system installation or migration	As needed
SAP system copy, clone, or refresh	Monthly/quarterly
DR test/drill	Yearly
SAP archiving	Monthly
SAP maintenance (including start/stop)	Monthly
OS patching/upgrades	Monthly/quarterly/yearly
SAP parameter optimization	Monthly/quarterly
SAP/OS password changes	Quarterly
SAP kernel update/host agent patching	Quarterly

Table 1.2 Examples of SAP Use Cases and Their Typical Frequencies

Activity	Frequency
Cloud systems snoozing	Daily
SAP jobs maintenance/reporting	Daily
HA/DR configuration	As needed
Any other business use case	As needed

Table 1.2 Examples of SAP Use Cases and Their Typical Frequencies (Cont.)

Automation can benefit an organization in different ways. As shown in Figure 1.3, you can use certain *lenses* when evaluating an automation solution or tool.

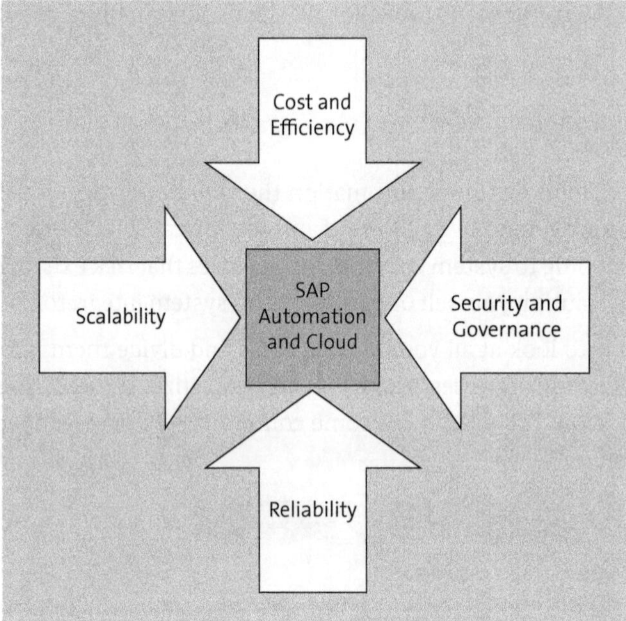

Figure 1.3 Automation through Different Lenses

These lenses include the following:

- **Cost and efficiency**
  Lower cost and increased team efficiency are the most commonly cited benefits for using an automation solution, but in our experience, this advantage is not realized immediately. Figure 1.4 shows how initial cost are higher and efficiency is lower, both of which improve with the introduction of automation and can reach a minimum efficient scale at which point long-run costs are minimized. If you keep trying to increase efficiency by automating non-recurring smaller tasks, the cost advantage may not be sustained. Therefore, to get the most out of investment in automation, we recommend focusing on larger tasks and recurring tasks.

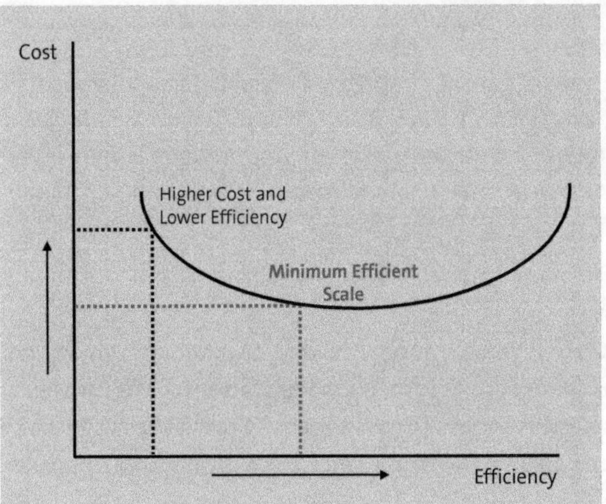

Figure 1.4 Cost and Efficiency Relationship with Automation

- **Security and governance**
  Centralized management of activities using automation improves governance, and you can apply that at scale for all existing deployments as well as new deployments. Similarly, elevated access can be controlled, and security aspects are integrated rather than going through a list of security controls to ensure all were met. Security and governance are not only part of deployment initially, but these can be easily audited (internally or externally).

- **Reliability**
  Earlier in Section 1.2.1, we discussed reliability in the context of the target architecture; still other aspects of reliability include *avoiding mistakes* and *drift detection*.

  The cost of unplanned downtime for business-critical applications, such as SAP, have been estimated from \$5,600 to \$10,000 per minute (see *http://s-prs.co/v5776137*), and given that a typical SAP landscape comprises of hundreds of VMs, mistakes such as shutting down the wrong system are not unheard of. Similarly, with the cloud, detecting any drift from the initial configuration is easy to configure, which also ensures that production system configurations are same across the board. Therefore, for example, the difference of a small patch or parameter doesn't break a critical business process.

- **Scalability**
  Scalability touches everything in the cloud, and automation adds multifold advantages there. Statistics such as the number of servers managed by an operating system (OS) admin or team sizes for SAP administration have evolved and consider the efficiency provided by the cloud and by automation. We could even argue that, given all the capabilities provided by cloud and managing DevOps in a large environment, scaling would be impossible without the right automation tools and technologies.

> **[+] Automation Solution Evaluation**
>
> If an automation solution meets one of the lenses discussed earlier but conflicts with another, this solution is not a good long-term solution. For example, if you use a tool to provision infrastructure resources but it doesn't meet all your governance requirements, any re-work negates whatever benefit it had in the first place.

1.2.4   Compliance and Security

Global businesses often deal with several compliance and regulatory frameworks, including region-specific such as General Data Protection Regulation (GDPR), as well as data sovereignty requirements. Gathering this requirement is probably easier since details are already well known and usually the compliance team is on top of things.

Security relates to risk reduction as part of your business requirements. Specifically, security practices protect organization and customer data. Security configurations vary across identity, encryption, network, data logging, and retention. Some questions you should ask to understand compliance and security requirements include the following:

- What specific compliance certifications are required?
- What type of operational compliance is needed?
- How often do audits happen, and what areas are covered?
- What is the specific data sovereignty for each system, and how does that impact any cross-region replication of data?
- What type of network security controls are required, and are the requirements different for cloud providers versus on-premise?
- What type of data encryption for data at rest, data in transit, and data in use is needed?
- How are access controls (identity) governed, and how is role-based access control managed?
- What type of logging is required, and what are the retention periods?
- What are the security requirements in the application (SAP) layer?

In the cloud, you'll often find out-of-the-box options or more visibility across all layers, which is a transparency generally lacking in on-premise environments. Figure 1.5 shows the building blocks for compliance and security in cloud environments.

These building blocks include the following:

- **Understand shared responsibility model**
  Security is everyone's responsibility, and understanding the shared responsibility model—you must perform some tasks and your cloud provider must perform

others—is the first step towards securing your infrastructure and applications with any cloud provider.

As a general rule, the *security of the cloud* is the cloud provider's responsibility; this area includes things such as protecting the hardware and software that runs the cloud-based services. *Security in the cloud* is your responsibility, as the customer, and this area includes items such as applying encryption, network security controls, configuring OS hardening, managing data, and configuring access controls.

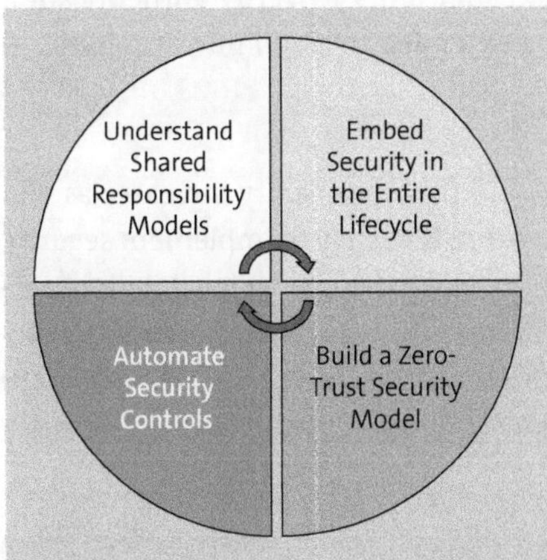

Figure 1.5 Building Blocks of Compliance and Security

Tools for Security in the Cloud

Even for security in the cloud, which is your responsibility as the customer, cloud providers often make encryption or network security tools available. You can bring your own security tools as well.

Cloud security breaches are often related to misconfigurations; thus, you should pay attention to securing each layer that comes under your responsibility.

- **Embed security in entire lifecycle**
  Infrastructure architecture point of view talks about security at each layer such as VMs, storage, network, etc., but you should also look at security from application lifecycle point of view as well. Security should be embedded as part of design to deployment to operations. Application viewpoint is crucial here because SAP does recommend security practices, which may not be applicable for other non-enterprise applications.

- **Build a zero-trust security model**
  Let's say you log on to your workplace device using your corporate credentials, but when you need to log on to your customer relationship management (CRM) system, you're asked to verify your credentials again, often using a biometric or physical

device such as a YubiKey. That's your organization's way of saying that they don't trust the initial logon for everything.

The idea behind a zero-trust security model is to continually verify a user's identity using different mechanisms. Just being in the corporate network doesn't grant you access to everything. In a sense, philosophically, the model assumes a data breach and protects the data leak rather than focusing completely on perimeter security. This is achieved using techniques such as micro-segmentation, using least-privilege access principles, and using different context indicators to verify authentication and authorization. The same principle can be extended to cloud infrastructures in addition to end-user access.

- **Automate security controls**

  SAP deployments on the cloud are complex, and this complexity increases as you start using more cloud-native services and features. If you try to implement security controls as a separate step and verify using a checklist, you'll be stuck in a never-ending process. Embedding controls in the automation process throughout the life-cycle is the way to go. Automation doesn't have to stop at the configuration of security controls—you can extend it to continuously monitor and remediate security gaps as well.

[»]

Compliance Certification on Cloud

Cloud providers offer compliance certifications (e.g., HIPPA, FedRamp, FIPS, etc.), and this shared responsibility model applies this that context as well. Cloud providers certify the cloud services, and *you* must ensure that your applications follow certification guidelines.

1.2.5   Cost

No doubt you've heard terms like resilient, secure, and scalable to describe architectures—let's now introduce you to *cost-aware architectures*. Every design decision you make in the cloud has an impact on cost. Think of it this way: When you buy an airline ticket, you get the seat, but on some airlines, you need to pay extra for snacks and, in some cases, seat selection as well.

The cost model for cloud services is different; not only do you pay the price for the service, but you also pay for data storage, replication, number of API calls, etc. You should start thinking about costs in terms of *run rate* and *long-term consumption*. Some things to explore as a starting point include the following:

- What components constitute the definition of costs on-premise? Examples include servers, licenses, datacenters, management, round-the-clock support, etc.

- What are the areas where optimization opportunities exist?

- How many different OS and database combinations exist? Is there an opportunity to lower the total number of permutations?

- What activities are good candidates for automation to streamline long-term costs?

- Which applications can be decommissioned or combined when you move to the cloud?

Moving from cable TV to streaming service subscription made us first think about what shows and movies we really wanted to watch and then pick the service that has those shows and movies. You should see cloud cost management from the same lens—decide which services you need and only pay for those. If not properly managed, costs can spiral out of control with everyone in your organization provisioning their own services and storing a ton of unnecessary data. Figure 1.6 shows the building blocks of a cost-aware architecture in the cloud.

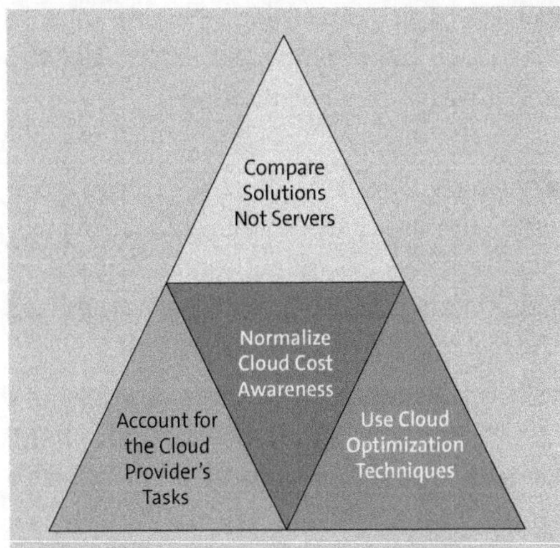

Figure 1.6 Cost-Aware Architecture Building Blocks

These building blocks include the following:

- **Compare solutions not servers**
  In our experience, every organization compares the cost of running SAP on-premise with that of the cloud; however, how the comparison is performed makes the difference in the long run. We recommend comparing the cost for the solutions and/or applications you are running (in the context of both peak usage and average usage), not the number of servers you currently have. Similarly, HA and DR configuration should be thought of in terms of uptimes and RTO/RPO not in terms of mapping the equivalent of what you currently do, but in the cloud.

  Cloud migration lets you reset a lot of things and provides an opportunity to think about processes you usually don't consider during day-to-day management. Let's look at a few examples:

- The CPUs that power old, on-premise servers are different from the CPUs used in the cloud. So, there is no point in comparing CPU/RAM to decide on a target configuration. Conducting an analysis using SAP Application Performance Standard (SAPS) is the right way to pick target systems.

- Do you really need the backups to be retained for 60 days, that you always had configured on-premise just because there was no incremental cost to it?

- You had factored the growth for next 5 years when on-premise hardware was ordered. The cloud operating model doesn't need that kind of planning and several months' notice for infrastructure availability.

- **Normalize cloud cost awareness**
  In a traditional on-premise setup, not all teams are aware of the cost. Technical teams are encouraged to optimize the system without much insight into whether a person, working on an initiative, costs more in terms of time than the infrastructure cost it saves. By normalizing cost conversations, you can enable each team to make informed decisions; everyone reviewing cloud costs and budget reports can also trigger optimization initiatives that can be used by other teams as well.

- **Account for cloud provider's tasks**
  When you think about total cost of ownership (TCO), in the cloud, it's easy to forget about things that cloud providers manage on your behalf. Anything related to physical security, data center buildings, maintenance, physical networking, infrastructure procurement and decommissioning, etc. are completely managed by the cloud provider. They do it at scale and pass the savings on to you.

  So, essentially, your cost for those tasks is baked into the services you buy. As a result, when calculating TCO, for correct comparisons, you must account for those costs on-premise. Add automation tools and personnel efficiencies, and you'll have some intangibles to consider as well.

- **Use cloud optimization techniques**
  A large part of SAP deployments rely on IaaS solution, which involve several cost optimization techniques that you should be aware of, such as the following:

  - Even though part of the popularity of the cloud is its pay-as-you-go model, all cloud providers provide discounts for sustained use. For example, AWS has different pricing plans (i.e., reserved instances and compute savings plan), which come at a significant discount in lieu of longer-term commitment. Long-term commitment may sound counterintuitive, but given that you're not going to get rid of your SAP systems anytime soon, fear of commitment seems a moot point. These plans also have some flexibility built in, so the commitment is not a sunk cost.

  - You can use the pay-as-you-go model for systems that are not always on and also use the compute savings plan for discounts on production systems (that are always up). You'll need to calculate the breakeven point of how many hours does a system must be down for the pay-as-you-go model to be efficient rather than going with the compute savings plan.

– Always look at costs in the context of performance (i.e., SAPS). A price-performance ratio is a better measure than just the price for a similar configuration. Often, newer-generation servers, storage, etc. have better performance metrics at the same or lower cost than older-generation ones.

– We already discussed right-sizing an instance, but monitoring its ongoing utilization is equally important. If not being used as expected, you can always size it down to a lower configuration. Scaling down is an underrated aspect of scaling in the cloud.

Cloud Costs

Certain cloud services are often advertised as free. However, these services are often bundled with another service that costs money. For example, a cloud provider can say that a certain monitoring tool is free; you just need to pay for the resources under the hood. What they mean is you may not pay a licensing fee upfront, but the cost is built for the resources you use. Thus, increases in consumption increase you overall costs. One analogy is when buying a phone: If the phone comes with headphones, it doesn't mean headphones are free. The price is bundled with the phone's price.

1.2.6   Innovation

Close your eyes and think of the word *innovation*. If technology was first in the list of things that came to mind, then keep going down the list until you find the business context. Technology in itself is just an enabler; what you do with it is a bigger part of innovation.

Cloud tools and technologies enable faster innovation cycles that lets organizations experiment with business models and improve the customer experience. A few things to explore when staring your innovation journey on cloud include the following:

- What business processes are currently supported by SAP? Can something be done better with cloud-native tools?

- What processes do you need to enable faster turnarounds when testing ideas?

- How can you unlock new business insights using machine learning and artificial intelligence (AI) models that wasn't practical on-premise?

- How can you ensure resiliency in IT operations and improve the customer experience with your applications?

Innovating on the cloud is not something to be considered only after migration and operational optimization; rather, innovation should be front and center of the cloud business case. According to Gartner, by 2027, the cloud will serve as key driver for business innovation (visit *http://s-prs.co/v5776130* for more information). Figure 1.7 shows the building blocks that you should consider when innovating in the cloud.

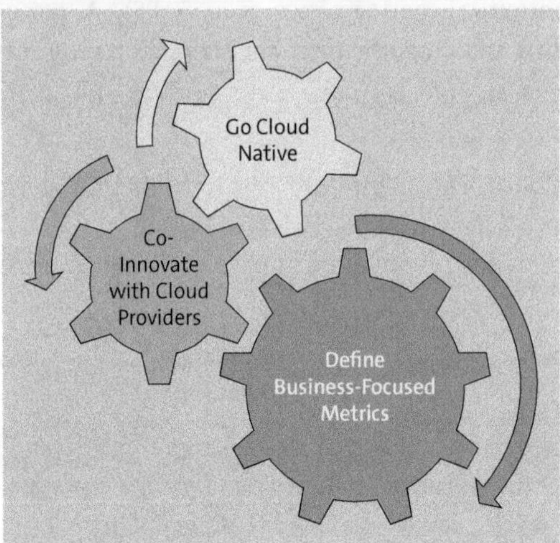

Figure 1.7 Building Blocks of Innovation in Cloud

These building blocks include the following:

- **Define business-focused metrics**
  Imagine buying a smart phone, thinking you can be more productive and instead looking for all the games you can now play—you're getting distracted from the main purpose. Similarly, focusing on all the fancy tools and services you can use to run a monthly report in half the time when nobody cares about it is using technology for fun, not for the business.

  Find the metrics that matter the most to business and optimize for those metrics; it may be as simple as trying to shorten the month-end close process or as complex as figuring out how your organization's customers make their buying decisions from analyzing disparate sources of data.

- **Co-innovate with the cloud provider**
  Instead of thinking of your cloud provider as a vendor, think of them as a partner and see what their business experience can bring to the table as well. For example, AWS is a subsidiary of Amazon, which has an optimized supply chain. Thus, AWS can potentially bring this business knowledge to bear and help you with your supply chains as well.

  Cloud providers often run SAP themselves (several Amazon subsidiaries, for example, run SAP), so they understand the challenges of migrating and optimizing SAP landscapes. This understanding could be a shared learning opportunity as well.

- **Go cloud native**
  According to the Cloud Native Computing Foundation (CNCF), "Cloud native technologies empower organizations to build and run scalable applications in modern, dynamic environments such as public, private, and hybrid clouds. Containers,

service meshes, microservices, immutable infrastructure, and declarative APIs exemplify this approach" (visit *http://s-prs.co/v5776128* for more information).

Faster innovation requires standing on the shoulders of giants and using modern, often cloud-native, technologies, and business-focused platform services can lower your investments when creating scalable solutions. Gartner also estimates that 95% of new digital workloads will be deployed on cloud-native platforms by the year 2025 (visit *http://s-prs.co/v5776129* for more information).

> **Low-Code/No-Code Solutions**
>
> Cloud providers also provide tools to help you create reports and to help with other business context solutions without coding know-how. These capabilities often involve some configuration steps rather than code or feature a visual drag-and-drop interface. Often called *low-code/no-code solutions*, this capability empowers business analysts to customize their own solutions without IT involvement. SAP calls this *citizen development*, and it's the next wave of solution innovation!

1.2.7   Integration

SAP systems are some of the most highly integrated enterprise applications out there—a claim substantiated by the fact that SAP Integration Suite features more than 2,200 prebuilt connectors! In our experience, organizations that have been running SAP for a while often find themselves making an inventory of all the integration points to their ERP systems as a part of migration project.

A few ways to think about integration requirements include the following questions:

- What type of interfaces currently exist, and how do they connect to SAP systems?
- Which interfaces are inbound versus outbound, and how do data flows impact network design?
- What type of authentication and authorizations are used with integration points?
- What type of data integration exists among different systems (SAP or otherwise)?
- Which integrations and interfaces will still be required when we move to the cloud?
- Do any consolidation, replacement, or decommissioning opportunities exist?
- What other SaaS solutions are connected to SAP?
- After migrating to the cloud, would any hybrid integration be required, such as on-premise to cloud?
- What is the frequency of communication among interfaces?

Not coincidentally, innovation and integration have a lot of overlap for cloud platforms. New integrations often offer innovative solutions to business processes. Figure 1.8 shows the building blocks of cloud integration through SAP.

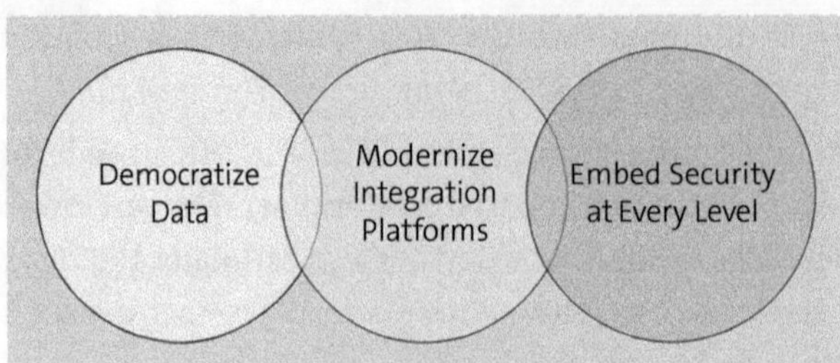

Figure 1.8 Building Blocks for Integration

These building blocks include the following:

- **Modernize integration platforms**
  While you're modernizing your overall SAP infrastructure and even your applications by moving to SAP S/4HANA, legacy integrations can hold you back with a restricted cloud-native architecture (since those integrations were not designed for the cloud) as well as the lack of comprehensive security controls.

  Along with the native services of cloud providers, you should also look into SAP's offerings for an integration platform as a service (iPaaS) and SAP Business Technology Platform (SAP BTP) as a part of your SAP integration scenario. Modern platforms not only accelerate your organization's journey, but also make it easier to adapt to future technological changes.

- **Democratize data**
  The evolution of low-code/no-code solutions has paved the way for self-service in analytics and business processes. Combine these capabilities with SAP data-driven insights, and you've enabled informed business decision-making. Data democratization provides access to the right data to everyone without the need for extensive knowledge about technical tools and data transformation. Democratizing your data is a prime candidate to include into your modern integration approach.

- **Embed security at every level**
  As you integrate data with different platforms, any business data or customer data is stored and processed at different providers, thus potentially exposing sensitive data. Thus, you must ensure that security is built into each and every step of the way: right from the infrastructure to the software to the data encryption throughout the integration flow.

  Make sure you look beyond data privacy and security and focus also on data residency. If your organization requires data to be stored in a particular boundary, that's an important consideration for your integration layers.

1.2.8   Sustainability

The United Nations (UN) defines sustainability as "meeting the needs of the present without compromising the ability of future generations to meet their own needs."

From this definition into practice, the goal is to reduce the environmental impact of your organization's business. As part of moving to the cloud, many organizations also extend their commitment to sustainability by tracking and reducing carbon emissions.

A few things that you should consider to get started include the following:

- How does your organization currently gather data and track sustainability efforts?
- What kind of reporting information do you need from your cloud provider, and how does this information integrate with your organizational reporting?
- Specific to IT infrastructures, what optimizations are currently done to reduce the environmental impacts?

Then, as shown in Figure 1.9, look for reporting provided by your cloud provider to track emissions and optimize your infrastructure and application architecture to reduce it.

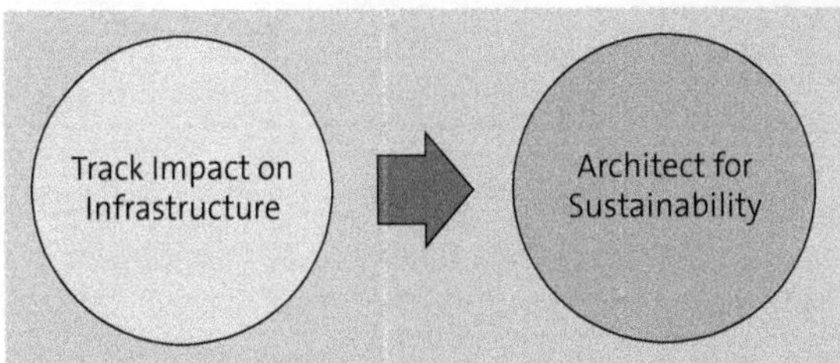

Figure 1.9 Sustainability Tracking and Target Architecture

Similar to security, sustainability comes under the shared responsibility model too, where the cloud provider is responsible for sustainability *of* the cloud, and you are responsible for sustainability *in* the cloud. AWS, for example, provides a customer carbon footprint tool in its billing reports so you can analyze the data and download the information for further use. We'll discuss architecture principles in detail in Chapter 3, but from a design point of view, architecting for sustainability means reducing waste by optimally utilizing cloud services.

1.2.9   Migration

We specifically saved this discussion of migration requirements to the end because, whether you are using the lift-and-shift approach or the transformational approach for your SAP systems, many considerations fall within the areas we've discussed throughout this chapter. Some migration-specific considerations include the following:

- How much downtime can your business afford? Also, you should differentiate between technical downtime and business downtime.

- What would be the operational and SAP development effort look like while the migration project is in progress? For instance, will some systems be in the cloud while others remain on-premise?

- What tools will be used, and are any costs associated with using these tools (both licensing costs and operational costs)?

SAP migration planning can be an involved process and take several iterations to optimize. Figure 1.10 shows the principles that you should keep in mind when migrating any application to the cloud.

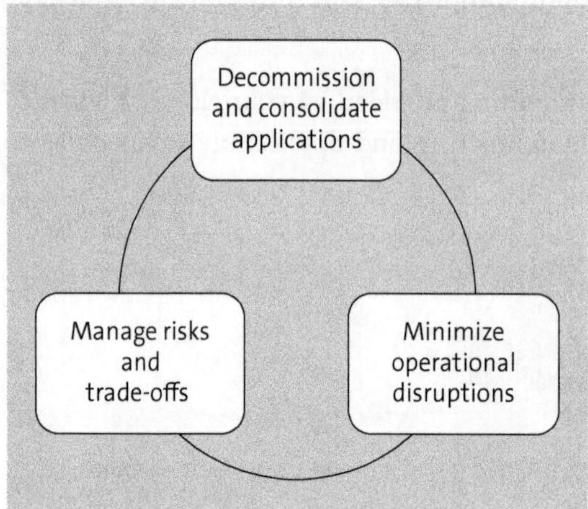

Figure 1.10 Principles Migrating to the Cloud

These principles include the following:

- **Decommission and consolidate applications**
  Think of last time you moved: Did you pack everything or first evaluated what's worth taking and either selling or discarding things that you didn't need? Migration to cloud is no different; it gives you opportunity to pause to evaluate, working with business, what use cases are important and which applications those belong to. Also, with using new cloud native and integration services, which use cases would have reduced footprint on existing applications, thus proving consolidation opportunity.

[+]   **Application Decommissioning**
Always start from a business process that's being used for an application rather than looking at the application function when picking candidates for decommissioning.

- **Minimize operational disruptions**
  SAP migration projects can span anywhere between months to years depending on the complexity, number of systems, and integration points. While you are moving

full steam with the migration planning and pilot, the business is still using production systems. Normal operations such as development, monitoring, maintenance, and patching, etc., are still being managed by the operations team. While some inconvenience is bound to happen, minimizing disruption in the operations flow and including business-critical dates in the schedule are key parts of a successful migration.

- **Manage risks and trade-offs**
  Any change in the existing landscape could introduce potential risk, such as security risk when connecting on-premise to cloud, regulatory or compliance risk, etc. Part of migration planning is to identify those risks and mitigate them with either a process, a technology, or both. For example, if your organization has certain compliance requirements, not only must you ensure the cloud provider meets these requirements, but in a shared responsibility model, you may also need to make changes to your application as well.

Similarly, you may need to make decisions during the migration while considering costs, performance, and downtime. For example, the lower downtime from using database replication would mean higher costs since you would be paying for both on-premise and cloud infrastructures simultaneously.

1.3 Summary

In the early days of the cloud, we saw a lot of organizations doing proofs of concept (POCs) to figure out how SAP and the cloud worked together. Some of the larger companies publicly announced their moves to the cloud, and since then, discussions have moved from proving that SAP works in the cloud to exploring how the target architecture better supports the business!

In this chapter, you learned what makes a public cloud stand out from hosting and co-location infrastructure services. We emphasized always starting from the business in mind, translating business use cases into cloud requirements, and using the right metrics to determine the best-fit solution. Our next chapter takes a tactical approach and explores various cloud deployment models, from on-premise to AWS connectivity, and discusses AWS-relevant products and features for SAP applications.

Chapter 2
AWS Service Offerings for SAP

Cloud providers have gone from focusing on technical use cases to application and business use cases as their offerings have matured. SAP is at core of many Fortune 500 companies, as evident by the fact that 87% of global commerce is generated by SAP customers (see http:// s-prs.co/v5776132). Amazon Web Services (AWS) continues to follow an application-focused approach and provides customized services for SAP deployment and management.

Consider an analogy: In the United States, grocery stores typically have around 40,000 items—but how many items did you actually buy the last time you went grocery shopping? It's no news that people buy ingredients only for what they plan to eat based on their preferences and restrictions. Likewise, at the time of writing, AWS has more than 200 fully featured services, but SAP only uses a subset of that.

In this chapter, we start with discussing various deployment models in the cloud and where SAP applications fit. Then we talk about options for connecting to AWS, followed by AWS services and features that are used for SAP landscapes and use cases.

2.1 Deployment Models

The design of SAP applications predates the cloud, so for obvious reasons, it's not a cloud-native design. While SAP is rearchitecting some of its applications, most deployments, at the time of writing, are still an on-premise version mainly installed on physical machines or virtual machines (VMs).

As cloud providers have matured their offerings in the SAP space, it's common for organizations to move their whole SAP estate rather than going for a hybrid model, which involves splitting systems between on-premise and cloud. In Chapter 1, we discussed the business view when moving to the cloud. You should not just lift and move systems; instead, you should evaluate the best way to utilize the cloud. This evaluation is where deployment models come into play, which we'll discuss in this section.

You may have seen a responsibility matrix outlining who is responsible for which part of an infrastructure component and how it relates to infrastructure as a service (IaaS) versus platform as a service (PaaS) versus software as a service (SaaS). Figure 2.1 shows another view of deployment models through examples of where organizations focus their efforts with each model.

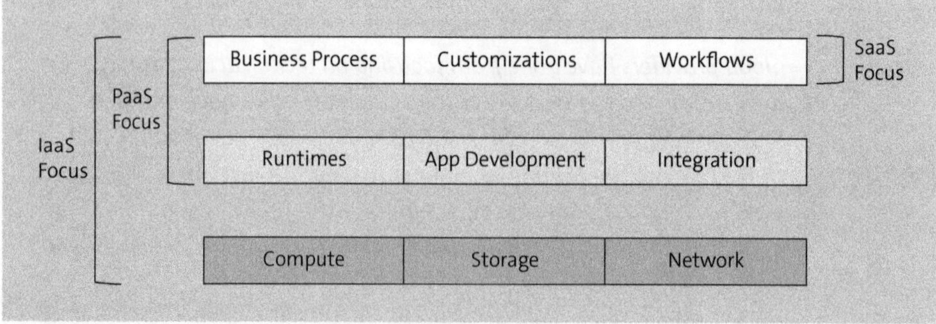

Figure 2.1 Deployment Model by Required Focus of the Organization

Often, what you gain in simplicity of management is lost in control; so, one is not necessarily better than the other, and it depends on the context. In the following sections, you'll learn more about these deployment models in terms of AWS services and how they relate to deploying SAP systems, including SAP's own program called RISE with SAP.

2.1.1   Infrastructure as a Service

From an SAP infrastructure point of view, think of IaaS as renting a house (or perhaps a data center) instead of buying or building your own. You're still responsible for managing everything inside the house, keeping it clean, etc., but you don't have to worry about electrical systems, water, or air conditioners. Back to cloud computing, as shown in Figure 2.1, if you're managing compute, storage, and network resources (the building blocks of any cloud infrastructure, along with your applications), you're using an IaaS. Cloud providers always manage the physical infrastructure, ensure security, and provide tools for the management of the IaaS.

Advantages of using an IaaS include better control over the infrastructure's configuration (such as RAM or CPU, parameters that must be fine-tuned to optimize the application) and the ability to migrate applications to cloud as-is, without much refactoring. Amazon Elastic Compute Cloud (EC2), Amazon Elastic Block Storage (EBS), Amazon Virtual Private Cloud (VPC), and Amazon Simple Storage Service (S3) are examples of IaaSs. From the user persona point of view, infrastructure architects and administrators would be the ones responsible for these components.

SAP on IaaS

SAP's primary on-premise version of products like SAP S/4HANA, SAP ERP, SAP Business Warehouse (SAP BW), etc. are deployed as IaaS models. Note that cloud versions of some of these products exist as well, but when you migrate your systems to IaaS on AWS, you're still using the on-premise flavor of these applications.

2.1.2 Software as a Service

If you've used an online travel portal to book a vacation, then you're familiar with what an SaaS model looks like. The top of Figure 2.1 shows the focus areas of the SaaS model, where application customization and business processes are your responsibility, and everything else is baked into the cloud provider's offering.

Advantages of using an SaaS include not having to worry about the infrastructure behind the scenes or about how to configure for resiliency. You can instead focus on business processes, while application providers focus on infrastructure. AWS Management Console (*https://console.aws.amazon.com*) is an example of an SaaS solution.

SAP on SaaS

SAP offers several SaaS solutions like SAP SuccessFactors, SAP Concur, SAP Ariba, etc. that can also be integrated with other IaaS SAP solutions.

2.1.3 Platform as a Service

Let's consider an example of how a PaaS works. Let's say Jill, our user persona, is an SAP developer; she doesn't want to manage any infrastructure (e.g., compute resources) to get started with development work. She just wants the correct libraries to get started. She can neither use IaaS nor SaaS for her work. Although she can work with an infrastructure admin to get the right resources provisioned, this type of dependency affects her productivity and increases total development time. Instead, she can use AWS Lambda to create a tool, using Python, for SAP deployment drift detection. This example shows how PaaS resides somewhere in between IaaS and SaaS with a focus on app development and runtime, as shown in Figure 2.1.

Advantages of using an PaaS include the ability to focus on development and management of your application (like Jill did) rather than on capacity planning and the freedom from managing the infrastructure.

Deployment Model Trade-Offs

As you can see, advantages exist for each of these models, but each also come with trade-offs. With SaaS, you lose control over the infrastructure (e.g., its resiliency, its configuration), and you must rely on the provider for those features. Any downtime impacts your business without being in control over the resolution.

On the flip side, some would argue that infrastructure work is undifferentiated, and you're better off focusing on business processes rather than on resiliency.

In the context of SAP, we do see a push to adopt more PaaS and SaaS solutions, but not all products are there yet.

2.1.4   RISE with SAP and SAP Business Technology Platform

If you've been around SAP for a while, you might be familiar with its former SAP HANA Enterprise Cloud offering, which was SAP's foray into hardware and SAP-managed services through partners. Think of RISE with SAP as the next generation of SAP HANA Enterprise Cloud with SAP incorporating lessons from SAP HANA Enterprise Cloud customer feedback! Similarly, SAP Business Technology Platform (SAP BTP) is an evolution of SAP's former PaaS offering, SAP Cloud Platform. We'll discuss both RISE with SAP and SAP BTP in the following sections.

RISE with SAP

This SAP offering (as opposed to a "product") launched in the beginning of 2021 to help organizations streamline the procurement and management of their SAP landscapes. Figure 2.2 shows the responsibility matrix for SAP on AWS directly versus RISE with SAP on AWS. While RISE with SAP still uses the power of the cloud, such as AWS among others, it simplifies SAP licenses by following a subscription model and takes the technical management away from you—all in a single contract. You still manage application and development aspects.

RISE with SAP supports you through phases like migrating your current on-premise version of supported SAP software to AWS, moving to SAP S/4HANA, business process transformation, and analytics so you can focus on your business and customers. Thus, SAP refers to RISE with SAP as "business transformation as a service."

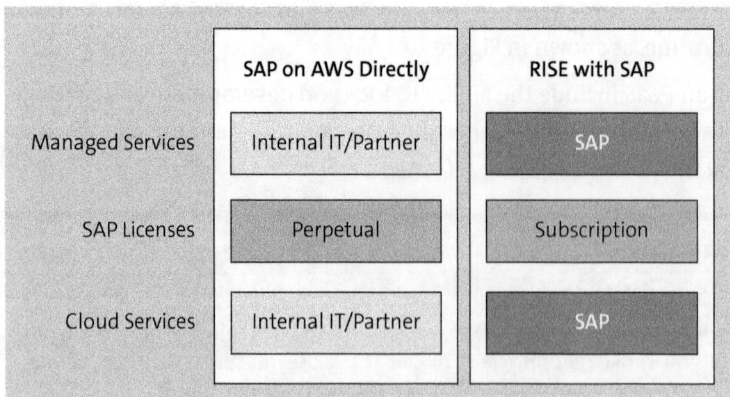

Figure 2.2 How Management and Licenses Differ between SAP on AWS Directly versus RISE with SAP

SAP provides several options for deployment under the RISE with SAP program, as shown in Figure 2.3. Since the destination is SAP S/4HANA (from SAP ERP and other on-premise SAP software), these offerings are called *SAP S/4HANA Cloud*.

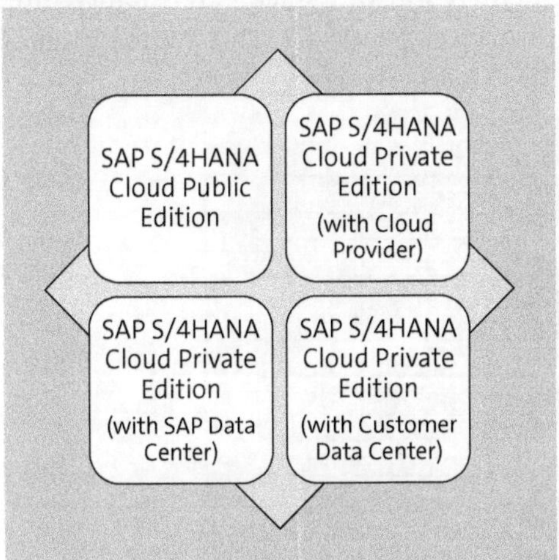

Figure 2.3 RISE with SAP: SAP S/4HANA Deployment Options

These deployment options include the following:

- **SAP S/4HANA Cloud Public Edition**
  This SaaS model includes preconfigured business processes data management solutions. This option is helpful for companies without complex customizations.

- **SAP S/4HANA Cloud Private Edition (with a cloud provider)**
  Migration and new implementation of SAP, with SAP S/4HANA journey, at the cloud provider of your choice (such as AWS, Microsoft Cloud, or Google Cloud). This option is helpful to make use of the scale and automation of the cloud.

- **SAP S/4HANA Cloud Private Edition (with an SAP data center)**
  With this option, you can run SAP software in SAP's own data center at select regions.

- **SAP S/4HANA Cloud Private Edition (with a customer data center)**
  Option to run SAP S/4HANA at your data center or co-location facilities on dedicated hardware. This option is helpful with data sovereignty, residency, and security requirements where the above-mentioned options are not available.

> **GROW with SAP**
>
> In April 2023, SAP launched an offering called GROW with SAP that focuses on mid-sized companies and that onboards customers to SAP S/4HANA Cloud Public Edition.

SAP Business Technology Platform

You're likely familiar with SAP's clean core strategy of following standard guidelines and less customizations for elements of core namely software stack, extensibility, integrations, data, processes, and operations, as shown in Figure 2.4. This approach helps improve management and lower the total cost of ownership (TCO).

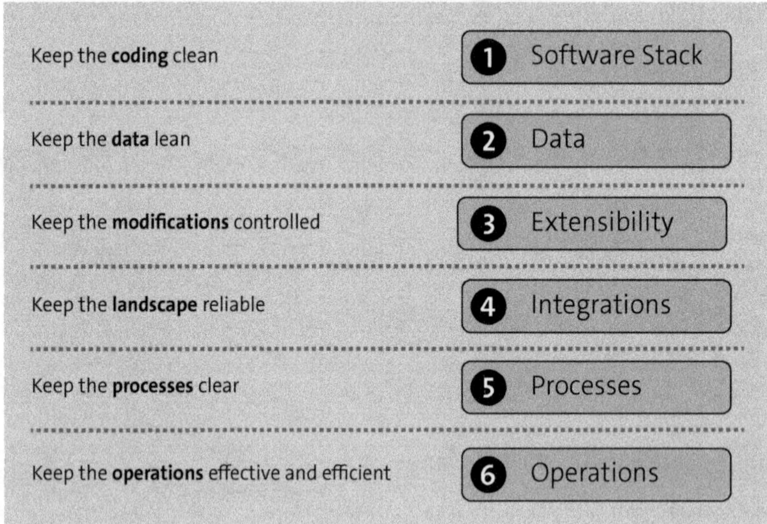

Figure 2.4 SAP's Clean Core Approach

SAP BTP supports the clean core approach, cloud mindset, and innovation by providing a common integration platform to create solutions by industry, line of business (LOB), etc. out of the box. SAP BTP provides a consistent experience across portfolios such as data, application integration, and analytics. Figure 2.5 shows the pillars across which SAP BTP services are categorized. You can find more details about SAP BTP services at SAP Discovery Center *(http://s-prs.co/v577601)*. SAP has also consolidated existing services under the common SAP BTP umbrella, including those from SAP Cloud Platform.

> **SAP BTP Missions**
>
> SAP provides architectural patterns and other guidance to help implement your use cases using SAP BTP in the form of "missions." These missions include SAP solutions as well as partner solutions. For more information about reference architectures and to embark on a mission, go to *http://s-prs.co/v577602*.

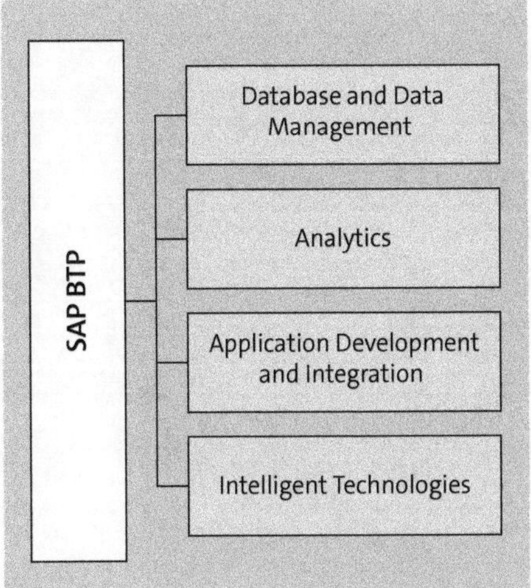

Figure 2.5 SAP BTP Technology Portfolio Pillars

SAP BTP Access

SAP BTP can be accessed either from RISE with SAP account or directly from cloud providers (AWS) using the HTTP layer.

2.2   Connecting On-Premise Infrastructures to AWS

One of the tenets of the cloud is the accessibility over internet. However, for business-critical applications such as SAP, many organizations do not allow internet connectivity for several reasons, such as security, latency, etc. This section discusses some other ways you can connect to AWS from your data center or office.

2.2.1   AWS Direct Connect

AWS Direct Connect is a network service using IEEE 802.1Q virtual local area networking (VLAN) that provides private, low-latency, and high-speed connections to AWS cloud. At the time of writing, AWS Direct Connect traffic is not encrypted out of the box; however, encryption can be implemented by your organization. You can either have a dedicated connection or use delivery partners through a hosted connection. Dedicated connections support up to 100 Gbps while hosted connection goes up to 10 Gbps.

The top of Figure 2.6 shows what a connection from an on-premise data center or office location would look like using AWS Direct Connect. Since global companies have multiple locations, the SiteLink feature of AWS Direct Connect lets you create a private network connection on an AWS backbone among the locations (also shown in Figure 2.6).

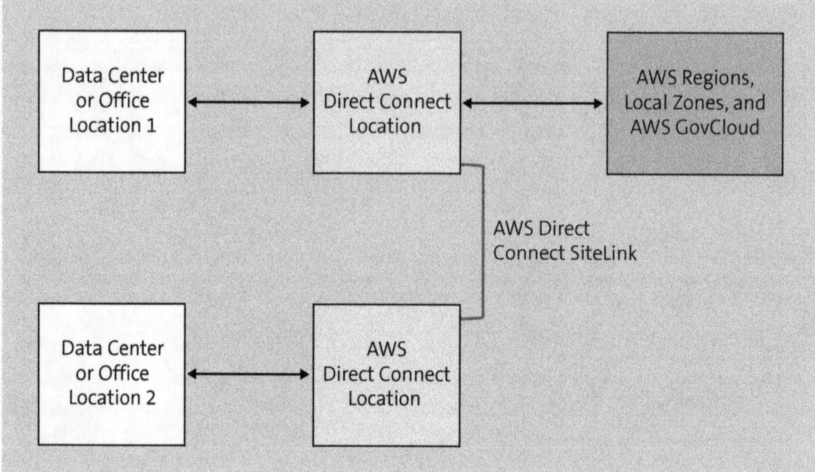

Figure 2.6 AWS Direct Connect to Cloud and Connection between Locations Using SiteLink

[»]

AWS Direct Connect and SAP Landscapes

In our experience, the majority of organizations moving their SAP landscapes to AWS use AWS Direct Connect for latency, throughput, and security reasons.

2.2.2   Virtual Private Networks

A virtual private network (VPN) uses Internet Protocol security (IPsec) to create a secure connection to AWS *over the internet*; all data over VPN is encrypted. As far as connectivity to AWS, you can use following VPN options:

- **AWS Site-to-Site VPN**
  This option offers VPN tunnel between AWS and the customer gateway used to connect to an office location for AWS. It has built-in redundancy with two tunnels and supports throughputs of up to 1.25 Gbps.

- **AWS Client VPN**
  This option connects an end user to AWS using an OpenVPN client and supports up to 10 Mbps bandwidth.

- **AWS VPN CloudHub**
  If you have more than one site-to-site connection, say, at multiple office locations, this option lets your office locations communicate with AWS cloud as well as with each other.

- **Third-party VPN appliance**
  This option involves bringing your own software model, where the software is installed on an AWS server similar to any other third parties and is not provided or maintained by AWS.

2.2.3   AWS Transit Gateway

From our AWS Direct Connect and VPN discussions earlier in this chapter, we hope you noticed a pattern for scaling: If you're still not convinced that scaling has met your organization's needs, we're at the connectivity option that you might be looking for. AWS Transit Gateway acts as a central hub for connectivity between your on-premise network and AWS cloud (specifically Amazon VPC, which we'll cover in detail in Section 2.3.6). Think of AWS Transit Gateway as a scalable cloud router that connects your VPNs and AWS Direct Connect instances to Amazon VPC instances, as shown in Figure 2.7, instead of establishing complex connections (such as *peering*) among Amazon VPC instances.

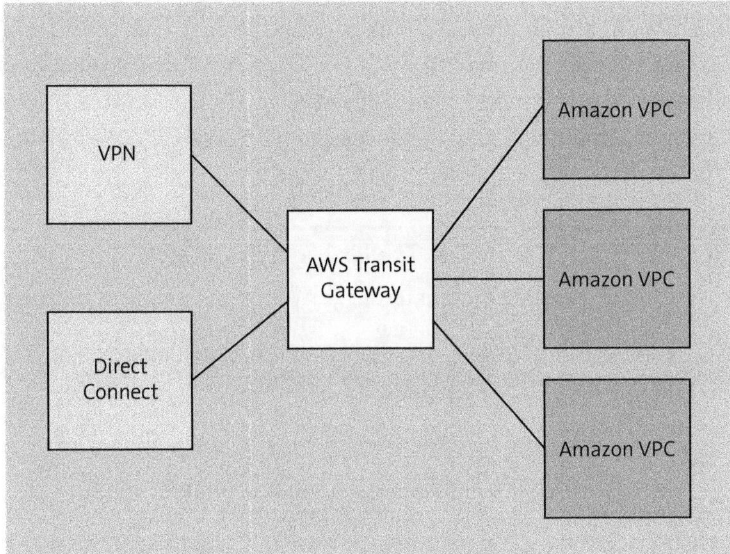

Figure 2.7 On-Premise to AWS Connectivity Example Using AWS Transit Gateway as the Hub

AWS Transit Gateway is a regional service that operates at layer 3 (network) of the open systems interconnection (OSI) model, and you can create more than one in a region as well. As your organization's AWS footprint increases, you can peer inter-region transit gateways to use the AWS network backbone.

2.3   AWS Services for SAP

Deploying SAP on AWS is similar to cooking a specialty food item by following a recipe. You acquire specific items in particular quantities and follow the instructions. For SAP deployments on AWS, you choose specific services according to SAP certification guidelines and follow SAP- and AWS-recommended practices. To that effect, even though we talk about generic services such as storage and databases, we'll keep most of our discussion focused on what's relevant for SAP.

In this section, we'll discuss the most commonly used infrastructure components and services provided by AWS, namely, its compute, network, storage, and management services.

2.3.1   Overview

If you've been around SAP, you know that an SAP Note exists for everything. Let's start with introducing you to two main SAP Notes related to AWS, which we'll refer to throughout this book:

- **SAP Note 1656250 - SAP on AWS: Support prerequisites** (*http://s-prs.co/v577603*)
  This SAP Note covers the criteria you'll need to meet to get support for SAP applications from AWS and SAP. At a high level, it spans network, storage, and tools from AWS and how to create support tickets with AWS. It's worth highlighting that you must be on a business or higher support plan from AWS, as shown in Table 2.1, for SAP environment support. In other words, the developer plan is not eligible for support.

Support Plan Name	Recommended for use cases
Developer	If you are experimenting or testing in AWS
Business	Minimum recommended tier if you have production workloads in AWS
Enterprise on-ramp	If you have production and/or business critical workloads in AWS
Enterprise	If you have business and/or mission critical workloads in AWS

Table 2.1 AWS Support Plans Overview

- **SAP Note 1656099 - SAP Applications on AWS: Supported DB/OS and AWS EC2 products** (*http://s-prs.co/v577604*)
  This SAP Note includes details about AWS infrastructure certification for SAP products. SAP's Product Availability Matrix restrictions still apply for supported combinations, but this SAP Note layers in AWS-specific details. It also lists all Amazon EC2 machines (compute resources) that are certified for SAP applications, databases, etc., as discussed in next section. According to this SAP Note, "Any conflicting sources are irrelevant," which essentially says that this SAP Note is the source of truth and should always take precedence over any blog or other website posting conflicting information.

[+]

Tip

Mark these SAP Notes as favorites in the SAP for Me portal to receive email notifications every time they are updated.

Next, we'll discuss the applications, operating systems (OSs), and databases supported by SAP. For the following lists, please refer to SAP Note 1656099 for more information on specific versions and kernels. At the time of writing, AWS supports the following SAP products:

- SAP NetWeaver ABAP and Java
- ABAP platform
- SAP BusinessObjects Planning and Consolidation on NetWeaver
- SCM Optimization, TREX, and SAP liveCache
- SAP BusinessObjects Business Intelligence (SAP BusinessObjects BI) and SAP BusinessObjects Financial Consolidation
- SAP Data Services and SAP Afaria
- SAP IQ and SAP Business One
- SAP Financial Information Management
- SAP Content Server 7.5
- SAP Convergent Charging

At the time of writing, AWS supports following OSs for SAP:

- SUSE Linux Enterprise Server (SLES)
- Red Hat Enterprise Linux (RHEL)
- Oracle Linux (OEL)
- Microsoft Windows Server

At the time of writing, AWS supports the following databases for SAP:

- SAP HANA
- SAP ASE
- SAP on IBM Db2 for Linux, UNIX, and Windows
- SAP MaxDB
- Microsoft SQL Server
- Oracle (only on Oracle Linux)
- Amazon Relational Database Service (RDS) (for a subset of applications as discussed in Section 2.3.7)

> **SAP 2027 Deadline**
>
> After 2027, SAP will stop mainstream maintenance for SAP Business Suite 7 to focus on SAP S/4HANA and the SAP HANA database. Thus, some application-and-database combinations we just listed will no longer be available. This deadline is also reflected in cloud offerings; new features are being released for SAP HANA database first.

> **Name Prefixes for Services**
>
> As we discuss each service, pay attention to naming conventions: Some start with *Amazon* (such as Amazon EC2) while others with *AWS* (such as AWS Outposts). While no official documentation exists on Amazon's naming prefix logic (at the time of writing), an unofficial AWS re:Post (see *http://s-prs.co/v5776133*) answer suggests that services starting with "Amazon" function independently and services starting with "AWS" work with other services.

2.3.2   AWS Global Infrastructure

Going global in minutes is another tenet of the cloud and, at the time of writing, AWS serves 245 countries and territories through 33 geographical regions, as shown in Figure 2.8, around the world (with 6 more regions coming soon).

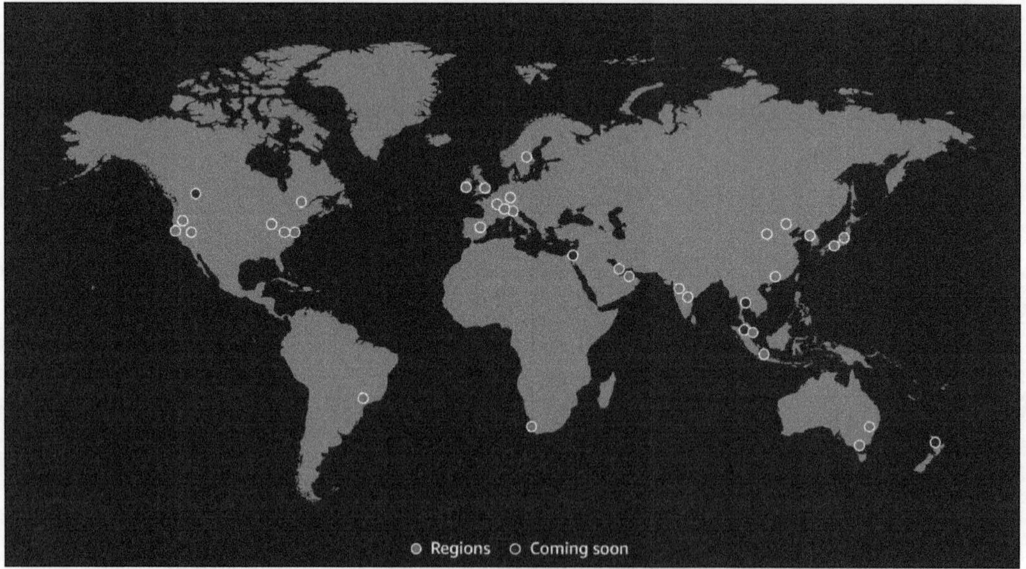

Figure 2.8 AWS Global Infrastructure Map Showing Regions

Breaking down the global infrastructure in pieces, a few definitions to note include the following:

- **Regions**
  Physically separate locations around the world with cluster of data centers, for example, the US East (Northern Virginia) region.

- **Availability zones (AZs)**
  Individually separated data centers with redundant power, cooling, networking, etc. AZs in an AWS region reside within 100 kilometers of each other but are separated by a meaningful distance, enabling easier application high availability configuration.

- **Local zones**
  A subset of AWS services, for common use cases, made available in bigger cities (with a large population or industry presence).

- **Wavelength zones**
  AWS infrastructure deployment at communication providers' locations so it uses 5G networks for low-latency mobile applications.

- **AWS Direct Connect locations**
  Locations (not necessarily regions) where you can access AWS Direct Connect services (135 locations at the time of writing). Once linked to an AWS Direct Connect instance, you can access all AWS regions.

- **AWS Outposts**
  Think of this service as an AWS cloud in any location you choose. It's available in server and rack deployments that lets you bring the AWS cloud-native experience to your data center for applications that can't be moved to an AWS region either for security or latency reasons.

- **AWS account**
  This concept is an administrative, security, and billing boundary that contains the AWS resources you provision. Therefore, sequentially, you create an AWS account before anything else.

AWS GovCloud (US)

In US East and US West, AWS has dedicated regions to support government and their partners with regulatory compliance requirements such as FedRAMP, the Department of Justice (DOJ)'s Criminal Justice Information Systems (CJIS) Security Policy, U.S. International Traffic in Arms Regulations (ITAR), etc.

2.3.3   Designing Regions and Availability Zones

Designing cloud infrastructures requires attention to several considerations, such as geographical and geopolitical isolation, land and labor availability, power cost and reliability, connectivity options, and *natural disaster risks*. AWS regions are geographically separated to withstand catastrophic disasters such as earthquakes, hurricane, etc.

Each region has three or more AZs, designed to withstand co-related failures with independent power, cooling, and connectivity. AWS doesn't publish the exact distances between AZs, but these are located within 100 km (60 miles) but meaningfully distant from one another so as to not be affected by the same event. You may have realized the common theme of *resiliency by design*. This approach applies to service updates as well: When AWS pushes updates, not all AZs are updated at the same time, thus limiting the impact of a potential issue. Figure 2.9 shows a logical representation of regions, AZs, and data centers depicting availability zone independence (AZI).

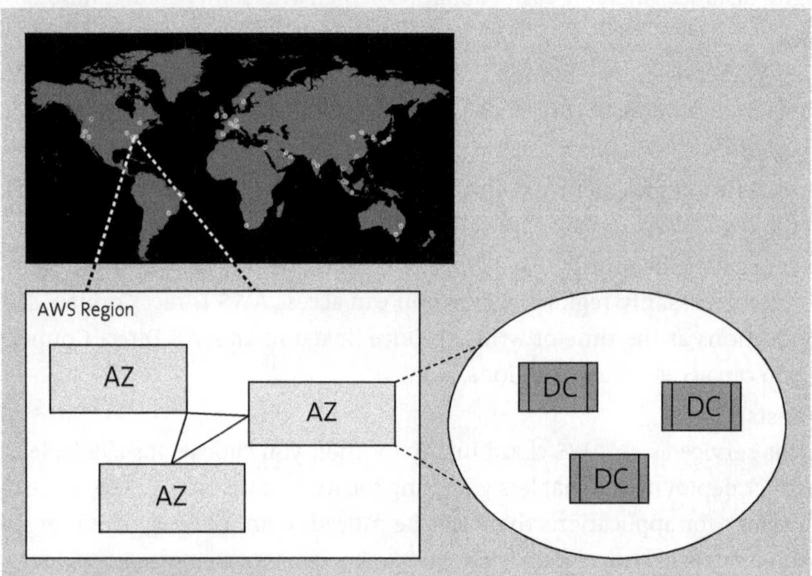

Figure 2.9 AWS Region, AZ, and Data Center (DC) Design Separation Representation

> **Application Resiliency**
>
> This section covers AZI and resiliency in infrastructures to showcase AWS's resilient design. In Chapter 5, we'll discuss in detail how to design for SAP application resiliency with high availability (HA) and disaster recovery (DR) patterns.

2.3.4   Compute

Compute, storage, and network are primary pillars of infrastructure deployment, and the next few sections focus on these topics. We realize sometimes we have circular dependencies, meaning you need to know some storage elements to understand compute resources better. We have tried to minimize these dependencies, but if and when that happens, read the relevant section then come back to original one.

First up is compute. We define compute as processing and memory capacity to run an application or a program. Thus, a VM would be a compute resource. Since VMs are called instances on AWS, we'll use that term going forward. The size of an instance is elastic in the cloud; as in, you can change the size without redeploying the application.

Amazon calls this resource an Amazon Elastic Compute Cloud (EC2) instance. These instances are subdivided by types of stock-keeping units (SKUs), offering different vCPU and memory combinations, and their generations (shown later in Table 2.3).

In this section, we'll talk about compute-related offerings from AWS that are most commonly used with SAP deployments.

Decoding an SAP-Certified Amazon EC2 Instance

Let's walk through how to decode an SAP-certified Amazon EC2 instance, which involves three steps:

1. **Naming convention**
   Figure 2.10 shows how AWS names the Amazon EC2 instance family, and Table 2.2 explains what these letters mean.

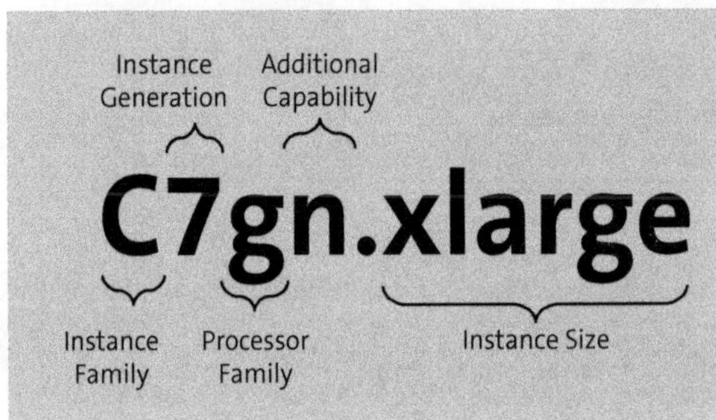

Figure 2.10 Amazon EC2 Instance Naming Conventions

Letter	Instance Family
C	Compute optimized
D	Dense storage
F	Field programmable gate arrays (FPGA)
G	Graphics intensive
Hpc	High performance computing
I	Storage optimized
Inf	AWS Inferentia
M	General purpose
P	GPU accelerated
R	Memory optimized

Table 2.2 Reference for Letters in the Naming Convention

Letter	Instance Family
T	Burstable performance
Trn	AWS Trainium
U	High memory
VT	Video transcoding
X	Memory intensive
Mac	macOS
Letter	**Processor Family**
a	AMD processors
g	AWS Graviton processors
I	Intel processors
Letter	**Additional Capabilities**
d	Instance store volumes
n	Network and Amazon EBS optimized
e	Extra storage or memory
z	High performance

Table 2.2 Reference for Letters in the Naming Convention (Cont.)

2. **Instance family**

   Table 2.3 shows a representative list of the current-generation Amazon EC2 instance family based on categories such as compute optimized or memory optimized, etc. (Refer to the AWS documentation for an up-to-date list at *http://s-prs.co/v577605*.) Note that Table 2.3 shows all offered instances; for SAP systems, you still need to refer to certified ones (which would be a subset of these instances, as highlighted in bold).

Compute Family	Type								
General purpose	**M5, M6a**	M6g	M6gd	**M6i**	**M6id**	**M6idn**	**M6in**	M7g	T4g
Compute optimized	**C5, C6a**	C6g	C6gd	C6gn	**C6i**	**C6id**	C6in	C7g	Hpc6a

Table 2.3 Categorization of Current-Generation Amazon EC2 Instance Types (Sample Set)

Compute Family	Type							
Memory optimized	Hpc6id	R6a, R6g, R6gd	**R6i, R6id**	**R6idn,** R6in, R7g	X2gd, **X2idn, X2iedn**	**U-3tb1, U-6tb1, U-9tb1**	**U-12tb1, U-18tb1, U-24tb1**	
Storage optimized	**I3en, I4i**	Im4gn	Is4gen					
Accelerated computing	G5g	Trn1						

Table 2.3 Categorization of Current-Generation Amazon EC2 Instance Types (Sample Set) (Cont.)

3. **SAP-certified Amazon EC2 instance**

   Figure 2.11 shows the SAP-certified instances from SAP Note 1656099. Notice that the last column also shows SAP Application Performance Standard (SAPS) rating, which is helpful since that's a hardware-independent metric. Now, let's pick an instance, say, r6i.8xlarge. This name indicates the following information:

   - R = Memory optimized (family)
   - 6 = 6th gen
   - i = Intel processor
   - 8x = 8 times extra-large configuration (r6i.xlarge has 4 vCPUs and 32 GiB RAM, so 8x is 8 times the vCPU/RAM configuration i.e., 32 and 256.)
   - 49,013 SAPS (Note that SAPS doesn't necessarily scale in a linear fashion.)

Instance Type	vCPUs	RAM GiB	2-tier SAPS
r6i.16xlarge	64	512	98,025
r6i.12xlarge	48	384	73,519
r6i.8xlarge	32	256	49,013
r6i.4xlarge	16	128	24,506
r6i.2xlarge	8	64	12,253
r6i.xlarge	4	32	6,127
r6i.large	2	16	3,063
r6id.metal	128	1,024	203,600

Figure 2.11 SAP-Certified Amazon EC2 Instances Sample (SAP Note: 1656099)

Tip

Notice that the RAM is published in gibibyte (GiB) and not gigabyte (GB); one gibibyte equals 1.074 gigabytes.

SAP Note 1656099 shows all certified Amazon EC2 instances for SAP application, SAP HANA database, and any other SAP-supported database—to refer to only SAP HANA ones, SAP publishes a list called "Certified and Supported SAP HANA Hardware," available at *http://s-prs.co/v5770606*. This list is a useful reference for comparison since it lists hardware from all vendors, not just AWS, and has additional information, as shown in Figure 2.12.

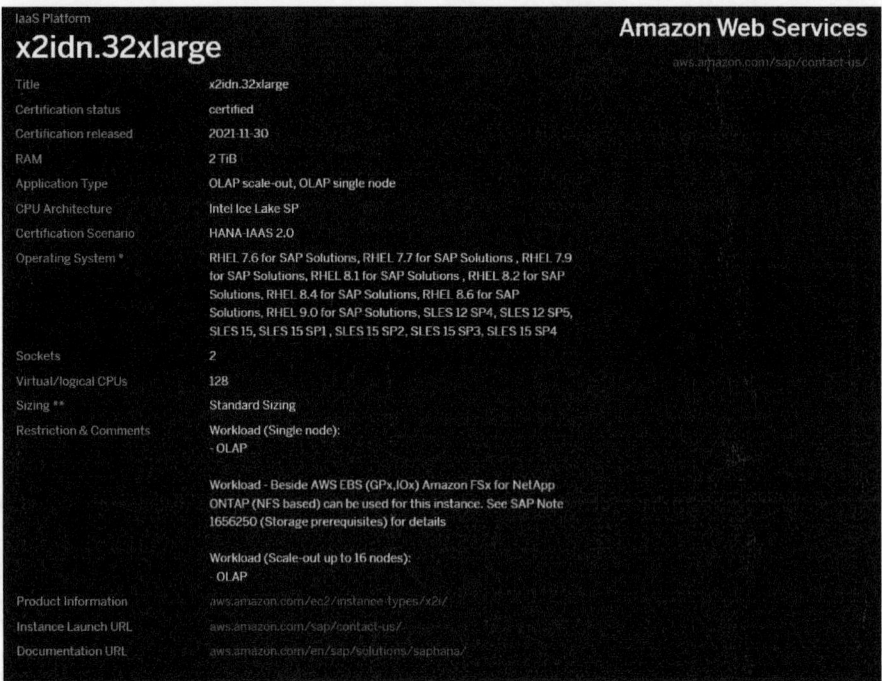

Figure 2.12 SAP HANA-Certified Amazon EC2 Instance and Additional Context from SAP HANA Hardware Directory

Other AWS Computes

Some other AWS compute resources include the following:

- **Bare metal**
  This option is supported and certified for SAP (e.g., r6.id.**metal** shown earlier in Figure 2.11).

- **Dedicated host**
  This option is also supported for SAP.

Note, however, that containers are not yet supported for SAP.

Nitro-Based Amazon EC2 Instances

AWS has a proprietary system called *Nitro*, which is a collection of hardware and software that the improves performance, availability, and security of its instances. The Nitro system includes a Nitro card, security chip, and lightweight hypervisor that eliminates overhead from virtualization and reduces the surface area of an attack. Figure 2.13 shows a list of virtual Amazon EC2 instances based on AWS Nitro. (Bare metal instances are built on Nitro as well.) Notice that SAP-certified instances are built on Nitro, and that's not an accident, given performance is important for SAP systems. For more information about the security design of a Nitro system, refer to the AWS whitepaper at *http://s-prs.co/v577607*.

General purpose: A1, M5, M5a, M5ad, M5d, M5dn, M5n, M5zn, M6a, M6g, M6gd, M6i, M6id, M6idn, M6in, M7g, T3, T3a, and T4g

Compute optimized: C5, C5a, C5ad, C5d, C5n, C6a, C6g, C6gd, C6gn, C6i, C6id, C6in, C7g, C7gn, Hpc6a, and Hpc7g

Memory optimized: Hpc6id, R5, R5a, R5ad, R5b, R5d, R5dn, R5n, R6a, R6g, R6gd, R6i, R6idn, R6in, R6id, R7g, U-3tb1, U-6tb1, U-9tb1, U-12tb1, X2gd, X2idn, X2iedn, X2iezn, and z1d

Storage optimized: D3, D3en, I3en, I4g, I4i, Im4gn, and Is4gen

Accelerated computing: DL1, G4ad, G4dn, G5, G5g, Inf1, Inf2, P3dn, P4d, P4de, Trn1, Trn1n, and VT1

Figure 2.13 List of AWS Nitro-Based Virtualized Instances

Features and Limitations of Amazon EC2 Instances

Some other things to note about Amazon EC2 instances include the following:

- Networking and storage features depend on instance type, so pay attention to these features.

- IPv6 is supported for all current-generation instances.

- I/O performance depends on the size of the instance. For example, a larger allocation of shared resources (CPU/RAM) would result in higher I/O performance.

- Each AWS account has an initial limit on the number of instances; you may need to request additional resources given SAP landscapes involve a large number of servers.

- The "T" instance family provides burstable CPU, which means it can use more CPU than baseline if required.

- AWS lets you enable termination protection in an instance. This feature prevents an instance from being accidentally deleted; we recommend enabling this feature for all production systems.

- Anytime an Amazon EC2 instance is shut down, you release that capacity to AWS. Unless you reserve capacity (discussed next), there is no guarantee that an Amazon EC2 instance will be available. Therefore, for DR scenarios, capacity reservation is recommended.

- All Amazon EC2 instances have autorecovery enabled by default. If a failure occurs, AWS would recover that instance and bring it up with the same properties. We recommend disabling this feature for SAP systems that already have HA configured using a Pacemaker cluster (more on this topic in Chapter 5).

- Some Amazon EC2 instances come with temporary storage called "instance store" volumes; for SAP landscapes, instance stores are not used.

- SAP requires all Amazon EC2 instances to have *AWS Data Provider for SAP* (more on this topic later in Section 2.3.10) installed for support. This component collects relevant data from the OS, storage, network, etc. Without a data provider, you wouldn't see the correct data in Transaction OS06 or Transaction OS07.

Amazon EC2 Purchase Options

The on-demand availability of resources is a tenet of the cloud, but that principle often comes at a high cost. Since SAP instances persist for a relatively long time, other purchasing options, as discussed in Table 2.4, may be more suitable. In our experience, most organizations use a *3-year savings plan* for SAP landscapes since an SAP system's lifecycle is long enough for 3 years (or more) of commitment.

Purchase Option	Description
On-demand instances	You pay for every second that the instance is up; shutting down the instance stops billing for that duration. Useful for extra capacity during month end close for example.
Reserved instances	Discount based on committed use, of an instance type and region, for 1 or 3 years. Note that this is completely a billing construct; the instance is NOT reserved for you.
Savings plan	Discount based on commitment for consistent amount of use, regardless of instance type or region, for 1 or 3 years.
Spot instances	Request unused Amazon EC2 instance at significant discount but be prepared to be evicted. Can't be used for business-critical application such as SAP.
Dedicated host	Buy a physical host fully dedicated to you; often used for per core software licensing model or to meet compliance requirements.
Dedicated instances	You pay by the hour for single tenant hardware (instead of shared tenant).
Capacity reservation	Pay for reserving Amazon EC2 capacity to ensure you always have an instance available. This is helpful for large scale maintenance event or DR where you need to ensure that servers are available to you (in a specific AZ).

Table 2.4 Amazon EC2 Purchasing Options

Service Level Agreement for Compute

At the time of writing, the AWS instance level SLA for an Amazon EC2 is 99.5%, and the region level SLA with multiple instances running concurrently across AZs is 99.99% (four 9s). Note that the region level SLA doesn't mean all instances SLA increase to four 9s; rather, the overall availability of at least one instance is considered for the SLA. More details about compute SLAs and service credits are available at *https://aws.amazon.com/compute/sla/*.

Amazon Machine Image

When you build an on-premise server for an SAP installation, you provision an OS and include other software configurations to make it ready for installation. Similarly, AWS provides an image that you can customize and use to launch an Amazon EC2 instance (or multiple instances, much like having a *golden image*, a concept that you're likely familiar with). As shown in Figure 2.14, Amazon Machine Image (AMI Catalog) allows you to select from among several options in AWS Console when provisioning an Amazon EC2 instance.

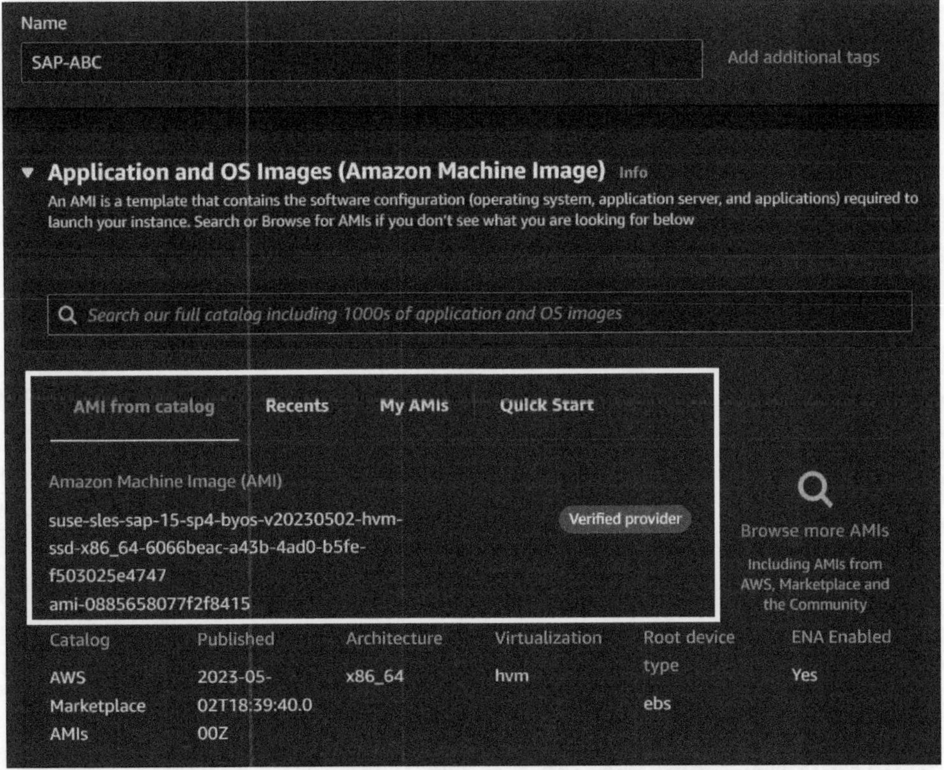

Figure 2.14 AWS AMI: Options When Provisioning an Amazon EC2 Instance

AWS AMI Options for SAP

You can bring your own image (and OS license), use AWS marketplace ones from partners such as RedHat and SUSE, or save one as a template after customizing your own AWS image.

AWS Lambda

AWS Lambda is a serverless compute service that lets you focus on executing code rather than managing infrastructures; think of this function as a service that can be event driven and that can be called from other AWS services as well as other SaaS solutions.

Some common applications include file processing, web application programming interface (API) calls, and Internet of Things (IoT) data processing. With SAP deployments, you could use AWS Lambda for activities such as starting or stopping SAP, Amazon S3 integrations, etc.

2.3.5   Storage

The next pillar is storage, and as you would imagine, several options exist for each block, file, and object storage category that aligns with SAP use cases, which we'll discuss in more detail in this section. Table 2.5 shows the different available storage types, their relevant AWS offerings, and SAP use cases for reference.

Storage Type	Feature	AWS Offering	SAP Use Case
Block store	Splits data into blocks and stores separately	■ Amazon Elastic Block Store (Amazon EBS) ■ Amazon FSx for NetApp ONTAP	■ SAP application storage ■ SAP database storage
File store	File level hierarchical data storage	■ Amazon Elastic File System (EFS) ■ Amazon FSx for Windows File Server ■ Amazon FSx for NetApp ONTAP	■ Shared storage such as for SAP mount (sapmnt), SAP transport (trans), interface, etc.
Object store	Flat storage of data as separate unit	■ Amazon Simple Storage Service (S3)	■ Database backup ■ Long term data archiving and backup retention

Table 2.5 AWS Storage Offering for SAP and Related Use Cases

Amazon Elastic Block Store

To deploy an application, you need file systems on a server—Amazon EBS backs those file systems on an Amazon EC2 instance. So, effectively you can create an Amazon EBS storage volume, attach that to Amazon EC2, and create a file system. It spans an AZ and provides similar SLAs as Amazon EC2 (i.e., 99.9% at volume level and 99.99% at region level). More details are available at *https://aws.amazon.com/ebs/sla/*.

Amazon EBS volumes can either be based on a solid state drive (SSD) or a hard disk drive (HDD) and offer different features and price points so you can decide the best option for your needs. Table 2.6 and Table 2.7 show the characteristics of SSD and HDD volume types, respectively. *GP3* is the most commonly used volume type for SAP.

Characteristic	General Purpose SSD Volumes		Provisioned IOPS SSD Volumes		
Volume type	gp3	gp2	io2 Block Express	io2	io1
Durability	99.8%–99.9% durability (0.1%–0.2% annual failure rate)		99.999% durability (0.001% annual failure rate)		99.8%–99.9% durability (0.1%–0.2% annual failure rate)
Volume size	1 GiB–16 TiB		4 GiB–64 TiB	4 GiB–16 TiB	
Max IOPS per volume (16 KiB I/O)	16,000		256,000	64,000	
Max throughput per volume	1,000 MiB/s	250 MiB/s	4,000 MiB/s	1,000 MiB/s	
Amazon EBS Multi-attach	Not supported		Supported		
Boot volume	Supported				

Table 2.6 Characteristics of SSD-Backed Volume Types

Characteristic	Throughput-Optimized HDD Volumes	Cold HDD Volumes
Volume type	st1	sc1
Durability	99.8%–99.9% durability (0.1%–0.2% annual failure rate)	
Volume size	125 GiB–16 TiB	

Table 2.7 Characteristics of HDD Backed Volume Types

Characteristic	Throughput-Optimized HDD Volumes	Cold HDD Volumes
Max IOPS per volume (1 MiB I/O)	500	250
Max throughput per volume	500 MiB/s	250 MiB/s
Amazon EBS Multi-attach	Not supported	
Boot volume	Not supported	

Table 2.7 Characteristics of HDD Backed Volume Types (Cont.)

Some features and limitations of Amazon EBS to note include the following:

- Elastic volumes feature of Amazon EBS lets you increase capacity dynamically, adjust IO and throughput, and change volume type without downtime; this helps with performance tuning of SAP storage.

- Amazon EBS supports (point-in-time) snapshots of volumes to Amazon S3.

- Amazon EBS volumes for SAP environments are usually attached to only 1 Amazon EC2 instance; provisioned IOPS SSD (io1 and io2) supports the multi-attach feature.

- Only SSD supports boot volume.

- SAP certification may have specific IOPS, and throughput listed, specifically for SAP HANA, so ensure your systems adhere to those guidelines (we'll learn more on this in later chapters).

- Amazon provides encryption options for Amazon EBS (for both boot and data volumes).

- Usually, you don't need to worry about configuring a redundant array of independent disks (RAID or striping) since you can adjust the performance characteristic dynamically. However, in some cases, such as > 16 TB GP3, > 16k IOPS, etc., a RAID 0 (striping) might make sense for SAP environments.

Some Amazon EC2 instance types provide temporary storage, included in the Amazon EC2 cost, called *instance store*. This storage is not used for SAP systems since the data in an instance store doesn't persist if an instance is terminated, hibernated, or stopped (although data on instance store survives Amazon EC2 reboot). So, you can't rely on any data stored here for long-term use.

Amazon EBS Purchasing Options

At the time of writing, AWS doesn't provide discount options for long-term commitment usage similar to those offered by Amazon EC2. Amazon EBS's pricing is determined by volume type, storage size, throughput, and IOPS. As is the case with technology, newer ones are often faster and cheaper than existing ones; for example, GP3 provides 3,000 baseline IOPS and 125 MiB/s baseline throughput at 20% lower cost.

Amazon Elastic File System

A *managed service* (i.e., a PaaS solution), Amazon EFS is a file store accessed via a network file system (NFS); thus, Amazon EFS is the ideal solution for shared storage use cases (/sapmnt, /usr/sap/trans, etc.) for SAP on *Linux*-based systems. Amazon EFS is also used as shared storage for HA configurations since its standard storage class has built-in resiliency that spans multiple zones with 99.99% SLA. Table 2.8 provides more details about each Amazon EFS type.

Amazon EFS Class	Durability	Availability	AZs
Standard	99.999999999% (11 9's)	99.99%	>=3
Standard–Infrequent Access (IA)	99.999999999% (11 9's)	99.99%	>=3
One Zone	99.999999999% (11 9's)	99.90%	1
One Zone-IA	99.999999999% (11 9's)	99.90%	1

Table 2.8 Amazon EFS Storage Classes Characteristics

Note

Amazon EFS, at the time of writing, is not supported for Windows machines since Amazon EFS supports only the NFS protocol.

Amazon EFS lets you mount the root of a filesystem to Amazon EC2 with all its data that resides underneath made accessible. *Access points* follow a different approach by providing application-specific entry points, making it easier to manage shared data. For example, in an SAP landscape, you can use the same Amazon EFS for multiple systems by creating application-specific access points for each system ID (SID). You can also differentiate access between the transport directory (/usr/sap/trans) and other shared directories.

Figure 2.15 shows a screenshot of a specific access point in AWS Management Console. You can also enforce different policies and authorizations for these access points.

Pricing

Billing is based on the average space used throughout the month and depends on the class of storage used, the throughput provisioned, access requests for infrequent access, etc.

Some other features and limitations of Amazon EFS to note include the following:

- It supports NFS version 4 (4.1 and 4.0).
- It is Portable Operating System Interface (POSIX) compliant.

- It supports mount targets only in one Amazon virtual private cloud (VPC) at a time.

- It supports replication to a different region, which is useful for SAP DR configuration.

- It supports intelligent tiering, which moves data to infrequent access storage class (and back) based on access patterns; this feature is not commonly used for SAP deployments.

- It doesn't support the `noconnect` mount option.

- You can also mount an Amazon EFS filesystem to your on-premise servers, given the connectivity is in place.

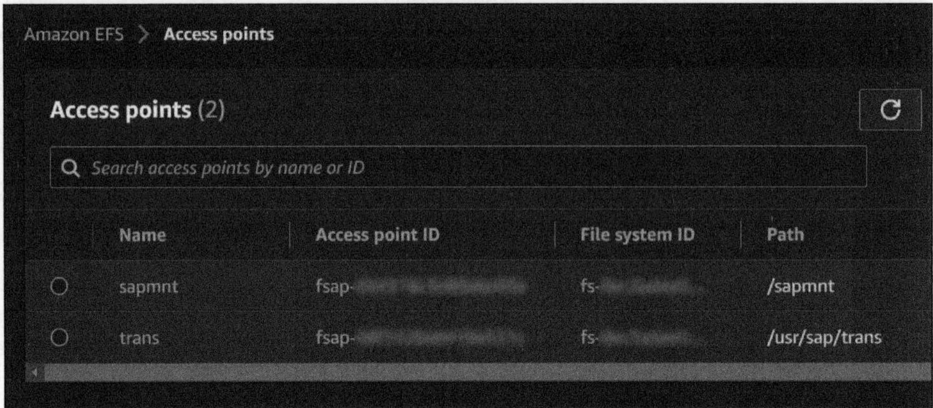

Figure 2.15 Amazon EFS Access Points Showing Different Entry Points for Application

Amazon FSx for NetApp ONTAP

Cloud service providers have commoditized hardware for a good reason, but what if you like certain technologies that provide additional features, enhance performance, or are just familiar technologies you currently use?

Cloud providers realize this need, and thus, the Amazon FSx Family supports widely used technologies such as OpenZFS, Windows File Server, Lustre, and NetApp ONTAP.

Amazon FSx for NetApp ONTAP is a managed service, *certified for SAP HANA*, based on ONTAP technology from NetApp that supports multiple protocols, namely, the Network File System (NFS), Server Message Block (SMB), and Internet Small Computer System Interface (iSCSI) protocols. Table 2.9 shows performance and scale characteristics for Amazon FSx for NetApp ONTAP.

Characteristics	Values
Latency	<1 ms
Max. throughput per file system	4-6 GB/s

Table 2.9 Performance, Availability, and Scale Characteristics for Amazon FSx for NetApp ONTAP

Characteristics	Values
Max. IOPS per file system	Hundreds of thousands
Maximum file system size	Virtually unlimited (100s of PBs)
Cross-region replication	Supported (NetApp SnapMirror)
Deployment options	Single and Multi-AZ
Availability	Multi-AZ: 99.99%, Single-AZ: 99.9%
Crash-consistent incremental backups	Supported
Inline instantaneous snapshots	Supported

Table 2.9 Performance, Availability, and Scale Characteristics for Amazon FSx for NetApp ONTAP (Cont.)

File and Block Store

Notice that, in Table 2.5 shown earlier, Amazon FSx for NetApp ONTAP is listed as both file storage and as block storage; that's because it supports shared block storage over the iSCSI protocol!

Pricing

Pricing is based on SSD storage capacity, provisioned IOPS, throughput, backup storage consumption, etc.

Some other features and limitations to note include the following:

- Supports data compression, deduplication, and compaction for reduced consumption, thus, helping control costs.
- Supports AWS management tools like AWS Console and AWS Command Line Interface (CLI) as well as NetApp tools like NetApp Cloud Manager and the ONTAP REST API.
- Supports common database features such as application-consistent snapshots, Flex-Clone, and continuously available SMB shares.
- Volumes are thin provisioned, which means that you consume storage capacity for the data stored in them.
- You can apply user quotas or group quotas to manage/restrict capacity for users and applications.
- Data at rest is automatically encrypted using AWS Key Management Service (KMS).
- Supports file access auditing.

Amazon FSx for Windows File Server

Built on Windows Server, Amazon FSx for Windows File Server is a managed service that supports the SMB protocol and integrates with Microsoft Active Directory (AD). Table 2.10 shows some of the performance and scale characteristics of this service.

Characteristics	Values
Latency	<1 ms
Max. throughput per file system	2-3 GB/s
Max. IOPS per file system	Hundreds of thousands
Maximum file system size	64 TiB
Deployment options	Single-AZ and multi-AZ
Availability	Multi-AZ: 99.99%, single-AZ: 99.5%
Crash-consistent incremental backups	Supported
Inline instantaneous snapshots	Supported
Cross region/cross account backups	Supported

Table 2.10 Performance and Scale Characteristics for Amazon FSx for Windows File Server

Pricing

Charges are based on storage and throughput capacity.

Some other features and limitations to note include the following:

- Integrates with on-premise Microsoft AD as well as AWS Microsoft-managed AD and is helpful for identity-based authentication
- All data is automatically encrypted in transit and at rest
- Supports file access auditing
- Supports continuously available file shares, which helps with HA configurations of Microsoft SQL Server
- Supports Windows Access Control Lists (ACLs)
- You can enable data deduplication and compression to reduce costs

OpenZFS and Lustre

We don't discuss OpenZFS and Lustre in detail since these options are not relevant for SAP on AWS; refer to the AWS documentation for more details at *https://aws.amazon.com/fsx/*. Also, a comparison of all Amazon FSx-supported storage options is available at *http://s-prs.co/v577608*.

Amazon Simple Storage Service

Amazon S3 is an object storage-managed service that can be used for several use cases such as data lakes, mobile apps, backups, archiving, etc., depending on your access, cost, and resiliency needs. Amazon S3 stores objects within resources called "buckets."

The most common use cases for SAP and Amazon S3 are for storing installation files and backups and for data transfers from SAP to Amazon S3. Transfer to Amazon S3 for reporting, analytics, and long-term data retention for compliance, among others. Amazon S3 offers several storage classes, and Table 2.11 shows some of the important performance and availability characteristics of Amazon S3 across those storage classes.

Characteristic	Amazon S3 Standard	Amazon S3 Intelligent-Tiering	Amazon S3 Standard-infrequent access (IA)	Amazon S3 One Zone-IA	Amazon S3 Glacier Instant Retrieval	Amazon S3 Glacier Flexible Retrieval	Amazon S3 Glacier Deep Archive
Durability	99.999999999% (11 9's)	(11 9's)	(11 9's)	(11 9's)	(11 9's)	(11 9's)	(11 9's)
Availability SLA	99.9%	99%	99%	99%	99%	99.9%	99.9%
AZs	≥3	≥3	≥3	1	≥3	≥3	≥3
Minimum storage duration	-	-	30 days	30 days	90 days	90 days	180 days
Retrieval charge	-	-	Per GB retrieved	Per GB retrieved	Per GB retrieved	Per GB retrieved	Per GB retrieved
First byte latency	Milliseconds (ms)	ms	ms	ms	ms	Minutes or hours	Hours

Table 2.11 Performance and Availability Characteristics of Amazon S3

Pricing

Amazon S3 charges are based on object sizes, storage classes, and duration of storage.

Some other features and limitations to note include the following:

- Several bucket naming rules must be followed, and prominent among these is that the name must be unique across all AWS accounts within a partition (not just your organization), that is, across the grouping of AWS regions such as standard (all commercial regions), China regions, and AWS GovCloud.

- Individual objects can have maximum of 5 TB.

- Supports batch operations (such as copying, tagging, etc.) to manage storage at scale.

- Supports version control and accidental deletion prevention, for instance, using multi-factor authentication (MFA).

- Supports cross-region replication (CRR) of data.

- Supports an object locking feature that lets you enforce write-once-read-many (WORM) policies.

2.3.6   Networking

Networking, the third pillar of cloud infrastructures, has several aspects such as connectivity to the cloud (as discussed earlier in this chapter), provisioning resources in the cloud network (Amazon VPC), load balancing, etc. This section focuses on these latter aspects and how compute, storage, and network all fit together for SAP deployments.

Amazon VPC

At this point, you know how to identify an SAP-certified Amazon EC2 instance and the storage that goes along with it. Now, you're ready to launch an instance—what is the first thing you need to do in networking? If you said assign a VPC, as shown in Figure 2.16, you would be right!

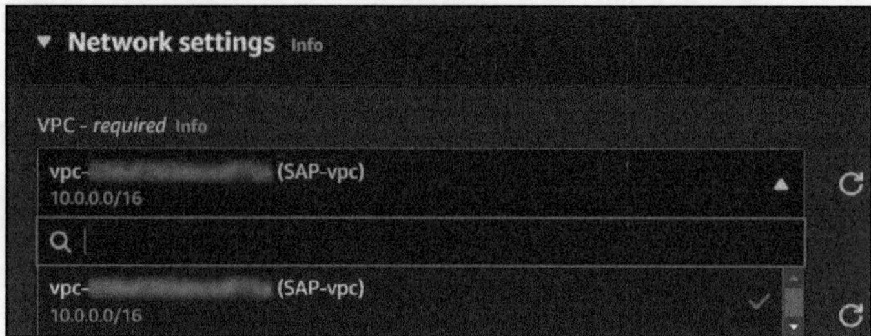

Figure 2.16 Network Settings in AWS Management Console while Launching an Amazon EC2 Instance

Bound to an AWS account, a VPC is similar to a virtual network that you're familiar with on-premise systems. By design, a VPC is an isolated network unless you establish *VPC peering* (defined later in this section). You would pick an IP Classless Inter-Domain Routing (CIDR) range for VPC, which is further subdivided into smaller ranges called *subnets*. Figure 2.17 shows an SAP VPC (in this case, a VPC named SAP) and its subnets in different AZs.

Note

Subnets are scoped to an AZ (i.e., each subnet is restricted to an AZ). If you're provisioning servers in a different zone, you must create another subnet there in the same VPC, which is scoped to a region.

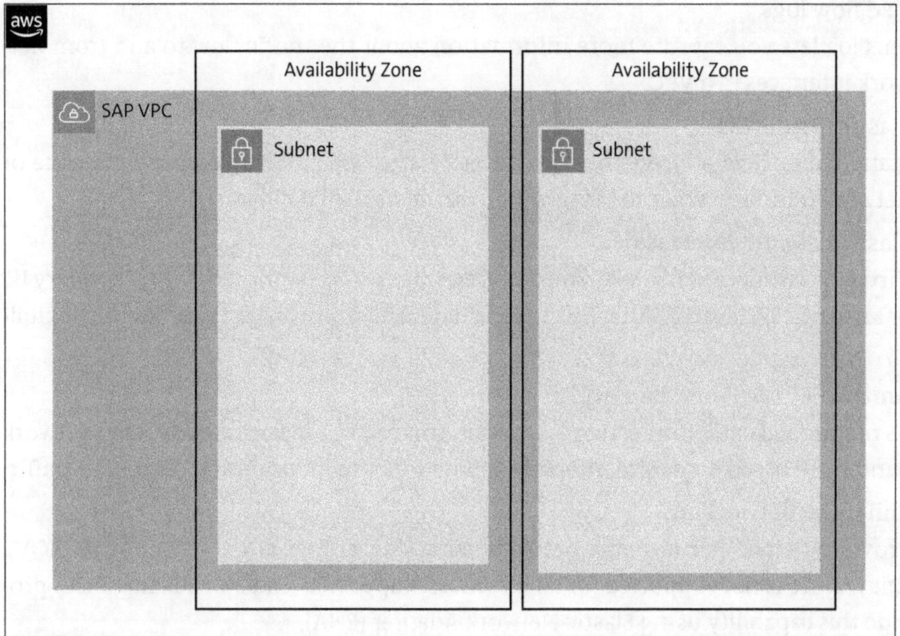

Figure 2.17 Block Representation Showing AWS Account, a VPC, AZs, and Subnets

Once you have the basic structure of a VPC and its subnets, you're ready to provision Amazon EC2, Amazon EBS, Amazon EFS, etc. The definitions of a few other related items include the following:

- **IP CIDR range**

  The CIDR notation describes a block of IP addresses. The open-source website *https:// cidr.xyz/* lets you visualize the range, IP count, usable IPs, etc.

- **Route table**

  This table contains a set of rules that governs where the network traffic is directed.

- **VPC peering**

  Network connection between VPCs that allows communication. Peering can be cross-account and cross-region as well.

- **Gateways**

  While peering lets you connect VPCs, gateways let you connect to VPC as well as to other networks. Examples include *internet gateway* (connect VPC to internet); *network address translation (NAT) gateway* (enable one way, namely outbound, connectivity from VPC to internet); and *transit gateway* (discussed earlier in this chapter).

- **PrivateLink**

  These default AWS service endpoints are public interfaces, but for most data flowing between VPCs and other services, we don't want to go through a public interface. PrivateLink can be enabled to provide private connectivity between VPCs and other AWS services. You can access any service supported by PrivateLink using VPC endpoints.

- **VPC flow logs**
  This log lets you capture more information about the traffic flow to and from network interfaces in a VPC.

- **Elastic IP Address**
  Static public IPv4 address that can be associated with an Amazon EC2 instance or network interface; you can bring your own *public* IP (BYOIP) as well.

- **Elastic network interface**
  Virtual network card in a VPC, this interface can include a primary IP, a secondary IP, or an elastic IP. It can be attached to (and detached from) an instance for full flexibility.

- **Amazon EC2 network bandwidth**
  As mentioned earlier in Section 2.3.4, each Amazon EC2 instance is assigned network bandwidth based on its size. This rule applies for both inbound and outbound traffic.

- **Enhanced networking**
  Providing a high-performance network using single root I/O virtualization (SR-IOV), this feature is not supported for all instances. Supported Amazon EC2 instances provide this capability using Elastic Network Adapter (ENA).

- **ENA express**
  Increases bandwidth and reduces latency within a subnet using an AWS technology called Scalable Reliable Datagram (SRD). SAP supports this technology for ENA express-supported, SAP-certified Amazon EC2 instances.

- **AWS Elastic Load Balancing (ELB)**
  As the name suggests, this component performs load balancing by distributing traffic to targets such as Amazon EC2, IP addresses, and containers by monitoring health and routing traffic to only healthy nodes. It works across AZs and is "elastic" in that it scales based on demand. It supports the *Application Load Balancer* (layer 7), the *Network Load Balancer* (layer 4), the *Gateway Load Balancer* (layer 3), and classic load balancer (layer 4 or 7; meant for Amazon EC2 classic network, which is not recommend anymore).

- **Route 53**
  This Domain Name System (DNS) service can be used to register domain names, for routing traffic (DNS routing), and to conduct health checks of resources (in any combination).

[»]

Default VPCs

AWS creates a default VPC (/16 CIDR) in each region to get deployments like websites up faster; however, with SAP systems, a lot of planning is involved, so default VPCs are not used. You can delete these VPCs if you like.

Placement Groups

In general, AWS determines where an Amazon EC2 instance is deployed across its hardware—the key principle used is minimizing potential failures. What if we had a use case where influencing (not necessarily controlling completely) the placement of *interdependent* instances would benefit the application? *Placement groups* solve that problem!

As shown in Figure 2.18, AWS provides the following grouping methods within an AZ:

- **Cluster placement group**
  Use this method to group instances more closely within the same AZ for better network performance among application nodes. At the time of writing, this grouping is used especially for *SAP HANA scale-out* deployments; it can be helpful for high performance computing (HPC) applications and can span peered VPCs.

- **Partition placement group**
  Use this grouping method to combine applications into groups and place the groups in different partitions so that hardware failures can be mitigated. This option can be useful for workloads such as Hadoop, Kafka, etc.

- **Spread placement group**
  Using this grouping method places instances into separate hardware, thus reducing related failures.

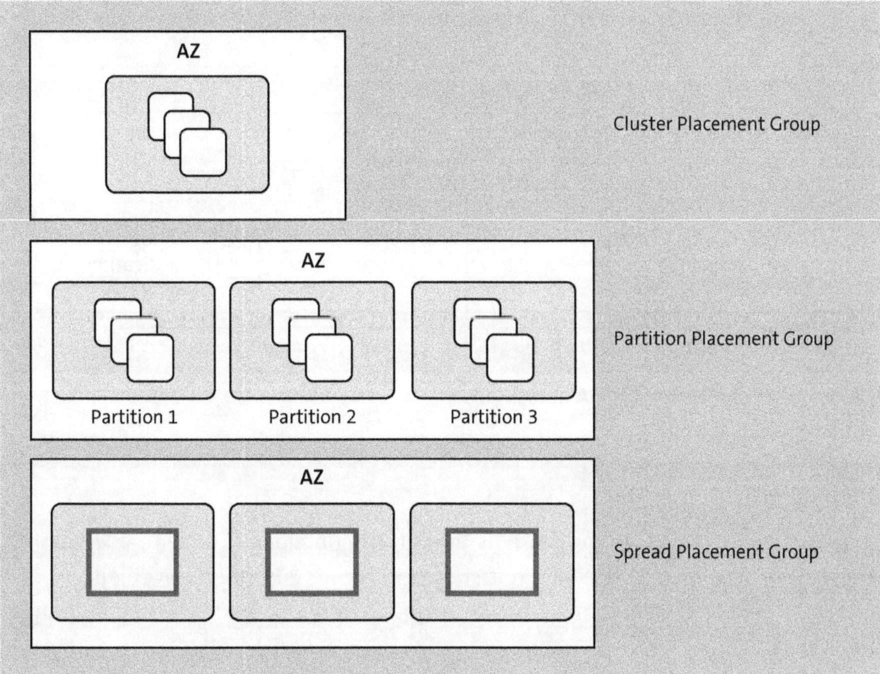

Figure 2.18 Placement Groups

> **Placement Group and Amazon EC2 SLA**
>
> The Amazon EC2 SLA remains the same regardless of whether you use placement groups or not. In other words, using placement groups does not provide a higher Amazon EC2 SLA.

Overlay IP

While an overlay IP is not a service per se, this concept is used in HA for SAP. Discussing this topic now will help you when we talk about HA in Chapter 5.

Since subnets are scoped to an AZ, when deploying an Amazon EC2 instance in a second AZ, you must create a different subnet in the same VPC. Thus, for SAP in an HA configuration with applications deployed across AZs, you need something that can route traffic to the secondary zone if the primary zone suffers a disruption. In terms of the architecture, you can use an IP address (the overlay IP), which is external to both subnets and alter route table info to redirect traffic to an active node, as shown in Figure 2.19.

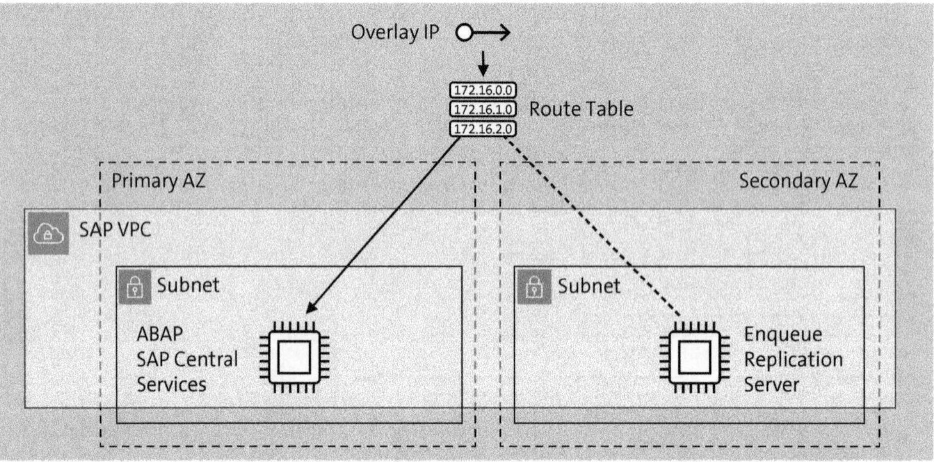

Figure 2.19 Use of Overlay IP for HA across AZs

> **End-User Traffic**
>
> For end-user connections, an overlay IP is used in conjunction either with a Network Load Balancer or a transit gateway (not shown in Figure 2.19; more on this topic in Chapter 5), where end users connect to SAP using the Network Load Balancer (pointing to the overlay IP).

> **Pricing**
>
> There is no additional price for creating VPC and subnets themselves. However, you'll pay for other VPC capabilities such as NAT gateway, network analyzer, etc. You also be charged for data transfers (per GB) across AZs and regions.

2.3.7   Databases

AWS offers several managed databases, called AWS Cloud Databases—more than 15 engines at the time of writing! Therefore, you don't have to worry about configuring resilience, fine-tuning parameters for performance, or creating extensive monitoring setups!

SAP's Product Availability Matrix lists all supported databases for SAP applications, and Table 2.12 lists all the databases offered by AWS. Note that some of these databases might support multiple engines; for example, Amazon RDS offers engines such as SQL Server, MySQL, Oracle, etc.

Database Type	AWS Offering
Relational	Amazon Aurora, Amazon RDS, and Amazon Redshift
Key value	Amazon DynamoDB
In-memory	Amazon ElastiCache and Amazon MemoryDB for Redis
Document	Amazon DocumentDB (compatible with MongoDB)
Wide column	Amazon Keyspaces
Graph	Amazon Neptune
Time series	Amazon Timestream
Ledger	Amazon Ledger Database Service

Table 2.12 Database Services Offered by Amazon (AWS)

How do SAP-supported databases fit into AWS database offerings? Most databases certified for SAP must be installed on an Amazon EC2 instance, making this landscape an IaaS setup. However, a subset of applications do support AWS databases, as shown in Table 2.13. SAP Note 1656099 discusses application versions and OS dependencies in detail.

Amazon Database	SAP Application
Amazon RDS (SQL Server engine)	SAP BusinessObjects Data Services
Amazon RDS (SQL Server and MySQL Engine)	SAP BusinessObjects BI

Table 2.13 Amazon Databases Supported for SAP Applications

> **Pricing**
>
> Prices depend on the database type and usage; several options are available, such as on-demand, and reserved instances offer a selection of instance types as well.

2.3.8   Key Management Services

We briefly mentioned encryption and bringing your own keys earlier in the chapter; now is where it all comes together! AWS KMS, a managed service, lets you create and manage cryptographic keys and provide audit capabilities for evaluating access. AWS KMS uses hardware security modules (HSM) and validates under the Cryptographic Module Validation Program. (China regions use an OSCCA-certified HSM.)

AWS KMS integrates with other services that uses encryption and works with either customer-managed keys (i.e., keys you create) or AWS-managed keys (keys that AWS services create). Auditing is facilitated by integration with AWS CloudTrail and Amazon CloudWatch (described later in Section 2.3.10), as shown in Figure 2.20.

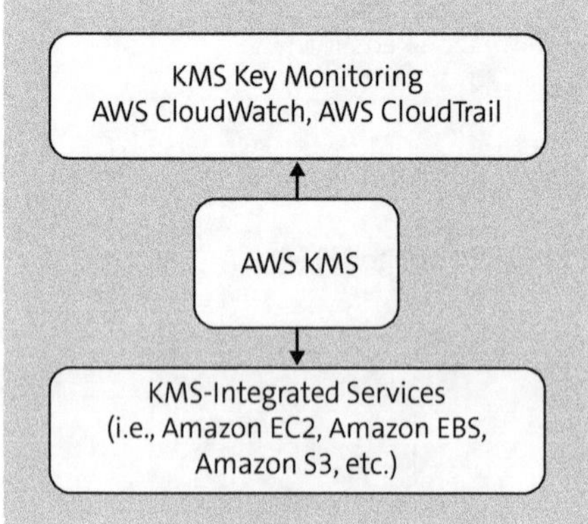

Figure 2.20 AWS KMS Integration

Some other features and limitations of AWS KMS to note include the following:

- It supports both symmetric and asymmetric keys; however, integrated AWS services use only symmetric keys.
- It supports key policies and IAM policies.
- It supports the automatic rotation of keys.
- It supports multi-region keys (copies of the same key in a different region).
- It supports external key stores (outside of AWS).

Pricing 

Prices are calculated by the number of keys required and the number of API requests made to AWS KMS.

AWS CloudHSM

AWS CloudHSM is an HSM that lets you use your own encryption keys, using dedicated FIPS 140-2 Level 3-validated HSMs, in AWS. It integrates with applications using APIs such as Java Cryptography Extensions (JCE), Microsoft CryptoNG (CNG) libraries, etc. and allows you to export keys to other HSMs (depending on the configuration).

2.3.9   Identity, Security, and Compliance

Identity is an integral part of overall security, and protecting user and application data is always a top priority in our experience. We discussed the concept of security *of* the cloud versus the concept of security *in* the cloud in Chapter 1. In this section, we'll focus on security in the latter sense.

AWS Identity and Access Management

AWS Identity and Access Management (IAM) helps you set boundaries to define *who* can access *what* AWS services or resources. When you create a new AWS account, your user, called root, gets full access for all services and resources. Only few tasks like changing account settings, closing accounts, enabling Amazon S3 MFA, signing up for AWS GovCloud access, etc. require root access. For everything else, you should create more users.

Tip

Use the root user only for activities that require root access. Create different users as necessary for any other activities and for day-to-day tasks.

Some other features of IAM to note include the following:

- It supports shared access of AWS account, which allows granting admin access to other folks.
- It supports granular permission such as read, write, billing access, etc.
- It supports integration with applications running on Amazon EC2; useful if application needs access to AWS resources such as Amazon S3 bucket.

- It supports multi-factor authentication (MFA).

- It supports identity federation such as with your corporate network.

- It supports logging of any requests using AWS CloudTrail.

- It supports Payment Card Industry (PCI) and Data Security Standard (DSS) compliance.

- It supports eventual consistency, which means the changes may not be immediately effective while it's replicating across multiple servers within AWS.

- It supports access via console, command line, software development kit (SDK), and APIs.

In IAM, a user is a specific person or application (the *principal* entity) versus role is an identity with specific permissions, which can be assumed by a person temporarily. IAM roles are useful in scenarios such as cross-account or cross-service access.

Policies list specific resources and access, which can be attached to principals. Once attached, principals get permission to perform those actions on specified resources since by default anything that's not explicitly allowed is denied. Expressed in JavaScript Object Notation (JSON) format, policies can either be *identity based* (attached to an IAM user, group, or role) or *resource based* (attached to AWS resources such as Amazon S3). Figure 2.21 shows an example of identity-based policy that allows read and write access to an Amazon S3 bucket.

```json
{
    "Version": "2012-10-17",
    "Statement": [
        {
            "Sid": "ListObjectsInBucket",
            "Effect": "Allow",
            "Action": ["s3:ListBucket"],
            "Resource": ["arn:aws:s3:::bucket-name"]
        },
        {
            "Sid": "AllObjectActions",
            "Effect": "Allow",
            "Action": "s3:*Object",
            "Resource": ["arn:aws:s3:::bucket-name/*"]
        }
    ]
}
```

Figure 2.21 AWS Policy (JSON) Example that Allows Read and Write Access to Amazon S3 Bucket

> **Policy for SAP Database Backups to Amazon S3**
>
> Brining SAP to the policy discussion, if you wanted to backup a database to an Amazon S3 bucket, for example, you would need to assign a policy to the Amazon EC2 instance for access to Amazon S3.

Attribute-based access control (ABAC) defines permissions based on attributes (also called *tags*) as opposed to *role-based access control (RBAC)*, where permissions are based on a job function or role. You can attach tags to IAM and AWS resources and create ABAC policies for IAM principals. Then, you can allow operations when the tags of the principal and those of the resource match. This matching is helpful in managing policies in rapidly growing environments as is often the case with new SAP implementations.

> **Pricing**
>
> AWS IAM doesn't have additional charges; you pay for other AWS services used by IAM users and ancillary services, such as the IAM access analyzer.

AWS IAM Identity Center

The AWS IAM Identity Center is a single-sign-on (SSO) service that enables you to manage SSO across AWS accounts and applications; it supports built-in Security Assertion Markup Language (SAML) integrations with business applications such as Salesforce, Microsoft 365, etc.

> **SSO with SAP**
>
> AWS IAM Identity Center supports SSO for several SAP applications such as ABAP platform (SAP S/4HANA), SAP NetWeaver, etc.

Other Related Services

Some other identity-, security-, and compliance-related services include the following:

- **AWS Directory Service**
  AWS Directory Service is a managed service for Microsoft Active Directory (AD) that helps you features such as group policy and SSO with AD, which supports SAP as well.

- **Amazon Cognito**
  This service provides an identity store and supports sign in with social media identity providers (such as Apple, Facebook, Amazon, etc.) for web or mobile applications.

- **Amazon Detective**

  This service offers prebuilt data aggregations that use machine learning and statistics on data collected from AWS resources to streamline security-related investigations.

- **Amazon GuardDuty**

  This service monitors potential threats and malicious activities within AWS accounts such as unusual API calls, unauthorized deployments, etc. Then, it uses anomaly detection (through machine learning) to report to services such as Amazon CloudWatch events and AWS Security Hub besides the Amazon GuardDuty console.

- **Amazon Inspector**

  This service scans AWS workloads (such as Amazon EC2, AWS Lambda, etc.) but not applications like SAP for software vulnerabilities.

- **Amazon Macie**

  This service scans *Amazon S3* for sensitive data types such as personal health information (PHI), personally identifiable information (PII), etc.

- **AWS Security Hub**

  This service aggregates security alerts from several AWS services (and supported partners) in an actionable format. Also integrates with Amazon GuardDuty, Amazon Inspector, Amazon Macie, etc. to help you with your security posture management.

- **AWS Artifact**

  This service provides security and compliance reports for AWS and third-party services on AWS Marketplace.

- **AWS Audit Manager**

  This service provides a prebuilt framework (and lets you create your own frameworks) for auditing AWS resources and fulfilling regulatory reporting requirements, such as General Data Protection Regulation (GDPR), Payment Card Industry (PCI), etc.

- **AWS Certificate Manager (ACM)**

  This service facilitates the creation and management of Secure Sockets Layer/Transport Layer Security (SSL/TLS) certificates to be used with AWS services.

- **AWS Firewall Manager**

  This service allows you to configure firewall rules centrally and enforce them across AWS accounts and AWS organizations.

- **AWS Network Firewall**

  This service lets you create firewall rules for fine-grained control across VPCs and integrates with AWS Firewall Manager.

- **AWS Resource Access Manager (RAM)**

  In a multi-account scenario, this service helps you share resources, such as subnets, transit gateways, etc., across accounts within AWS organizations.

- **AWS Secret Manager**
  This service helps you manage the lifecycles of secrets (such as credentials, API keys, etc.) and automates secret rotation policies.

- **AWS Shield**
  As you may have guessed, this service shields your applications from distributed denial of service (DDoS) attacks. AWS Shield Standard, which protects against network layer DDoS, is an automatic benefit at no additional cost, but AWS Shield Advanced, which protects against attacks in applications running on Amazon EC2, AWS ELB, etc., require a subscription.

- **AWS Web Application Firewall (WAF)**
  This service helps you create security rules to protect your web applications against common exploits such as SQL injection and cross-site scripting. It also supports Captchas to force users to go through a challenge (or puzzle) before their request is allowed, thus helping block bot traffic, similar to what you may see on social media sites.

Security and Amazon EC2

Since a large part of SAP deployment is on Amazon EC2 (at least until SAP convinces everyone to adopt SaaS), we wanted to focus this section on security in Amazon EC2. Thinking about a (virtual) machine, you would want to secure it at different layers such as network, OS, logging, access, and storage, etc.

Consider following ways to manage network traffic to and from Amazon EC2:

- **Network Access Control List (NACL)**
  Stateless and scoped to a subnet, NACL is an allow/deny list for network traffic (using IP and port). So, you can allow only IPs from your on-premise network and SAP-related ports for an SAP subnet, for example.

- **Security group**
  Going from subnet to instance level, a security group is a stateful and allowed list (anything that's not allowed explicitly is denied) for a particular resource, such as Amazon EC2. For an SAP deployment, let's say that web layer (such as SAP Web Dispatcher) should not be able to talk to database or only application servers can talk to database. In this case, you can attach a security group, to database Amazon EC2 instances, that is configured to deny any traffic that doesn't originate from application servers. We recommend implementing both NACL and security groups for SAP environments. Figure 2.22 shows how this works together at different layers (i.e., subnet and instance).

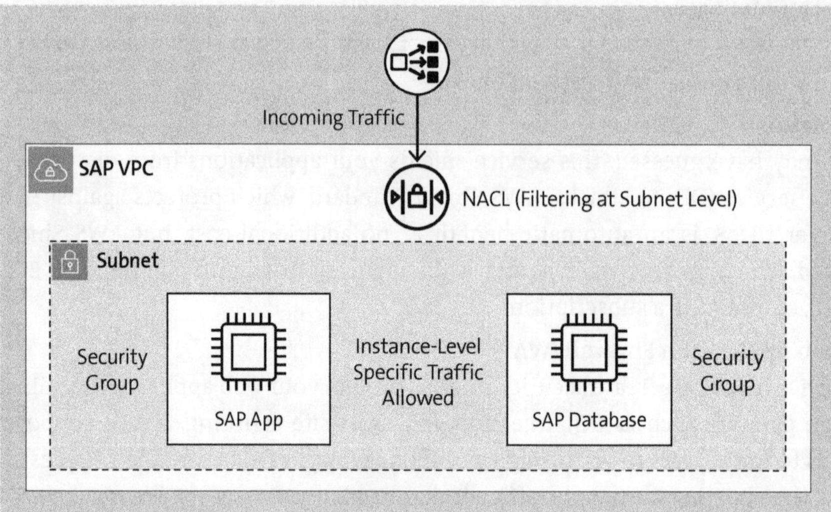

Figure 2.22 Network ACL and Security Groups Representation

Amazon EC2 Instance Connect Endpoints

You can use an Amazon EC2 Instance Connect (EIC) endpoints to connect to an instance either using Secure Shell Protocol (SSH) or remote desktop protocol (RDP) without enabling public IP for the instance (as opposed to creating bastion hosts), as shown in Figure 2.23. An Amazon EIC endpoint uses IAM-based access control combined with security groups and logs everything in AWS CloudTrail.

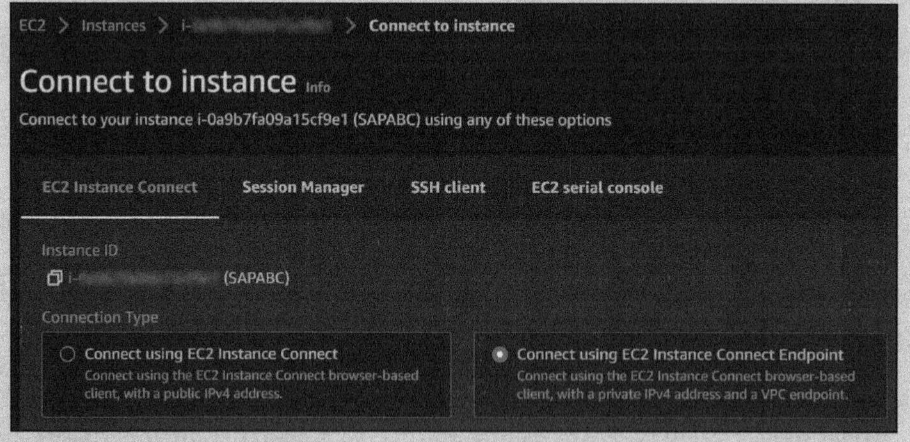

Figure 2.23 AWS Management Console Screenshot Showing EIC Endpoint Connection Option for Amazon EC2

In our experience, every organization has its own policies around OS hardening, including agents that run on the OS, updates, and security patches etc. The Center for

Internet Security (CIS) has OS hardening guidelines (and images at AWS Marketplace) for cyberdefense as well. You can also use Amazon Inspector for security vulnerability assessments on Amazon EC2 instances.

Note

When hardening OS, be sure you follow SAP's guidelines (SAP publishes security guides for OSs, including for database servers) and don't block anything that could disrupt SAP installation and operations.

Configure the right IAM roles and policies based on principle of least privilege (i.e., provide only the minimum access required to perform specific tasks). You should enable the logging of all API calls using AWS CloudTrail, which records the identity of the API caller, the time, the source IP, etc. and maintains history, which is useful for any security analysis as well as audit and compliance requirements.

Access Notification

You can also enable notification for SSH access to Amazon EC2 either using Amazon Simple Notification Service (SNS) or a third-party service for an additional layer of security.

Security discussions are never complete unless we mention encryption! Consider both encryption for data at *rest* as well as for data in *transit*. A few ways of doing so include the following:

- **Amazon EBS volumes**
  For Amazon EBS encryption, AWS KMS keys are used to encrypt the volumes on the host Amazon EC2 instance; thus both data at rest and data in transit between Amazon EC2 and Amazon EBS are encrypted with minimal effect on performance and latency.

- **Amazon FSx**
  All Amazon FSx file systems are automatically encrypted using AWS KMS at rest. Both Amazon FSx for Windows File Server and Amazon FSx for NetApp ONTAP support the encryption of data in transit.

- **Amazon S3**
  Amazon S3 automatically encrypts all data by design using server-side encryption using Amazon S3 managed keys (SSE-S3). On the client side, you're responsible for managing the encryption process and keys. You can use either Secure Socket Layer/ Transport Layer Security (SSL/TLS) or client-side encryption to protect data in transit to Amazon S3.

2.3.10   Management, Governance, and Operations

The struggle between moving fast and maintaining good governance is not new—we have seen both ends of the spectrum, from organizations moving so fast they don't know what would break if they make a change to organizations tied into central governance mechanism so much that they often lose independence of deploying infrastructure for standing up SAP landscape quickly. Having the right framework and long-term thinking (such as for putting operational procedures in place) helps avoid rework and carves a path for the quicker adoption of the new features frequently launched by cloud providers!

In this section, we'll discuss services related to management, governance, and operations that help you create the right governance map for deploying SAP landscapes in AWS throughout the stages of cloud adoption, as shown in Figure 2.24.

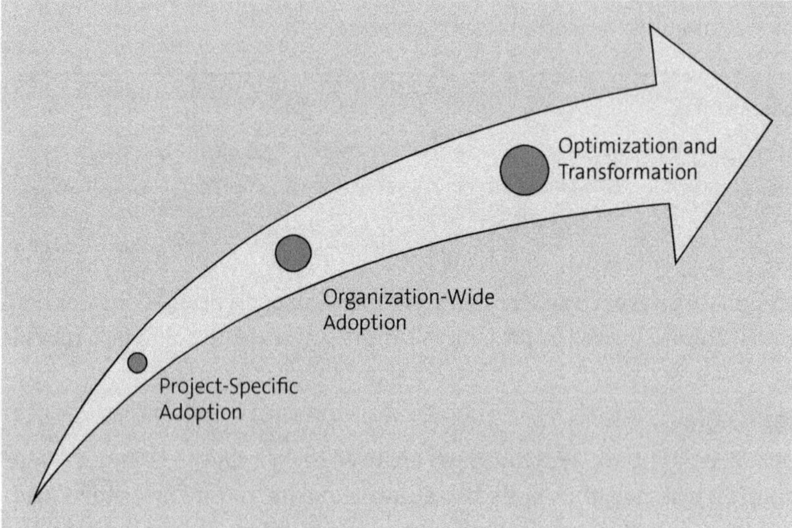

Figure 2.24 Cloud Adoption Stages for an Organization

AWS Organizations

As your organization's footprint on AWS grows (with different stages as shown in Figure 2.24), you'll almost always end up creating more than one AWS account, which is our recommendation as well, for one or more of following reasons:

- Separation of different applications
- Separation of business units
- Separation of production versus nonproduction workloads
- Compliance, governance, and security requirements of different workloads

AWS Organizations helps you manage multiple AWS accounts through a structure called the *organization*. It provides for the central management of all accounts, billing, and governance using organizational units (OU). You can read a more detailed discussion in Chapter 4.

Service Control Policies

Service control policies (SCP) define sets of rules to allow or deny services (or actions) at the organization level that would be enforced to all accounts under that OU. For example, for SAP-related accounts, if you wanted to restrict folks to be able to create only SAP-certified Amazon EC2 instances, that would be defined in an SCP.

AWS Control Tower

With SAP deployments, using multiple AWS accounts is fairly common, usually for reasons such as resource segregation, compliance, security, or organization policy.

Multi-account environment setup and management add to the complexity; that's where AWS Control Tower comes into play and helps with governance for things such as Landing Zone (discussed in detail in chapter 4), logging, policy enforcement, data isolation, etc. across AWS accounts.

AWS Config

AWS Config is a service that enables the continuous monitoring and recording of resource configurations of AWS resources. It tracks changes to configurations and the relationships between AWS resources, thus allowing you to assess overall resource inventory, track compliance with configurations, and diagnose causes of changes.

For SAP systems running on AWS, AWS Config can help establish governance to validate that SAP deployments meet architectural guidelines by providing configuration history. AWS Config rules can also check for unused security groups, unrestricted public access, etc.

AWS Service Quotas

When you create a new account, this account is restricted in terms of the number of services, operations, and Amazon EC2 machines you can create. You can request an increase in service quotas (for allowed services using templates from AWS Management Console), as shown in Figure 2.25, and use it as a tool to prevent unnecessary provisioning of AWS resources.

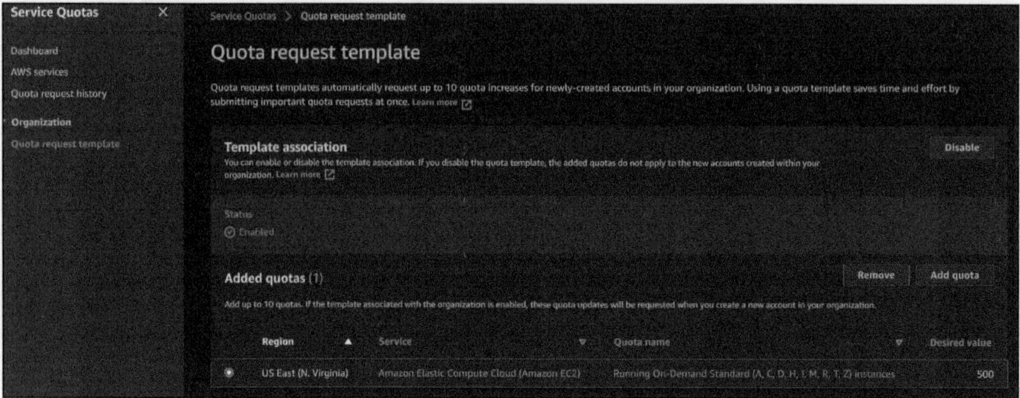

Figure 2.25 Requesting Higher Service Quota for an AWS Account

AWS CloudFormation

If your organization likes to manage infrastructure as code (IaC) instead of say, the console, using continuous integration/continuous deployment (CI/CD) tools, AWS CloudFormation fits right in with supporting both YAML Ain't Markup Language (YAML), a human-readable data serialization language, and JavaScript Object Notation (JSON). You can also use one of many AWS-provided templates as a starting point and develop from there.

AWS Auto Scaling

For non-predictable workloads, if you wanted to set utilization threshold and scale up and down AWS resources based on those thresholds, you'd use the AWS Auto Scaling feature. At the time of writing, it is supported for Amazon EC2, Amazon DynamoDB, Amazon Elastic Container Service (ECS), and Amazon Aurora.

Amazon EC2 Auto-Scaling

Although useful for managing cost and to some effect resiliency, Amazon EC2 auto-scaling is NOT supported for SAP at the time of writing.

AWS Resource Groups

Resource groups allow you to manage AWS resources as a single unit. For example, for a particular SAP system ID (SID), you could group all associated resources such as Amazon EC2, storage, etc., into a single resource group.

AWS Service Catalog

Let's say you want your organization's SAP group to self-service AWS resources but don't want them to use things that don't work for SAP applications (such as Amazon

EC2 auto-scaling), or you want them to use services with certain compliance supported. You would create AWS Service Catalog portfolios and grant access using roles. This approach also allows you to assign different services to separate groups and supports both AWS CloudFormation and open-source Terraform templates as IaC.

AWS CloudTrail

If you use any tool such as AWS Management Console, AWS SDK, AWS CLI, or AWS CloudFormation that makes an API call, it is logged into AWS CloudTrail. You can use this information for audits, to ensure compliance, or any type of analysis by answering questions about actions performed, which user IDs performed which actions, and when the action performed.

The information logged includes event name, time, user, source, resource type, resource name, request, and response parameters, etc. some of which are shown in Figure 2.26. Logs up to 90 days have no additional charge.

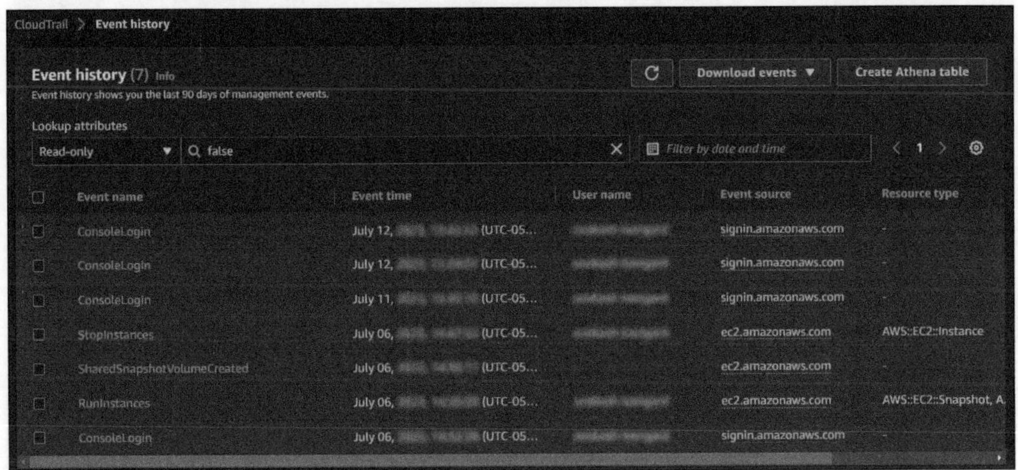

Figure 2.26 Event Log History in AWS CloudTrail

Amazon CloudWatch

If you wanted to know how an SAP system is doing, you would usually collect relevant data, which is combination of metrics (point-in-time measurements such as CPU utilization) and logs (details of an event such as system crash). Amazon CloudWatch collects metrics and logs from services and lets you create dashboards, generate alerts, and analyze root cause of issues.

> **Amazon CloudWatch Application Insights**
>
> The Application Insights feature of Amazon CloudWatch takes monitoring to the application level for the bigger picture of *observability*. At the time of writing, this tool supports SAP HANA databases, SAP NetWeaver systems, and ABAP platform systems.

AWS Compute Optimizer

If a downside exists to cloud computing, in our view, it's the ability provision resources without traditional barriers such as availability of infrastructure at scale. Without right governance, costs can get out of hand. In this context, tools like AWS Compute Optimizer can be useful. It uses machine learning on historical metrics to analyze and recommend optimal configuration for following resources:

- Amazon EC2 instances
- Amazon EBS volumes
- AWS Lambda functions
- Amazon Elastic Container Service (ECS) on AWS Fargate

SAP-Certified Amazon EC2 Instances

At the time of writing, AWS Compute Optimizer is *not* SAP aware; thus, the service recommendation is completely based on metrics. As a result, the solution may suggest instances that are not SAP certified or may not consider that SAP HANA is an in-memory database.

AWS Trusted Advisor

While AWS Compute Optimizer, as the name suggests, makes recommendations only related to compute, AWS Trusted Advisor takes you to the next level with recommendations across multiple factors such as performance, security, fault tolerance, service quotas, and cost optimization, as shown in Figure 2.27.

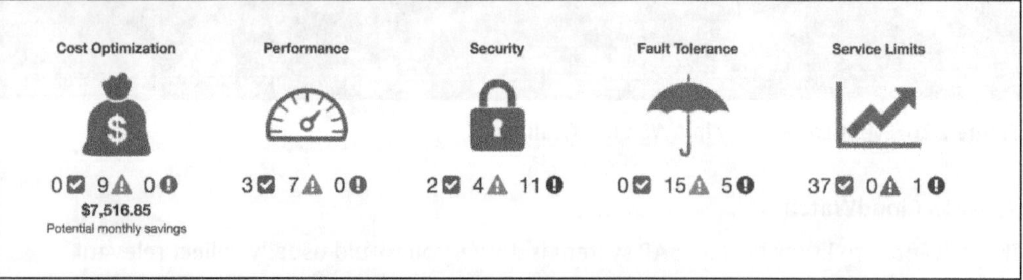

Figure 2.27 AWS Trusted Advisor Makes Recommendations across These Factors

AWS Health

AWS Service Health Dashboard (*https://health.aws.amazon.com*) is a publicly available site that shows the general health status (and past 12 months interruptions) of AWS services, the AWS Health Dashboard has *personalized* availability and performance information within your account for services you're using.

This solution avoids unnecessary noise and allows you to focus on the services important to you (related to SAP systems since SAP doesn't use all AWS services). You can also trigger alerts based on custom rules and receive relevant information in operational tools (such as Slack channels, Microsoft Teams, or email) using AWS Health Aware (AHA) framework.

AWS Cost Management

As the name suggests, AWS Cost Management lets you manage the cost aspects such as budgeting, forecasting, reporting, and *anomaly detection*. This solution also provides right-sizing recommendations across Amazon EC2 instance families, which is powered by AWS Compute Optimizer.

AWS Pricing Calculator

From a lifecycle point of view, AWS Cost Management comes in play after the applications (and thus AWS resources) are deployed. To access the potential cost view of a new deployment, AWS Pricing Calculator is helpful. This web tool, accessed at *https://calculator.aws*, shows you the public pricing (your organization's negotiated discounts are not reflected in the public calculator) that you can customize based on your configuration and save/ share as uniform resource locator (URL) or in comma-separated values/ portable document format (CSV/PDF).

Tip

Use the bulk import feature of AWS Pricing Calculator to import Amazon EC2 and Amazon EBS in a Microsoft Excel file so you don't have to input the configuration of hundreds of instances for an SAP landscape.

AWS License Manager

If you're bringing licenses from software vendors such as Microsoft, Oracle, IBM, etc. that have restrictions related to CPU cores, dedicated hosts, user-based licensing (to name just a few aspects), you can create custom licensing rules using AWS License Manager. Custom rules let you emulate and enforce licensing terms, thus reducing the risk of non-compliance and increasing ease of reporting.

AWS Marketplace

Deploying SAP requires several different types of licenses such as application, database, OSs (or any other third-party software in use), and your organization may not necessarily have a relationship with any other vendor. AWS Marketplace allows you to procure licenses at standard terms from vendors such as RHEL for OS and SLES for SAP. AWS

Marketplace also has a professional services section where you can access offerings from service providers related to support, assessment, training, and managed services.

AWS Launch Wizard

From an application point of view, if you're trying to deploy, say, SAP S/4HANA, you need to either learn about infrastructure (such as SAP-certified instances, storage options, performance and connectivity requirements, etc.) or work with someone who understand AWS services in SAP context. To bridge the gap, AWS Launch Wizard provides a guided experience for sizing, configuring, and deploying AWS resources required for supported applications behind the scenes.

Figure 2.28 shows applications supported by AWS Launch Wizard, and for SAP, the following deployment use cases are supported (at the time of writing):

- SAP HANA database on Amazon EC2 (single instance, HA, and multi-node scale out)
- SAP NetWeaver and ABAP platform (standard, distributed, and HA deployments)
- RHEL and SLES Pacemaker cluster setup as part of HA deployment
- Adding new nodes to already deployed SAP HANA scale-out system and application server for distributed setup

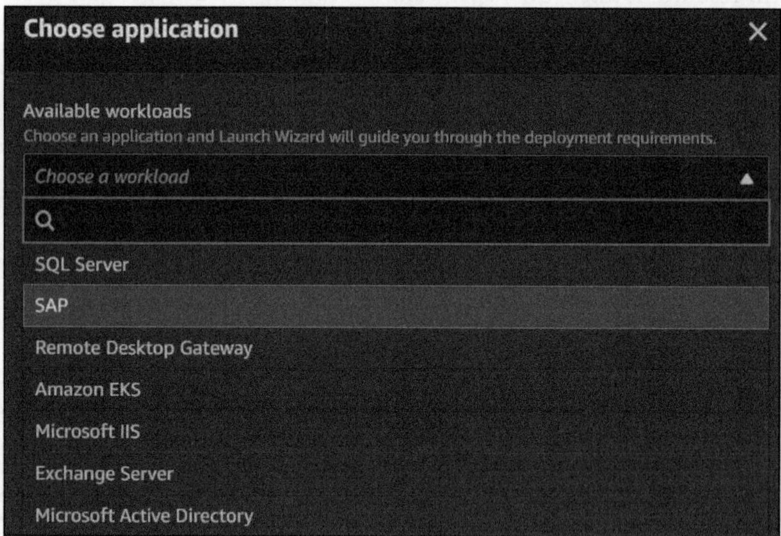

Figure 2.28 Application/Workloads Supported by AWS Launch Wizard

AWS Launch Wizard also creates AWS CloudFormation templates that you can use as a baseline for any new deployment and it installs AWS Data Provider for SAP on Amazon EC2 instances.

AWS Data Provider for SAP

AWS Data Provider for SAP collects performance data (such as OS, storage, network, etc.) and configuration data from several AWS sources (such as Amazon EC2 and Amazon CloudWatch) and makes this information available to SAP. You can access this information via Transaction OS06 and Transaction OS07.

> **Note**
>
> Installing AWS Data Provider for SAP is a support prerequisite from SAP as highlighted in SAP Note 1656250. If you don't use AWS Launch Wizard for SAP deployments, ensure you have this step captured in your list of deployment tasks.

Amazon Simple Email Service

SAP systems often send emails, and usually, an organization uses their own Simple Mail Transfer Protocol (SMTP) system. To simplify things on AWS, you can use a cloud email service called Amazon Simple Email Service (SES) that also supports sender statistics and an email deliverability dashboard.

Amazon Simple Notification Service

The Amazon SNS, as the name suggests, sends notifications in two different ways:

- Application to application (A2A): This option can be used for distributed applications that are not tightly coupled to send a notification that the target would act on based on publish-subscribe (pub-sub) model.
- Application to person (A2P): This option can be used to send notification directly to a recipient using short messaging service (SMS on phone number), mobile push notification, and email.

> **AWS Launch Wizard and Amazon SNS**
>
> AWS Launch Wizard for SAP supports Amazon SNS notifications that you can use to monitor or be notified about any failure event, completion, etc.

AWS Elastic Disaster Recovery

The AWS Elastic Disaster Recovery (DRS) service provides a block-level replication mechanism at the server level and can be used to migrate systems to AWS or for DR setup. Since SAP databases usually have a high rate of change, we recommend using AWS DRS for non-database Amazon EC2 instances in DR setups.

AWS Systems Manager

Often, the discussion focuses on the architecture and migration to AWS, but once on AWS, operating is equally important! (We discuss SAP operations in detail in Chapter 7.) AWS Systems Manager acts as an operations hub for multiple AWS services and some aspects of SAP applications as well. This solution is divided into four pillars, as shown in Figure 2.29:

- **Operations management**

  Supports capabilities such as incident management, Amazon CloudWatch dashboards, and custom resource dashboards.

- **Application management**

  Supports capabilities such as grouping resources into applications, viewing operational data, and integration with AWS KMS for database strings and password management.

- **Change management**

  Supports capabilities such as change workflows, the automation of common operational aspects, and the scheduling of maintenance windows.

- **Node management**

  Supports capabilities such as server fleet management, compliance dashboards, and patch management.

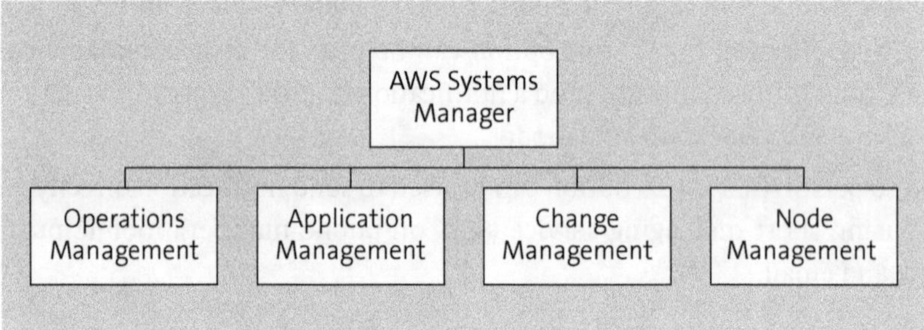

Figure 2.29 AWS Systems Manager Pillars

AWS Well-Architected Framework

You may have noticed already the number of resource and service options and optimizations to be done for SAP deployments and management in AWS. The AWS Well-Architected Framework brings these together with best practices framework for application architecture and tool that lets you track progress over time. AWS Well-Architected Framework's pillars include security, reliability, performance efficiency, operational excellence, cost optimization, and sustainability (discussed in detail in Chapter 3). The tool aspect is part of AWS Management Console, which is based on the framework and provides guidance related to these pillars.

2.4   Summary

Application deployment and management requires learning all aspects of the cloud such as security, resiliency, cost, and performance rather than just service offerings. To that extent, we started this chapter with a focus on applications by discussing deployment models and walked you through several AWS services and offerings for SAP landscape; we also set the context of how these pieces fit together for application architectures. In the next chapter, you'll learn about application architecture frameworks with a focus on recommendations for configuring these services and best practices as it relates to the pillars of the AWS Well-Architected Framework. We'll also take the SLA discussion from a technical definition to a more business-focused conversation.

Chapter 3
Framework for Designing Solutions on AWS

There comes a time during every SAP migration project when you ask yourself, "Are we ready to cutover?" In other words, can our business folks start using the system? We'd like you to ask yourself one other question first: Are we well architected? A systematic approach to considering all aspects of an architecture such as performance, cost, reliability, etc. in the form of a framework is a prerequisite for a successful deployment, and this chapter walks you through these design considerations.

So far, we've discussed how to look at cloud services from a business requirement lens and Amazon Web Services (AWS) services for SAP deployments. Now, we'll look at the framework for designing solutions on AWS. Think of this task like building a house. You've decided on a location, picked a builder, and got all the materials to get started. Now, you don't have to be in the construction business to realize that the industry must follow certain guidelines to ensure the quality of work and to accelerate construction; Trade-offs may be needed involving costs, the speed of construction, and managing project risks. Coming back to SAP, what if you had access to a design framework and principles for solutions, such as SAP, based on AWS's experiences; not only could you ensure you cover every aspect of the architecture but you can also provide a systematic approach to decision-making.

In this chapter, we'll discuss the framework for SAP solution design on AWS and walk you through the decision-making process. We'll conclude this chapter by talking about how service level agreements (SLA) play a role in all this!

3.1 AWS Well-Architected Framework

The AWS Well-Architected Framework acts as a bridge between your requirements and your target architecture in terms of decision-making and trade-offs. For example, you know that the application (SAP in this instance) is critical to business and requires that the cloud architecture provide the same or better reliability as on-premise—connecting that requirement to application recovery, high availability (HA), and disaster recovery (DR) is what the framework helps you focus on!

The AWS Well-Architected Framework is built around six pillars: reliability, security, performance efficiency operational excellence, cost optimization, and sustainability, as shown in Figure 3.1. We'll cover each pillar, as well as more general design principles, in the following sections. The AWS Well-Architected Framework also has the following aspects:

- **Domain-specific lens**
  Recommendations pertaining to specific industry or technology domain such as SAP. The SAP Lens is discussed later in Section 3.2.

- **Hands-on labs**
  Experience the recommendations by performing related tasks in an AWS account. Lab instructions can be accessed online at *https://www.wellarchitectedlabs.com/*.

- **Tool in AWS Management Console**
  An AWS service (called the AWS Well-Architected Tool) that lets you document and measure your workloads (such as SAP) against the framework's best practices.

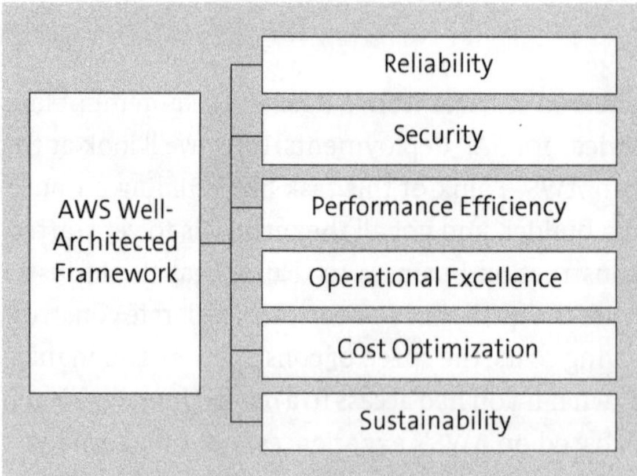

Figure 3.1 AWS Well-Architected Framework Pillars

Note

At the time of writing, the AWS Well-Architected Tool does not have automation aspect to it, meaning you can't run a script to measure your application's compliance to best practices; it's more of a guidance and measurement tool that highlights the risk (and provides improvement ideas) when you go through a questionnaire and provide inputs regarding which recommendations were followed versus not.

3.1.1   General Design Principles

As we go through each pillar, we'll discuss design principles, some of which may not apply to SAP landscape directly, hence, AWS has a guide called *SAP Lens* complementing the AWS Well-Architected Framework that provides context to SAP related nuances. As you go through system deployments, feel free to experiment with different patterns (within the constraints of SAP certifications and AWS supported architecture) and measure things that can be improved in the next iteration; think of it in terms of continuous improvement lifecycle (i.e., experiment, measure, optimize), as shown in Figure 3.2. You can then add your own questions based on your learnings, organization's policies, and operational procedures as part of the *custom lens* to the AWS Well-Architected Framework.

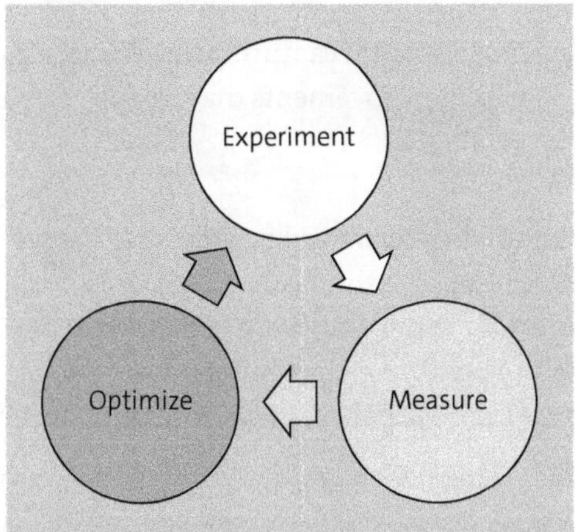

Figure 3.2 Components of Continuous Improvement Lifecycle

Figure 3.2 also shows the crux of the AWS Well-Architected Framework's general design principles, which include the following:

- **Test systems at production scale**
  Simulating a full-scale environment and peak usage is made easier by the cloud since you can provision resources quickly and pay for what you use (on-demand pricing). You can test HA/DR configurations and simulate year-end peak loads in a non-production system for SAP, for example.

- **Don't speculate on long-term capacity needs**
  The cloud enables scaling up and down for resources making adapting to workload needs easier and doesn't require 3- to 5-year capacity forecast. You can gradually change the Amazon Elastic Compute Cloud (EC2) instance type as your SAP HANA database grows instead of maintaining spare capacity.

- **Automate and experiment with architecture**
  Creating new systems and replicating your current ones using automation enables faster deployment and training cycles, especially for use cases such as SAP system copies and creating parallel landscapes for upgrade testing or training. You can also experiment and measure the impact on cost and performance for adopting new features to achieve the optimal architecture.

- **Move towards evolutionary architectures as standard**
  As your organization's needs evolve, you'll need to re-evaluate your original architectural decisions and make changes to adapt. In a cloud environment, this adaptation is a fluid process where you make use of new features and innovations through automation. SAP architectural changes (such as moving to a standalone enqueue server 2) also become easier to implement when you think of the architecture in terms of continuous improvement instead of a series of static decisions.

- **Make data-driven decisions**
  Every choice you make affects some behavior of a system; capturing the effects (e.g., on availability, resiliency, cost, etc.) lets you make improvements that are supported by data over time.

- **Improve through game days**
  Game days (or learning by gamification) simulate unexpected events or failure scenarios, usually in non-production systems, to ensure the team has procedures in place to minimize the impact to the business if the problem were to happen in production. This approach also helps you identify possible architectural improvements or gaps from previous decisions across areas such as operations, performance, security, cost, etc.

> **Note**
>
> This chapter focuses on the design principles of the AWS Well-Architected Framework pillars rather than on providing prescriptive guidance. You'll learn more about how these principles are used for SAP on AWS architecture in Chapter 4 and Chapter 5.

3.1.2   Reliability

AWS defines reliability as the ability of a workload to perform its intended function correctly and consistently when it's expected to, making it a combination of availability, performance, and resiliency. Figure 3.3 shows the five design principles of the reliability pillar:

- **Automate recovery**
  Availability is a central part of resiliency and depending on how the application handles data consistency and user connections, using monitoring and automation to recover the system is more efficient than manual intervention. Always work with your business context and key performance indicators (KPIs) for recovery automation since technical recovery doesn't necessarily measure business impact. For example, recovering a non-production system using Amazon EC2 auto-recovery has a lower impact than doing so for production system, especially when HA is configured.

- **Test recovery procedures**
  Having a recovery plan and executing it are not the same things. On-premise environments provide limited opportunity to validate failure scenarios and recovery process, and that's where the flexibility of the cloud is instrumental in letting you conduct scenario planning and chaos engineering (randomly injecting a failure and ensuring/test recovery).

- **Scale horizontally**
  Horizontal scale is having larger number of less powerful servers rather than fewer but more powerful ones. In an SAP context, having a larger number of SAP application servers is a more resilient design pattern, since it statistically contributes to increased availability, than having fewer, with same total SAP Application Performance Standard (SAPS). You may need to make some trade-off with operations and management, but automation takes a lot of complexity away from that.

- **Manage capacity based on utilization**
  We discussed not guessing about your long-term capacity needs as general principle, but even for short-term needs, you should take the data-driven approach. Monitor the utilization, peak usage, etc. and combine that information with other known business needs such as year-end close to adjust capacity! We mean adjusting down as well as up, which helps with service quota constraints as well.

- **Automate change management**
  Managing change often has an application context such as patching, upgrade, etc., but you must remember to apply that idea to automation as well, so everything is tracked and audited. In case issues arise, problems are easier to identify and roll back as well.

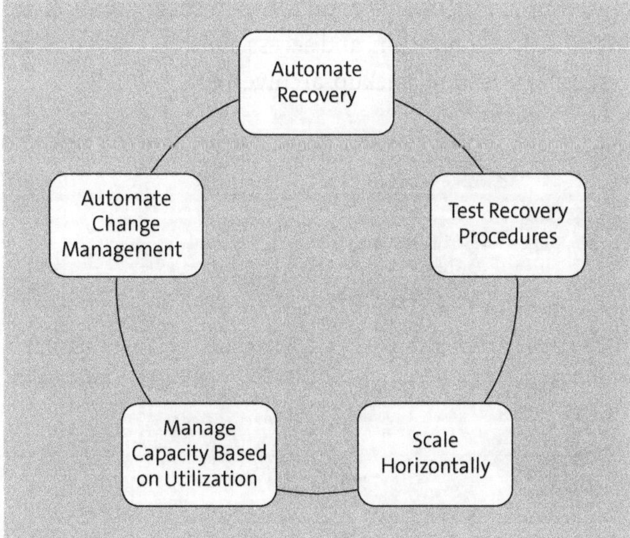

Figure 3.3 Design Principles for Reliability in Cloud

3.1.3   Security

The security pillar focuses on strengthening of security in cloud and covers application and data protection, encryption, and identify related principles, as shown in Figure 3.4:

- **Implement foundation for identity**
  Identity and security go together, so having a strong identity foundation is essential to any security model. Implementing the least privilege model, separation of duties enforcement with correct authorization, and centralized identity management are key foundational aspects of security.

- **Maintain traceability**
  Having logs is not only an audit thing but also helps with investigation and accountability. Always configure traces and enable application programming interface (API) logging for actions in the cloud and for application-related logs for SAP.

- **Apply security at all layers**
  This principle is like having a deadbolt on your door and still getting a security system, sometimes also called the *belt and suspenders approach* (strict adherence). Apply security controls wherever applicable such as the network layer, in your virtual private clouds (VPCs), Amazon EC2, operating system (OS), SAP, etc.

- **Automate security practices**
  Automation never goes out of style! Automating security controls and processes have less friction while implementation and provide effective long term governance structure as well. Version controlled code to implement controls provides the ability to trace and rollback. For example, you could mandate that every Amazon EC2 instance have security groups associated with it during creation.

- **Protect data at rest and in transit**
  Start by classifying the data according to sensitivity and use tools for encryption (for both data at rest and data in transit), tokenization, and access control. Also, consider your data at every level: database, filesystem, backup, archives, etc.

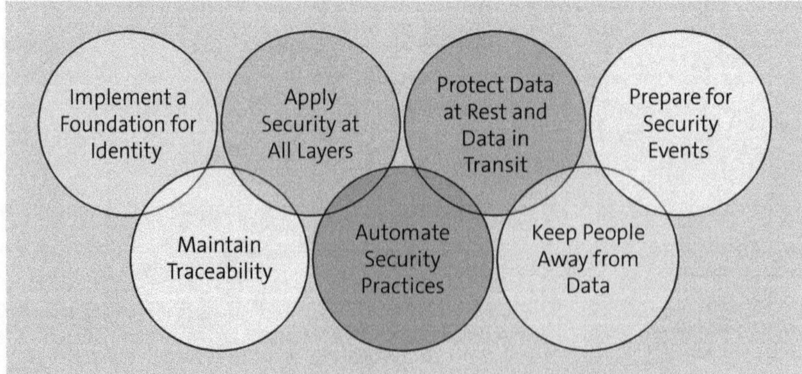

Figure 3.4 Design Principles for Security in Cloud

- **Keep data away from people**
  This principle may sound counterintuitive at first, but the best way to prevent mistakes and reduce risks is to prevent access, which doesn't mean nobody gets access, of course. The idea is to eliminate the *need for the direct access* and any manual processing.

- **Prepare for security events**
  Always assume a security breach and implement controls for better response and investigation of security events. Run simulations for various detect, investigate, and recovery processes with different breach scenarios.

3.1.4   Performance Efficiency

For good reason, this pillar is called *performance efficiency* and not shoot for the *best performance*—trade-offs will always need to be made, and to that extent, this pillar focuses on meeting your business requirements by using your resources efficiently. For example, your organization may prefer acceptable performance at a lower cost over spending a lot for the best performance, depending on the application's criticality.

Figure 3.5 shows the design principles for the performance efficiency pillar:

- **Democratize advanced technologies**
  Newer and more advanced technology implementation requires huge investments in terms of learning curve and time to maturity. You're better off letting cloud providers deal with the complexity and maturity and enabling teams to use the technology as a cloud service. A few examples include NoSQL databases, artificial intelligence (AI), media transcoding, etc.

- **Think and go global**
  SAP customers often have global footprint; if your organization does too, then think about latency from different regions and potentially use multiple regions for better customer experience. For SAP landscape, it's not always possible to use multiple regions at the same time because of the architecture, but picking the right region is still important.

- **Go serverless**
  In theory, everything runs on servers but from user experience point of view, a lot of the complexity is hidden behind the service offering. So going serverless means you just focus on utilizing the service for your use case rather than burdened by managing servers as well. Most of the SAP systems rely on you using Amazon EC2, because of the application architecture, but services such as Amazon Simple Storage Service (S3), Amazon Elastic File System (EFS), Amazon Elastic Block Storage (EBS), Amazon Elastic Load Balancing (ELB), etc. are still utilized and simplifies overall management.

- **Experiment often**

  The ease of deploying systems and automation make it easier to validate your architecture using data points. Make note of important KPIs and experiment around those when new features and services are released by the cloud provider.

- **Consider mechanical sympathy**

  Mechanical sympathy, in this instance, refers to understanding the technology, intended use, and consumption models better. For example, even though you can mount Amazon S3 to an Amazon EC2 instance, that doesn't make Amazon S3 suitable for database files since Amazon S3 is an object store. For SAP systems, AWS already has a recommended pattern that guides you about what service to choose, while still allowing you to experiment with IOPS, throughput, etc., for system-specific performance.

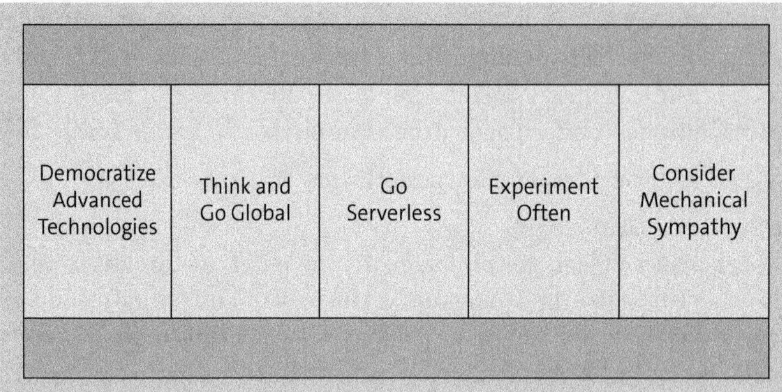

Figure 3.5 Design Principles for Performance Efficiency in Cloud

3.1.5   Operational Excellence

We've discussed the importance of keeping operations in mind right from the beginning of a cloud migration project. Otherwise, you may make decisions that sound right for the migration but have ripple effect during operations. The operational excellence pillar focuses on efficient operational procedures, gaining visibility, and continuous improvement. Figure 3.6 shows the design principles of the operational excellence pillar:

- **Perform operations as code**

  Infrastructure as code (IaC) helps automate a lot of things, and SAP also supports use cases such as silent application install, system copy, and replication, which can be managed using DevOps processes and tools. Not only does it help with scalability, but also it avoids mistakes that can potentially affect business-critical systems.

- **Make frequent, small, and reversible changes**

  The scale of automation can go either way—if a mistake arises, that problem would be enhanced as well! So, the idea behind this principle is to restrict the potentially large-scale impacts of something going wrong. Making small changes frequently that can be reversed quickly avoids chaos. For example, moving small changes to an SAP production system using transports every week is better than a monthly transport window because it limits the impact and because smaller changes are easy to revert.

- **Refine operational procedures often**
  With the ease of adopting new architectures and services, operational procedures must be updated and refined as you use them. For example, if you decide to use Amazon FSx for NetApp ONTAP instead of Amazon EBS for your SAP HANA database, remember to not only update the process docs saying so but also include the advantages you were hoping to get and the operational use cases that you hoped would be easier going forward.

- **Anticipate failure**
  If something is always true for the cloud, it is "failures happen!" Prepare to mitigate failure scenarios either through operational procedures (such as monitoring and taking action based on that) or design (such as configuring HA for business-critical systems). Always test your procedures either through game days or during a maintenance window.

- **Learn from operational failures**
  Although this principle applies on-premise as well, it's even more important in a cloud environment. Sharing these lesson not only improve your operational procedures, but they can also act as data points to make better architectural decisions. For example, if a database restore fails during an SAP system copy, you might need to improve your backup architecture as well.

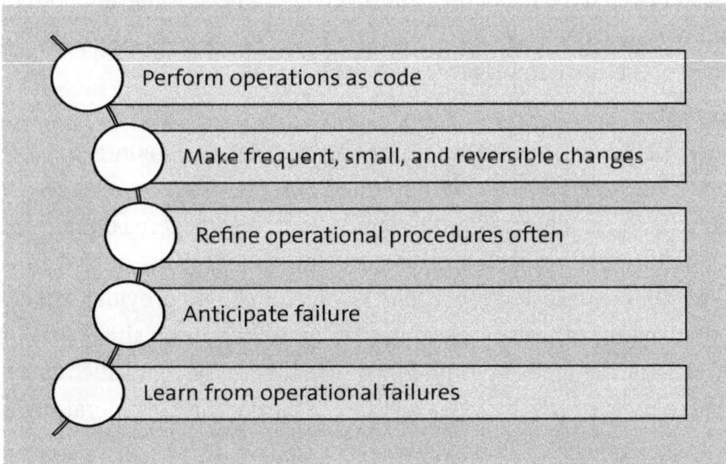

Figure 3.6 Design Principles for Operational Excellence in Cloud

3.1.6   Cost Optimization

With on-premise servers, you hardly think about optimizing cost after the initial deployment; we've seen terabytes of storage with software, exports, old interface files, etc. sitting without any focus to delete those since it's in nobody's priority list (because it doesn't save money). However, cost optimization in cloud is an iterative and continuous process. Just like you try to optimize your electricity bill by switching off fans not in use, turning everything off when nobody is at home, etc., cloud cost can be optimized using several techniques. The cost optimization pillar focuses on principles, as shown in Figure 3.7, that let you deliver the same business value at the lowest cost:

- **Implement cloud financial management**
  Cloud cost is fairly transparent, and you can see cost of each component such as the load balancer, public IP address, etc. Add the various cost models and flexibility of service deployments in the cloud, and now you need pay more attention to planning, controlling, and optimizing the cost. That's the purpose of cloud financial management. If your organization doesn't already have a structure in place, think about implementing cloud financial management right from the beginning of a project. A lot of what you hear about cloud costs getting out of control stems from the lack of a correct governance model for managing costs.

- **Adopt a consumption model**
  Think of systems in terms of being used versus not and how you're paying for it. (We discussed Amazon EC2's pricing model in Chapter 2.) Cloud flexibility is easy to interpret as you can deploy a new SAP system when you need to. Often forgotten, however, is the part where you can shut down the system when not in use (for systems with on-demand Amazon EC2 pricing). A sandbox system that is used only once a month doesn't need to be up all the time. Even if you need something out of the expected window, you can always bring one up or create one using automation, thus allowing better operational management as well.

- **Focus on business value**
  When you consider costs, not only the cost of the server, but also maintenance of data centers, operations, spare capacity planning, and other costs are involved. So, if your organization sells shoes, all of this infrastructure can be considered *undifferentiated* work that doesn't help you sell more shoes. You should focus on business projects and use the right service and cost model) from your cloud provider, and by doing so, your organization can make more money by selling more shoes than it pays for the cloud service. Using the right pricing model for SAP systems, such as the compute savings plan for systems that are always on, is part of this consideration as well.

- **Analyze and attribute expenditure**
  In our experience, a lot of organizations do not know the true cost of running SAP system on-premise because often the infrastructure and maintenance costs are shared among all IT systems. Since the cost of the cloud is easy to classify, *you should*

create a workload-based approach to map costs to business value. In this way, you'll know how much of your business runs on SAP systems or on a particular module. This approach also helps with chargebacks in some organizations where each business unit pays for their own infrastructures.

- **Measure efficiency**
  Once you can associate your SAP landscape cost to business value, you can use that information as an indicator for the value from cost savings (i.e., efficiency). For example, if your $100,000 spend on SAP system supports $1 million of business and you save $2,000 by making a change, that doesn't add as much value as if you can save $10,000 with an architecture change and support more business users. The idea here is to make decisions that promote efficiency over raw numbers.

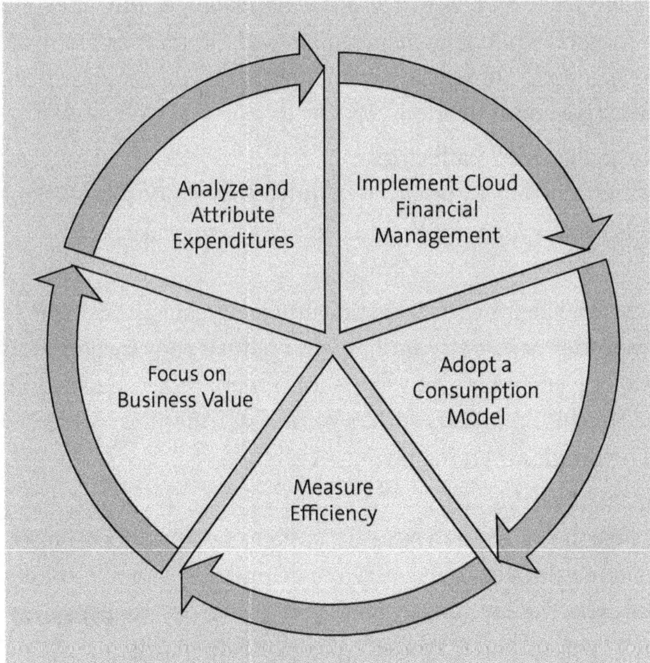

Figure 3.7 Design Principles for Cost Optimization in Cloud

3.1.7   Sustainability

Sustainability is rarely top of mind when you're involved in an SAP migration project, in operations, or on related projects. Start by looking at your organization's goals rather than a project goal—how committed is the overall organization towards sustainability, what is being done in other groups, and how collaboration and cross learning would help the goal. The sustainability pillar focuses on efficiency and reducing environmental impact through better cloud resource utilization and reducing waste, which reduces total cost of ownership (TCO) and carbon emissions. Figure 3.8 shows the principles of this pillar:

- **Understand the workload impact**

  Following this principle might not be a straightforward task since there are moving pieces, but the idea is to measure the impact of the workload (say SAP), including the impact of customers using the product. From this information, you can try to improve productivity while reducing the overall footprint and impact.

- **Establish sustainability goals**

  In the long run, your goals would be something like reducing Amazon EC2 or storage for every SAP transaction or per business user (i.e., doing more by optimizing resource and utilization).

- **Maximize utilization**

  If you have five SAP application servers running at, say, 40% utilization each, you're not using all your available resources. Aim for efficient resource utilization by reducing the number of servers (while still ensuring HA and DR metrics are met). Don't keep too much free space as a storage allocation, for instance. As a side effect, this principle also helps with cost optimization.

- **Adopt efficient hardware and software offerings**

  The key idea is flexibility and utilizing automation to move applications to efficient hardware (as and when released by AWS) and software that you may use.

- **Use managed services**

  Using managed services not only helps with operating efficiently, but also (since the infrastructure behind the scenes is shared) can help you reduce your carbon footprint. For example, instead of using your own servers for Grafana, you can use Amazon's managed Grafana offering. Similarly, other serverless offerings could be, depending on use case, a better pick then managing servers.

- **Reduce downstream impact of workloads**

  Pay attention to the user side, that is, the downstream impact of your applications as well. Users needing to upgrade devices to use services or utilizing more resources with inefficient design impacts the amount of energy or resources used. You can understand these potential impacts better by testing using device farms.

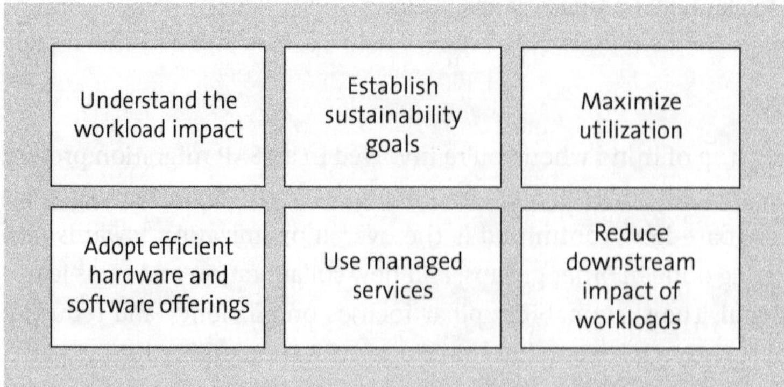

Figure 3.8 Design Principles for Sustainability in the Cloud

3.2   SAP Lens

At this point, if you're thinking that being "well architected" sounds great, but since SAP is special, would it apply there, AWS' answer is: yes, with few tweaks. As we alluded to in Section 3.1.1, SAP related recommendations collection is captured in SAP Lens of the AWS Well-Architected Framework.

To put it in context, let's say you know how to drive and bought a new electric vehicle (EV), which not only looks different, but also has buttons and controls at different places (compared to your older car); you already know that maintenance requirements for EV is probably different from a gas (petrol) powered one and you need to learn all of these. The fact that you already know how to drive a car doesn't change but you still need to know the nuances of EVs and particular make/model you bought. Similarly, you need to learn SAP architectural and operational specific recommendations in addition to general cloud and well architected principles to ensure that your SAP workload is well-architected; thus, SAP Lens for the AWS Well-Architected Framework was born!

SAP Lens

The SAP Lens does not replace the AWS Well-Architected Framework, but is supplemental content that clarifies how to interpret and adopt the general best practices of the AWS Well-Architected Framework into SAP workload designs.

Don't think of this as something that needs to be done after the deployment or go-live, rather it should be in your radar right from the planning stage, as various teams collaborate to define the target architecture and migration pattern and in your operational procedures as a periodic review item at the following times:

- **Early design phase**
  Reviewing your workloads during the design phase ensures that there are no one way door (i.e., irreversible) decisions being made. This helps in laying the right foundations for workloads using guidance from all five pillars.

- **Before going live**
  Reviewing your workloads before going live ensures that the decisions, trade-offs, and the application architecture is as intended. It also provides an opportunity to test items such as HA/DR procedures, making broader teams familiar with cloud architecture, and potentially fix any issues identified as part of the review.

- **Architecture and business evolution**
  Over time your architecture might consume newer AWS services and evolve based on new features released from SAP and AWS. So, use SAP Lens review when you make changes to your architecture, and periodically as and when business requirements evolve. For example: business might determine that an SAP system that wasn't critical last year is now important, which might require HA architecture for that landscape.

Table 3.1 shows the topic areas that align with the pillars of the AWS Well-Architected Framework. Both the AWS Assessment tool (*https://a2t.accelerate.amazonaws.com*) or the AWS well architected tool in AWS Management Console (*https://console.aws. amazon.com/wellarchitected*) are good places to start and provide best practice ratings.

AWS Well-Architected Framework Pillar	Topics
Operational excellence	<ul><li>Prepare</li><li>Manage SAP changes</li><li>Operate</li><li>Evolve</li></ul>
Security	<ul><li>Identity and access management</li><li>Detective controls</li><li>Infrastructure protection</li><li>Data protection</li></ul>
Reliability	<ul><li>Foundations</li><li>Change management</li><li>Failure management</li></ul>
Performance efficiency	<ul><li>Selection</li><li>Review</li><li>Monitoring</li><li>Trade-offs</li></ul>
Cost optimization	<ul><li>Cost-effective resources</li><li>Matching supply and demand</li><li>Expenditure awareness</li><li>Optimizing over time</li></ul>
Sustainability	<ul><li>Sustainability-centric design decisions</li><li>Improvements for infrastructure</li><li>Monitor and measure</li></ul>

Table 3.1 AWS Well-Architected Framework Pillars and the Related Topics for SAP Lens

Using the AWS Well-Architected Framework tool, not only can you import SAP Lens, but you can also create your own custom lens for your organization, according to your compliance and operational requirements.

To use this tool, you must first define the workload and import the JavaScript Object Notation (JSON) file for SAP Lens. SAP is considered a custom lens in the tool, at the time of writing. Following are high level steps that you need to execute:

1. Download the SAP Lens JSON file, which available at *http://s-prs.co/v577609*.

2. Create a custom Lens in the AWS Well-Architected Framework tool in the AWS Management Console using the downloaded JSON file.

3. In the AWS Well-Architected Framework tool, under custom lenses, as shown in Figure 3.9, select the SAP Lens and choose **Publish lens**.

4. Apply the SAP Lens to your SAP workloads. You can also share the lens with all accounts in the organization or in an organizational unit (OU).

Figure 3.9 Importing SAP Lens as from Custom Lenses Menu in AWS Well-Architected Framework Tool

Figure 3.10 shows an example of how you can review the SAP Lens against recommendations. At the end of the review, you can generate a PDF report and also examine an improvement plan, if any, in the tool itself, as shown in Figure 3.11.

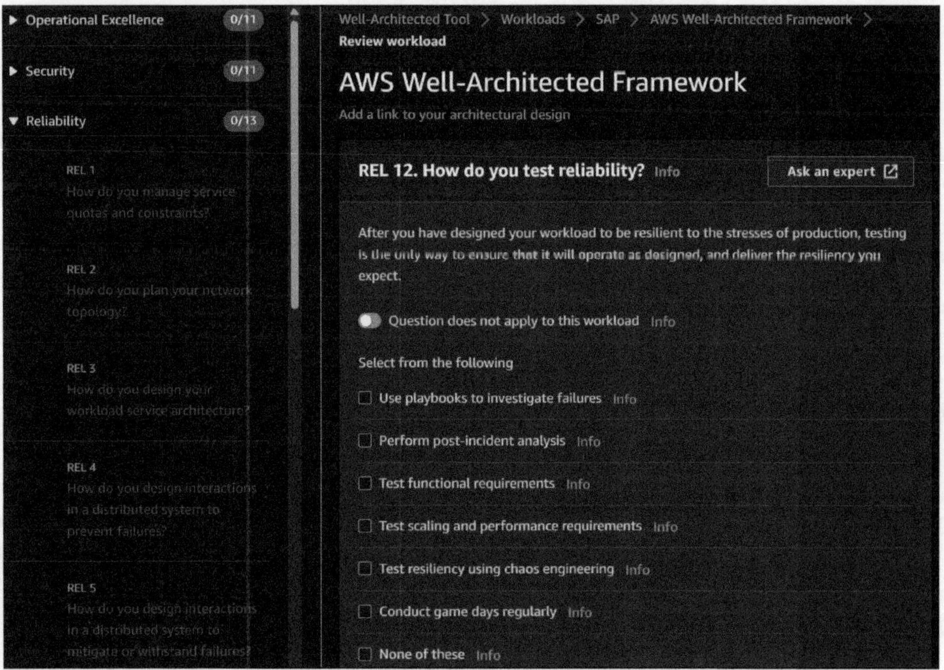

Figure 3.10 Reviewing SAP Lens

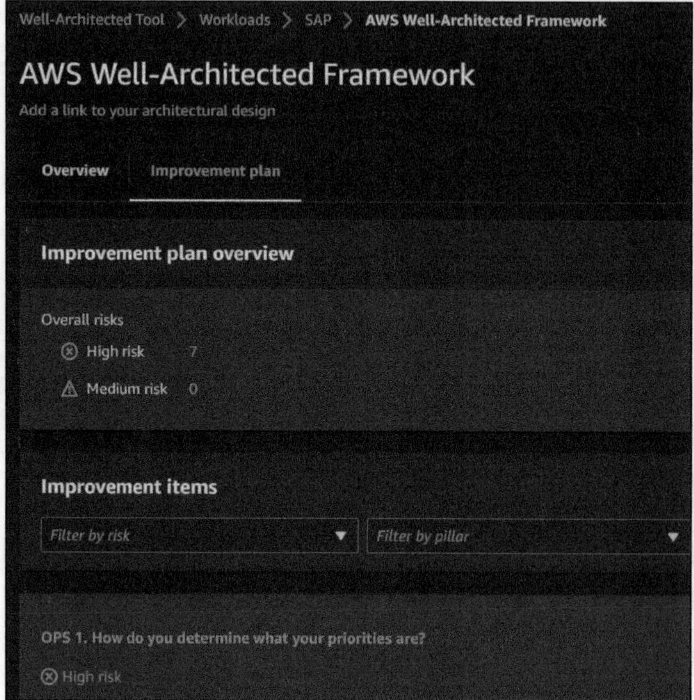

Figure 3.11 Improvement Plan Based on Your Answers during the Review

3.3   Cloud Service Level Agreement

An SLA is a contract that outlines the terms of service, how to measure metrics, and the actions to be taken if the service terms are not met. For example, cloud providers have availability SLAs for compute.

In the context of daily life, when you buy a new phone, you get some coverage through the warranty. When your favorite game is streaming, you expect internet to be available and fast enough (for worry-free gaming) so you can focus on the game (and hosting the party!) Similarly, when you use a cloud service, you expect certain assurances in terms of availability, reliability, and performance, which are governed by SLAs. If you work with a system integrator, you're already familiar with how they use SLAs to maintain the quality of service, such as ticket response/resolution time, SAP performance metrics, etc.

Suffice to say, SLAs are important and highlight not only the cloud provider's responsibilities but also your organization's responsibilities, as a customer. Some common elements of an SLA include the following:

- Service overview and definitions of metrics such as what does "error rate" or "uptime" mean in the context of that service

- Service level objectives (SLOs) such as uptime percentage

- Security standards and joint responsibilities between the cloud provider and your organization as the customer

- Penalties, which can be financial, if the SLA is not met and the claim process

- The process for termination or SLA expiration

- Exclusions from the SLA

- Data residency and industry compliance such as compliance with the Health Insurance Portability and Accountability Act (HIPPA) and the Payment Card Industry (PCI), as applicable

In this section, we'll discuss how to understand SLAs in context of AWS and what it means for SAP deployments.

SLA as Financial Constructs

One could argue that "an SLA is purely a financial construct, since the penalties are often financial in nature and thus not a reliable metric for architectural decisions." We certainly have come across that sentiment; however, cloud providers do strive to meet SLOs and defined KPIs the with often more stringent internal SLO or design target and back the claim with money. So, in a sense, the argument of not making decisions based on SLA is like saying we should measure distance in "minutes" instead of "miles," since we talk about distance in terms of how long it takes to travel all the time! In addition to officially published SLA, if available, you should also look for actual downtime statistics.

3.3.1   Understanding AWS Service Level Agreements

As you may have guessed already, AWS provides an SLA for all paid, generally available (GA) services. You can find uptime SLA of AWS services from the public page at *http://s prs.co/v577611*. As discussed earlier, when looking at an AWS service, you should pay attention to the span of the service. (For example, a subnet resides within an AZ and doesn't span zones.) The SLA is another important metric you should look at.

Let's explore this topic with an Amazon EC2 SLA as an example, which you can find at *http://s-prs.co/v577612*. The SLA covers the following topics:

- **The definition of the service and scope**
  The Amazon EC2 SLA applies for each account separately and also includes "any Amazon Elastic Graphics, Amazon Elastic Inference, and Elastic IP Address resources purchased with the relevant Amazon EC2 instance(s)." Also included is what happens when a conflict arises between the terms of the SLA and the customer agreement.

- **SLA scope**

  The region (concurrent deployment across AZs in same region) and the instance level (single Amazon EC2 instance).

- **Metric and penalty**

  Uptime as a monthly percentage and the service credit percentage if the uptime goals are not met, as shown in Figure 3.12, which is an example Amazon EC2 uptime SLA.

- **SLA credit and request procedure**

  How the credits are calculated as a percentage of the monthly bill and how to submit a claim (if not an automated refund).

- **SLA exclusion**

  The SLA doesn't apply to certain things such as the suspension of, the termination of, and performance issues in Amazon EC2. Also included is what happens if you fail to respond to resource health concerns.

- **SLA definitions**

  Defines terms such as AZs, monthly uptimes, service credit, etc.

Region-Level SLA

For Amazon EC2 with all running instances deployed concurrently across two or more AZs in the same region (or at least two regions if there is only one AZ in a given region), AWS will use commercially reasonable efforts to make Amazon EC2 available for each AWS region with a Monthly Uptime Percentage of at least 99.99%, in each case during any monthly billing cycle (the "Region-Level SLA"). In the event Amazon EC2 does not meet the Region-Level SLA, you will be eligible to receive a Service Credit as described below.

Monthly Uptime Percentage	Service Credit Percentage
Less than 99.99% but equal to or greater than 99.0%	10%
Less than 99.0% but equal to or greater than 95.0%	30%
Less than 95.0%	100%

Instance-Level SLA

For each individual Amazon EC2 instance ("Single EC2 Instance"), AWS will use commercially reasonable efforts to make the Single EC2 Instance available with an Instance-Level Uptime Percentage of at least 99.5%, in each case during any monthly billing cycle (the "Instance-Level SLA"). In the event any Single EC2 Instance does not meet the Instance-Level SLA, you will be eligible to receive a Service Credit as described below.

Instance-Level Uptime Percentage	Service Credit Percentage
Less than 99.5% but equal to or greater than 99.0%	10%
Less than 99.0% but equal to or greater than 95.0%	30%
Less than 95.0%	100%

Figure 3.12 Amazon EC2 Uptime SLA

Hourly Single Amazon EC2 Unavailability

Besides the monthly uptime, the SLA also states that "AWS will not charge you for any Single Amazon EC2 Instance that is Unavailable for more than six minutes of a clock-hour." This credit is automatic, as in you don't have to specifically request these refunds.

Note

A region-level SLA defines the availability of service for a region, not an individual instance; it translates into: if you deploy 2 instances in different AZs, at least one of the instances will be available 99.99% of the time).

Now that we know single Amazon EC2 uptime SLA is 99.5% and 2 or more Amazon EC2 instances deployed across zones provide the SLA of 99.99% let's correlate this to SAP architecture and HA: for any system that's business critical, such as SAP S/4HANA production, you should deploy the components in multiple AZs. This provides the infrastructure level (Amazon EC2) SLA of 99.99% for that component and mitigates single point of failure (SPOF) to enable SAP HA. *Individual components* and *HA architecture* are discussed in detail in Chapter 5. For a non-production system or an SAP system that's not business critical, where availability is not as important, the architecture could have all Amazon EC2 instances in a single AZ, which provides Amazon EC2 SLA of 99.5%

Using the website *https://uptime.is* Figure 3.13 shows the potential downtime for Amazon EC2 uptime SLAs. (It doesn't necessarily mean that a single instance is down for **3h 37m 21s** every month, just that if it is, that's still within the defined SLA; hence the importance of choosing right resiliency architecture for application, which is also consistent with AWS Well-Architected Framework principles).

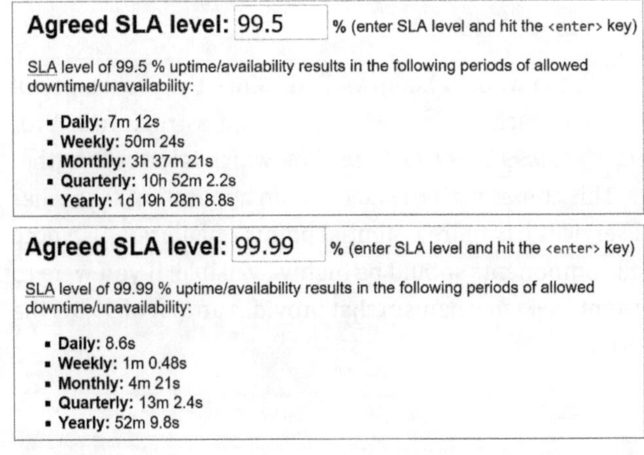

Figure 3.13 Downtime Shown in Terms of Time for 99.5% and 99.99% Uptime SLA

> **[»]**
>
> **SLA for Other AWS Services**
>
> Similarly, you should review SLAs of other services used in SAP architecture, such as Amazon EBS (region level versus volume level), Amazon EFS (standard versus one zone), Amazon Elastic Load Balancing (multi-AZ versus single load balancer), etc. and not just assume that the Amazon EC2 SLA follows for every other service.
>
> Note that some actions do not have an associated SLA, such as Amazon EC2 time to terminate, time to provision, etc. and you should pay attention to these things as well. You are not guaranteed that an Amazon EC2 instance will be available for you to use unless you use an on-demand capacity reservation (ODCR).

3.3.2   Workload Service Level Agreements

Consider a *workload* as a collection of resources that work together to serve a business function. For example, SAP is a workload with multiple components, including several cloud services that enables deployment, HA, DR, etc. So, from a broader workload uptime perspective, you should consider SLAs for all services supporting the application, which is called a *composite SLA*.

> **[»]**
>
> **Note**
>
> This section discusses SLAs in the context of uptime of services and considers uptime as a large part of the reliability of an application.

Consider an SAP system deployed in AWS in a HA configuration, as shown in Figure 3.14. Further detailed discussions about architecture follow in Chapter 5.) All the components are installed on an Amazon EC2 instance across AZs, and Amazon EFS is shared. At the time of writing, the AWS SLAs for these services included the following:

- Amazon EC2 regional SLA: 99.99%
- Amazon EFS standard storage class SLA: 99.99%

If any of the SAP components fail, the whole system will fail. Since the probability of each service failing is independent of each other, the composite infrastructure SLA for the architecture is *99.99% * 99.99% * 99.99% * 99.99% = 99.96%,* which is lower than that of the individual components. This comes with the fact that an application that relies on multiple components and services has more potential points of failures. Also note that, for an HA architecture, all components should be highly available: If you were to use a different network file system (NFS) mechanism that provides an SLA of 99.9% the overall SLA would then be 99.87%.

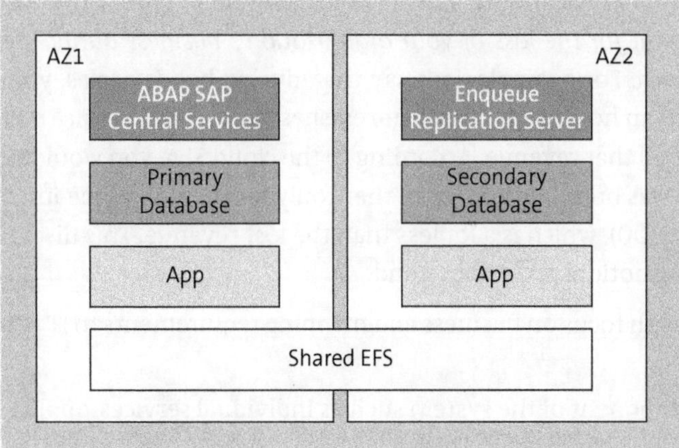

Figure 3.14 Example HA Deployment for SAP

3.3.3 Business Uptime and Infrastructure Service Level Agreements

We mentioned earlier that you should know your organization and its requirements such as uptime, performance, etc. of its SAP systems. From the previous section, you learned how to think of application uptimes in the context of the failure probability of individual components. Since AWS just provides the infrastructure, the SLAs we discussed account only for *infrastructure components*. If the application goes down because of a misconfiguration, that's an additional downtime you need to consider in your overall calculations.

Let's say that your organization's SAP S/4HANA system is critical to your business and IT has agreed with business users that the application should be up 99.95% of the time, measured monthly. As shown in Figure 3.15, your organization can thus tolerate an application downtime of 21m 44s every month without significantly impacting your customers. Now, based on that information, you can choose the right architecture, choose the right cloud services, determine application downtime probability and mean time to recover, etc.

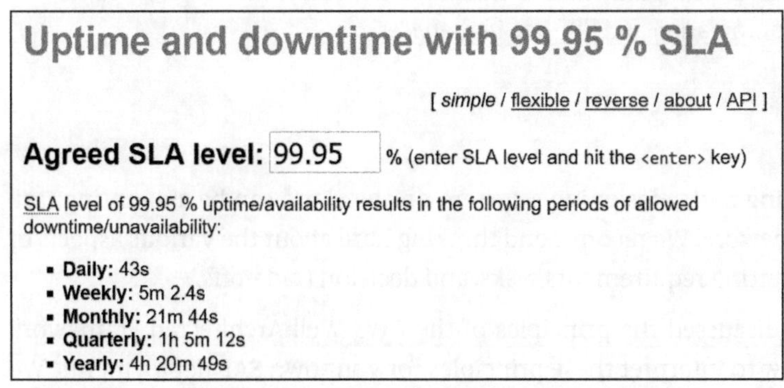

Figure 3.15 Potential Downtime with 99.95% Uptime SLA

However, looking at the SLA's details, the SLA credit mechanism only refunds the cost of service and *doesn't account for the loss of your organization's business during the period when service was down*. For example, let's say that, during holiday sales, your company's revenue is $1000 an hour, and the website crashes because of a service outage. Your company has lost all that revenue. According to the cloud SLA, you would be eligible for credit *if* the SLA was breached and, even then, only for the cost of the infrastructure service itself (say $100), which is a lot less than the lost revenue. This discussion highlights the following notions to keep in mind:

- You should always start with focus on business and mapping requirements to IT systems.

- Pay attention to each component of the system such as individual services, application, etc.

- HA applies to each component; you can't just have HA for the frontend and expect a resilient system, for example.

- Overall application uptimes depend on your team's mean time to recover applications.

- Cloud SLAs govern the services you use, not necessarily the revenue dependency your business has on its IT systems.

SLA and RISE with SAP

Since RISE with SAP combines infrastructure and application management, the published SLA is for applications. When you see a 99.7% or 99.9% SLA commitment in RISE with SAP, it's not the AWS service SLA, but the uptime you get from application or composite SLA, as described in Section 3.3.2.

Availability, Resiliency, and Automation

The higher the availability retirement metric, the more important resiliency configuration and automation becomes. For example, achieving > 99.9% application uptime is improbable without HA configuration and even more difficult to have higher availability target without automation and self-healing services.

3.3.4   Summary

With so many moving parts, designing a system on the cloud can be a daunting task without a good framework. We recommend thinking hard about the various aspects of design without forgetting requirements, risks, and decision trade-offs.

In this chapter, we discussed the principles of the AWS Well-Architected Framework and showed you how to interpret these principles for your own SAP deployments. We also covered how to read and interpret SLA docs, apply principles to your architecture, and how to think about business risk in the context of cloud service SLAs. With these principles in mind, you'll next learn how to set up landing zones and related services in the next chapter. We'll also further discuss architectural patterns for application resiliency in the cloud.

Chapter 4
AWS Landing Zones

For any SAP, non-SAP, or enterprise workloads being deployed or migrated to Amazon Web Services (AWS) you need a secure, scalable, and compliant foundation. In this chapter, we'll talk about the AWS landing zone and its purpose. We'll show you how AWS tools/services can help you quickly set up a landing zone before you start provisioning any AWS services for running SAP workloads.

You've done an assessment on the benefits of migrating your SAP systems to AWS. Based on the business value proposition from AWS, you've received the "go" decision from leadership to move your SAP workloads to AWS. The SAP team and the IT team are ready to go with migrating that first SAP sandbox environment! AWS services are available for you to consume, and you were promised the agility that the AWS cloud provides. So, are you ready to start? Well, not yet! Think of your house and the way it was built. It all started with the design and then blueprinting of the foundation—the floor layout, the material/equipment used for building it from the ground up, the wiring for the electricity, the plumbing pipes, the gas lines, the door locks etc. Eventually, when all the house is finally built, now you're ready to move in with all your furniture.

Similarly, on AWS, you need a solid foundation before you can start migrating any enterprise-grade workloads such as SAP. You must design for your AWS account structure (the floor layout), networking topology (the electric wiring and plumbing), and the most importantly the security guardrails (the door locks) including user authorizations. From an agility perspective, AWS offers native services (the material/equipment) that you can use to accelerate the foundational implementation before you migrate and run your SAP workloads on AWS. Once your foundation is built, you're ready to start migrating your SAP workloads (your furniture, new or old) to AWS.

To summarize, an AWS landing zone is a multiple-AWS account environment with a foundation related to networking and with security guardrails defined based on your requirements and governance model to house your enterprise workloads such as SAP. In this chapter, you'll learn key concepts behind using AWS Organizations, AWS Control Tower, and AWS Service Catalog, and we'll show you how these AWS services can help you build scalable, secure, and resilient AWS landing zones for SAP systems on AWS.

AWS Landing Zones

Based on our experience, building a baseline landing zone on AWS can take 4 to 6 weeks. You can accelerate the implementation process using the AWS Partner Network and/or AWS Professional Services.

After building the baseline landing zone on AWS, you'll follow an iterative process to continuously monitor and improve on your configurations related to the landing zone. For example, you might need to address a particular security and/or a compliance configuration that was missed during the baseline implementation or modify the configuration to account for a new requirement later.

4.1   AWS Landing Zone Architecture

Each organization is unique with respect to its culture, size, business units, IT, and SAP team structure. Its security and compliance requirements are unique. For these reasons, no single go-to architecture exists for an AWS landing zone. In subsequent sections, we'll look at the baseline structure that should be implemented as a bare minimum based on AWS best practices.

An AWS account, which is a 12-digit number (you can create a friendly alias as well), is a container within which you'll provision your resources, such as Amazon Elastic Compute Cloud (EC2), and on top of which you'll have your SAP workloads running. No costs are associated with creating new AWS accounts. You only pay for the resources or services that you consume within those AWS accounts.

AWS Account

An AWS account is different from your AWS user account. An AWS user account could be created and restricted to a particular AWS account. An AWS user account is an identity that will be assigned roles and permissions to define what a user is allowed to do in an AWS account.

When it comes to the number of AWS accounts that you need, there is no one-size-fits-all answer. Any enterprise moving or implementing any workload in the cloud will end up with multiple accounts, as shown in Figure 4.1.

Let's consider each reason for the need for multiple accounts, as shown in Figure 4.1:

- **Security controls**
  An organization with different lines of business (LOBs) could have different security and compliance requirements. One AWS account could host the SAP systems belonging to a particular LOB while another AWS account could host SAP systems

for another LOB within the enterprise. Account segregation could also be based on application types and their respective security requirements. Example SAP applications could go into one account and non-SAP could into another. Tasks become easier from an auditing perspective as well.

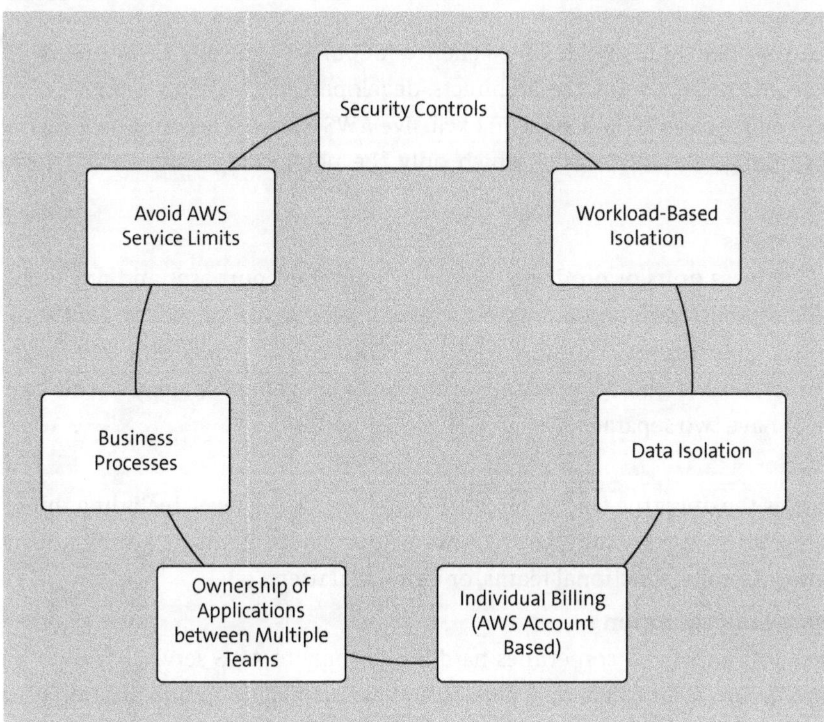

Figure 4.1 Why Multi-Account on AWS?

- **Isolation**
  An account is a container or a boundary with your resources related to an infrastructure such as compute, storage, and networking. In this context, the concept of *blast radius* comes into play. If for any reason your non-production account is compromised, you'll still have your production account isolated, thus reducing your attack surface. A financial institution running its SAP workloads on AWS could have a requirement of segregating production accounts from non-production accounts as an audit and compliance mandate. You may have different security requirements that are workload based that require you to isolate one account from one another based on the application type (e.g., SAP versus non-SAP).

- **Data isolation**
  An air-gapped or a data bunker AWS account that contains your immutable backups. This account will be an isolated account with access given to the chosen few, with limited interconnectivity with other accounts, and restricted programmatic or

application programming interface (API) access. This account is specifically used for data protection and for isolating your data against ransomware attacks. Another example is a separate account for an SAP data lake on AWS for use in data isolation and for segregating access between the SAP team and the data lake team.

- **Multiple teams**
  The IT team or the cloud center of excellence (CCoE) are typically structured with multiple teams such as solutions architects, developers, networking team, DevOps team, etc. You probably would need an exclusive AWS Network account hosting the AWS networking related services, which only the networking team should have access to.

- **Business process**
  Different business units or products might have different purposes and processes. You should establish different accounts to serve business-specific needs. For example, consider an insurance company with two LOBs: a finance planning sub-ledger (on SAP) and another 360-degree user view (on SAP). The best strategy in this case would be to have two separate SAP-specific AWS accounts.

- **Billing**
  An account is the only true way to separate items at a billing level, including things like data egress charges. Multiple accounts help separate items at a billing level across business units, functional teams, or individual users.

- **AWS service limit allocation**
  Sometimes soft limits and sometimes hard limits apply to AWS services within the AWS accounts. These limits are on a per-account basis. Imagine an organization with hundreds of applications, including SAP, spread across thousands of virtual machines (VMs). In this scenario, the ideal approach is to segregate the workloads into multiple accounts to prevent hitting your limits.

Now that we've established the relevance of implementing multiple AWS accounts, let's dive into the AWS landing zone architecture. First and foremost, no right or wrong answers exist when it comes to an AWS landing zone architecture. However, AWS-recommended architectures should be considered first when you're designing one. Figure 4.2 shows a typical AWS landing zone that is a good starting point for any organization planning to run their SAP workloads on AWS.

Let's define the important terms related to the implementation of this multi-account strategy, as shown in Figure 4.2:

- **Organization**
  An organization is the logical container that you'll create to group all of your AWS accounts together. The architecture shown in Figure 4.2 is the organization starting from root.

- **Root**
  The parent node is a container that has your organization's hierarchical structure defined based on your requirements.

- **Organizational unit (OU)**
  Under this logical construct, you'll have your accounts assigned. These accounts could be new accounts or existing accounts that you invite to your organization and eventually to the relevant OU.

- **Policy**
  A policy is a document using which you can control what activities are all allowed in the AWS accounts that are created. A policy can be assigned to an OU, which is then applied to all the accounts within the OU.

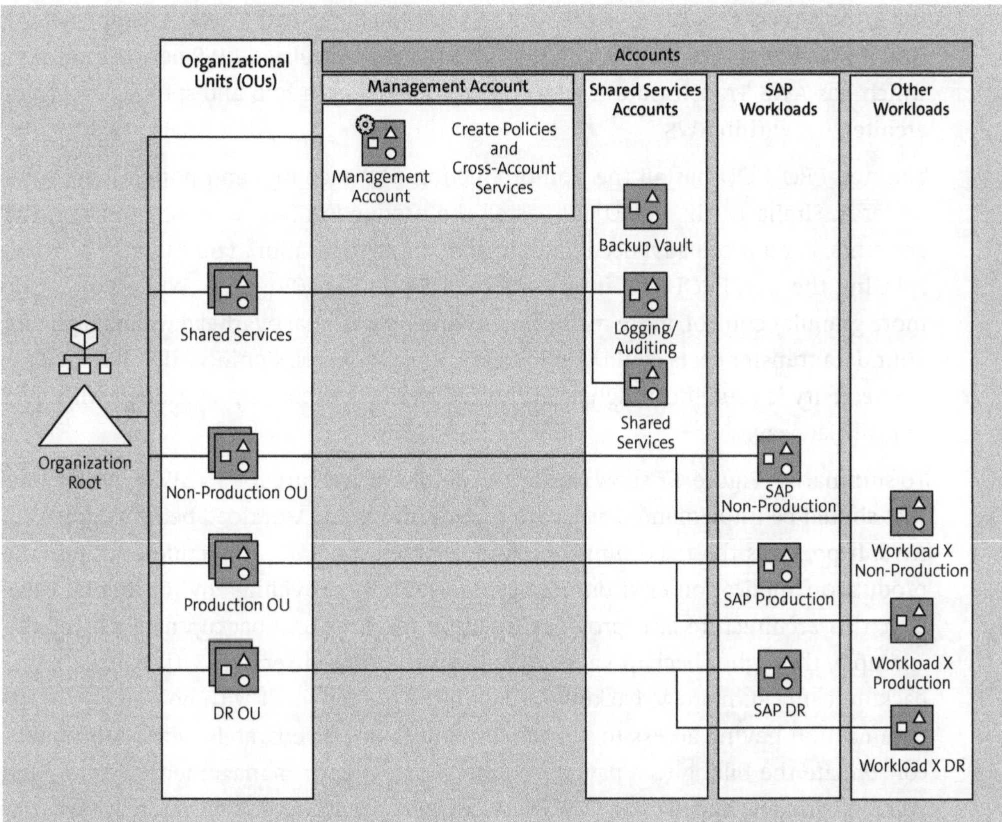

Figure 4.2 AWS Landing Zone Architecture

Now that we've established some important concepts and terms, let's look more closely at the architecture shown in Figure 4.2. The organizational root has the management account, and all the OUs are defined. The management account is the core account that will house all your custom code pipelines to set up landing zones and/or AWS services/ tools to help you manage and operate according to this multi-account strategy. The root account further has OUs under it, and further OUs can be segregated based on environment types or the LOBs in your enterprise. If you're considering splitting your OUs based on LOB, you can have nested OUs containing different environment types (e.g., LOB1 containing non-PROD and PROD OU, LOB2 containing non-PROD and PROD OU, respectively).

As shown in Figure 4.2, you can segregate OUs based on the application environment types, which are shared services, non-PROD, PROD, and disaster recovery (DR). Within these OUs, you'll have the respective accounts assigned. The Shared Services OU has a common centralized logging/auditing, backup vault, and shared services accounts. A Shared Services account within the Shared Services OU will typically house all the solutions/tools/AWS services (e.g., the backup vault) that are shared between applications. Under the shared services OU, you may also have a centralized AWS network account, which has AWS Transit Gateway to help you provision a hub-and-spoke networking architecture within AWS.

The non-PROD OU has all the non-production SAP accounts and non-SAP accounts under it. Similarly, the PROD OU has all the production SAP accounts and non-SAP accounts. For a more advanced landing zone implementation, you can also consider splitting the non-PROD OU into sandbox, DEV and QA/Pre-Prod. While you'll gain more granular control, more management and operational overhead will be incurred. Your data transfer costs could also increase, in some cases, significantly high if all the connectivity is routed through a centralized Network account provisioned with AWS Transit Gateway.

To summarize, Figure 4.2 shows a starting point architecture for an AWS landing zone that should be implemented as a bare minimum for any workload being migrated to AWS. It provides the maximum isolation between the SAP production and the non-production application environment types. Not only providing environmental isolation, this architecture also provides isolation for your SAP backup data as well, for example, through a backup vault account in the shared services OU. Your storage/ backup team can manage backups stored in the backup vault with no one else in the organization having access to it. Each account is billed separately with an option to consolidate the billing to a payer account, which is your management account, and you're no longer capped by the AWS service limits per account. You can create security policies that automatically trickle down to the accounts assigned under an OU, giving you the authority to implement security controls based on your application environment type.

Figure 4.3 shows the options available when building and implementing an AWS landing zone:

- AWS-native services such as AWS Organizations and AWS Control Tower including AWS Service Catalog.
- Custom-built AWS landing zones built by your system integrator using their go-to infrastructure as code (IaC) tool, such as AWS CloudFormation, Terraform, Pulimi, etc. This strategy is typically followed by customers who have unique security and compliance requirements that cannot be met by AWS-native services such as AWS Control Tower.

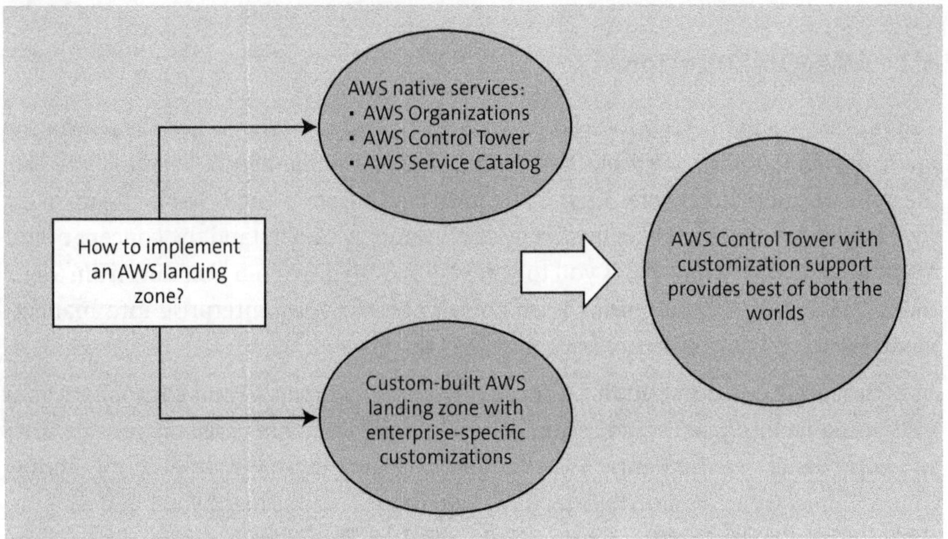

Figure 4.3 How to Implement an AWS Landing Zone

Note

In subsequent sections of this chapter, we'll cover AWS-native services like AWS Organizations, AWS Control Tower, and AWS Service Catalog that are used to build a multi-account AWS architecture. A heavily customized or a custom-built landing zone using self-owned code pipelines are beyond the scope of this book.

Typically, cloud architects or your CCoE platform team, a partner, and/or AWS professional services team, or some combination thereof, implement your AWS landing zones. For example, AWS Professional Services could build the AWS landing zone for an enterprise, and the enterprise could own it after its implementation is complete.

If you don't have much or any in-house expertise and/or are new to cloud, AWS highly recommends you leverage a certified partner from their partner network and/or leverage AWS Professional Services to build your landing zone.

Some enterprises have non-SAP applications or data lake/analytics solutions running on AWS, and because of this reason, they already have an AWS landing zone built on AWS. Unless your security requirements are different, in these scenarios, we highly recommend leveraging an existing landing zone by spinning up new AWS accounts for SAP workloads. With this approach, you can significantly cut down on the timelines for migrating your SAP systems to AWS.

4.2   AWS Organizations

You're done designing your AWS landing zone, and as you have reached the realization phase of your landing zone project, it's time to build or implement it. Whether you take the approach of manual versus an automation landing zone build, you'll end up using AWS Organizations. So, it becomes imperative for us to understand the concept of this AWS native service which helps you to organize and manage your accounts from a central location. AWS Organizations is an entry point for your enterprise into management and governance of a multi-account AWS world.

AWS Organizations is a centralized service to help you organize and consolidate your AWS accounts into your organization and eventually into OUs based on your security and compliance requirements. As in our earlier example of a multi-account landing zone architecture in the previous section, AWS Organizations helps you build a hierarchical account strategy with security as a top priority. By centrally governing and controlling your AWS resources, you can easily *innovate* (spin up multiple controlled AWS accounts for your developers to explore the art of possible) and *scale* (seamlessly add additional accounts based on existing or new workloads being migrated to cloud).

In this section, we'll cover the capabilities and benefits of AWS Organizations and its key concepts and show you how to implement it in your AWS account.

4.2.1   Capabilities and Benefits

Some key capabilities and benefits of AWS Organizations include the following:

- **Organize, manage, and control all your accounts from a central location**
  You can create new accounts using the AWS Console, API, or AWS CLI, and combine them into an OU using AWS Organizations. You can also invite existing standalone accounts into an organization. Let's consider an example: Company A and Company B have SAP systems running on AWS and are using their respective AWS Organizations. Company A acquires Company B. One option is to consolidate both the SAP systems, but we all know consolidating multiple SAP systems is not that straightfor-

ward and could become a year-long (or more) project. So, for now, Company A can invite Company B's AWS accounts housing the SAP systems to its organization. In this scenario, Company B would, in some cases, need to comply with the security and compliance policies defined at the OU level. Alternatively, Company B could be a separate entity under the same organization but in a different OU. To summarize, you can quickly scale by adding new or existing AWS accounts, thus ensuring security policies are applied based on your organization protocols.

- **Share resources and simplify access management**
  Using AWS Organizations, you can share AWS resources such as your networking components like your virtual private clouds (VPCs), subnets, Transit Gateway, etc. at the organization, OU, or account level, which reduces the need to set up multiple subnets or to duplicate resources. In this case, you would use AWS Resource Access Manager (RAM), which is integrated with AWS Organizations.

  For access management, you can use AWS Organizations to integrate your identity provider or store with AWS IAM Identity Center and thus streamline user access to the AWS environment based on your requirements.

- **Secure and audit your environment for compliance**
  We discussed earlier how to apply common security policies across your organization, OU, or account to ensure you meet your security and compliance standards. In addition, AWS Organizations can help you log all the security-related events such as the following:

  - API calls being called within your organization using the integrated AWS CloudTrail service. This service helps you identify "who did what" in your AWS environment at an organization, OU, or account level.

  - Identify security deviations from your standard configuration, which you define based on your organization protocol. This task is accomplished using another integrated service: AWS Config.

  - Actively monitoring and reducing security threats across your organization using integrated services such as Amazon GuardDuty.

- **Organize costs and identify cost-saving measures**
  First and foremost, zero costs are incurred to use AWS Organizations. You don't pay for defining your organizational structure within the service; however, you do pay for the resources you activate. A good example is when you activate Amazon GuardDuty using AWS Organizations; you'll incur a cost for using Amazon GuardDuty.

  AWS Organizations provides your FinOps team with a consolidated bill, which may result in quantity discounts based on the number of resources used within your organization. You can also view and optimize costs with the findings identified at a particular service level, such as your total spend on Amazon EC2 instances for SAP workloads in a specific AWS account.

Figure 4.4 shows the benefits of AWS Organizations.

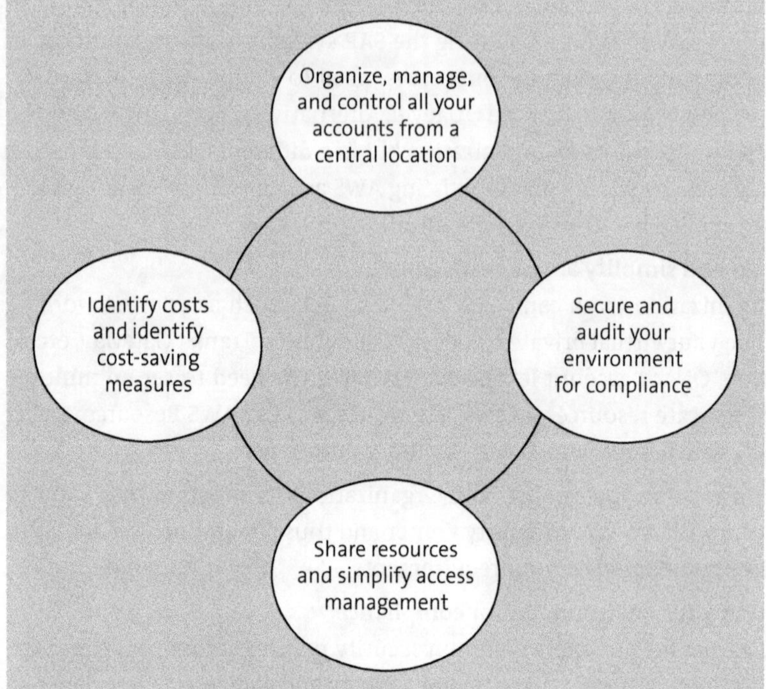

Figure 4.4 Benefits of AWS Organizations

4.2.2   Key Concepts

Figure 4.5 shows a multi-account strategy built using AWS Organizations. This example has the organization root with the AWS Org management account and OUs with the member accounts assigned to them. In the previous section, we defined some key terms and concepts, such as the organization root, the organization, the OU, and security policies.

With AWS Organizations, as shown in Figure 4.5, a few additional concepts that you should know include the following:

- **AWS Org management account**
  This account is your main or the starting point AWS account and houses all the continuous integration/continuous deployment (CI/CD) pipelines or native AWS services to build your AWS landing zone. In other words, this account is used to create/build AWS accounts in your organizations. You can have one and only one management account. This account is also your payer account, where consolidated billing happens. You also have the ability to remove and add accounts within organizations. One of the most important functionalities of the management account is to apply security policies across the root, within an OU, or directly at the account level. Another important functionality is its integration with other supported AWS services.

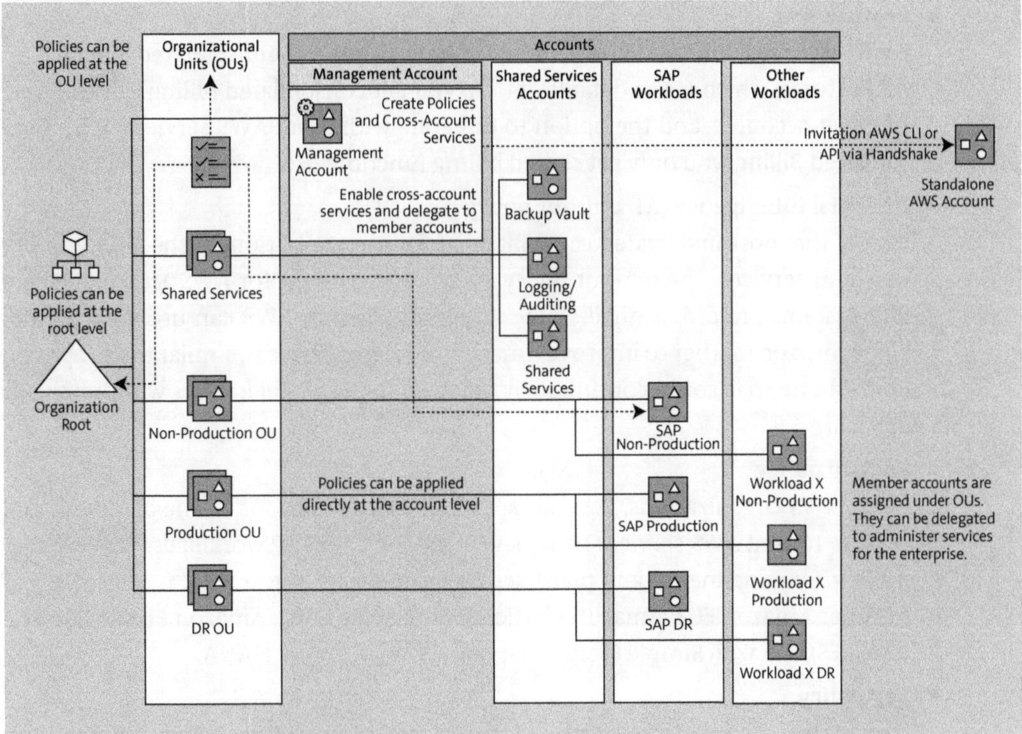

Figure 4.5 AWS Organizations Key Terms

- **Member accounts**

  Member accounts are the AWS accounts assigned to an OU. You could have multiple member accounts, depending on your requirements. As shown in Figure 4.5, you can have SAP and non-SAP AWS accounts grouped under their respective OUs. An account can be a member of only one organization at any given point in time. You can also attach security policies directly to member accounts. For example, you might have a backup policy to help you build immutable SAP database backups across the member accounts where your SAP systems are provisioned.

- **Invitation and handshake**

  Using AWS Organizations, you can invite standalone AWS accounts or accounts in other organizations by using invitations. However, keep in mind that, at any given point in time, an AWS account can only be part of one organization. The process that handles invitations is called a *handshake*. If you use AWS Command Line Interface (CLI) or the AWS software development kit (SDK) handshake—a multi-step process is being used behind the scenes. A handshake is also initiated with consolidated billing when the management account requests for the spending from the member accounts. As shown in Figure 4.5, the management account initiates the invite and the handshake process.

- **Feature sets**

  AWS Organizations has two feature sets—**All Features** and **Consolidated Billing**. With **All Features**, which is the default option, you get consolidated billing, control over member accounts, and the option to integrate with other AWS services. With **Consolidated Billing**, you only get shared billing functionality.

- **Artificial intelligence (AI) opt-out policy**

  One of the most underrated capabilities of AWS Organizations is the ability to opt out of AI services. The opt-out policy is typically relevant for AWS accounts where SAP systems are provisioned. Some AI services within AWS can use your data in some form or another to improve Amazon AI-related services, similar to the apps on your phone that could continuously track your usage unless you withdraw your consent.

- **Backup policy**

  Using AWS Organizations, you can apply a backup policy for all the compute- and storage-related AWS services being leveraged for your SAP workloads. This capability helps you define backup plans across compute resources, that is, regardless of whether Amazon EC2, Amazon Elastic Block Storage (EBS), Amazon Elastic File System (EFS), Amazon Simple Storage Service (S3), and/or SAP HANA.

- **Tag policy**

  Tags are key/value pairs to help you identify AWS resources, manage authorizations, and control the costs associated with these resources. They can also be used for automation. An example would be automatically starting or stopping SAP systems tagged "non-prod" during non-business hours. Consider a tagging requirement for an SAP system deployed on an Amazon EC2 instance, which should have "SAP-Environment-Type" as the key and "non-prod or prod" as the value. First , you'll create a tagging policy using AWS Organizations and apply it to the OU that contains the AWS SAP accounts. Another option is to directly apply the policy to the AWS SAP accounts. Thus, for example, a resource like an Amazon EC2 instance should contain this tag whether you build the Amazon EC2 instance manually or using IaC. For monitoring whether you're compliant or not with your tagging process, you'll have use to Amazon EventBridge.

[+]

AWS Organizations Tag Policy

When you're using IaC tools such as AWS Launch Wizard for SAP to build and install your SAP systems on AWS, some tags will be generated by these tools and/or services. Ensure that you update the tags for the AWS resources that these tools and/or services deploy based on the tag policies defined at your organization or at the SAP AWS account level.

- **Security control policies**

  Centralized security control policies (SCPs) are the most important security policies that you define within AWS Organizations. Let's consider an example where you have hundreds, even thousands, of AWS accounts, and you need to set up an Amazon S3 policy that restricts the deletion of any bucket from your production AWS accounts. Some of these production accounts also house your SAP systems' backup data in the Amazon S3 buckets. One approach is to set up roles and permissions and assign them to each user in all these accounts using Identity Access and Management (IAM), or you could also use a Microsoft Active Directory (AD) group policy, which can be integrated with AWS IAM. This first layer of defense controls access to users. However, what if your production account is compromised? In this case, the bad actor or hacker could have admin rights to the accounts, and they could modify/delete/steal your data. Now, let's bring AWS Organizations into play. With its centralized security policies applied at the OU level trickling down to the AWS accounts, you can restrict the deletion of any Amazon S3 bucket. Even an administrator with a full admin access within AWS cannot modify, update, and/or delete the contents of the Amazon S3 bucket in an AWS account.

SCPs are written in plaintext file structured in JavaScript Object Notation (JSON). In terms of syntax, these files are similar to IAM policies or Amazon S3 policies.

Let's consider another example where you have a dedicated production SAP AWS account and a non-production SAP AWS account. By now, you know that, within AWS, only certified and supported instance types can be used. Listing 4.1 shows an example SCP policy that denies the launch of any t3.micro instances, which are typically not used for SAP systems.

```
{
  "Version": "2012-10-17",
  "Statement": [
    {
      "Sid": "RequireMicroInstanceType",
      "Effect": "Deny",
      "Action": "ec2:RunInstances",
      "Resource": [
        "arn:aws:ec2:*:*:instance/*"
      ],
      "Condition": {
        "StringEquals": {
          "ec2:InstanceType": "t3.micro"
        }
      }
```

```
     }
   ]
 }
```

Listing 4.1 Example SCP Policy to Deny Launch of Non-Certified Resources on AWS

- **Delegated administrator for AWS Organizations**
  Another important concept in AWS Organizations is the delegated administrator. All policies, including SCPs, backup policies, tags, and/or AI opt-outs are controlled and assigned via the AWS Org management account within AWS Organizations. With a delegated administrator, you can give control to a member account to create and maintain their own security polices according to their specific requirements and protocols. Returning to our previous example of an OU with an SAP workloads account, with this feature, the SAP team can control and operate their own SAP AWS account-specific SCP policies based on their requirements, which could be unique compared to other workloads within the enterprise.

> **[+]**   **AWS Organizations OU**
>
> Do not define an OU in AWS Organizations based on your internal enterprise organization structure or on application owners. Don't use your AD structure either. This structure should typically be based on your organization's workload types or LOBs.

AWS Organizations supports integration with multiple AWS-native sources such as AWS Account Management, AWS Config, Amazon GuardDuty, AWS Control Tower, AWS CloudTrail, etc. AWS keeps adding new services to this list. Thus, for the latest comprehensive list, visit the AWS documentation at *http://s-prs.co/v577613*. Any AWS service-to-service integration is allowed via roles that are part of AWS IAM. Similar to trusted remote function calls (RFCs) in an SAP system where you establish trust between two SAP systems with the correct authorizations/permissions but without any user authentication, AWS Organizations uses trusted access to create a service-linked role to build the trust between AWS Organizations and other AWS services. The trusted service can then perform tasks in your organizations and its accounts are used on your behalf. The trusted service does this asynchronously as needed, not necessarily in all accounts in the organization at the same time. The service-linked role has pre-defined IAM permissions that allow the trusted service to perform only specific tasks within that account. In general, AWS manages all service-linked roles, which means that you typically can't alter the roles or the attached policies. Example, if you integrate AWS Config with AWS Organizations, you'll use trusted access to enable AWS Config—the trusted service to perform tasks in your organizations and make calls on your behalf.

> **AWS Organizations Integration with Other AWS Services**
>
> AWS highly recommends you enable and disable trusted access using the other service's console, API, or CLI and not using AWS Organizations. There are dependencies such as initialization or clean-up which are better handled by the other service's APIs when you enable and disable the trusted access. Service-linked roles are exempt from any SCP restrictions.

4.2.3   Implementation

Now that we have established the "why" and "what" of AWS Organizations, let's look at some of the options for you on "how" to implement it:

- **Using AWS Management Console**

  AWS Management Console is a web browser-based user interface (UI) and the most straightforward way to implement AWS Organizations. Figure 4.6 shows the landing page for AWS Organizations in AWS Console. AWS Organizations is available under the **Management & Governance** section under **Services**.

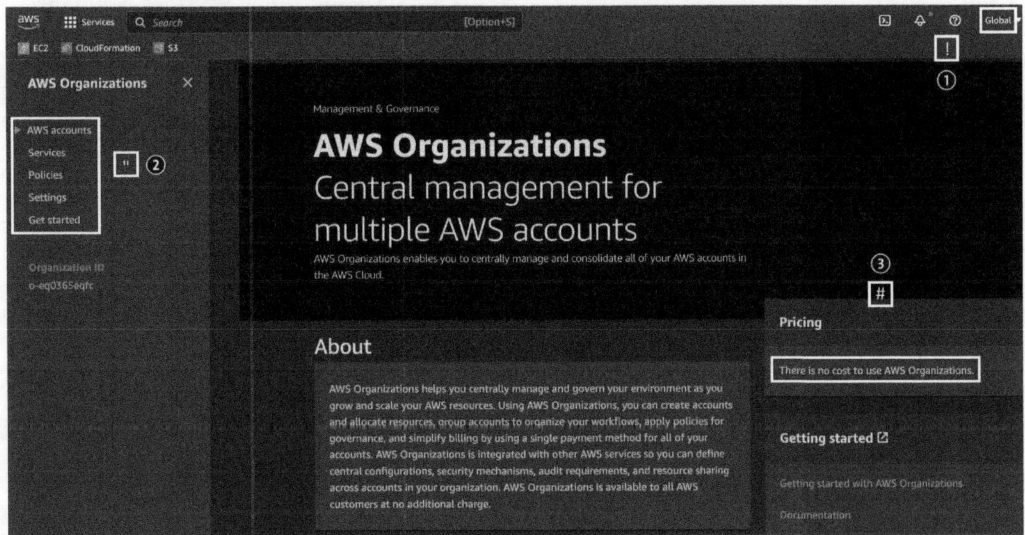

Figure 4.6 AWS Organizations Using AWS Management Console

As shown in Figure 4.6, using AWS Management Console, you can use AWS Organizations which is a ❶ a global AWS service to ❷ create and/or invite multiple AWS accounts, enable trusted access to AWS Services, create security guardrails (policies), and create delegated administrator for AWS Services and organizations using Settings. Lastly, ❸ AWS Organizations is free of cost.

- **Using AWS CLI**

  You can also use AWS CLI to build your multi-account strategy using AWS Organizations. This approach is typically used for managing and controlling AWS services and can help automate tasks related to these services through scripts.

  Listing 4.2 shows an example of an SAP AWS account being moved to the SAP OU.

```
aws organizations move-account \
--account-id "<SAPProdAccountId>" \
  --source-parent-id "<RootOUId>" \
        --destination-parent-id "<SAPOUId>"
```

Listing 4.2 AWS Account Moving from One OU to Another OU Using AWS CLI

- **Using AWS SDK**

  When you're building your custom-built AWS landing zone, your CloudOps team or the DevOps team will end up using AWS SDK. The AWS SDK provides APIs, modules, and libraries in your preferred programming language to connect to AWS services such as AWS Organizations, as shown in Listing 4.3.

```
def attach_policy(policy_id, target_id, orgs_client):
    """

    Attaches a policy to a target. The target is an organization root,
    account, or
    organizational unit.
    :param policy_id: The ID of the policy to attach.
    :param target_id: The ID of the resources to attach the policy to.
    :param orgs_client: The Boto3 Organizations client.
    """

    try:
        orgs_client.attach_policy(PolicyId=policy_id, TargetId=target_id)
        logger.info("Attached policy %s to target %s.", policy_id, target_id)
    except ClientError:
        logger.exception(
            "Couldn't attach policy %s to target %s.", policy_id, target_id)
        Raise
```

Listing 4.3 Python Code to Attach a Policy to a Target Root, an OU, or an AWS Account

- **Query requests using HTTPS APIs**

  Query APIs are used to perform actions related to the service through actions—the operations to be performed. AWS Organizations supports GET and POST for all actions. AWS Organizations has a single global HTTPS-based API endpoint, which is mandatory, that is hosted in the US East (N. Virginia) region.

  When using APIs, you must ensure that you're signing requests using AWS Signature Version 4 using an access key and secret access key. Listing 4.4 shows an example of an action called AttachPolicy, an operation that can be performed using APIs.

```
POST / HTTP/1.1
X-Amz-Target: AWSOrganizationsV20161128.AttachPolicy
{ "TargetId": "ou-examplerootid111-exampleouid111", "PolicyId":
"p-examplepolicyid111" }
```

Listing 4.4 Performing Action of AttachPolicy Using APIs

AWS Organizations HTTPS Query APIs

When you're using AWS CLI or AWS SDK, these tools sign their API requests for you, unlike HTTPS query APIs. Thus, we recommend using AWS SDKs instead of making direct calls to AWS Organizations APIs.

4.3 AWS Control Tower

You typically have three options available to you when building a landing zone in the AWS world:

- *Option 1*: You work with AWS Partners, or you purchase ready-to-go solutions available in AWS Marketplace to build your landing zone on AWS.

- *Option 2*: You work with an AWS Solutions Architect, they bring in all the code, automation scripts, etc., to deploy a landing zone based on your requirements.

- *Option 3*: You use AWS native services with both automation and customization capabilities to deploy a landing zone for your enterprise.

Option 1 and Option 2 are more geared towards building a heavily customized or a custom-built landing zone on AWS. Option 1 is still widely used by enterprises who usually bring in a partner of their choice, or an incumbent partner in the organization or can request AWS to bring in their professional services or a partner to help them build the landing zone.

Tip

"Partner" implies system integrators (SIs), global system integrators (GSIs), or any boutique consulting company who have the AWS competency and are part of the Amazon Partner Network (APN).

With Option 2, enterprises had the option of deploying their landing zone free of cost with the help of an AWS solutions architect. These solutions architects would come in with some baseline code available in GitHub, but you could modify/customize the code based on your organization's requirements. You could have a custom-built ready-to-go landing zone for your SAP or non-SAP workloads in no time, in theory. Option 2 was called "AWS Landing Zone" (Note the upper-case L in Landing and Z in Zone).

Option 2 really worked well. However, as you can imagine, this solution was difficult to scale with thousands and thousands of customers flocking towards AWS because the burden was entirely on the solutions architects to help enterprises build the AWS Landing Zone for them. Interest arose within AWS to build a native AWS service that could automate landing zone deployment to help customers deploy a ready-to-go landing zone quickly based on the experience gathered from implementing thousands of landing zones and based on best practices from AWS, customers, and partners. In essence, we are describing the automation of AWS multi-account account implementations, enabling controls across your environment, and monitoring at *scale*—exactly what AWS did with Option 3.

In 2019, AWS introduced a service called AWS Control Tower, a free service to help your IT, cloud, or DevOps team build multi-account AWS environments with out-the-box security guardrails and manage, operate, and monitor the entire landscape based on AWS best practices. AWS Control Tower is a service version of Option 2 or perhaps even its successor. If the out-of-the-box capabilities of AWS Control Tower do not meet your requirement, you can modify your deployment using the customization feature within the service. In the beginning of this chapter, we mentioned that you might need 4 to 6 weeks to build a baseline landing zone on AWS. Out of those 4 to 6 weeks, the most time should be spent in planning and designing your landing zone. The realization or implementation of that plan generally takes less than a day or two using AWS Control Tower.

Tip

After the introduction of AWS Control Tower, SIs/GSIs, or any boutique consulting company and/or AWS Professional Services started leveraging AWS Control Tower to set up a multi-account structure for enterprises.

As shown in Figure 4.7, the heart of AWS Control Tower is AWS Organizations, which means AWS Control Tower calls AWS Organizations APIs to quickly and securely build your multi-account landing zone on AWS. In fact, AWS Control Tower shares lot of common ground with AWS Organizations, such as the AWS account, organization root, OU, member accounts, etc.

Behind the scenes, AWS Control Tower uses IaC through AWS CloudFormation—an AWS-native IaC service to create and apply controls to enforce policies and detect violations in the accounts that it creates. Controls are rules to govern your multi-account AWS environment, and these controls are called *guardrails*. AWS Control Tower further integrates with other trusted and reliable AWS services such as AWS Lambda to orchestrate the application process of guardrails. Guardrails come in the following three types:

- **Preventive guardrails**
  In AWS Organizations, we introduced the term "policies," especially SCPs. Preventive

guardrails or controls are set up using SCPs. For example, you can disallow changes to Amazon S3 buckets for a member account (SAP non-prod and prod) under an SAP OU.

- **Detective guardrails**
  These guardrails are implemented using AWS Config rules to check for deviations against the standard or desired configurations for an AWS resource. For example, let's say you need to ensure the encryption of data at rest for Amazon EBS volumes used for SAP systems: A strongly recommended control within AWS Control Tower helps detect whether an unencrypted Amazon EBS volume is attached to an Amazon EC2 instance in your landing zone.

- **Proactive guardrails**
  Proactive controls are implemented using AWS CloudFormation hooks to invoke custom logic to inspect resource configuration before provisioning. Thus, proactive guardrails are used to detect whether resources are compliant based on your protocols and enterprise policies. For example, a security-related hook might verify security groups for the appropriate SAP HANA database inbound and outbound traffic rules for your Amazon Virtual Private Cloud (VPC).

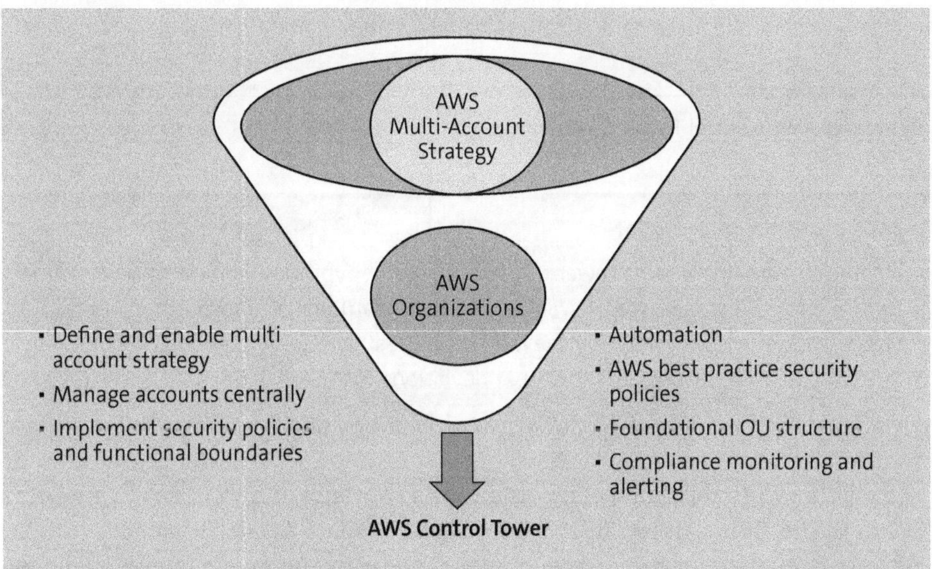

Figure 4.7 AWS Control Tower Uses AWS Organizations to Build a Multi-Account Structure

As shown in Figure 4.7, another important feature of AWS Control Tower is compliance monitoring and alerting. So, AWS Control Tower becomes your one-stop shop, a single pane of glass service to monitor your entire landing zone environment. Figure 4.8 shows a dashboard within AWS Control Tower to monitor your multi-account environment with an overview of OUs and accounts, including the recommended actions for guardrails that are in a non-compliant status.

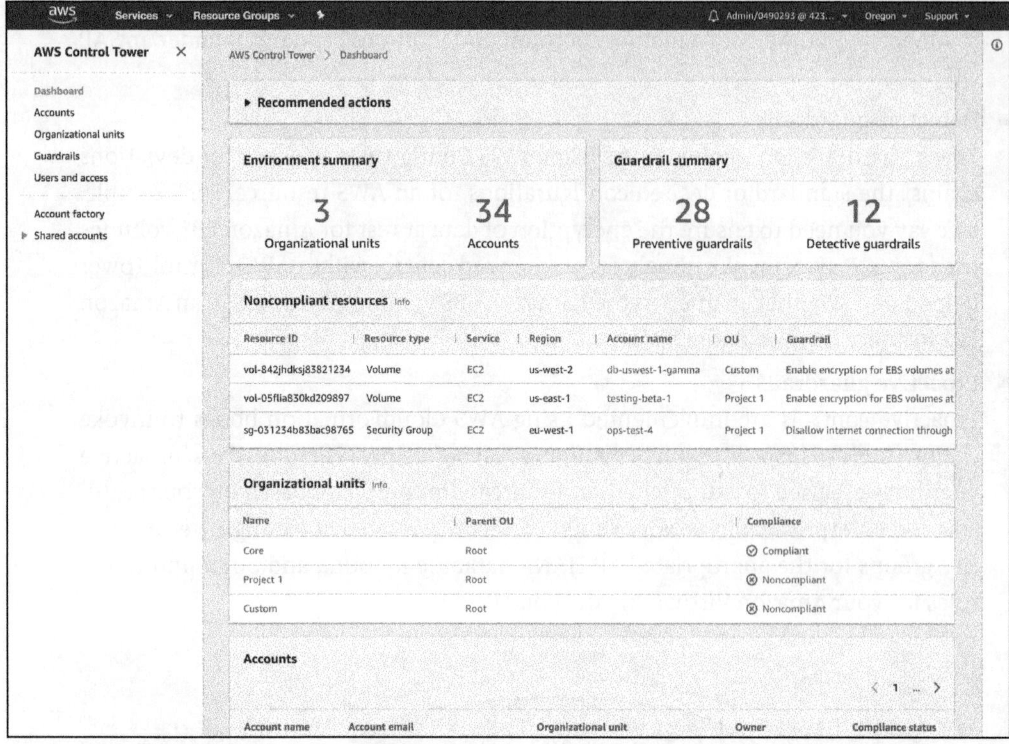

Figure 4.8 AWS Control Tower Dashboard

AWS Control Tower Updates

AWS releases AWS Control Tower updates regularly. Typically, your cloud administrators/IT who will ensure that AWS Control Tower is updated and patched regularly. These updates not only give you the latest and greatest release of AWS Control Tower but can also help resolve any drift (i.e., deviations from the defined controls in your organization).

Now that we've established the importance and relevance of AWS Control Tower, let's look at the baseline architecture that it helps you build via AWS Management Console, AWS CLI, or AWS SDK. The easiest way to get started with AWS Control Tower is by using AWS Management Console. Figure 4.9 shows an example of an enterprise who started their journey of migrating their SAP workloads to AWS by building an AWS landing zone using AWS Control Tower.

As shown in Figure 4.9, some important actions that AWS Control Tower performs as part of its automation process include the following:

- Starting with the AWS Org management account in your AWS home region, AWS Control Tower uses AWS Organizations to create two OUs: a security OU and a custom OU, which are contained within the organization's root.

- AWS Control Tower creates a log archive account and an audit account in the security OU, and these accounts are shared with your other AWS member accounts.

- AWS Control Tower also creates a cloud-native directory with AWS IAM Identity Center (previously called AWS Single Sign On) with preconfigured groups and single sign-on. You also have the option to integrate with your corporate identity provider such as Microsoft Entra ID (formerly Azure AD), Okta, Google Workspaces, etc. as long as your identity provider supports Security Assertion Markup Language (SAML) 2.0.

- AWS Control Tower applies all the configured/enabled guardrails across OU and the respective accounts except to the management account.

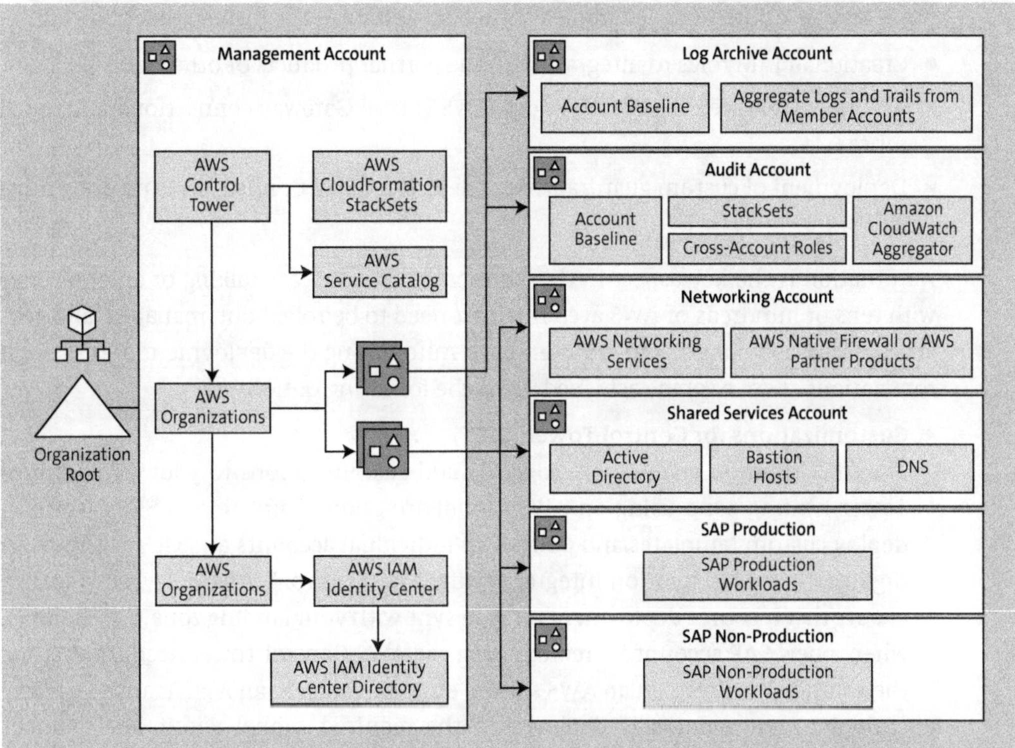

Figure 4.9 AWS Control Tower Architecture

All these actions are part of your out-of-the-box baseline multi-account AWS environment that is setup using AWS Control Tower. But how do you provision AWS member accounts such as network services, shared services, SAP production, and SAP non-production as represented in Figure 4.9? How do we set up an AWS account baseline and a networking baseline for these accounts?

The answer to this question is AWS Control Tower Account Factory, also known as the *account vending solution*. If we zoom in on the management account shown in Figure 4.9, you'll notice AWS Service Catalog being called by AWS Control Tower. AWS Service

Catalog is another trusted AWS service that AWS Control Tower uses to create and provision your member accounts based on the default guardrails that it offers. We'll look at the AWS Service Catalog service in detail in Section 4.3. You have other options as well such as using AWS CLI or using the AWS Control Tower console. However, AWS Service Catalog is the standard way to provision new accounts or to enroll existing accounts into your AWS Control Tower landing zone.

The Account Factory is a great feature to roll out new accounts based on the default baseline. However, you might have a requirement to roll out customizations based on your code pipelines, and you might have unique controls. You might need to deploy AWS resources and/or services (or third-party products) in an automated way into these new accounts. Some examples of these customizations include the following:

- Creation of IAM roles to integrate with the partner products or other tools

- Additional features like VPC flow logs, AWS Transit Gateway connections, and much more

- Deployment of custom guardrails, both detective (AWS Config rules) and preventive (SCPs)

Automation is the key aspect in this context because we are talking of an enterprise with tens or hundreds of AWS accounts that need to be rolled out, managed, and governed based on your custom requirements. Automating the deployment of these customizations at scale can be achieved using the following options:

- **Customizations for Control Tower (CfCT)**
  The CfCT solution enables you to easily add customizations to your AWS Control Tower landing zone using an AWS CloudFormation template and SCPs. You can deploy custom templates and policies to individual accounts and OUs within your organization. This solution integrates with AWS Control Tower lifecycle events to ensure that resource deployments stay in sync with your landing zone. For example, when a new SAP account is created using the AWS Control Tower Account Factory, the solution ensures that an AWS service endpoint such as an AWS Transit Gateway endpoint is automatically deployed in the required subnet within the member account's VPC. This endpoint might be required for on-premise connectivity via AWS Transit Gateway.

> **Customizations for Control Tower**
>
> CfCT is targeted towards customizations related to accounts and OUs in your landing zone by deploying resources that you specify after the account has been built using the Account Factory.

- **Account Factory Customization (AFC)**
  AFC allows you to customize accounts at creation time, automatically deploying the AWS services you choose on new accounts. For this task, you'll leverage AWS Service

Catalog Product, which runs behind the scenes on AWS CloudFormation templates that you can use later to deploy new resources on new accounts. We highly recommend setting up a separate AWS account called Service Catalog Blueprint Hub and then delegate Account Management API operations for this hub account in the organization. For example, when a new SAP account is created using the AWS Control Tower Account Factory, this feature can help you define a non-overlapping VPC CIDR range (not overlapping with other VPCs or on-premise ranges) based on your requirement.

- **Account Factory for Terraform (AFT)**
  In our experience, Terraform is being used extensively by many enterprises for its multi-cloud support capabilities. You can use Terraform with AWS Control Tower to vend a new AWS account. AFT sets up a Terraform pipeline that helps you provision and customize your accounts in AWS Control Tower. AFT follows a GitOps model to automate the processes of account provisioning in AWS Control Tower. You'll create an account request in a Terraform file, commit the file to the repository, which provides the necessary input that triggers the AFT workflow for account provisioning. After account provisioning is complete, AFT can run additional customization steps automatically.

At this point, you might wonder if, for an enterprise, AWS Control Tower can set up everything including customizations, then why would an enterprise invest time and effort in a custom-built landing zone? This question along similar lines to whether you should invest in an SAP S/4HANA Cloud Public Edition SaaS versus opt for SAP S/4HANA Cloud Private Edition. With SAP S/4HANA Cloud Private Edition, you're allowed to configure complex customizations including modifying source code. Along similar lines, AWS Control Tower does offer customizations that might or might not meet your requirements, whereas a custom-built AWS landing zone will provide you with greater flexibility when it comes to more advanced or complex customizations.

With that in mind, let's look at some of differences between AWS Control Tower and a custom-built landing zone, as shown in Table 4.1.

AWS Control Tower	Custom-Built Landing Zone
A managed service to setup multi-account environment in AWS	You, a partner, or AWS Professional Services builds all the environment components from scratch
AWS provides regular updates to the service	You or a partner are responsible to manage, operate, upgrade, and maintain the solution
Automated provisioning on AWS accounts	Write your own code to provision AWS accounts or manually use AWS Organizations

Table 4.1 AWS Control Tower versus Custom-Built AWS Landing Zone

AWS Control Tower	Custom-Built Landing Zone
Built-in preventive and detective guardrails	Build your own guardrails and write code to automate application of policies
Dashboard to monitor compliance	Build your own dashboard using Amazon CloudWatch, AWS's monitoring service
Flat OU structure	Nested OUs
Out-of-the-box centralized logging and auditing	Need to create, and then automate, centralized logging and auditing
Not available in all AWS regions but there are options/workarounds	Available in all AWS regions
Low- to medium-level customization available	Can be highly customized

Table 4.1 AWS Control Tower versus Custom-Built AWS Landing Zone (Cont.)

With all these capabilities and features, including support for customizations, we highly recommend you use AWS Control Tower before you consider a custom-built landing zone on AWS. Only if the already available features and customizations do not meet your requirements, then should you pivot towards a custom-built landing zone, while still keeping in mind the differences shown in Table 4.1.

[»] **AWS Landing Zone with RISE with SAP**

In RISE with SAP, SAP takes responsibility for building the landing zone. By default, all systems are deployed into a single Amazon VPC per AWS region.

For cross-region DR, an additional AWS account is used in the DR region. Per AWS account, there is only one VPC. Keeping in mind AWS landing zone best practices and recommendations, if you require network isolation/separation (e.g., of PRD and non-PRD systems), you have two options:

- **Separation into multiple subnets**
  You can request to deploy SAP systems into different subnets, for example, subnet-1a for PRD systems and subnet-1b for non-PRD systems. A separation by multiple VPCs is not possible. This option is only available for tailored approaches and does not apply to standard RISE with SAP S/4HANA Cloud, private edition.

- **Separation by multiple contracts**
  You could sign different RISE with SAP contracts for the SAP landscape to separate systems into different AWS accounts/VPCs. This contract could have a legal impact and must be considered carefully before signing off on the RISE with SAP contract.

4.4   AWS Service Catalog

Let's look at a simple example of provisioning an Amazon S3 bucket in an AWS account. One option is to do it manually using the AWS Console—you enter the name of the bucket, ensure that public access is disabled, ensure that encryption is enabled, and make sure your tags are configured based on your organization's standards. If you forgot to configure tags, and depending on your AWS Control Tower guardrails, your IT/CloudOps team might start getting alerts or notifications. Instead of an end user doing this end-to-end, which is error prone and might not adhere to the organization's standard, the IT/CloudOps team has the option of creating a product, let's call it "S3 Bucket Creation" with all the other parameters related to security and compliance defined already. The end user in this case just needs to enter the name of the Amazon S3 bucket and launch the product. The end result is an Amazon S3 bucket based on your organization's security and compliance standards.

Let's consider another example of a deployment of an SAP system on AWS. You have two options—manual or automation (IaC) to provision the infrastructure, and then eventually install the SAP system on the provisioned infrastructure resources. Now let's assume you're using automation using CloudFormation template for provisioning the infrastructure. As part of this automation, you still need to input the VPC CIDR range, availability zones (AZs), subnets, IP addresses for Amazon EC2 instances, golden AMI for the OS, tags, and then eventually all the SAP related parameters such as System ID, SAP Instance number, number of App Servers, etc. What if the IT administrator or CloudOps team creates a product which has all the predefined values related to VPC, AZs, Subnets, IP, golden AMI for the OS, tags and an SAP BASIS team member simply needs to input the SAP related parameters such as SAP System ID, SAP instance number and the number of app servers. This is exactly what AWS Service Catalog does! In our example, the cloud administrator needs to create an SAP on AWS-related product by importing IaC files/templates, which is one of the many products in your catalog, with predefined networking/DNS/AD related parameters and values including enforcement of tagging, and make it available to the SAP BASIS admin to roll out an SAP system on AWS. It makes the life of your IT administrators/CloudOps team easy, avoids back and forth between the teams, helps to deploy a consistent SAP system much faster, and most importantly helps the IT/CloudOps ensure AWS resources are deployed based on internal procedures and security protocols.

A native service within AWS to help you build products and make them available to your end users or developers is AWS Service Catalog. This service helps to deploy and manage infrastructure and applications in the form of products that are based on your organization's security and operational policies. These products are consumed typically by developers, SAP administrators, data scientists, and other operations teams in a self-service manner. As shown in Figure 4.10, a few key terms to note when it comes to AWS Service Catalog include the following:

- **Product**

  An IT service or AWS resource written in JSON, YAML Ain't Markup Language (YAML), or Terraform.

- **Portfolio**

  A collection of AWS products.

- **Constraint**

  Security, governance, and deployment control.

- **Product lists**

  List of products that an end user can launch.

- **Provisioned products**

  Enable users to update or perform AWS service actions.

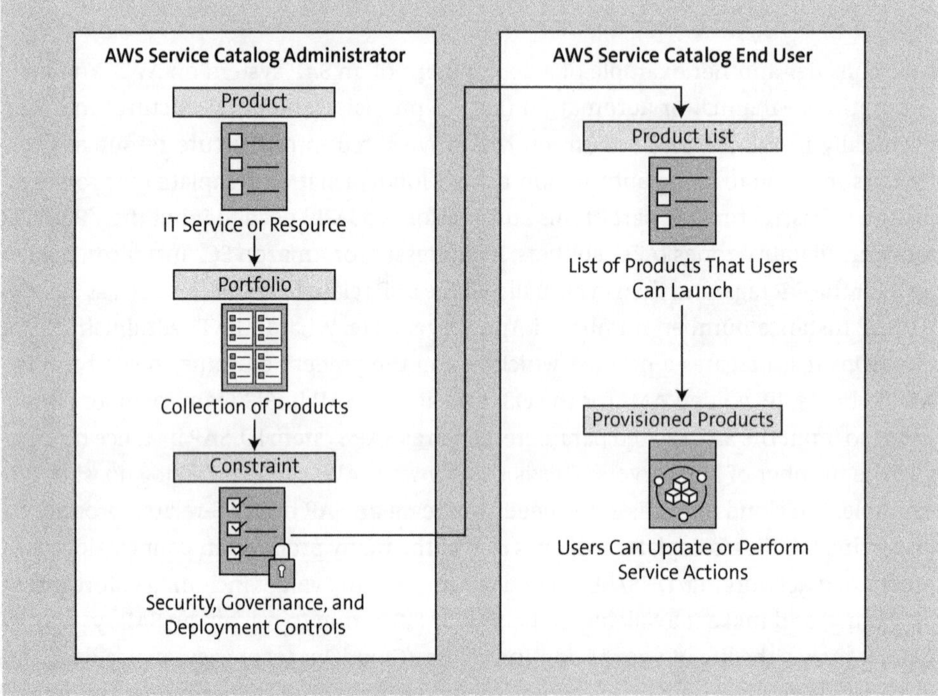

Figure 4.10 Key Terms: AWS Service Catalog

Additionally, AWS Service Catalog can be used for SAP workloads with GxP requirements, which holds relevance typically in the healthcare and life sciences industry. Your cloud administrator can define an AWS Service Catalog product for SAP on AWS using AWS CloudFormation templates. Typical SAP products include preapproved configurations for SAP application servers and databases that are deployed using AWS resources like an Amazon EC2 instance. After your security administrator defines the GxP application, in this case, SAP using an AWS CloudFormation template as a framework, they can publish it for SAP administrators in the AWS Service Catalog. SAP administrators can use this framework to further enhance the template based on the application's requirements.

The AWS Service Catalog is used heavily within AWS Control Tower; in fact, after AWS Organizations, it is the next most important service that you should learn. As a trusted and integrated service with AWS Control Tower, AWS Service Catalog is useful when you're building your landing zones in the following two ways:

- **Account Factory**
  Other than creating or enrolling accounts using AWS CLI or AWS Management Console, AWS Service Catalog is the most standard, and the recommended, service to roll out new accounts when you're using AWS Control Tower. Using APIs, AWS Control Tower builds an AWS Service Catalog product, which administrators or developers can use to launch new accounts. This product within the AWS Service Catalog console is called AWS Control Tower Account Factory. In the AWS community, this approach is popularly known as an account vending solution. New accounts are launched based on out-of-the-box guardrails that you enabled during the setup of AWS Control Tower defined at the OU or at the organization root level.

  With AWS Control Tower, you can customize both new and existing AWS accounts when you provision their resources from the AWS Control Tower console. However, this task is more of a manual process. Let's consider an example with two AWS accounts: SAP Production and SAP DR. Within these accounts, you must define the VPCs within which your SAP systems will be provisioned. To establish network connectivity between these two VPCs and other VPCs in your landing zone, you'll need to ensure the VPCs are using a non-overlapping CIDR range. So, you would manually set up the IP range (non-overlapping) in Account Factory before enrolling each account. As you can see, creating these accounts is a manual process first. Eventually, you can ensure that VPC CIDR ranges are non-overlapping. The manual process works out great specifically when you have only few (5 to 8) workload accounts to manage.

- **Account Factory Customization (AFC)**
  AWS introduced AFC to combat the manual effort needed to create and/or enroll tens or hundreds of customized AWS accounts according to your specific requirements of security guardrails, and AWS services or resources that need to be deployed within these accounts. This feature allows you to customize accounts at creation time all based on automation by building the pipelines for you. This capability still uses AWS Service Catalog in the background; however, now you have an additional blueprint/template, or the product is customized based on your specific requirements ideally in a separate "AWS Service Catalog Blueprint Hub Account." The blueprints are available as AWS managed or from the AWS Partner network. In our previous example of ensuring non-overlapping CIDR ranges, you can use an AWS-managed blueprint, tweak, or customize it or use an out-of-the-box blueprint provided by an AWS partner to ensure that your VPCs created have separate CIDR block

ranges so that there are no routing conflicts. Figure 4.11 represents an overall Account Factory customization flow:

❶ Within the AWS Control Tower console, you have the option to create new accounts using Account Factory. You must provide account details (e.g., email, account name); access configuration (AWS IAM Identity Center user email and name), and OU.

❷ Account Factory calls AWS Service Catalog product "AWS Control Tower Account Factory," which creates baseline account and applies all the guardrails to the account to the OU that the account is associated with.

❸ After creating the baseline account, through AWS Control Tower lifecycle events, in this case, CreateManagedAccount, Account Factory calls another AWS Service Catalog Product, which is deployed in a separate AWS Service Catalog Blueprint account. This product is deployed based on your customization requirement, which is built based on your own, AWS-managed, or partner-managed blueprints.

❹ The product in the AWS Service Catalog Blueprint account deploys all the AWS services or any third-party products in the target member accounts based on your unique customization requirements or security controls.

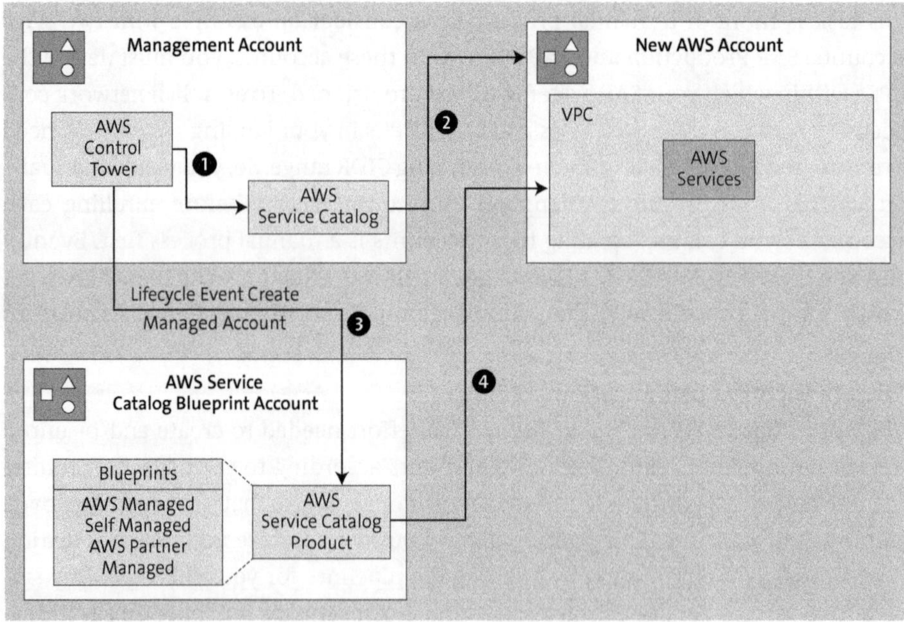

Figure 4.11 AWS Control Tower Account Factory Customization (AFC) Process Flow

4.5   Resilient Architecture Patterns

For SAP workloads, the *resiliency* of an SAP system is its ability or the capacity to recover quickly from a computer hardware or a software failure. SAP systems, for most organization, are deemed mission critical to the extent that they are sometimes referred to as "crown jewels" of an enterprise. For an enterprise, SAP systems are revenue generating and/or reporting, external facing to your suppliers/vendors, and/or it could be your core system around, which all the business functionalities revolve. Typically, all the business processes and core business transactions revolve around the SAP ERP or SAP S/4HANA system with loose or tight coupling with other systems. For these reasons, it becomes imperative to protect your critical SAP systems, specifically your production systems against, all failures both hardware and software.

Additionally, as part of your landing zone design, you should always consider AWS regions and the related AWS services or any third-party products related to networking and/security that are in scope. Therefore, the first step in your journey towards building a reliable SAP system is to ensure that you have a landing zone designed to accommodate a single region with multiple AZs and/or multi-regions with multiple AZs, which can support all the relevant AWS services and tools needed to support the SAP system. Figure 4.12 shows resiliency, its important components, and the objective of each of its components, which we'll discuss further in the following sections.

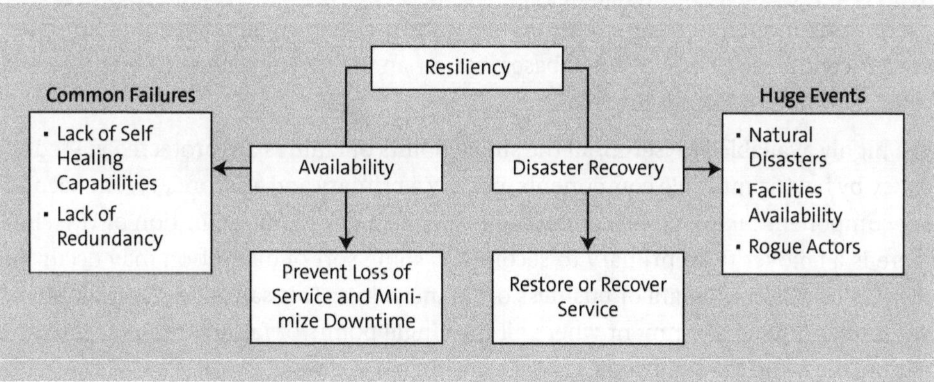

Figure 4.12 What Is Resiliency?

Reliability versus Resilience

"Reliability" is the outcome, and "resilience" is the way you achieve it. An SAP system that can withstand failures (be resilient) will eventually lead to a consistently available, a reliable system.

4.5.1   High Availability

An important concept to understand is the difference between *fault tolerance* and *high availability (HA)*. Consider an example: An airplane has two engines so that, even if one engine fails, the plane will continue to operate normally. The ability of an IT system to continue its operations normally, without any business or data loss, is known as *fault tolerance*. On the other hand, a car usually has a spare tire in the trunk that can be used to replace a flat—however, it can take a few minutes or hours to replace the tire and get back on the road again. Similarly, the ability of an IT system to continue its operation after minimum or low disruption with low or ideally no business or data loss is known as *HA*.

To ensure business as normal and prevent data loss, SAP systems have fault tolerant capabilities, such as the following:

- An SAP system or instance comprises of multiple dialog work processes. An individual SAP dialog work process has the ability to restart after a failure to ensure business continuity without any noticeable impact or outage to the system.
- SAP HANA services have the ability to restart on failure on their own using the watchdog feature.
- An SAP HANA host has an auto-failover feature, which automatically fails over from the crashed host to a standby host in the same system.

In an SAP system, fault tolerant capabilities at the software level and/or hardware are usually not enough to ensure business continuity. Other single points of failure, such as SAP central services, your databases, etc. in an SAP system need to be protected, which is where HA comes into play!

In a highly available SAP setup, all the single points of failures are protected via redundancy by having multiple components, usually a primary and a secondary. The secondary component takes over in case of failure. Going back to our definition of HA, since there is a failover from primary to secondary, some sort of disruption may occur but with no or minimum data or business or financial loss. This safeguard is similar to an on-premise SAP environment where all the single points of failures are protected via redundancy, and the same logic is applicable on AWS as well.

On AWS, the components related to data centers such as networking, power, cooling, and other resources have already been designed and implemented with redundancy to ensure static HA, and in some cases, these components are fault tolerant. Let's look at an example: Within an AWS region, different AZs are connected via *transit centers*. At a minimum, two transit centers are always deployed in an AWS region with both always up and running. In case of hardware or software failures with one of these devices, the one that is still healthy can continue to serve it purpose, which is to ensure that instances from one AZ can connect to instances in other AZ. Other than data center components like a transit center, which AWS owns and manages, you still must ensure that you are protecting your SAP systems against hardware and software failures that your enterprise or partner is responsible for.

Figure 4.13 shows the single points of failure applicable for an SAP system running on AWS.

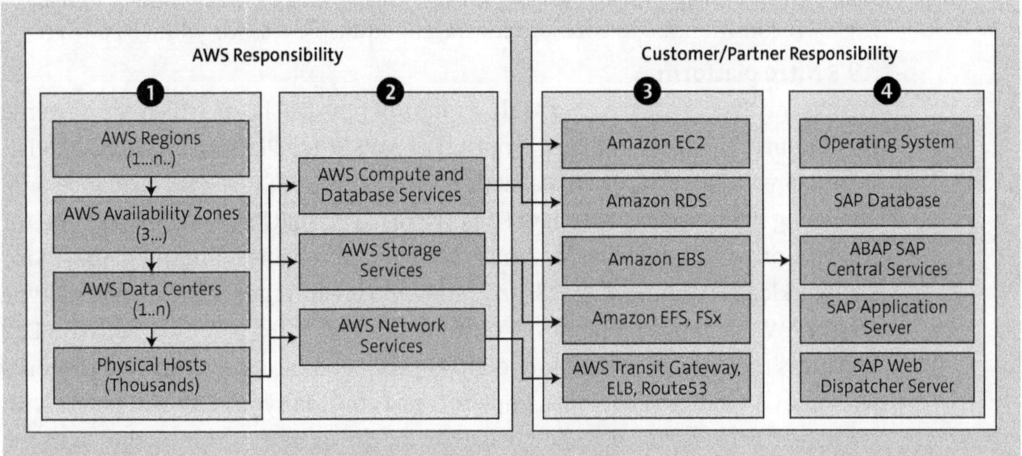

Figure 4.13 Single Points of Failure for an SAP System on AWS

Let's take a closer look at each of these points:

❶ **AWS regions, AZs, AWS data centers, and the physical hosts running on the AWS Nitro platform**

AWS does the heavy lifting when it comes to the resiliency, fault tolerance, and HA of its global physical infrastructure:

– **Regions**

AWS has multiple physical geographic locations spread across the world which are called regions. The main reason to have multiple locations around the globe is redundancy to ensure business continuity for your applications specifically during a disaster (physical or logical) across your primary region. Using multiple regions is typically used to set up a DR strategy for SAP workloads.

– **AWS AZs**

Each AWS region comprises of multiple—three or more AZs. An AZ is a logical construct. The AZs are built of one or more distinct data centers each with their own networking, power, and cooling. Before rolling out any region and the AZs within the region, AWS does thorough analysis, past several years of historical data, to ensure that the zones are not impacted simultaneously by floods, earthquakes, fires and/or tornadoes. Each of AZs are connected via transit centers and distance between any two zones is not more than 100 miles or 60 kms. Because of their design, and low latency networking capabilities, leveraging multiple AZs in an ideal architecture for an SAP HA setup.

- **AWS data centers**

  Within the logical boundaries of an AZ AWS has its data centers. These are usually one or more within an AZ. Since data centers are a single point of failure hence it's highly recommended to consider leveraging multiple AZs for your HA design.

- **AWS Nitro platform**

  Within the data centers AWS has thousands of physical hosts which are powered by in-house, custom built, and proprietary AWS Nitro Platform. AWS Nitro Platform helps AWS perform virtualization with minimum (~1%) performance penalty using lightweight AWS Nitro Hypervisor. Nitro platform is the backbone for AWS for providing high performance, secure, and reliable compute platform for SAP workloads running on AWS. Any latest and greatest generation instance type within AWS's compute services is running on AWS Nitro platform. Amazon EC2 instances that you provision for your SAP workloads are bound to data center(s) within an AZ hence it becomes imperative to consider your SAP workloads with HA spread across multiple AZs in case of a failure or intermittent issues with the compute service in an AZ.

❷ **AWS compute, storage, and networking services**

As shown in Figure 4.14, AWS offers core services related to compute, storage, and networking for running your SAP workloads on AWS. Some of these services are Amazon EC2, Amazon EBS, Amazon EFS, Route 53 etc. AWS builds highly vetted and secure services keeping HA in check. Most AWS services are split into the logic of "control plane" and "data plane." These terms come from the networking world of routers. Routers data plane distribute networking packets based on rules however routing policies have to be defined somewhere, and that's where control plane comes in. Simply put, when you launch an instance on AWS using console or APIs—this is a control plane operation, and running these instances itself, reading and writing to a volume—this is a data plane operation. AWS ensures HA of control, and data plane by segregating them and treating them as distinct components.

[»]

AWS Service Level Agreements: Control Plane versus Data Plane

AWS does not offer a service level agreement (SLA) related to control plane operations but does so for data plane operations. While designing a reliable SAP system specifically for DR and failover scenarios it's important to consider data plane actions that the service can provide in case the control plane operations are not available due to region failure.

❸ **Service level agreements**

AWS provides an SLA for availability for most AWS services. However, it's your responsibility or a partner's responsibility on how you leverage and set up these services to ensure HA based on your SLAs or mean time to resolution (MTTR) requirements. Due to the criticality of SAP systems for an enterprise, 99.9%, 99.95%, and

99.99% of uptime requirement annually is common in the SAP world. The general rule is to deploy the AWS services across multiple AZs to achieve multiple 9s of availability. Some AWS services are serverless and/or fully managed which, by default, are deployed across AZs and offer 4 9s of availability SLA (e.g., Amazon EFS):

- **Amazon Elastic Compute Cloud (Amazon EC2)**
  This is the most used compute service which offers virtualized or bare-metal instances/servers for running SAP workloads on top of it. A single Amazon EC2 instance deployed in a single AZ offers 99.5% availability SLA for every. Hence when you're deploying an SAP system which needs less than 52 minutes (4 9s availability) of unplanned downtime you must consider deploying SAP components across multiple Amazon EC2 instances spread across the AZs.

- **Amazon Relational Database Service (RDS)**
  This fully managed database service supports Oracle, SQL Server, MySQL, MariaDB, and PostgreSQL. There are only few SAP applications such as SAP BusinessObjects, which are supported on Amazon RDS, offering 99.95% SLA for multi-AZ database instance, and 99.5% SLA for single-database instance.

- **Amazon Elastic Block Storage (EBS)**
  For persistent block storage for SAP application servers and databases, Amazon EBS offers 4 9s of availability SLA when deployed across AZs and 99.9% for single Amazon EBS volume in an AZ. However, Amazon EBS volume types such as io2 or io2 block express do offer 4 9s availability SLA for a single volume in a single AZ.

- **Amazon Elastic File System (EFS)**
  EFS is a fully managed, set-and-forget, and a highly scalable shared storage service. For shared file system specifically relevant in a HA setup for SAP, EFS offers 4 9s of availability SLA by default as it is deployed across multiple AZs.

- **Amazon FSx**
  AWS offers file system service Amazon FSx where x can be for Windows File server, NetApp ONTAP, OpenZFS, and Lustre. For SAP workloads, the most used and certified services are Amazon FSx for Windows File Server, and Amazon FSx for NetApp ONTAP. Both are fully managed shared storage offering 4 9s of availability SLA for a multi-AZ setup.

- **Amazon Simple Storage Service (S3)**
  Amazon S3 is an object storage service which is specifically used for SAP database backups, SAP Data Lakes, and/or archiving old SAP data. Amazon S3 offers various storage classes, and by default Amazon S3 standard offers 4 9s for availability SLA.

- **AWS Transit Gateway**
  AWS Transit Gateway is a central router which helps you setup hub and spoke architecture related to networking and is used to connect multiple VPCs across your AWS accounts. It is offered as a fully managed service which offers 4 9s availability SLA by default.

- **Amazon Elastic Load Balancer (ELB)**
ELB is a fully managed service used for load balancing specifically relevant in the SAP Fiori world. A multi-AZ deployment offers 4 9s whereas a single AZ deployment offers 3 9s of availability SLA.

> **HA SLA**
>
> The AWS Services SLA is different from the Managed Service Provider (MSP) SLA. AWS service availability SLAs are at the service level, whereas MSP availability SLAs are at the application/database uptime level. Thus, you should treat these two SLAs differently.

❹ SAP single points of failures
In an SAP system, multiple points of failures exist, such as the following:

- The OS
- SAP Central Services message server and enqueue server
- SAP databases
- SAP application servers
- SAP Web Dispatcher
- Shared storage, for instance, network file storage (NFS)
- Network load balancers

OS, SAP Central Services, databases, application servers, and SAP Web Dispatcher are all protected using additional Amazon EC2 instances on AWS. Before Amazon EFS or Amazon FSx was introduced, a common approach was to have NFS servers running on Amazon EC2 instances, which meant you were responsible for management and operations and for ensuring HA for these file systems. However, today, the go-to solution for shared storage is Amazon EFS (for Linux) or Amazon FSx for Windows File Server (for Windows) because they are fully managed serverless shared storage with 4 9s of availability SLA offered by default. On load balancers, you can bring any third-party solution or product typically deployed on Amazon EC2 instances in an active-active mode. However, due to similar challenges that we saw with NFS servers on Amazon EC2, the recommended solution is Amazon ELB, which is fully managed by AWS offering 4 9s of availability SLA with a multi-AZ deployment.

Another aspect of availability SLAs are composite SLAs, which result from combining different AWS service offering SLAs. Your SAP application architecture will dictate whether your composite architecture can provide higher or lower uptime values. For example, consider an architecture with a single SAP application server that writes to an SAP HANA database configured with replication to a secondary database running on an Amazon EC2 instance in another AZ. At the time of writing, you could acquire the following SLAs from AWS services:

- SAP Application Server, SLA of 99.5% (single Amazon EC2 instance in a single AZ)
- SAP HANA Database, SLA of 99.99% (two Amazon EC2 instances spread across AZs)

In this case, due to hardware and/or software issues, if SAP Application Server fails, the whole application fails. However, in case of primary database failure, it can failover to the secondary instance in the second AZ either manually or using automation. So, the composite SLA for this architecture is 99.99% × 99.5% = 99.49%, which is lower than expected because the application relies on multiple services, which has more potential failure points.

4.5.2   Disaster Recovery

Ensuring business continuity for your mission-critical SAP systems is a must, and to build a resilient SAP production landscape, you need a DR strategy. A resilient DR solution should encompass getting your SAP business data and operations out of a failed infrastructure and running again on a new infrastructure. DR is not about saving the infrastructure but about saving the business. Disaster can include physical or natural calamities such as earthquakes, floods, tornadoes, etc. or can be logical (human induced), such as ransomware or malware attacks.

DR is the ability of your SAP systems, most often the production systems, to recover from physical or logical failures, which are widespread across multiple critical SAP applications with some or minimum data loss. The time it takes to recover these systems is usually hours. Let's go back to our example of a car and its spare tire. What if all the tires went flat? The chances are slim, but it could happen. In that case, your best option is to get a replacement car and send the primary car to a mechanic to get it fixed. The time it takes to get a replacement car is usually a few hours, while the time to get the primary car fixed could be a day or two. Along these lines, though the chances are slim or highly unlikely, an AWS region can fail due to a physical or a logical disaster, so it becomes imperative to protect the SAP systems against an AWS region going down. The strategy that you choose to protect the SAP systems against a disaster mainly depends not only on recovery point objective (RPO) and recovery time objective (RTO) but also on recovery consistency objective (RCO) keeping in check the impact to the business.

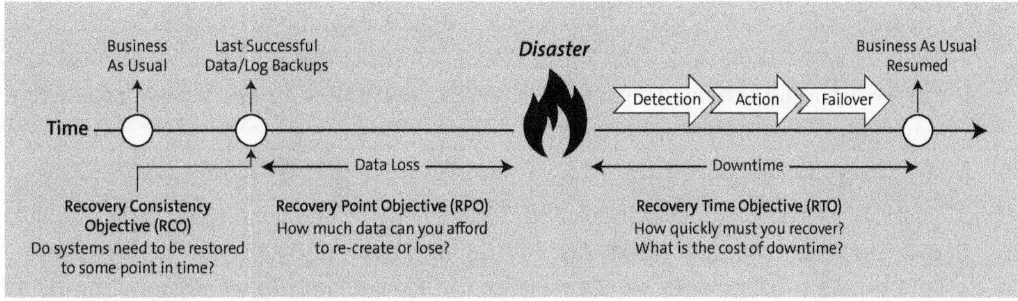

Figure 4.14 RPO versus RTO versus RCO

As shown in Figure 4.14, let's look at the terms related to DR in detail:

- **RPO**

  The data loss that an SAP application will suffer when a disaster strikes. The unit of measurement is time. Depending on the criticality of the SAP application, an enterprise typically has an RPO requirement of minutes or hours per SAP application, segregated based on tiers. For example, at tier 1, an SAP ERP or SAP S/4HANA system might have an RPO of 15 minutes, whereas at tier 3, SAP Solution Manager might have an RPO of 2 hours.

- **RTO**

  The time it takes to bring the SAP applications back up and making it available to your end users. The unit of measurement is time. Depending on the criticality of the SAP application, an enterprise typically has an RTO requirement of hours or days per SAP application, segregated based on tiers. For example, at tier 1, SAP ERP or SAP S/4HANA might have an RTO of 2 hours, whereas, at tier 3, SAP Solution Manager might have an RTO of 6 hours.

- **RCO**

  An uncommon but a still important term, RCO is a measurement of the consistency of business data distributed among interlinked systems. In a typical customer environment, SAP systems are tightly integrated, and data is frequently exchanged between these systems, like SAP ERP or SAP S/4HANA, SAP Business Warehouse (SAP BW) or SAP BW/4HANA, SAP Customer Relationship Management (SAP CRM), SAP SRM, SAP GTS, etc. This group of tightly integrated systems is called a *system group*. In case of DR failover, you may have a zero RCO requirement within the system group. In other words, in case of DR failover, all databases within the SAP system group must be recovered to the same point in time.

Another important term when designing for DR is the maximum tolerable downtime (MTD). This value is the RTO plus some additional time kept as buffer to accommodate for any untoward incidents during a recovery. The motivation behind introducing this term is to avoid downtime that breaches acceptable thresholds.

RPO and RTO are usually system specific with applications segregated into tiers. For example, SAP ERP with an RTO of 2 hours and SAP Solutions Manager with a 6 hours mean 6 hours of total RTO within which SAP ERP should be recovered in the first 2 hours. On the other hand, if the RTO is 2 hours for the entire SAP production landscape, then you should be able to recover both SAP ERP and SAP Solution Manager production environments within those 2 hours.

At an enterprise level, an important consideration is that three parties—business users, IT leadership (C-level suite), and IT teams—all be aligned on RPO, RTO, and RCO SLAs on DR for your SAP workloads. Additionally, if one of the primary drivers of moving to AWS is to improve on SLAs related to DR, then a thorough analysis should be performed between your current SLAs versus possible future SLAs on AWS with proper communication to both business and IT leadership.

On AWS, a failure at the software or hardware level can occur at an AZ level or at a regional level. So, when designing a resilient SAP architecture, an imperative step is to consider a multi-region approach, keeping in mind data residency requirements as well as minimum distance requirements for the DR location. Typically, on-premise, you would have one physical location segregated into two hosting facilities (often separate rooms or halls), and this setup is used to deploy a highly available SAP system protecting its single points of failures. This approach often maps to a multi-AZ architecture. The other physical location for DR situated at a certain distance, let's say 500 miles away, often maps to another region on AWS. No right or wrong answers exist when it comes to a single region or a multi-region resiliency design, but your business uptime requirements should drive the architecture.

4.5.3 Single Region

A common tactic among SAP customers is to leverage a single-region pattern when designing HA and/or DR into their SAP workloads. This choice is most often driven by two factors: data residency and minimum distance requirements for the DR location. Let's consider an example: Enterprise A may have stringent data residency requirements that data never cross a country's perimeter. If AWS has a region within the country, then it becomes an easy choice for the enterprise to go with an AWS region in that country.

Because of the way AZs are constructed, in this scenario, the enterprise sets up what is known as HA/DR. Two AZs are used to set up a resilient SAP system protected against compute and storage failures in a single AZ. As a single region has, at a minimum, three AZs; the third AZ can become the DR site as long as the minimum distance requirement is less than 60 miles (100 kms). Enterprise A, depending on the RPO and RTO requirements, can also consider having standby compute resources up and running specifically for databases in the third AZ. Another option is to deploy a non-production system in the third AZ to ensure capacity in case both production AZs fail.

With a single-region approach, you're not only avoiding costs related to the setup of foundational services such as networking, security, and auditing, and the related management and operation overheads; you're also avoiding the potential network latency to your on-premise or end-user locations that usually comes with multi-region approach. With a single-region pattern, you have the advantage of using synchronous databases or storage replication for HA/DR, which means you can achieve zero RPO and avoid any kind of data loss in case of failover event. Additionally, to optimize costs while keeping your RTO in check, a smaller sized standby production database or backup/recovery can also be considered as a HA/DR strategy. You must consider the time it takes to re-size the instance during a DR event.

[+]   AWS Single-Region Pattern

If an enterprise has data residency requirement for a particular country and AWS does not have a region in that country, one option is AWS Outposts where AWS sends a rack (of servers) to the enterprise's on-premise location. However, since AWS Outposts are linked backed to a "mother" AWS region, the enterprise must do their due diligence on whether it meets its country's cloud compliance standards. SAP on AWS Outposts is not supported yet; however, AWS Outposts can be leveraged for non-SAP workloads depending on the application's feasibility and supportability.

[»]   AWS Single-Region Pattern Network Latency

For synchronous database replication between a primary and a secondary database, SAP recommend you have a network latency of <=0.1 ms. You can use AWS Network Manager, a service using the AWS Console, to check the latency between the AZs within a region. As SAP components are spread across AZs in a HA setup and are latency sensitive, it becomes imperative for you to conduct this exercise of selecting the AZs within a region with lowest possible latency.

Within AWS, data egress costs will be incurred, and thus, with a single region pattern, you can avoid costs related to data transfers across the regions. However, you're still incurring any intra-region costs (e.g., cross-AZ data transfer costs). For SAP on AWS, you might have an application server in AZ 1 transferring network packets to the primary database in AZ 2, and vice versa. Tier 0 services or your core services such as DNS, AD, and/or LDAP or any third-party ISV solutions related to security and audit only need to be deployed in one AWS region.

For application servers and databases, you should always consider having a backup of a backup. Thus, at a minimum, to prevent failures specifically against logical corruption, malicious activities, or human error, you should consider Amazon S3 (for database backups) with intra-region replication from one Amazon S3 bucket to another. Also consider object locking, a feature within Amazon S3, to build immutable backups. You can also consider a data bunker or an air-gapped backup solution in another AWS account within the same region.

The easiest, most convenient, and standard way to govern and confine your deployments of your SAP workloads to a single region is using AWS Control Tower. AWS Control Tower has a region deny control that applies at the landing zone as a whole rather to any specific OU. With this control, you won't be allowed to deploy resources in the denied regions. This feature can be changed at any time.

We highly recommend you conduct, at a minimum, one to two HA/DR drills or simulations by defining runbooks and documenting procedures. Figure 4.15 shows HA/DR architecture options with a single-region pattern.

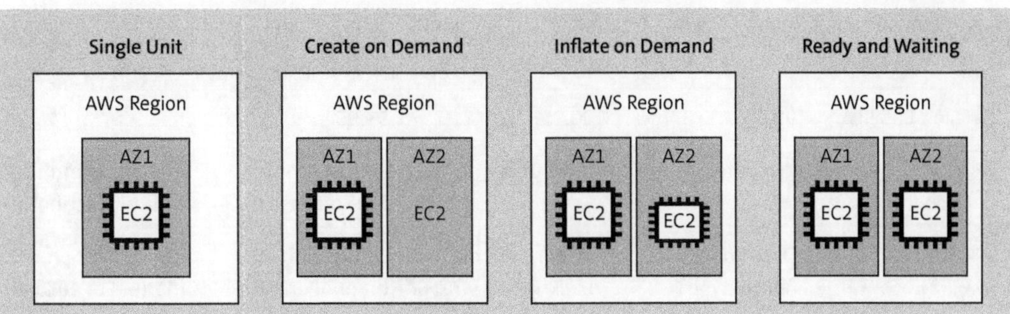

Figure 4.15 Single Region HA/DR Resiliency Patterns

Table 4.2 summarizes the HA/DR architecture patterns with an Amazon EC2 SLA, an Amazon EBS SLA, RPO, RTO (MTTR for HA), and resiliency levels against region, AZ, hardware, and logical failures.

Pattern	Single Unit	Create on Demand	Inflate on Demand	Ready and Waiting
Description	<ul><li>Single Amazon EC2 with auto-recovery</li><li>Single AZ</li></ul>	<ul><li>Primary Amazon EC2 with secondary Amazon EC2</li><li>Secondary Amazon EC2 created on demand</li><li>Dual AZs</li><li>Manual failover (AWS CLI or AWS Management Console Operations)</li><li>Logs/data stored in cross-AZ storage (Amazon EFS, Amazon FSx)</li></ul>	<ul><li>Primary Amazon EC2 with secondary Amazon EC2</li><li>Secondary Amazon EC2 smaller than primary</li><li>Secondary Amazon EC2 inflated on demand</li><li>Dual AZs</li><li>Manual fail over (AWS CLI or AWS Management Console Operations), can be automated.</li><li>Data synchronously shipped across AZs</li></ul>	<ul><li>Primary Amazon EC2 with secondary Amazon EC2</li><li>Secondary Amazon EC2 same sized as Primary</li><li>Dual AZs</li><li>Automated failover using auto-scaling groups or OS clustering</li><li>Logs/data synchronously shipped across AZs</li></ul>
Amazon EC2 SLA	99.5%	99.5%	99.99%	99.99%

Table 4.2 Summary of Single Region HA/DR Resiliency Patterns

Pattern	Single Unit	Create on Demand	Inflate on Demand	Ready and Waiting
Amazon EBS SLA	99.9%	99.9%	99.99%	99.99%
RPO	~Minutes	~Minutes	~0 (Last committed transaction)	~0 (Last committed transaction)
MTTR/RTO	Dependent on failure type (minutes to hours)	Dependent on backup/log strategy Dependent on manual or automated process (<4 hours)	Dependent on manual or automated process (<2 to 4 hours)	Dependent on manual or automated process (<20 minutes)
Resiliency against hardware failure	Yes	Yes	Yes	Yes
Resiliency against AZ failure	No	Yes (only if cross-AZ backup services/logs available)	Yes	Yes
Resiliency against an AWS region failure (Risk of occurrence: Very low)	No	No	No	No

Table 4.2 Summary of Single Region HA/DR Resiliency Patterns (Cont.)

4.5.4 Multi-Region

Continuing with our example of Enterprise A, what if the minimum distance requirement is 500 miles? In that case, one option is to go with another AWS region within the country's borders. If Enterprise A has data residency requirements to remain in the US, then they can pick another region within US that is more than 500 miles apart because AWS has multiple regions that are more than 500 miles apart.

But what if there is no other AWS region available in that country? In this scenario, a common approach is to follow a multi-cloud strategy or use a private cloud in a co-location coming into play as the DR site. In May 2023, the only Google Cloud Platform (GCP) region in France was flooded with water and was down for more than 3 weeks. Think of enterprises with data residency requirements and who did not consider another DR site within the country. Thus, an imperative task is to consider another DR site within the country! For mission-critical systems like SAP, you cannot afford this long of a downtime. Let's look at another scenario where Enterprise A, based out of the US, does not have any data residency requirements. In this case, it can select from multiple AWS regions across the globe. In this scenario, Enterprise A might choose a region based out of Canada for their DR site, or vice versa. (If Enterprise A has its primary region in Canada, then a region in the US can be the DR site since AWS has only one region in Canada, as of Q4 2023.)

When you're designing an architecture for SAP workloads on AWS with high levels of resiliency (i.e., aggressive MTTR, RPO, RTO, and/or RCO), typically a multi-region approach is considered. So, while designing your AWS landing zone environment, you should consider a multi-region landing zone setup, which can help you build that foundation, not only in your primary region but also in the DR region of your choice. AWS Control Tower has the option wherein you can select multiple regions where you plan to deploy your workloads including all the compliance, security, and audit related services.

With a multi-region pattern, you must consider costs related to setting up foundational services (e.g., for networking, security, and auditing) and related management and operation overheads but also the potential network latency to your on-premise or end-user locations. Because of geographic distance and laws of physics, with a multi-region pattern, you must consider asynchronous databases or storage replication for DR, which means you can achieve an RPO of minutes to hours. Therefore, the potential of some data loss or the best-case scenario is losing only in-flight transactions during a failover in a DR event. Additionally, to optimize costs while keeping your RTO in check, a smaller sized standby production database or backup/recovery system can also be considered as a DR strategy. You must consider the time it takes to re-size an instance during a DR event; in this case, the RTO will increase versus having a same-sized DR instance as primary production instance.

You must also consider data egress costs related to multi-region connectivity, specifically costs for database replication, shared storage replication, and/or database backups being copied from region to another. Additionally, you might be using AWS Transit Gateway or Amazon VPC peering to stitch the two regions together, and thus, those costs must be accounted for as well.

AWS Data Transfer

All data transfers over a VPC peering connection that stays within an AZ is free. You can use the AZ ID to uniquely and consistently identify an AZ across different AWS accounts.

Tier 0 services, or your core services such as DNS, AD and/or LDAP, or any third-party ISV solutions related to security and auditing must be deployed or should be planned for failover to the DR site in the second AWS region.

The easiest, most convenient, and standard way to govern and confine your deployments of your SAP workloads to multiple regions is using AWS Control Tower. AWS Control Tower has a region deny control that applies at the landing zone as a whole rather to any specific OU. With this control, you won't be allowed to deploy resources in the denied regions. This feature can be changed at any time.

For application servers and databases, you should always consider having a backup of a backup. Thus, at a minimum, to prevent failures specifically against logical corruption, malicious activities, or human error, you should consider Amazon S3 (for database backups) with intra-region replication from one Amazon S3 bucket to another. Also consider object locking, a feature within Amazon S3, to build immutable backups. You can also consider a data bunker or an air-gapped backup solution in another AWS account in another AWS region other than the DR region. We highly recommended you conduct, at a minimum, one to two DR drills or simulations by defining runbooks and documenting the procedures.

AWS Compute Capacity Planning

All data transfers over a VPC peering connection that stays within an AZ is free. You can use the AZ ID to uniquely and consistently identify an AZ across different AWS accounts.

Figure 4.16 shows DR architecture options with a multi-region pattern.

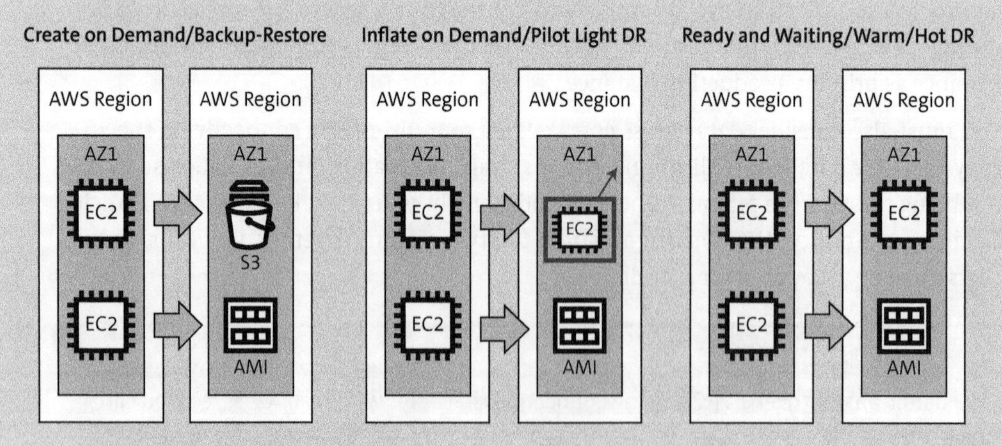

Figure 4.16 Multi-Region DR Resiliency Patterns

Table 4.3 summarizes the DR architecture patterns with RPO, RTO, and resiliency levels against region, AZ, hardware, and logical failures.

Pattern	Create On Demand/ Backup-Restore	Inflate on Demand/ Pilot Light DR	Ready and Waiting/ Warm/Hot DR
Description	■ Data backed up on Amazon S3 and via cross-region Amazon S3 replication migrated to another region at regular intervals ■ Application server Amazon Machine Image (AMI) replicated to secondary region	■ Only critical core systems of SAP (e.g., database) is replicated (usually small-scale) in the secondary region ■ Application server AMIs replicated to secondary region	■ Only critical core systems of SAP (e.g., databases) are replicated to a same-sized instance in secondary region ■ Application server AMIs replicated to secondary region ■ SAP application servers are up and running in another region (hot DR)
RPO	Minutes, using Amazon S3 Replication Time Control (RTC); 99.99% of objects are replicated within 15 mins	Minutes	Minutes
RTO	Dependent on backup/log Strategy	Dependent on manual or automated process	Dependent on manual or automated process
	Dependent on manual process	<2 to 6 hours	<2 hours
	(Double-digit hours)		
Resiliency against hardware failure	Yes	Yes	Yes
Resiliency against AZ failure	Yes	Yes	Yes

Table 4.3 Summary of Multi-Region DR Resiliency Patterns

Pattern	Create On Demand/ Backup-Restore	Inflate on Demand/ Pilot Light DR	Ready and Waiting/ Warm/Hot DR
Resiliency against an AWS region failure (Risk of occurrence: Very low)	Yes	Yes	Yes
Resiliency against logical corruption, malicious activity, and human error	Yes	Yes	Yes

Table 4.3 Summary of Multi-Region DR Resiliency Patterns (Cont.)

Multi-Region HA

We still recommend you leverage a multi-AZ architecture within a single region for an HA setup for your SAP systems on AWS. The second AWS region serves the purpose of a DR site leveraging cold, pilot-light, warm, or hot DR strategies depending on your RPO and RTO.

Multi-Region Data Bunker

We also highly recommend you consider another AWS region for your SAP-related backups. By securing backups in an AWS account that is isolated from the primary copy of your data, either directly or using replication, it's possible to minimize the risk of a compromised system also impacting your data recovery mechanisms. The secondary account can be viewed as a "data bunker" with access requirements aligned to this use case. For backups using Amazon S3, additional controls might include an Amazon S3 object lock to store objects using a write-once-read-many (WORM) model or multi-factor authentication (MFA) delete.

Disaster Recovery on AWS Flexibility

With the flexibility that AWS provides, start off with the architecture that meets your requirements, which should be driven by RPO, RTO, and RCO. You can pivot to any DR strategy at any point, which could end up in further improving RPO, RTO, and/or RCO, or you could find new cost savings with a strategy that still meets your RPO and RTO requirements.

To summarize, Figure 4.17 shows a decision tree that you can follow to select whether a single-region or a multi-region approach meets your requirements.

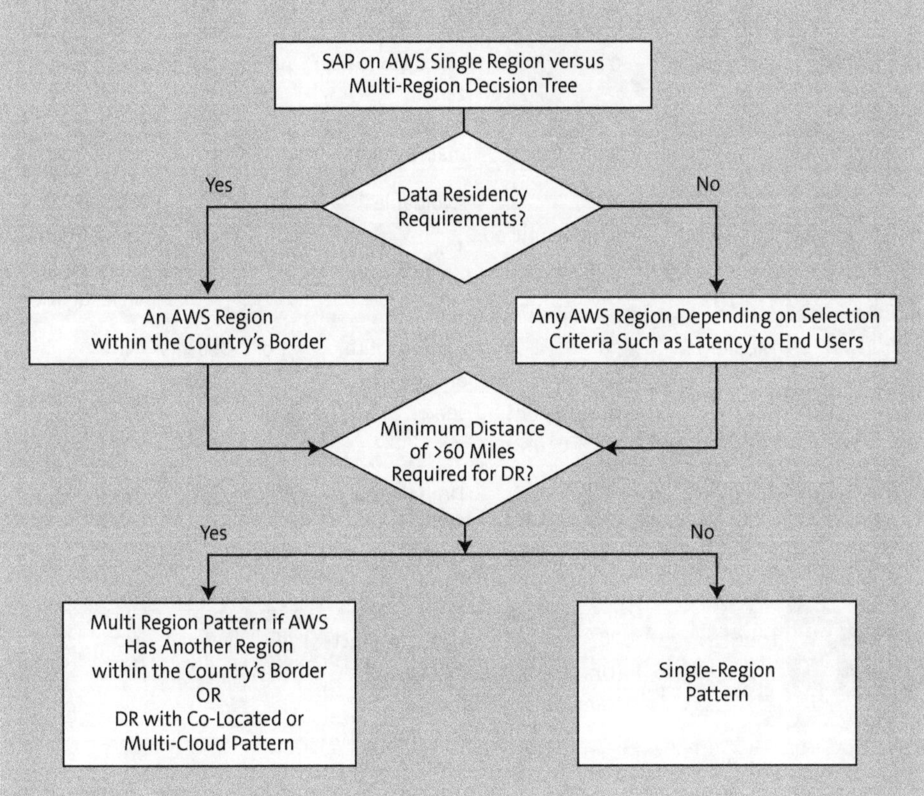

Figure 4.17 Decision Tree Single-Region versus Multi-Region

The decision tree shown in Figure 4.17 is only a recommendation, and there is no mandate to follow any rules to decide between a single-region or a multi-region approach. An enterprise with no data residency requirements and no minimum distance requirements can also go with a multi-region pattern; however, it should also consider single-region pattern as within a single region AWS offers high levels of resiliency, as shown in Table 4.4. Additionally, a single region pattern can be more cost optimized since you won't need to deploy services related to security, auditing, and networking in the other region. This approach eventually leads to less operational overhead of managing a multi-region landing zone and could potentially help you save on the cost of a managed service provider (MSP) as well.

	Comparative risk of occurrence	Data loss expected	Maximum outage	User impact	Requires operator input for service continuity
Planned/Controlled Maintenance	Planned	0	Managed to near zero for hardware related maintenance	All	Yes
Resource Exhausted (CPU Maxed/ Filesystem Full/Out of Memory)	Medium	~0 (Uncommitted transactions)	Avoidable	All	Focus is on prevention but may require investigation
Distributed Component Failure (e.g., App Servers)	Medium	~0 (Uncommitted transactions for % of users)	Reconnect to another app server.	% of GUI users / background jobs.	No
Single Point of Failure (Database/SAP Central Services)	Medium	~0 (Some uncommitted transactions)	Time to detect failure and failover (automated)	All	No
Data Center/AZ/ Network Failure	Low	~0 (Uncommitted transactions)	Time to detect failure and failover (automated)	All	No
Core Service Failure (DNS/AD/ EFS/API Calls)	Low	0	None expected	Dependent on failure	No
Corruption/ Accidental deletion/ Malicious attack/ Faulty Code Deployment	Low	Last consistent restore point before failure	Time to detect failure and failover or restore (manual)	All	Yes
Region Failure	Very low	Replication delay	Time to detect failure and make a decision to invoke DR plus takeover	All	Yes

Table 4.4 AWS High Levels of Resiliency within a Single Region

4.6   Summary

In the chapter, we went over the relevance and importance of an AWS landing zone. The multi-account setup, and its governance, and management can be implemented using custom-built code pipelines or using AWS-managed services like AWS Control Tower. In both options, core AWS services like AWS Organizations, and AWS Service Catalog are leveraged behind the scenes to manage, control, and operate your landing zones. AWS Control Tower Factory uses the Account Factory feature to build new accounts. This "account vending" solution leverages AWS Service Catalog to launch new accounts based on AWS CloudFormation templates without maintaining or managing code pipelines.

With an AWS landing zone ready to go, you also must ensure that your SAP workloads, specifically your production systems, are designed for resiliency. A resilient architecture can be unique to your enterprise based on the business uptime and recovery objectives. We looked at fault tolerance, HA, and DR and the importance of these terms, specifically with respect to an SAP workload on AWS. For SAP workloads on AWS, HA is typically implemented across AZs in a single region, and DR, across multiple regions. Thus, you must ensure that your landing zone can support both single HA in a single region and multi-region DR requirements. To summarize, we highly recommend you use AWS Control Tower for new as well as existing enterprises to deploy, operate, and govern a multi-account AWS environment across AWS regions to meet the security and resiliency requirements for your SAP workloads on AWS.

Chapter 5
Reference Architecture

Amazon Web Services (AWS) solutions for SAP are typically driven by reliability metrics, which eventually dictate your architectural patterns as well as the management and operations of your SAP applications on AWS. In this chapter, first, we'll look at solutions supported and certified for SAP on AWS; then, we'll cover architectural patterns and trade-offs based on costs, availability, and performance efficiency for your SAP application and database stack.

Because almost 90% of the world's finance and goods transactions touch an SAP system, a vested interest exists among technology hardware vendors to get their platforms certified for SAP (*http://s-prs.co/v577614*). The certification of hardware platforms is the basis for receiving support from SAP. In the on-premise world, you have numerous options for running SAP products on supported hardware platforms developed by vendors such as IBM, HPE, Dell, Fujitsu, Oracle, etc. On top of the physical host layer, you often have a virtualization layer using vendor-recommended virtualization technologies. Going further up the stack, you have non-modified or modified (from virtualization perspective) operating systems (OSs) on top of which you install your SAP systems.

Enterprises can run and operate SAP systems on virtualized or non-virtualized hardware on a certain OS because they have been certified and thus are supported by SAP. All hardware vendors are required to get their hardware platform components for the use of SAP systems on Microsoft Windows, Linux, and the SAP HANA database certified from SAP. Similar to the on-premise world, AWS is considered a hardware platform vendor that offers computing services. Hardware providers like AWS must get sign-off from SAP on what SAP products and applications and the operating system/database (OS/DB) combinations are supported on their platform.

Once you've selected what AWS solutions or services are supported for the relevant SAP applications and products, you can start building an architecture based on your requirements and around reliability, which are typically different for production and non-production systems. Architectural patterns will dictate costs related to infrastructure as well as the of management and operations. However, with the flexibility that AWS provides, you have the option to pivot from one architectural pattern to another based on your requirements without needing to procure or manage infrastructure components such as data centers, server racks, servers, networking, power cooling, and virtualization technologies. Before we dive more deeply into architectural patterns, let's look at SAP applications and products supported to run on AWS.

5.1   Support for SAP and Related Products

SAP offers its enterprise resource planning (ERP) software via multiple products such as SAP ERP, SAP Business Suite for SAP S/4HANA, SAP Customer Relationship Management (SAP CRM), SAP Supply Chain Management (SAP SCM), etc. AWS offers core services for compute, storage, and networking including Amazon Elastic Compute Cloud (EC2), Amazon Elastic Block Storage (EBS), AWS Elastic Load Balancing (ELB), etc. As mentioned in the introduction to this chapter, a critical component of certification is compute (i.e., the CPU type). Thus, you must approach designing and architecting a fully supported SAP solution on AWS from two angles: first, what SAP products are supported to run on AWS and, second, what AWS services specifically related to compute and storage are certified and supported to run SAP applications and databases on AWS. Additionally, support for any OS/DB combination should also be considered from two angles as well: first, from SAP's Product Availability Matrix and, second, from the AWS perspective (e.g., the release of a new OS version from a vendor for SAP on AWS). SAP support on AWS implies that SAP will provide services related to any issues, concerns, queries, or consulting services, which depends on the licensing agreement with SAP.

With hundreds of SAP products, let's try to simplify things with respect to the prerequisites that must meet to get support from SAP as well as from AWS. In this way, you can get support from SAP and AWS while also running a system with optimal performance.

SAP on AWS Support

The connection of an ABAP system as a source via remote function call (RFC) is not a close coupling and can therefore also be operated across cloud boundaries, with due consideration to network quality. For example, an SAP Business Warehouse (SAP BW) system can run on AWS, connected via RFC with SAP ERP or SAP S/4HANA still running on-premise. This architecture is fully supported by both AWS as well as SAP.

Through an example, let's further break down the support prerequisites. Let's assume you're running an SAP system that you plan to migrate as-is, or rehost, or lift and shift onto AWS. In this example, your on-premise data center is configured in the following way:

- SAP version: SAP ERP EHP6 SPS12
- SAP NetWeaver Stack: 7.52
- SAP Kernel: 7.52 PL1100
- Database: SAP HANA 2.0 SP4

- OS: Windows Server 2012 for SAP Application Servers and SLES 12 SP4 for SAP HANA database

Figure 5.1 shows the sequence of steps to ensure this configuration is fully supported by SAP and AWS.

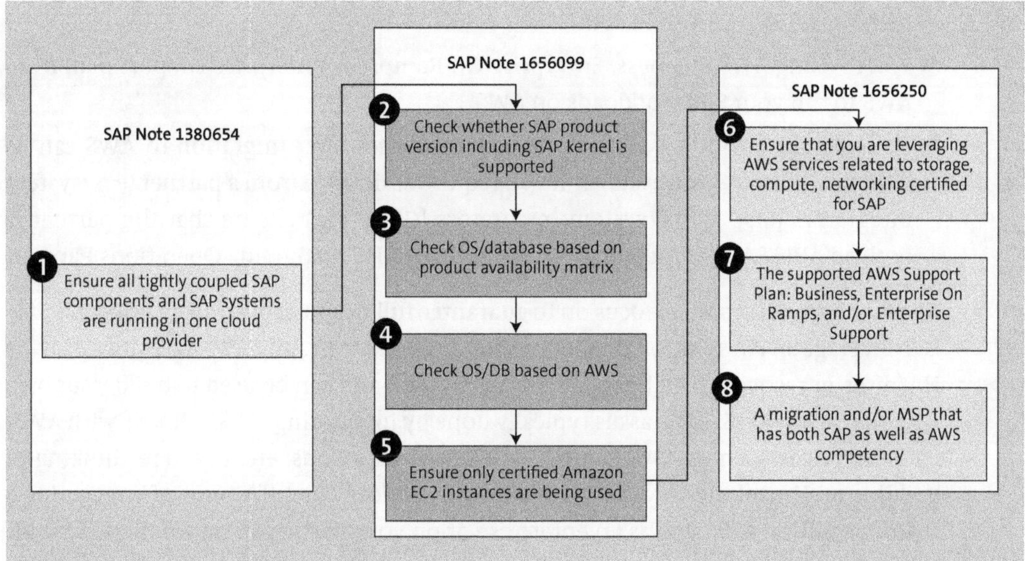

Figure 5.1 SAP on AWS Support Prerequisites

Let's look at each of these steps in more detail:

❶ Ensure that all the runtime components of an SAP system, in this case, the application servers and database for SAP ERP, are running and operating on a single cloud provider, which is AWS.

❷ Check whether SAP ERP EHP6 SP12 using SAP NetWeaver 7.51 including the SAP kernel is fully supported on AWS via SAP Note 1656099. A quick check by referring to this SAP Note reveals that support is available based on the statement "Applications running on the Application Server ABAP and/or Java as part of SAP NetWeaver 7.1 or higher and 7.21 kernel levels or higher."

❸ Check whether OS/DB combinations are supported in SAP's Product Availability Matrix.

❹ Check whether OS/DB combinations are supported on AWS using SAP Note 1656099. So, based on this SAP Note, for SAP application servers, Windows 2008, 2012, 2016, 2019, and 2022 are fully supported on AWS. SAP HANA 2.0 SPS00 or higher with SLES 12 SP3 or higher is supported. Therefore, you can perform a lift and shift of your source SAP system into AWS without changing or modifying the OS/DB combination, which is fully supported by both SAP and AWS.

❺ Ensure that you use only certified and supported Amazon EC2 instances as mentioned in SAP Note 1656099 and with latest and greatest SAP on AWS documentation at *http://s-prs.co/v577615*.

❻ Based on SAP Note 1656250, ensure that you're leveraging only SAP- and AWS-supported AWS services related to storage, networking, monitoring, backups, and high availability (HA) solutions.

❼ You're using AWS Business, Enterprise On-Ramps, or Enterprise support plan from AWS to run your SAP workloads on AWS.

❽ The migration to AWS, and the managed services after migration to AWS can be either handled in-house and/or if you require assistance from a partner (e.g., system integrators [SIs], global system integrators [GSIs], etc.) ensure that the partner is both an AWS SAP competency partner and an SAP Outsourcing Operations Partner.

Another important area to focus on to guarantee full supportability from AWS and SAP, is to leverage an HA solution that is certified by SAP as well as by AWS. SAP does provide HA/disaster recovery (DR) hooks written in Python and can be used to build your own HA solution on AWS. This task is typically done by integrating HA/DR hooks with AWS-native services such as AWS Lambda, AWS Step Functions, etc. If you're thinking of building an HA solution by using AWS-native services, then the onus of supporting a custom solution is on you as an enterprise or on your partner or on whomever builds this solution for you. Another option, and the recommended one, is to leverage an SAP partner-certified HA solution on AWS offered by Red Hat Enterprise Linux (RHEL), SUSE, Veritas, SIOS, Microsoft Windows, and NEC ExpressClusters. In this scenario, instead of building a solution, you're investing in an HA solution offered by the independent software vendor (ISV), which is certified by SAP as well as by AWS and is supported by the cluster software vendors themselves, not by SAP.

SAP on AWS HA Solution Support

At the time of writing, solutions like HPE Serviceguard and Oracle RAC are not supported on AWS. If you have an existing SAP environment using any of the on-premise HA solutions that are not supported on AWS, consider pivoting to a solution that is supported and certified by SAP as well as AWS.

For business continuity, specifically DR, in the on-premise world, a common practice is to use storage-based replication by protecting your application servers and database file systems using a vendor solution. On AWS, the same approach can be followed using AWS-native services such as the AWS Elastic Disaster Recovery (DRS) service, which is a continuous data protection tool based on block-level storage replication. AWS DRS can orchestrate and automate the failover process apart from the storage-based replication feature. From a supportability perspective, similar to on-premise vendor solutions, no additional certification is required from SAP for storage-based replication solutions.

The solutions are typically supported by the vendor themselves, and SAP will not typically intervene from a support perspective in this scenario. Instead of exclusively using a storage replication-based ISV solution or an AWS service, you can also pivot towards a combination of storage-based replication and a database-native replication solution. For example, application servers could be protected using a storage-based replication service—AWS DRS. Meanwhile, your databases are protected using database-native replication such as SAP HANA System Replication, Oracle Data Guard, SQL Server Log Replication, SAP ASE service replication server, etc. This architecture is fully supported by SAP as well as AWS.

SAP on AWS Storage-Based Business Continuity Support

Because AWS DRS is a storage-based replication capability offered by AWS as a service, the primary point of contact for support is the AWS Support team. Because AWS DRS operates at the OS and at the storage level, no additional certification is required, and thus this solution is fully supported as a DR solution for SAP application servers as well as databases. For SAP databases, ensure that the proper tests/drills are conducted that meet your recovery point objective (RPO), recovery time objective (RTO), and recovery consistency objective (RCO) requirements, if any.

Once you have nailed down on your HA and DR architecture leveraging SAP- and AWS-supported solutions, also ensure that you're sizing your systems appropriately, especially the SAP HANA database, by not only selecting the Amazon EC2 instances according to the publicly available SAP HANA-certified hardware directory but also by ensuring that your workloads meet the sizing requirements listed in the **Restrictions & Comments** for each Amazon EC2 instance listed in that directory. Let's go through an example of SAP HANA sizing on a certified Amazon EC2 instance (**x2iedn.32xlarge**) listed in the SAP HANA hardware directory. As shown in Figure 5.2, in the **Restrictions & Comments** section, there is a workload-based sizing restriction. Click **Read More** for more details about standard versus workload-based sizing from a supportability perspective.

What you're seeing is the difference between SAP HANA Tailored Data Center Integration (TDI) v4 and TDI v5. SAP has gone through many iterations of what a certified SAP HANA configuration can be. Initially, only appliances approved by SAP could be utilized to run supported SAP HANA workloads on. SAP then changed this limitations to allow their customers to conduct a tailored data center integration, whereby they could put together their own servers as long as they met certain hardware and performance requirements. TDI v4 had a strict core-to-memory ratio. While this rule worked for smaller SAP systems, as SAP HANA systems grew, traditionally online analytical processing (OLAP)-based systems would utilize the memory and cores they had, but online transaction process (OLTP)-based systems would utilize only a fraction of the CPU. SAP customers and hardware vendors asked SAP if they could relax the core-to-memory ratio to save on unnecessary CPU costs, and thus, TDI v5 was born. For x2iedn.32xlarge, 2.7 TiB is the SAP HANA TDI v4 spec. The 4 TiB is the SAP HANA TDI v5 spec. With SAP HANA TDI v4, you're guaranteed to meet the SAP HANA certification out of the box. With SAP HANA TDI v5, your enterprise must run its workload on the box and ideally should have SAP validate that they are comfortable supporting it.

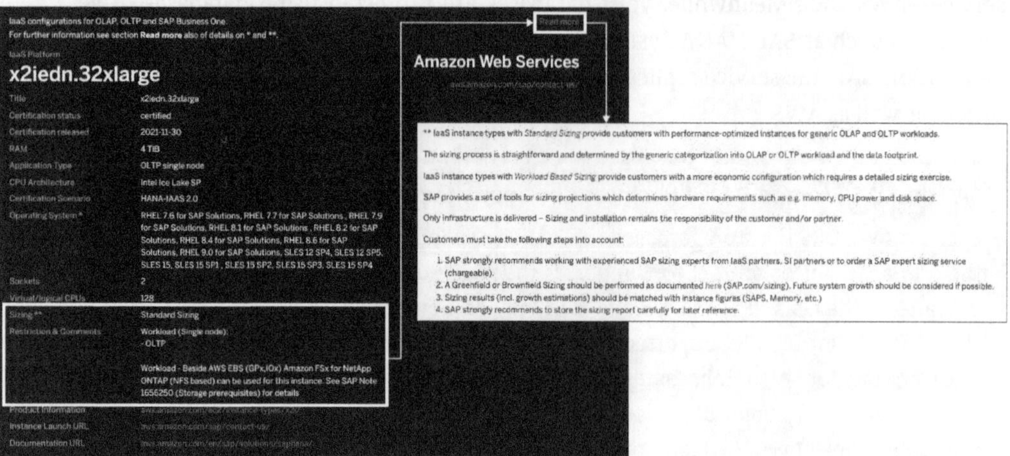

Figure 5.2 SAP on AWS Support for SAP HANA-Certified Amazon EC2 Instances: Standard versus Workload-Based Sizing

SAP S/4HANA or an SAP ERP on SAP HANA is largely an OLTP system, so it's a good candidate to use the full 4 TiB. If you start using more than 2.7 TB, you're going outside of the stricter TDI v4 standard and now must ensure your workload isn't going to be bound by CPU constraints. So, while you may not be out of support from SAP, SAP could absolutely request you to increase their Amazon EC2 instance size because it may have enough memory but not enough CPU to address the workload. Usually, for OLTP workloads, additional cores are not required, but some enterprises do run code that is not optimized. You can use your workload sizing results to determine the optimal number of cores and memory, which means that the scalability limits are no longer tied to the number of CPU cores available on the system. With TDI v5, you can scale up your workloads up to the maximum RAM size available on the server assuming your workload characteristics, thus allowing you to put more data per core than currently supported. Even with workload-based sizing, the recommendation is to consider the SAP Application Performance Standard (SAPS) or the Quick Sizer or SAP services (paid), data growth, non-optimized code, and preloading off.

As we've looked at support prerequisites for SAP products, let's pivot towards support criteria for other non-SAP systems in the landscape. An enterprise's SAP landscape typically will include non-SAP applications or ancillary or tertiary systems/applications to enhance or extend their industry-specific business process functionality. An example could be a non-SAP archiving solution or a non-SAP warehouse management system etc. integrated with an SAP ERP system. In this scenario, we highly recommend reaching out to the software provider to check the supportability of the solution on AWS. Similar to SAP, most of the time, the solution will be technically feasible to run on AWS. However, an imperative goal is that the solution should meet support prerequisites to guarantee full support from the software provider as well as AWS. Additionally, quite commonly, you'll see vendor solutions moving towards a software as a service (SaaS) offering on AWS, typically following a subscription-based licensing model. This case is a good opportunity for your enterprise to consider migrating to an SaaS offering, especially considering the amount of time and effort required for code remediation. In some scenarios, due to license restrictions and/or platform restrictions, it might not be possible to migrate your non-SAP systems to AWS. For instance, for a data center exit, you can consider an AWS partner offering in a co-location with premises in close proximity to an AWS region. Or, if you still have on-premise data center presence, then you can consider where your systems are and then integrate these systems with dedicated private network connectivity between AWS and on-premise networks to ensure lower latency using AWS Direct Connect.

Non-SAP Product Support

Depending on your support prerequisites, you might have a hard requirement to have an SAP and a non-SAP system located in close proximity with each other. In these scenarios, we highly recommend considering tightly coupled SAP and non-SAP systems as part of the same wave during a migration to AWS to avoid any network latency specifically if the integration is real time.

5.2   SAP Architecture

In SAP products like SAP ERP or SAP S/4HANA are a software delivered by SAP to manage and operate your business functions related to finance, sales and distribution, materials management, etc. Similarly, you could be leveraging other SAP products like SAP Process Integration and SAP Process Orchestration, SAP BW, SAP Supply Chain Management (SAP SCM), etc. to operate other business functionalities within your enterprise. An SAP product within an enterprise is typically built from multiple environments (e.g., sandbox, development, quality assurance, and production). Each of the environments consist of one or more SAP systems.

In this section, we'll look at the SAP architecture related to a three-tier SAP design, a distributed and standard deployment, and an SAP application technology stack.

5.2.1   Three-Tier Architecture

Each SAP system is built of, in principle, three tiers—a database tier, an application tier, and a presentation tier. Figure 5.3 shows an overview of the three-tier architecture of an SAP system. The presentation layer consists of the SAP GUI or a web browser (specifically SAP Fiori) through which users—developers, administrators, and the end users access an SAP system. The application tier is used to process business logic based on the user's request. The application tier comprises of message server, which consists of two critical services—the SAP central services and the enqueue service. The database tier is the data store that contains all your enterprise data as well as the program code, which is used to process business logic by the application tier. For you get full support from SAP as well as from AWS, the database and application tiers must be deployed in a single cloud provider, in this case, AWS.

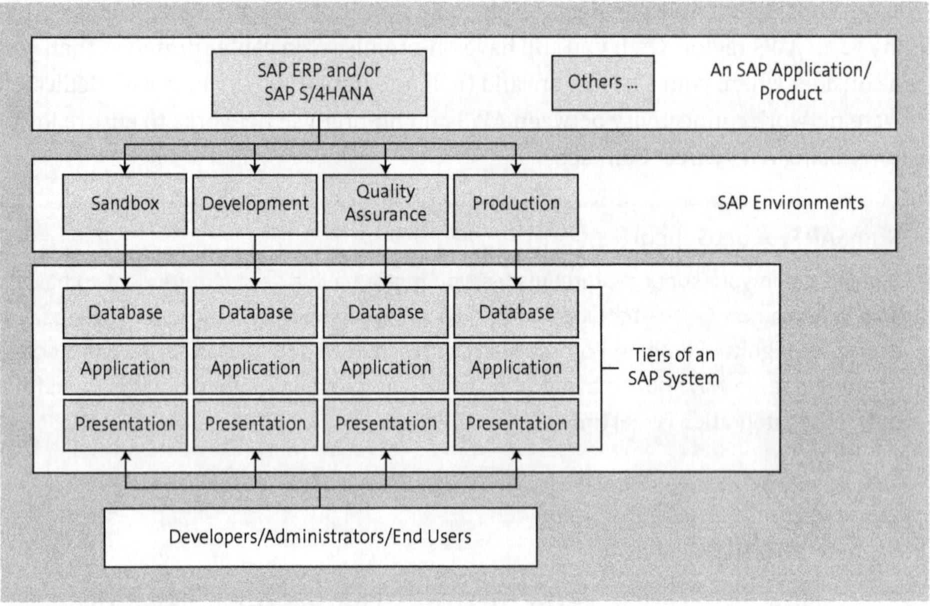

Figure 5.3 Three-Tier Architecture of an SAP System

From an installation and a deployment perspective, the application and database tiers are deployed on a certified and supported hardware platform, which could be on-premise data center, a co-location, or a cloud service provider like AWS. The presentation layer is deployed on the end-user machines.

5.2.2   Distributed versus Standard Deployment

Let's zoom in further into the application and the database tier and their deployment patterns, specifically on AWS. Depending on your performance and cost requirements, SAP systems are typically deployed in two ways:

- **Standard deployment**
  This option is a single-host deployment where all the components—application tier including ABAP SAP Central Services (ASCS)/SAP Central Services, the Primary Application Server (PAS) and database tier—are installed on a single physical/bare-metal machine or a virtual machine (VM). As shown in Figure 5.4, in AWS, a standard deployment of an SAP system is installed on a single SAP-certified Amazon EC2 instance in a single availability zone (AZ) within an AWS region.

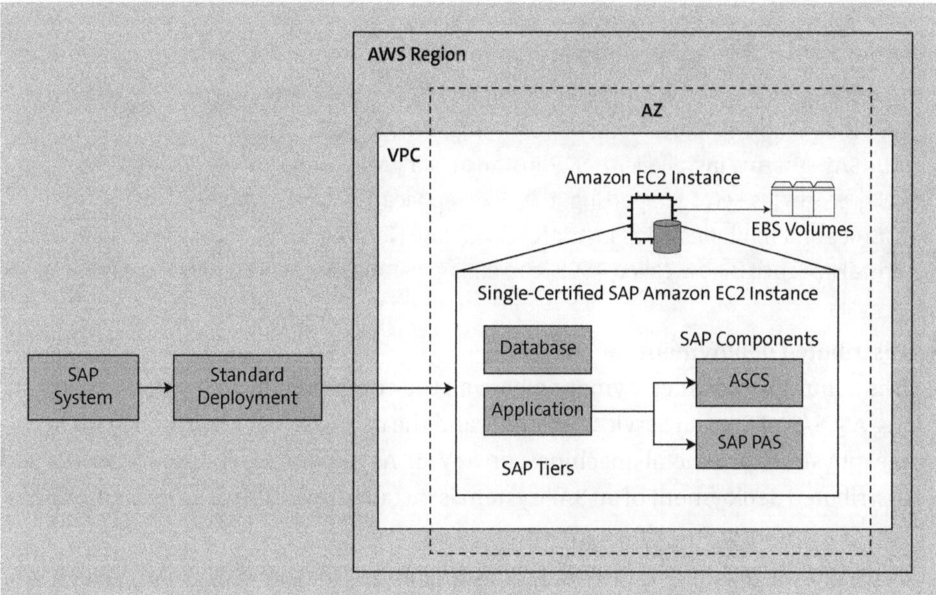

Figure 5.4 SAP Standard Single-Host Deployment on AWS

The risk of failure for this type of architectural pattern is high because, if the system goes down, everything goes down from an SAP tier perspective. In the event of AZ failure or disruption, it may take hours or days to restore operations. Thus, these patterns are suited best for applications and workloads that can tolerate such a downtime. While these patterns are commonly used for non-production systems, such as development and sandbox activities, they may be considered for production systems where the mean time to resolution (MTTR), RTO, and RPO, as defined by your enterprise allow for several hours or days of downtime for restoration of operations in a failure or DR event.

The single points of failure in this pattern are the SAP application, the ASCS, database, the OS software, the Amazon EC2 instance, and your Amazon EBS volumes. You'll be bound to single instance Amazon EC2 service level agreement (SLA) for a single AZ, which is 99.5%. You could achieve higher SLAs by having Amazon EC2 auto-recovery enabled; however, AWS does not provide SLAs on Amazon EC2 auto-recovery. Typically, this configuration is used for the following scenarios:

- SAP environments, for example, development, sandbox, proof of concept (POC)/ proof of technology (POT)/mock, QA, and test systems.

- SAP systems that can be stopped for maintenance activities during business hours.

- Non-critical SAP systems that can withstand the downtime needed to recover SAP application and database components from backups in case failures at the software or the physical layer arise.

[+] **An Amazon EC2 Instance and SAP Instance**

An Amazon EC2 instance is different from an SAP instance. An Amazon EC2 instance in AWS is a virtual or a bare-metal server on top of which you have the OS and eventually your SAP systems installed. An SAP instance is a group of resources such as memory, work processes, etc. supporting the SAP application and comprised of a central instance and multiple dialog instances. With SAP NetWeaver 7.3 releases and later, the central instance is now called ASCS, and dialog instances are called PAS and AAS.

- **Distributed deployment**

It is a multiple-host deployment where all the components application tier including ASCS/SAP Central Services, the PAS, and the database tier are installed on multiple physical/bare-metal machines or a VM. As shown in Figure 5.5, in AWS, a distributed deployment of an SAP system is installed on multiple SAP-certified Amazon EC2 instances in a single AZ within an AWS region.

Typically, the database is installed on a standalone Amazon EC2 instance, and ASCS/ SAP Central Services and the PAS are installed on another Amazon EC2 instance. The risk of failure for this type of architectural pattern is lower than in a single-instance architecture because of decoupled SAP tiers. However, you still need to protect your single points of failure. In the event of AZ failure or disruption, you may need hours or days to restore operations. Thus, these patterns are suited best for applications and workloads that can tolerate such a downtime. While these patterns are commonly used for non-production systems, such as development and sandbox activities, they may be considered for production systems where the MTTR, RTO, and RPO, as defined by your enterprise, allow for several hours or days of downtime for the restoration of operations in a failure or DR event.

The single points of failure in this pattern are the SAP application, ASCS, the database, the OS software, the Amazon EC2 instance, and your Amazon EBS volumes. You'll be bound to a single-instance Amazon EC2 SLA for a single AZ, which is 99.5%. You could achieve higher SLAs by having Amazon EC2 auto-recovery enabled; however, AWS does not provide SLAs on Amazon EC2 auto-recovery. Typically, this configuration is used for the following scenarios:

- SAP environments, for example, development, sandbox, POC/POT/mock, QA, and test systems.

- SAP systems that can be stopped for maintenance activities during business hours.

- Non-critical SAP systems that can withstand the downtime needed to recover SAP application and database components from backups in case failures at the software or the physical layer arise.

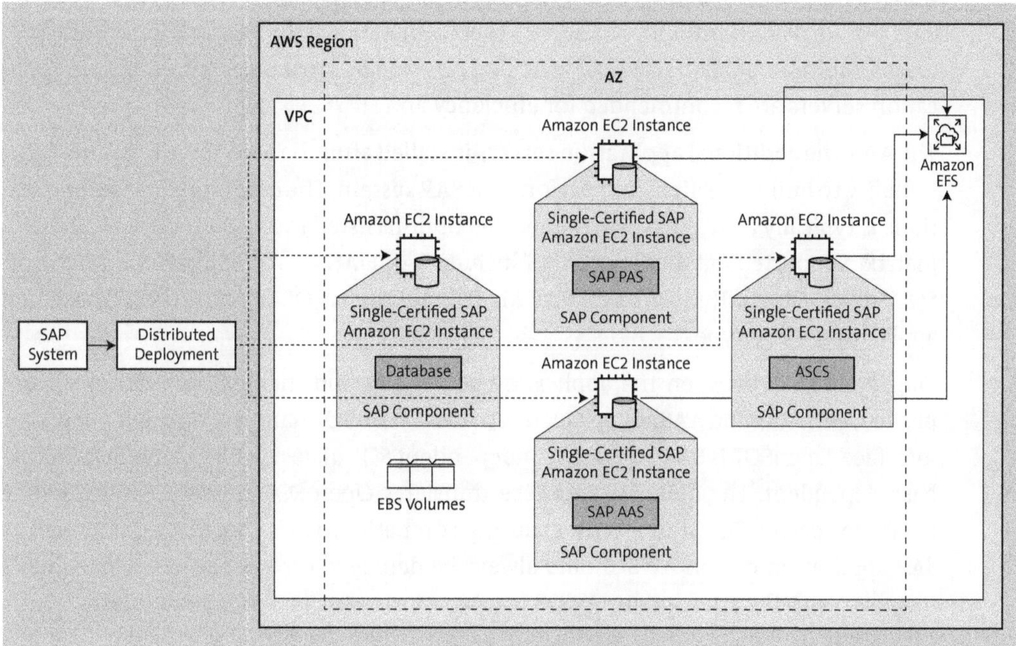

Figure 5.5 SAP Distributed Multiple-Host Deployment on AWS

RISE with SAP on AWS Deployment and Architectural Pattern Considerations

RISE with SAP is fully supported on AWS. The deployment and architectural patterns related to your SAP systems constructed by RISE with SAP depend on your contract with SAP and are fully controlled and managed by SAP. The deployment and architectural patterns highlighted in this chapter related to SAP NetWeaver, ABAP platform, and SAP HANA are applicable for RISE with SAP as well. The final reliability SLAs depend on your contract with SAP.

5.2.3 SAP Application and Technology Stacks

This section provides an overview of the client-server architecture of an SAP system, consisting of the presentation, application, and database tiers. The SAP system's technology stack involves application servers processing business functionalities and a database storing transactional and analytics data, which may be an OLTP and/or an OLAP relational database management system (RDMS). To ensure support from SAP and AWS, the right OS/DB combination must be used.

The SAP application server, comprising ASCS/SAP Central Services, PAS, and AAS, operates on SAP's NetWeaver technology stack. The ASCS/SAP Central Services, housing critical services like the message and enqueue servers, acts as a central communication channel and handles SAP locks. For HA, an additional passive standby enqueue replication server (ERS) is installed in another AZ within an AWS region. The PAS is based on SAP NetWeaver technology, handling user connections through various work processes. Multiple smaller-sized SAP-certified Amazon EC2 instances with multiple application servers are recommended for efficiency and HA.

The AAS, the additional application servers installed after the PAS, serve the same functionality to build a resilient and performant SAP system. The data store for SAP applications is typically an RDBMS, with supported options listed in SAP's Product Availability Matrix. Supported databases on AWS include SAP HANA, SAP ASE service, Microsoft SQL Server, Oracle, IBM Db2, and SAP MaxDB, serving as OLAP and/or OLTP based on application usage type.

The interaction between the application server tier and the database tier is accomplished, with PAS and AAS, using Open SQL and native SQL handlers through a database interface. OpenSQL is SAP's database-independent SQL dialect, while native SQL is database dependent. The database interface translates Open SQL statements into native database-specific SQL statements, ensuring compatibility. The database tier as well as SAP application tier on AWS should always be deployed on a supported Amazon EC2 instance, with the appropriate OS/DB combination based on SAP's guidelines.

In the next section, we'll explore application- and database-specific architectures and provide a comprehensive look at the compatibility of SAP products and applications with multiple databases.

SAP Technology Stack

Apart from SAP NetWeaver ABAP and Java, SAP also uses other technology stacks such as standard Java-based application products like SAP Hybris Commerce, SAP Billing and Revenue Innovation Management, etc. Both SAP Hybris and SAP BRIM are SAP-acquired products from other vendors; thus, SAP follows the same architectural design principles from their initial product designs.

5.3   Application- and Database-Specific Architecture

At this point, we've built a foundation related to the SAP application and technology stack. In this section, we'll cover SAP application- and database-specific architectures related to SAP HANA, Microsoft SQL Server, Oracle, SAP ASE service, SAP MaxDB, and IBM Db2 on AWS.

5.3.1   SAP Application Server Architecture

As we saw earlier, an SAP application is built up from multiple components, namely, the PAS, the AAS (for HA and load balancing), ASCS, and the SAP replication enqueue server (for HA). With respect to the architecture, in the following sections, we'll look at compute, storage, HA, and DR for the SAP application layer and how it is typically deployed on AWS.

Compute

Each component (i.e., PAS, AAS, ASCS, and the SAP ERS) are deployed on an SAP supported and certified Amazon EC2 instance. Since AWS gets new Amazon EC2 instances certified regularly from SAP, we highly recommend referring to SAP Note 1656099 and the AWS documentation online for the latest list of SAP-supported and -certified instances that you can use for SAP application servers.

The latest and the greatest generation on Amazon EC2 instances is built on the AWS Nitro platform with Intel- or AMD-based processors. With respect to sizing, CPU type, and mapping of on-premise severs to Amazon EC2 instances, typically you would look at SAP Early Watch Alerts reports generated from SAP Solution Manager for existing SAP systems. Alternatively, you would use the Quick Sizer for greenfield SAP implementations. You can also use AWS native discovery tools, such as AWS Migration Evaluator, as well to review information related to vCPU, memory, and storage requirements. With this information, you can map your requirements against a supported and certified Amazon EC2 instance.

SAP Applications on Amazon EC2 Compute

Because AMD-based Amazon EC2 instances could be cheaper and provide similar or more SAPS (as compared to Intel-based Amazon EC2 instances), consider AMD-based Amazon EC2 instances for SAP application servers for a cost-optimized architecture. Based on our experience, many enterprises are more used to Intel, in which case you can consider Intel-based Amazon EC2 instances for production and a production-like environment and for others (e.g., DEV, sandbox, etc.). Then, you could use AMD-based Amazon EC2 instances to lower your total cost of ownership (TCO).

AWS has a custom-built, proprietary ARM-based chipset (CPU processor)—AWS Graviton. At the time of writing this book, Amazon EC2 instances based on AWS Graviton are not certified and supported to run SAP workloads on AWS. For SaaS solutions like SAP HANA Cloud on AWS (different from a customer- or partner-managed SAP HANA instance running on Amazon EC2 on AWS), in some scenarios, SAP does use AWS

Graviton-based Amazon EC2 instances to gain economic benefits over Intel- and AMD-based instances.

Operating System

SAP application servers can be deployed on a Linux or Microsoft Windows OS on AWS. For Linux, vendor products from SUSE Linux Enterprise Server (SLES), Red Hat Enterprise Linux (RHEL), and Oracle Enterprise Linux (OEL) are certified and supported on AWS. Within the realms of SAP's Product Availability Matrix, refer to the latest documentation from AWS as well as SAP Note 1656099 for OS versions that are supported for AWS for SAP.

For SAP systems running on Oracle databases, a mandatory step is running your SAP application servers on the Oracle Linux OS. For up-to-date information refer to SAP Note 2358420 - Oracle Database Support for Amazon Web Services Amazon EC2. AWS offers licenses for sale and also supports bring your own license (BYOL). We highly recommend using the OS flavors tailored specifically for SAP such as RHEL for SAP and SLES for SAP. Over the standard, these flavors include additional features such as extended lifecycle support, live-patching, auto-configuration of OS parameters based on SAP products, and high availability extension (HAE) packages.

Storage

In a standard SAP installation, Amazon EBS (gp3 or gp2) is typically used as storage volumes to house the root, and local SAP binaries, which include your SAP kernel, trace, and log files. Amazon EBS volume types io1, io2, and io2 block express are supported and certified for SAP application servers but however are typically not used due to the uplift in costs. gp3 and gp2 provide a balance of costs and performance for running SAP applications on AWS.

For a distributed installation or a HA SAP installation, we recommend using a shared file system service like Amazon Elastic File System (EFS) for sapmnt if you're using the Linux OS and Amazon FSx for Windows File Server for the Windows OS. For an SAP transport file system, we highly recommend using a shared file system like Amazon EFS or Amazon FSx for Windows File Server depending on the OS. For enterprises with a mix of OS within their landscapes and/or who are more familiar with tools related to NetApp storage solutions can pivot to Amazon FSx for NetApp ONTAP, a native AWS service that provides fully managed shared storage in the AWS Cloud with the popular data access and management capabilities of ONTAP.

High Availability

In this section, we'll cover an HA architecture on AWS for SAP NetWeaver-based applications, SAP Cloud Connector, SAP Content Server, and SAP BusinessObjects.

SAP NetWeaver-Based Applications

For critical SAP systems with minimum downtime requirements, it becomes imperative to consider an HA architecture. The single points of failure for an SAP application server built up of SAP NetWeaver ABAP/Java are ASCS, PAS, and shared storage.

As shown in Figure 5.6, in this configuration, SAP applications and databases are deployed based on a highly available architecture in multiple AZs in a single AWS region. Primary systems are set up in the first zone, and the secondary systems are set up in another AZ. Potential single points of failure include the SAP application server, ASCS from a message server and enqueue server perspective, and the database and shared file systems (/sapmnt). Other points of failure include networking components like SAP Web Dispatcher and AWS ELB used specifically for SAP Fiori.

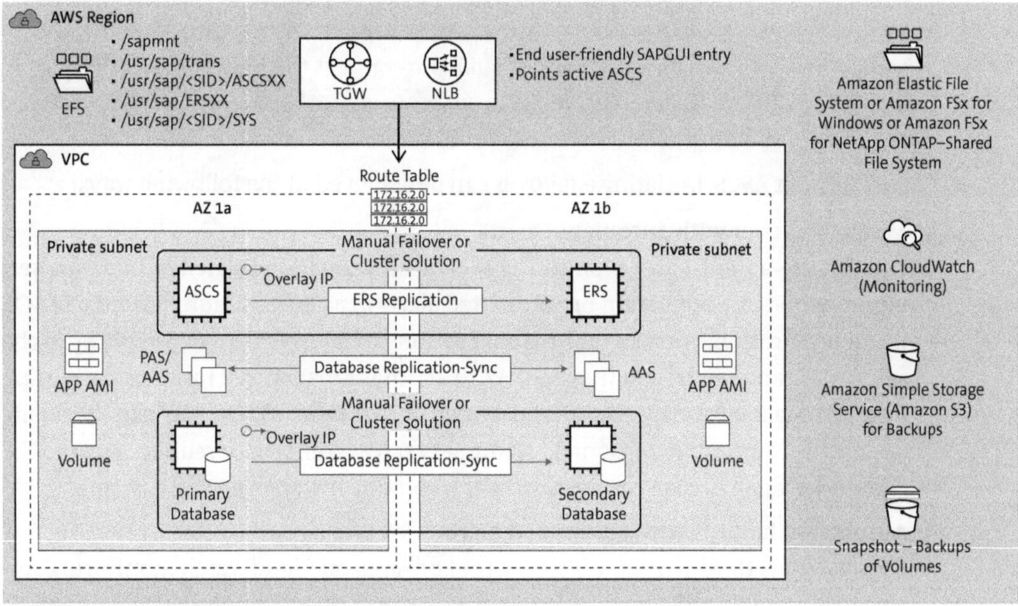

Figure 5.6 Multi-AZ HA for SAP Application on AWS

By design, the work processes in PAS are fault tolerant, which means that, if a work process fails, another work process within PAS can take the load over automatically without any human intervention. However, what if the OS or underlying hardware running PAS fails? Thus, having a single SAP PAS is an single point of failure. You can protect PAS running on an Amazon EC2 instance using Amazon EC2 auto-recovery which could help recover PAS within minutes. However, some downtime will arise. To ensure continuous business operations, you can protect PAS using additional SAP application servers across multiple AZs, which are called Additional Application Servers (AASs), which have the same functionality as PAS. As shown in Figure 5.6, we have multiple application servers—PAS and AAS spread across two AZs. Additionally, having multiple application servers helps balance incoming user loads more efficiently, leading to an optimally performant and reliable SAP system.

SAP on AWS HA Multi-AZ Single Region

Since, in a highly available architecture, multiple AASs are spread across two AZs, the background jobs might run longer due to the latency between the AZs since an AAS in one AZ communicates to the primary database in another AZ. Thus, we recommend using SAP job server groups to ensure all the critical jobs are running in the same AZ where the active primary database is.

SAP on AWS HA Multi-AZ Single Region

Since, in a highly available architecture, multiple AAS are spread across two AZs connecting to a database in one AZ, you should also consider the network costs when an application server from one AZ connect to database in another AZ.

Specifically, for ASCS, initiating a failover can be achieved in the following ways:

- **Manual failover with automated alerts**
  You can use a third-party monitoring tool or Amazon CloudWatch monitoring service to initiate a notification or automated alerts in case of a failure related to ASCS. Upon notification, you can take manual actions to failover the ASCS services in the same AZ or another (recommended) AZ to another Amazon EC2 instance built using Amazon Machine Image (AMI) and stored in Amazon S3. Or you can manually update the route tables or Domain Name Server (DNS) entries to redirect users to the failed-over node in case you already have a standby instance up and running.

- **Automated failures with automated alerts**
  You can use a third-party monitoring tool or Amazon CloudWatch monitoring service to initiate a notification or automated alerts in case of a failure related to ASCS. In this scenario, depending on the OS/DB and SAP application combination, you might completely rely on cluster solutions developed by vendors (e.g., SUSE, RHEL, Microsoft [Windows], Veritas, SIOS, NECCluster, etc.) to orchestrate the failover process for you. These cluster solutions, upon detection of a software and a hardware failure will trigger a failover, recover the ASCS services, and ensure they are up and running without any loss of data, in this case, SAP locks.

SAP on AWS HA Multi-AZ Single Region

As a best practice, you should always have a replica of production with a QA or Pre-Prod system, not only to perform quality assurance, but also from an HA solutioning perspective. Whether a custom-built solution using AWS services or an ISV product, as a best practice, the solution should be implemented in one of the lower environments to test any updates or features with the HA solution before implementation in production. To optimize compute costs, leverage the flexibility of AWS by using on-demand instances to shut down standby instances (e.g., the SAP ERS and database instance) when not in use.

In on-premise systems, you might be familiar with the concept of a *floating IP address*, also called a VIP (virtual) IP address. This kind of IP address can move across the active SAP servers that are typically part of the network Classless Inter-Domain Routing (CIDR) IP range, which spans across the data center or across different data centers within a wide area network (WAN). In other words, the subnets to which the ASCS and SAP ERS are assigned can span across data centers. For end users, a DNS entry for a floating IP address with a friendly name is configured in SAP GUI. A VIP address is used most cluster solutions for on-premise SAP systems.

In AWS, the IP address attached via an elastic network interface to an Amazon EC2 instance cannot be failed over to another Amazon EC2 instance in another AZ because subnets do not span across AZs. For this reason, an overlay IP is used, which is an IP outside of your VPC CIDR range, to access the active instance. Overlay routing allows you to use a non-overlapping RFC 1918 private IP address that resides outside the VPC CIDR range. You can then direct the SAP traffic to any instance setup across the AZ within the VPC by changing the routing entry in the route tables. To point to the active or the failed-over SAP component, the change in the routing entry is automated when using a cluster solution like SUSE's HAE, RHEL HA Add-On, SIOS, or Veritas to protect your SAP HANA database or SAP NetWeaver application.

For the networking configuration, the overlay IP is configured to point to the active ASCS instance whether it is the primary or the secondary node. The overlay IP is not reachable over AWS Direct Connect or AWS Site-to-Site VPN because its IP address is outside the VPC range. Therefore, you can use a Network Load Balancer through AWS ELB or Amazon Route 53 that points to the overlay IP. With this setup, end users connect to the Network Load Balancer, which points to the ASCS instance that is active, as shown in Figure 5.7. You can also use AWS Transit Gateway as a central hub to facilitate the network connection to an overlay IP address from multiple locations including Amazon VPCs, other AWS regions, and on-premise systems using AWS Direct Connect or AWS Site-to-Site VPN.

Figure 5.7 summarizes the network connectivity flow triggered from an on-premise network and involves the following steps:

❶ End-user/RFC connection to SAP.

❷ A Network Load Balancer with target group pointing to the overlay IP identifies the active ASCS elastic network interface serving the overlay IP using route table lookup.

❸ Once the elastic network interface is identified, the connection request is forwarded by the Network Load Balancer to the active ASCS elastic network interface.

❹ The active elastic network interface service's ASCS/SAP Central Services overlay IP sends the connection requests to the Message Server running on ASCS/SAP Central Services.

❺ The message server provides the application servers in AZ1a and AZ1b, which can server the requests (via SAP Logon Load Balancing).

❻ Application servers in AZ1a and AZ1b connect to the active primary database server serving the primary overlay IP (identified via route table lookup).

❼ Application server connection request is now connected to primary SAP HANA database instance serving the RFC/SAP GUI connection request.

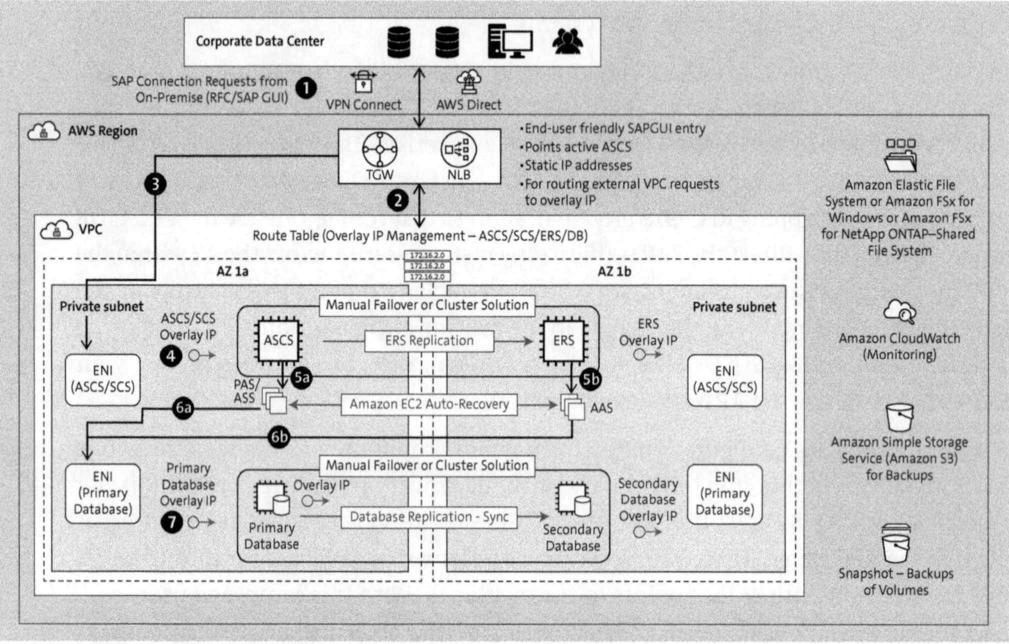

Figure 5.7 SAP End-to-End User Network Connectivity for Highly Available Multi-AZ Systems

SAP on AWS HA Overlay IP

A question that frequently comes up is "Does an overlay IP consume any of the customer IPs?" Yes, because the overlay IP is allocated from a non-overlapping CIDR range (external to where your SAP systems are provisioned) that is outside of the VPC CIDR range and does consume IP addresses from the CIDR range that you define for the overlay IPs. A single SAP system with an ASCS/ERS and an SAP database (e.g., SAP HANA) will consume a total of three private overlay IPs from the RFC1918 CIDR range (one each for ASCS, ERS, and SAP HANA. So, if you have clusters for multiple critical systems such as SAP ERP/SAP S/4HANA, SAP Process Orchestration/SAP Process Integration, SAP SCM, SAP Global Trade Services (SAP GTS), etc., then you should consider the number of IP address that will be consumed for the overlay IPs. The easiest way to track this would be to use an IP address manager solution like AWS VPC IP Address Manager (IPAM) or a third-party partner solution.

For some SAP components like ASCS and ERS running on Windows, one solution certified and supported on AWS is Windows Server Failover Clustering (WSFC). To protect shared file systems, `sapmnt`, `trans`, and `interfaces`, use Amazon FSx for Windows File Server, which is a fully managed shared storage for Windows with 4 9s of availability when deployed across AZs with full support for the Server Message Block (SMB) protocol, Windows NTFS, Microsoft Active Directory (AD) integration, and Distributed File System (DFS). For SAP systems running on Windows, you can also use other partner solutions, such as SIOS, Veritas; refer to the list available at *http://s-prs.co/v577616*.

To ensure HA in an SAP system, components like SAP Web Dispatcher and AWS native or third-party load balancers need to be protected as well. These components are primarily used for end-user connections coming in via SAP Fiori, which is SAP's web-based user interface (UI) accessed typically via web browser on PC, Mac, or mobile devices. With SAP S/4HANA, SAP is leaning into its strategy of enabling SAP Fiori as the primary UI/UX for their customers. Thus, an imperative task is to protect SAP Web Dispatcher and load balancers in an SAP environment.

- **SAP Web Dispatcher**
  SAP Web Dispatcher is a license-free tool with executables but no database that is deployed on Windows or Linux on a supported and certified Amazon EC2 instance. This solution acts as a reverse-proxy middleware between both external or internal end users and backend SAP systems. SAP Web Dispatcher helps distribute loads across SAP application servers and supports various functionalities such as multiple SAP systems support, URL filtering, web caching, URL rewriting, and Secure Sockets Layer (SSL) termination or re(encryption). You can protect SAP Web Dispatcher in three ways: HA on the same host using `RESTART` parameter to protect against any software issues; HA with a standby SAP Web Dispatcher instance on another host (passive) with an HA solution such as Pacemaker on Linux and WSFC on Windows; and the preferred and recommended approach, to have to multiple SAP Web Dispatcher instances to ensure HA and balance incoming loads more efficiently than either using a load balancer or with DNS load balancing.

- **Load balancers**
  On AWS, we highly recommend using fully managed services wherever possible. For load balancing, whether on layer 4 or layer 7 of the network open systems interconnection (OSI) model, the go-to native service is AWS ELB. For SAP applications specifically SAP Fiori, AWS ELB comes in two flavors: First, the Application Load Balancer,

which operates at layer 7 and supports HTTP/HTTPS, SSL termination, and path-based routing. Second, the Network Load Balancer operates at layer 4 and supports TCP/UDP/TLS, and end-to-end SSL. The Network Load Balancer can handle millions of requests per second. In the SAP Fiori world, which is a browser-based HTTP(S) access, the Application Load Balancer is leveraged whereas for any TCP connectivity such as SAP GUI, a Network Load Balancer is preferred. AWS ELB is a fully managed service that can scale on demand without any human intervention based on AWS's hyperplane (horizontal scaling) technology. AWS ELB provides an SLA of 4 9s of availability when deployed across multiple AZs.

SAP Cloud Connector

SAP Cloud Connector is a non-SAP NetWeaver, no-database application with a collection of executables deployed on supported and certified Amazon EC2 instances to connect SAP Business Technology Platform (SAP BTP) services with SAP systems or non-SAP systems/applications running on AWS or on-premise. As shown in Figure 5.8, for HA, another, secondary instance is deployed on another Amazon EC2 instance in another AZ. SAP Cloud Connector has inherent HA capabilities where the primary node pushes its configurations to the secondary node on a frequent basis to keep the nodes synchronized. No external HA solution is required in case of SAP Cloud Connector. If the master system is not reachable for a while, the shadow system tries to take over the master role and, if successful, establishes the secure tunnel to SAP BTP.

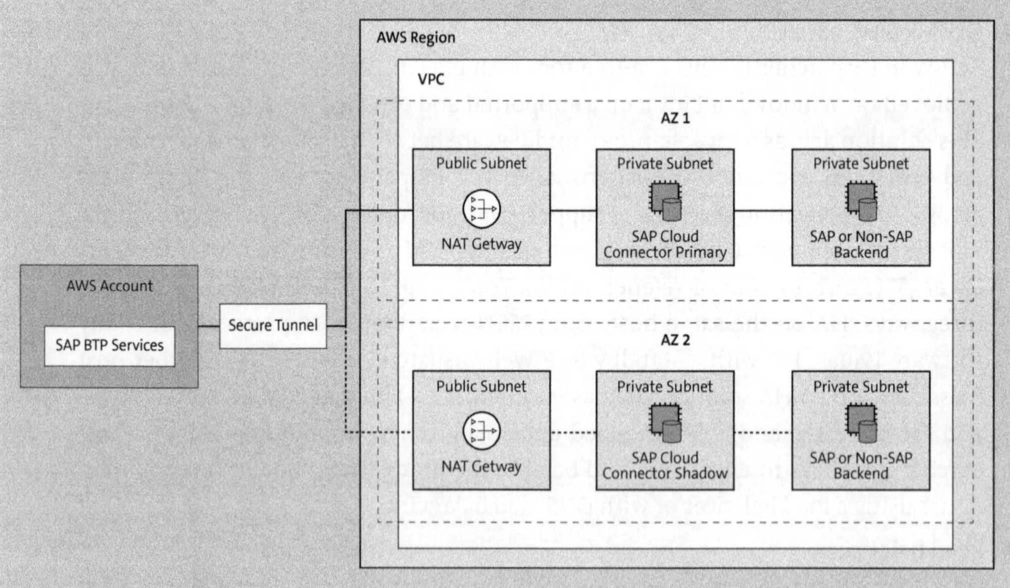

Figure 5.8 SAP Cloud Connector Multi-AZ HA

Another option for HA is using Amazon EC2 auto-scaling, which ensures that another instance is brought online using an AMI of the primary. As this approach/architecture is not supported by SAP, this scenario falls under your responsibility or your partner's responsibility.

SAP Content Server

As part of your archiving strategy, you might be using SAP Content Server to store large volumes of electronic documents of any format and with any content. This standalone component is tightly integrated with the SAP Business Suite of products such as SAP ERP. SAP Content Server holds critical data (documents), which can be saved in one or more SAP MaxDB instances running on a supported and certified Amazon EC2 instance. Another option is to store data in a shared file system such as Amazon EFS. Sales invoices, purchase orders, salary slips, emails, and architecture diagrams are examples of electronic documents commonly stored in SAP Content Server.

A standard non-HA deployment is created using a single Amazon EC2 instance with Amazon EBS volumes for storage in an AZ. To match the SLA for SAP Content Server with that of your OLTP-based SAP systems on AWS, SAP Content Server is deployed with the same HA as many core and critical SAP systems like SAP ERP. SAP Content Server consists of an application layer and a storage layer, which are its single points of failure as well. As shown in Figure 5.9, we are using SAP MaxDB database as the storage layer. The application layer is managed by the SAP's `sapcontrol` service.

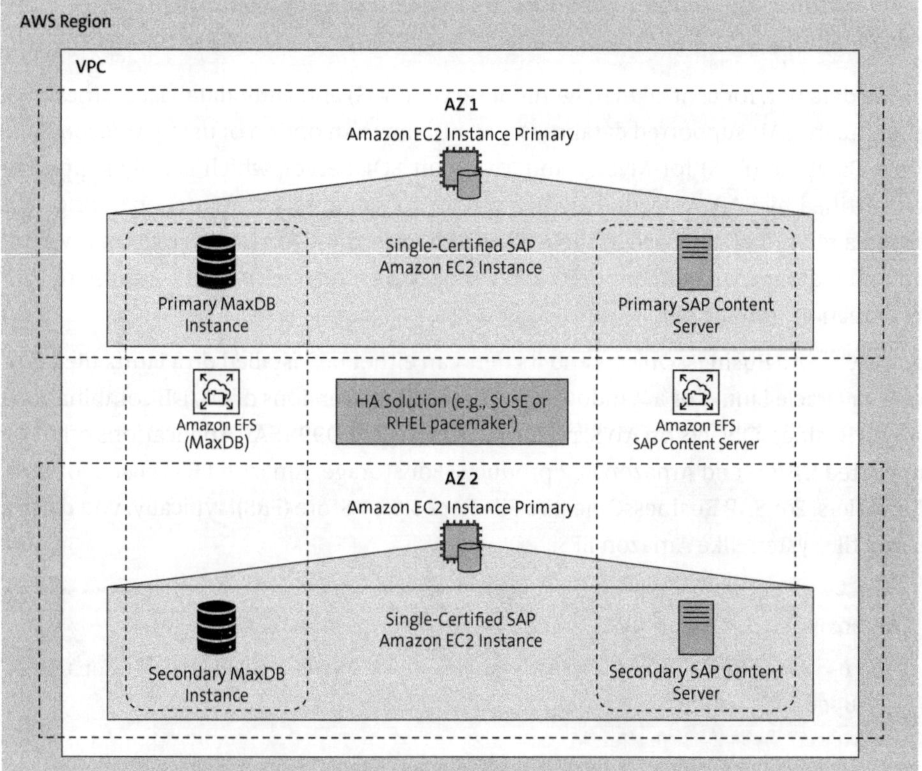

Figure 5.9 SAP Content Server HA Architecture

With your OS's HA solution, for example, SUSE or RHEL, you can manage both application and storage layer resources and build an automated, highly available solution.

SAP BusinessObjects

Next, from an SAP application server perspective, let's look at SAP BusinessObjects and its related architecture. SAP acquired BusinessObjects in 2007 and, along with it, its flagship product BusinessObjects XI with no intention of changing its design. Thus, even today, SAP BusinessObjects is a series of virtual tiers similar to any traditional web app. SAP BusinessObjects helps enterprise users in the reporting, querying, analyzing, and visualization of business data. At its core, this visualization tool is part of SAP's business intelligence suite of products or, in simpler words, part of SAP's analytics portfolio of products.

From an installation and architecture perspective, similar to SAP NetWeaver-based deployments, you have several options:

- Standard deployment: All tiers installed on a single SAP supported and certified Amazon EC2 instance.

- Distributed deployment: Application and data tier get installed in their individual SAP supported and certified Amazon EC2 instances.

- HA deployment: Redundant application and data tier are installed on SAP supported and certified Amazon EC2 instances to provide resiliency against hardware or software failures.

For the data tier, for central management server (CMS) and audit databases, other than the standard SAP-supported databases, you also have an option of using Amazon Relational Database (RDS) for MySQL and Microsoft SQL Server, which is fully supported and certified by SAP. Amazon RDS is a service to facilitate setting up, operating, and scaling a relational database in the AWS cloud. Amazon RDS can take over many difficult and tedious management tasks such as backups, software patches, automatic failure detection, and recovery.

For the OS, SAP BusinessObjects and its tiers can either be installed on a Linux (Red Hat, SUSE, or Oracle Linux) or a Windows OS. For the latest versions of OS/DB combinations for SAP BusinessObjects on AWS, refer to SAP Note 1656099 - SAP Applications on AWS: Supported DB/OS and Amazon EC2 products. For storage, Amazon EBS is used for individual tiers. For SAP BusinessObjects's File Repository Store (FRS), typically, you'd use a shared file system like Amazon EFS.

SAP BusinessObjects on AWS

If you use Amazon RDS as the database, application tiers must be installed on separate SAP-supported and -certified Amazon EC2 instances.

Figure 5.10 shows an HA architecture for SAP BusinessObjects with redundant tiers deployed across two AZs.

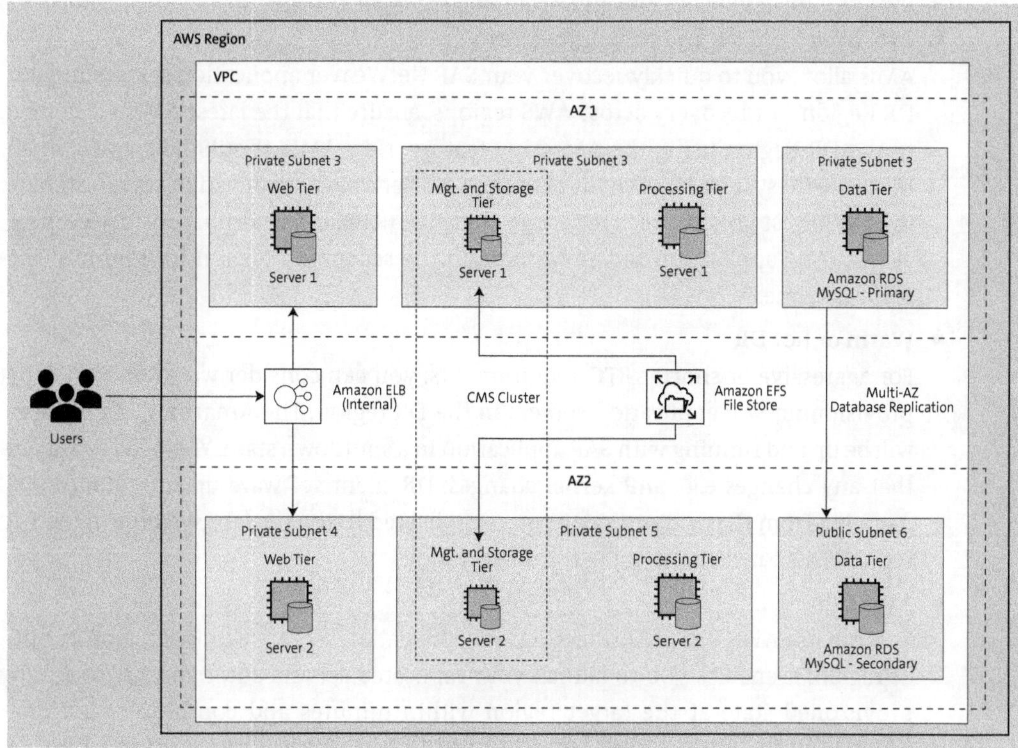

Figure 5.10 SAP BusinessObjects with HA Multi-AZ Architecture

SAP BusinessObjects on Oracle on AWS Considerations

On AWS, SAP BusinessObjects Business Intelligence (SAP BusinessObjects BI), version 4.2 SP4 or higher and version 4.3 or higher, are supported with Oracle databases and Oracle client software according to SAP Note 3015613 - Oracle Database support for BI Platform on AWS. For a detailed list of Oracle support restrictions on AWS, refer to SAP Note 2358420 - Oracle Database Support for Amazon Web Services Amazon EC2.

Disaster Recovery

A DR strategy for SAP application servers on AWS should be architected based on your RPO and RTO requirements. The options could be cold, warm, or hot or it could be a combination of warm and hot using native AWS DRS service built specifically for DR. Let's look at some of the options in detail:

- **Shared file system backup**
  NFS or SMB mounted shared file system is typically used for HA systems on AWS.

Shared file systems are also used for critical interface files for exchanging data between SAP and non-SAP systems. Amazon FSx or Amazon EFS file systems support backup of data both within a region and out-of-region.

- **AMI copy (cold DR and backup/restore)**
  AMIs allow you to quickly recover your SAP NetWeaver application servers in your DR Region. For recovery across AWS regions, ensure that the latest AMIs are copied to the DR Region using the AMI copy feature. New AMIs should be created when there are filesystem level changes to the SAP NetWeaver application servers. If having the lowest possible recovery time objective (RTO) is a priority, consider keeping at least one application server running in the secondary Region to minimize the recovery time.

- **Warm or hot DR**
  For aggressive or shorter RTO requirements, you can consider warm or hot DR by provisioning SAP application servers in the DR region. The Amazon EC2 instances will be up and running with SAP application in a shutdown state. You need to ensure that any changes e.g., SAP kernel changes, DB client software updates, and/or OS patching from the primary systems are replicated to the tertiary systems in the DR region. This can be done using third-party tools.

- **AWS DRS**
  You can use AWS DRS for protecting SAP application servers within a region or out-of-region. It enables you to launch your replicated servers automatically to a fully provisioned state at the target region within minutes and continue operations, which basically translates to a low RTO. It can be faster and more effective than a manual approach. The RPO of SAP application servers is dependent on time taken by the source system to send changes to the staging area. This is further impacted by the volume of transactions on the source system. Other factors include network throughput and latency, source and replication server performance, etc. These factors should be measured to calculate the potential amount of data loss during a DR event. Since the change rate on the source volumes on the SAP application servers is comparatively low as compared to large databases, DRS is a perfect fit for enterprises looking for a cost-effective and shorter RPO/RTO requirement.

Table 5.1 provides a comparison between cold, warm/hot, and AWS DRS DR for achievable RTO/RPO.

DR Architecture	Strategy	Costs	RPO/RTO
Cold	AMIs/snapshots	Low	High/high
Warm/hot	SAP app servers up and running	High	Low/low
AWS DRS	Storage-based replication (within or out-of-region)	Medium	Low/low

Table 5.1 DR Strategy for SAP Application Servers on AWS

5.3.2   Common Architectural Patterns for High Availability and Disaster Recovery for Databases

For an SAP system running on SAP HANA or any other database, database is the most critical component and is a single point of failure. If software or hardware issues arise a single point of failure, such as database, global operations supporting end-to-end business processes (e.g., sales or order entry or continuous manufacturing) will come to a halt. In Chapter 4, we covered how the requirement for a highly available architecture is driven by availability SLAs and by MTTR.

> **SAP on AWS: Database HA/DR Architectural Patterns**
>
> SAP database HA/DR architectural patterns with native database technologies are covered from Section 5.3.3 to Section 5.3.8.

For SAP HANA (or any database), following are some of the common architecture patterns related to HA and DR. The following patterns are applicable to all the supported databases for SAP applications on AWS:

- **Amazon EC2 auto-recovery**
  Amazon EC2 auto-recovery works at the compute layer (hypervisor and below) within AWS, which could use either simplified automatic recovery (default), Amazon CloudWatch action-based recovery, or dedicated host recovery techniques to recover an instance on a host which is deemed healthy. The overall concept of how recovery works for the three techniques remains the same. The automated recovery is initiated when the underlying database or SAP HANA physical host has software or hardware issues (e.g., hypervisor software issues or a loss of system power or network connectivity). A recovered instance is identical to the original instance, including the instance ID, private IP addresses, elastic IP addresses, and all instance metadata. If the impaired instance is in a placement group, for example, an SAP HANA scale-out, the recovered instance runs in the placement group. During instance recovery, the instance is migrated as part of an instance reboot, and any data that is in-memory is lost. No migration of data is required because, in the AWS world, all the services are application programming interface (API)-connected including Amazon EC2 and Amazon EBS services, which are segregated. When you create the Amazon CloudWatch alarm using AWS Management Console, associate it with Amazon Simple Notification Service (SNS) to receive email notifications. Alternatively, you can set up Amazon SNS notifications after creating the alarm.

  When you combine Amazon EC2 auto-recovery feature with the database auto-restart capability, you can have a robust HA solution without any human intervention. Typically, an instance is failed over in minutes. Specifically, for SAP HANA, consider the time it takes to load the data from disks into memory. If you're considering

Amazon EC2 auto-recovery as your HA solution, then consider that it does not offer a fixed Amazon EC2 SLA on the time it would take to move the failed instance to a healthy host. Additionally, Amazon EC2 auto-recovery is not application aware, which means it won't be triggered if database services such as the SAP HANA index server or the name server crashes. Amazon EC2 auto-recovery is AZ bound, which implies that it cannot protect Amazon EC2 instances against an AZ failure. To summarize, with Amazon EC2 auto-recovery, you get a 99.5% single-instance Amazon EC2 SLA and a higher MTTR with no SLA on the time to migrate to a healthy host.

If you have SAP HANA or another database configured in HA using cluster solutions like Pacemaker, then enabling Amazon EC2 auto-recovery could cause conflicts between both the solutions and could lead to unexpected behaviors that could cause data loss. Thus, we highly recommended you have Amazon EC2 auto-recovery enabled for instances where you do not have cluster solutions or dedicated host recovery enabled, for example, non-production SAP HANA environments such as DEV, SBX, or QA.

[»]

Amazon EC2 Auto-Recovery

The automatic recovery process attempts to recover your instance for up to three separate failures per day. If the instance is not recovered then, you should stop and start the instance manually.

Amazon EC2 simplified auto-recovery does not work when Amazon EC2 instance store volumes (i.e., the volumes physically attached to the instance) are in use or mapped to the instance. Do NOT use instance store volumes for Amazon EC2 instances that have SAP running on them.

Figure 5.11 shows the overall process of how Amazon EC2 auto-recovery works:

- **Amazon EBS snapshots**
  Snapshots are point-in-time copies of Amazon EBS volumes, which are crash consistent and not application consistent. Let's consider a scenario where the underlying storage service (e.g., Amazon EBS, Amazon FSx for NetApp ONTAP, or Amazon EFS) has failed, in that scenario, if it's the boot volume you could consider Amazon EC2 auto-recovery. However, what if it's database log and data volumes? In this case, one option that you can consider is to recover the database using snapshots if you do not have a secondary instance in another AZ. After file-based database backups, Amazon EBS snapshots are typically used as a second layer of defense when it comes to protecting data within a region or cross-region. You can restore Amazon EBS volumes from the snapshot or create a new volume out of a snapshot in the same or different AZ within a region or a different region and launch Amazon EC2 instances.

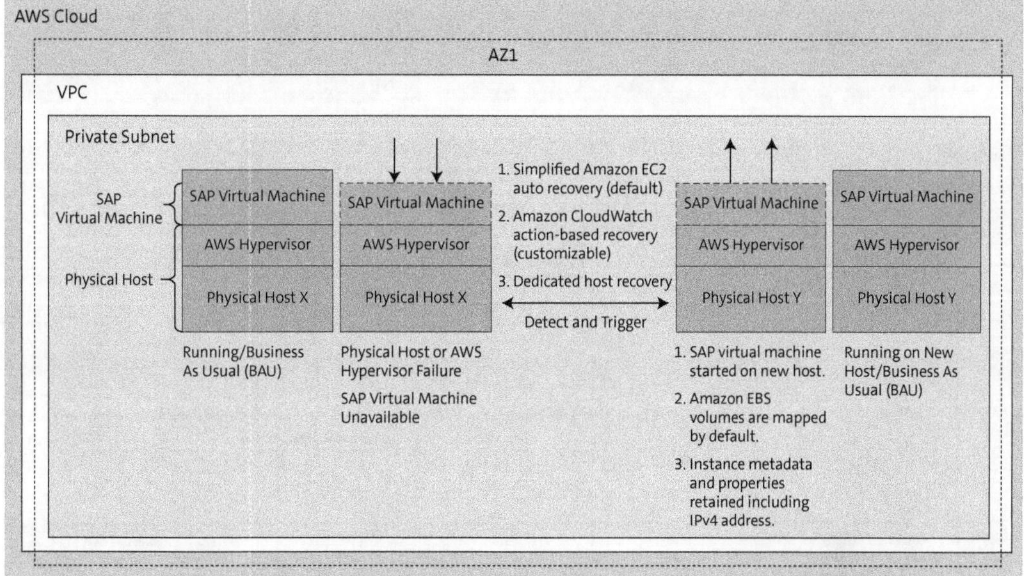

Figure 5.11 Amazon EC2 Auto-Recovery for SAP HANA on AWS

For HA, it's *not* common to have a snapshot-based HA strategy because it could lead to higher data loss unless you schedule snapshots at regular intervals (e.g., every 5 mins), or it could lead to higher MTTR with the time it takes to deploy and recover the database. Further, it is a crash-consistent backup and not an application consistent backup, which means that Amazon EBS snapshots are not application aware. In other words, it does not copy in-memory data or pending I/O transactions, which could lead to application inconsistency unless you put databases into storage snapshot mode or quiesce I/O operations on the disks. Significantly higher costs may arise with snapshots if proper housekeeping is not in place, and for lower RTO, you might have to consider Fast Snapshot Restore (FSR), which could further bump up the costs. Amazon EBS snapshots are typically used in DR strategies by replicating snapshots to another AWS account within the same region or another region to protect data against malware or ransomware attacks.

Figure 5.12 shows a snapshot-based HA/DR strategy to recover instances from failures related to storage components within AWS.

SAP HANA Amazon EBS Snapshot-Based Recovery

You can consider automating the deployment and installation of SAP HANA (or any database) using IaC based on AWS CloudFormation (e.g., AWS Launch Wizard for SAP HANA) or third-party tools like Terraform to further reduce the MTTR or RTO. Additionally, if you have Amazon EC2 instances running SAP HANA, you can automate the creation and retention of application-consistent Amazon EBS snapshots of the Amazon

EBS volumes attached to those instances using scripts via AWS System Manager Documents for SAP HANA only.

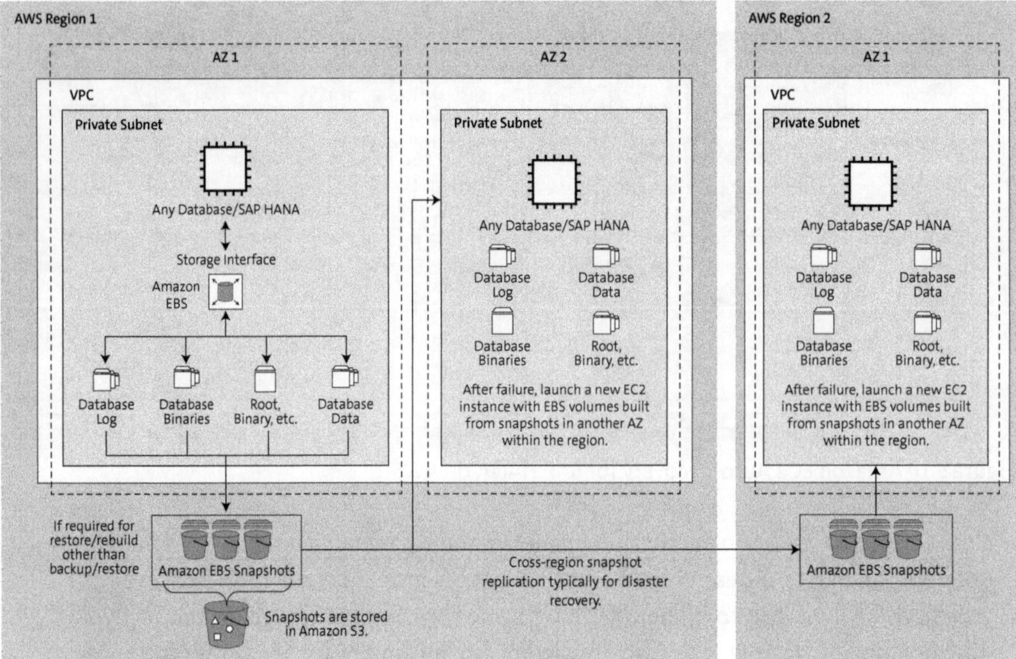

Figure 5.12 Snapshot-Based HA/DR Strategy for SAP HANA (or Any Database)

- **Backup/restore SAP HANA or any other database**
  Using database backups, you can recover SAP HANA (or any other database) running on an Amazon EC2 instance in another AZ within the same region or in another region when the AZ in which your primary database was running becomes unavailable. You could take your backups directly on an Amazon EBS st1 volume type, but we highly recommended you make backups of your backups or directly trigger backups to a storage location available during an AZ failure. The perfect service storage service for this scenario, providing a balance of cost and performance, is Amazon S3. As mentioned earlier, to remove the middle layer of storing backups on an Amazon EBS volume and then running scripts to move the backups to Amazon S3, AWS introduced AWS Backint Agent for SAP HANA. AWS Backint Agent runs as a standalone application that integrates with your existing workflows to back up your SAP HANA database to Amazon S3 and to restore it using SAP HANA cockpit, SAP HANA Studio, and SQL commands. For any database, you can user third-party tools to trigger backups directly to Amazon S3. For Oracle you have the option to use Oracle Secure Backups (OSB) Cloud Module or for other databases you can use Syntax's Emory solution.

As shown in Figure 5.13, SAP HANA is running on an Amazon EC2 instance that has auto-recovery enabled to protect against hardware issues within the AZ. For backups, SAP HANA full, differential, incremental, and log backups are stored in Amazon S3 using AWS Backint Agent. In case of an AZ failure, you should launch an Amazon EC2 instance with a IaC template and then restore/recover SAP HANA with backups stored in Amazon S3 using AWS Backint Agent.

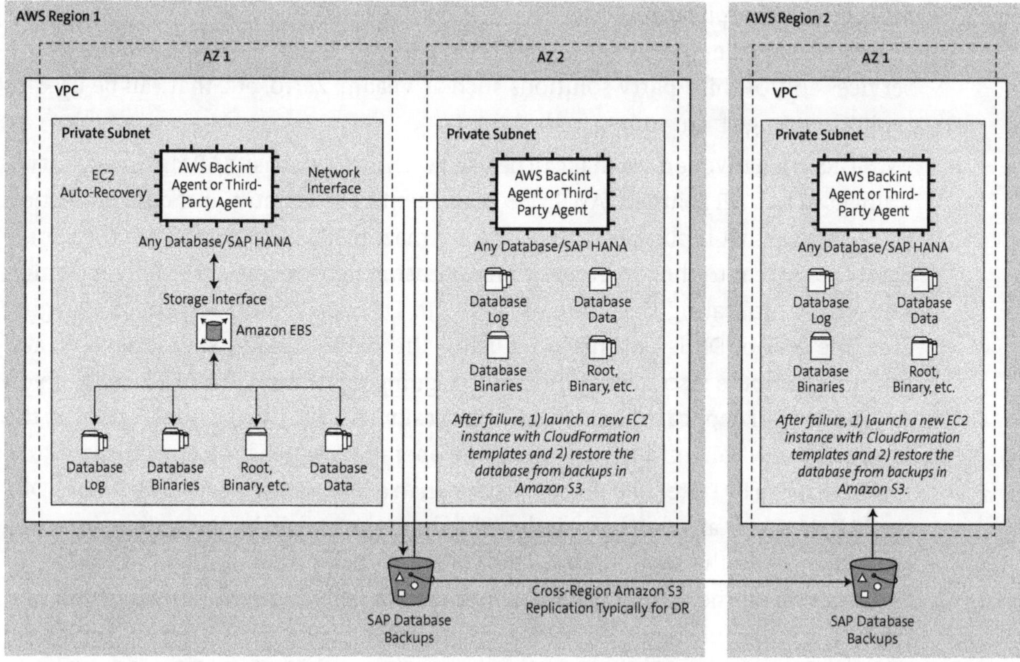

Figure 5.13 SAP HANA/Any Database Backup/Restore HA and DR Strategy

This low-cost solution does not have any compute provisioned for the standby instance on the secondary AZ. Numerous recovery and post-recovery steps must be performed leading to longer MTTR (for HA) and RTO (for DR). Data loss (or RPO) is greater than zero since it's dependent on how frequently you store your SAP HANA backup files in Amazon S3 preferably using AWS Backint Agent. Additionally, you can quickly restore from the backup files in Amazon S3 without creating custom scripts to manually copy your SAP HANA backup files to and from Amazon S3.

Also, as shown in Figure 5.13, there are two important things to call out—the network interface and storage interface (in this case, for Amazon EBS). Each SAP-certified and -supported Amazon EC2 instance will have these characteristics and the maximum amount of storage and network throughput it can support. Keeping this in mind, some key aspects consider to ensure optimal performance in the time required to backup and restore your data (which will dictate your MTTR or RTO) include the following:

- Database size
- Amazon EC2 instance type (its network and storage throughput)
- Amazon EBS volume type (its throughput)
- Network throughput supporting the communication channel with Amazon S3 (ideally using Amazon S3 gateway endpoints)
- Available CPU resources on the instance type

- **Storage based-replication**

  As described earlier in Section 5.1, similar to on-premise, on AWS, you can use native services and/or third-party solutions such as Veaam, Zerto, etc. that can be used to replicate storage from source to the target.

  AWS DRS is a native service that can be used with any database SAP HANA and is typically used as a DR solution. AWS DRS is an agent-based continuous data protection (CDP) service that replicates storage at the Amazon EBS block level from the source database system to the target environment using lightweight AWS DRS replication servers. By replicating the source systems to DRS replication servers in a staging area, the cost of DR is optimized by using affordable storage, shared servers, and minimal compute resources to maintain ongoing replication. AWS DRS enables you to launch your replicated servers automatically to a fully provisioned state at the target region within minutes and continue operation, which basically translates to a low RTO. It can be faster and more effective than a manual approach. The RPO of DRS for SAP HANA (or any database) is dependent on time taken by the source system to send changes to the staging area. This value is further impacted by the volume of transactions on the source system. Other factors include network throughput and latency, source and replication server performance, etc. These factors should be measured to calculate the potential amount of data loss during a DR event.

> **[»]**
>
> **SAP HANA (or Any Database) with AWS DRS Consideration**
>
> An SAP HANA database (or any database) can have large disk sizes and high change rates. We highly recommend conducting tests to ensure that your reliability SLA requirements are met in such events. Additionally, you must ensure that the primary and target databases are in sync during peak activity cycles.

You can perform non-disruptive tests, known as *drills*, to confirm that your AWS DRS implementation is ready for a DR scenario. AWS DRS automatically converts your servers to boot and run natively on AWS when you launch instances for drills or recovery. The service also automatically creates point-in-time snapshots of your server state as it replicates. If you need to recover applications, you can launch recovery instances on AWS within minutes, using the latest snapshot or an earlier point-in-time snapshot. Once your applications are running on AWS, you can choose to keep them there, or you can initiate data replication back to your primary site when the issue is resolved. You can fail back to your primary site with AWS DRS tools, such as the Failback Client.

As shown in Figure 5.14, the source servers running SAP application components, such as SAP HANA database, are replicated using AWS DRS within a single region, across regions, or from outside AWS such as on-premise to an AWS region.

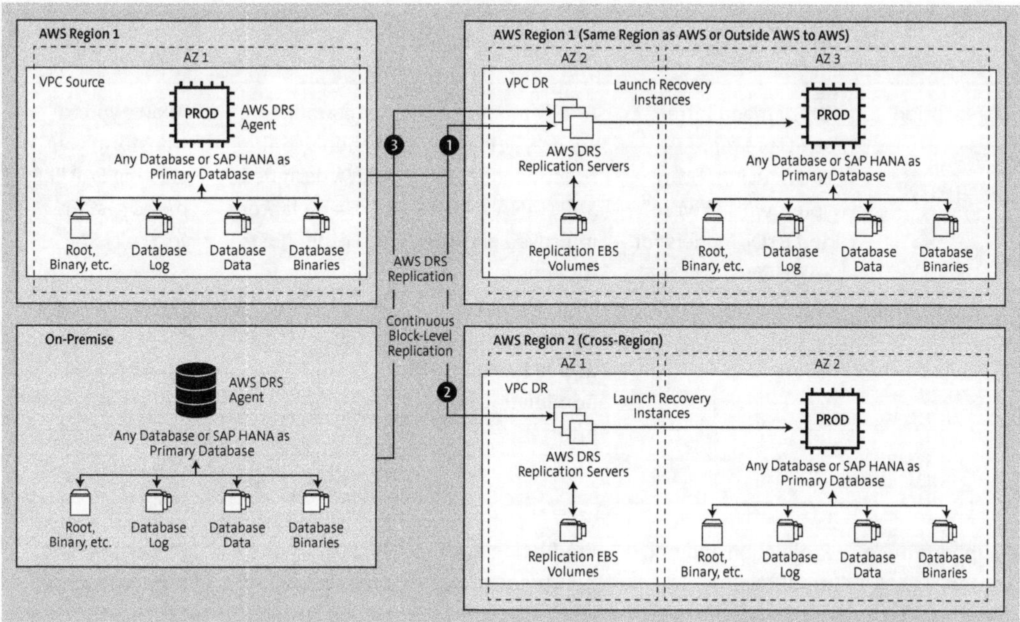

Figure 5.14 AWS DRS Disaster Recovery Scenarios for SAP HANA (or Any Database)

Based on our experience following are the important key takeaways when you're considering HA and DR strategies for any database or an application:

- Choose strategy based on your business drivers and budget.

- Align with all the key stakeholders (including business users, IT teams, C-suite execs) on the reliability SLAs.

- By using flexibility that AWS provides switch between options based on changing business requirements.

- Leverage third-party solutions for automated failover (e.g., SLES for SAP HA extensions) or create your own custom scripts and hooks.

5.3.3 SAP HANA

As every vendor's hardware platform needs to be certified and supported specifically for SAP HANA, AWS has collaborated with SAP to certify the AWS platform so that companies of all sizes can fully realize all the benefits of the SAP HANA in-memory computing platform on AWS. For SAP HANA architecture on AWS, let's explore our deployment options, as well as compute, OS, and storage and how they are typically deployed on AWS.

Deployment Options

SAP HANA can be deployed in multiple ways on AWS. Table 5.2 provides an overview of the four main SAP HANA offerings on AWS.

Offering	SAP HANA (BYOL)	SAP HANA Cloud	SAP HANA, Express Edition	RISE with SAP on AWS
Description	On-demand infrastructure for SAP HANA using "bring your own software" and BYOL models for SAP HANA	A fully managed service, get the functionality of SAP HANA with improved elasticity, consumption, subscription-based pricing, and reduced administrative effort	A streamlined version of SAP HANA designed to run on laptops and other hosts	Allows you to transform your existing SAP ERP-based systems to a fully SAP S/4HANA on AWS system or implement a new SAP S/4HANA ERP on AWS that is technically managed and operated by SAP
Supported use cases	Production and non-production	Production and non-production	Typically, non-production but can be used for Production	Production and non-production
Supported SAP HANA scenarios	<ul><li>Native SAP HANA applications</li><li>Data marts, analytics, and big data</li><li>SAP S/4HANA</li><li>SAP BW/4HANA</li><li>SAP Business Suite on SAP HANA</li><li>SAP Business Warehouse (SAP BW) and SAP Business Planning and Consolidation on SAP HANA</li><li>SAP Business One on SAP HANA</li></ul>	<ul><li>Native SAP HANA applications</li><li>Data marts, analytics, and big data</li><li>Extensions to SAP S/4HANA and other SAP applications</li></ul>	<ul><li>Native SAP HANA applications</li><li>Data marts, analytics, and big data</li><li>BYOL and microservices</li><li>Predictive analytics</li><li>Machine learning algorithms</li><li>Geospatial processing</li></ul>	<ul><li>SAP S/4HANA</li><li>SAP BW/4HANA</li><li>SAP Business Suite on SAP HANA</li><li>SAP BW and SAP Business Planning and Consolidation on SAP HANA</li><li>SAP Business One on SAP HANA</li><li>SAP HANA Cloud</li><li>SAP DataSphere</li><li>SAP BTP</li></ul>

Table 5.2 SAP HANA Offerings on AWS

Offering	SAP HANA (BYOL)	SAP HANA Cloud	SAP HANA, Express Edition	RISE with SAP on AWS
Licenses	BYOL	Subscription- and consumption-based licensing for the SAP HANA service (including required AWS infrastructure) procured through the SAP Business Technology Platform (BTP)	Free SAP HANA license, up to 32 GB. On-demand Licenses for 64 GB–128 GB memory sizes available for purchase	■ Subscription-based licensing model which includes infrastructure, software subscription, software support, and technical managed services ■ Relief from SAP annual maintenance fee (typical for perpetual licenses)
Memory	■ OLTP and OLAP scale-up to 24 TB ■ SAP S/4HANA scale out-up to 48 TB ■ OLAP scale-out up to 100 TB	Scale-up, up to 5,970 GB memory/ 1,000 PB cold (data lake)	<=32 GB free 33-128 GB licensed	■ OLTP and OLAP scale-up up to 24 TB ■ SAP S/4HANA scale-out up to 48 TB ■ OLAP scale-out up to 100 TB

Table 5.2 SAP HANA Offerings on AWS (Cont.)

Compute

In Section 5.1, you saw how it is imperative to select SAP-supported and -certified Amazon EC2 instances specifically for SAP HANA. Since AWS gets new Amazon EC2 instances certified regularly from SAP, we highly recommend you refer to SAP Note: 1656099 and the AWS documentation online for the latest list of SAP-supported and -certified instances that you can use for SAP HANA. The latest and the greatest generation on Amazon EC2 instances for SAP HANA are built on AWS Nitro platform. As SAP HANA in only certified for Power or x86 CPU architecture, on AWS, SAP HANA runs on Intel-based Amazon EC2 instance types. In the AWS world, it is important to know that Amazon EC2 instance offerings of 6 TB (DRAM) and above fall into the category of high-memory instances. These SAP-certified instances are available both virtually as well as bare-metal running natively (AWS's data centers), which are also powered by the AWS

Nitro platform. Some common instances families used for SAP HANA on AWS include r5, r6i, x1, x1, x2idn, x2iedn, and u-*.

With respect to the sizing, CPU type, and mapping of on-premise SAP HANA servers to Amazon EC2 instances, you should consider the following tools:

- SAP EarlyWatch Alert reports generated from SAP Solution Manager for existing SAP systems.

- Using AWS native discovery tools such as AWS Migration Evaluator as well to get to the information related to vCPU, memory, and storage requirements and use it to map against a supported and certified Amazon EC2 instance.

- Quick Sizer for greenfield SAP implementations.

- ABAP reports or the SAP S/4HANA readiness check based on the source systems to get to the memory and CPU sizing required on AWS. This tool is applicable for both SAP S/4HANA system conversion or a database migration from SAP HANA (or any database).

SAP HANA on AWS Cost Optimization

For SAP HANA on AWS, there are additional Amazon EC2 instances (on top of the production-certified instance types) available when used only for non-production. These Amazon EC2 instances are NOT SAP-certified for production. These Amazon EC2 instances are part of the x2iedn instance family for non-production, and you should evaluate if the relaxed core-to-memory ratio will meet their performance requirements for that non-production environment.

Operating System

SAP HANA can be deployed on a Linux OS on AWS only on an x86 based CPU architecture. For Linux, vendor products from SLES and RHEL are certified and supported on AWS. Within the realms of SAP's Product Availability Matrix, refer to the latest documentation from AWS as well as SAP Note 1656099 for OS versions supported for AWS for SAP. AWS supports both BYOL and licenses purchased from AWS. We highly recommend you use OS flavors tailored specifically for SAP, such as RHEL for SAP and SLES for SAP. Over the standard, they include additional features such as extended lifecycle support, live-patching, autoconfiguration of OS parameters based on SAP products, and HA extension (HAE) packages.

Storage

Though SAP HANA is an in-memory database, it still must persist data on disks to prevent data loss specifically to recover from a crash, from planned downtime, and from unplanned downtime. The changed data from memory is flashed to disks every 5 minutes (which is configurable) via a "savepoint" process. Transactions must be persisted to disk quickly to ensure transaction integrity. Since the data stored in DRAM is volatile, after every restart of SAP HANA or the OS or the Amazon EC2 instance, SAP HANA loads data from disk into memory—another reason you need disks with minimum KPIs defined by SAP and AWS for SAP HANA to operate efficiently during the data preload process. The right throughput and IOPS on your disks can ensure that SAP HANA is performant, not only during normal operations, but also during preloading data onto to the disks after restarts, backups, and restores. Table 5.3 provides an overview of AWS storage services that are certified and supported for SAP HANA on AWS.

AWS Storage Service	SAP HANA Scale Up	SAP HANA Scale Out
Amazon EBS	Yes	Yes
Amazon EFS	No	Yes (only for /hana/shared and backups)
Amazon FSx for NetApp ONTAP	Yes	Yes
Amazon S3	Yes (only backups)	Yes (only backups)

Table 5.3 AWS Storage Services Overview for SAP HANA Scale Up and Scale Out

gp3 with Non-AWS Nitro-Based Amazon EC2 Instances
gp3-based configurations are only supported in production for AWS Nitro-based instances (u, x2, r5/r6), not for Xen based instances (x1, r4, etc.).

Figure 5.15 shows an example from AWS's technical guide on storage configuration, in this case, related to an x2iedn.32xlarge Amazon EC2 instance.

General-purpose SSD gp3 volumes balance price and performance IOPS as throughputs can be increased from a high baseline value on running instance without any downtime. io2 Block Express volumes provide the highest performance consistently and resiliency for mission-critical applications, with the lowest (below a millisecond) latency specifically for SAP HANA production systems. As shown in Figure 5.15, notice the 2 × 2.4 TB for gp3 or io2 block express, which translates to two volumes of 2.4 TB, each configured in RAID 0 (striping) get more throughput and IOPS out of these combined volumes.

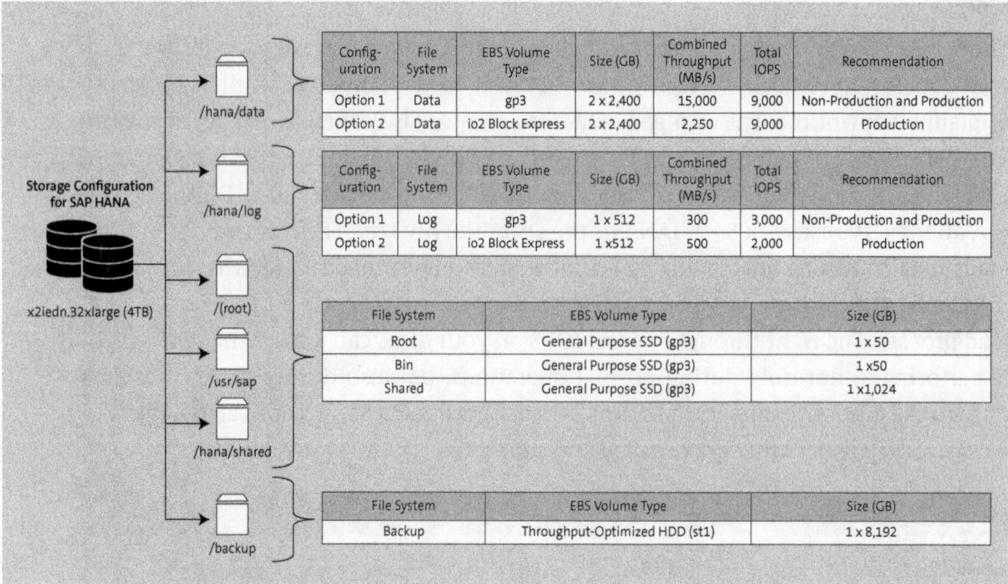

Config-uration	File System	EBS Volume Type	Size (GB)	Combined Throughput (MB/s)	Total IOPS	Recommendation
Option 1	Data	gp3	2 x 2,400	15,000	9,000	Non-Production and Production
Option 2	Data	io2 Block Express	2 x 2,400	2,250	9,000	Production

Config-uration	File System	EBS Volume Type	Size (GB)	Combined Throughput (MB/s)	Total IOPS	Recommendation
Option 1	Log	gp3	1 x 512	300	3,000	Non-Production and Production
Option 2	Log	io2 Block Express	1 x512	500	2,000	Production

File System	EBS Volume Type	Size (GB)
Root	General Purpose SSD (gp3)	1 x 50
Bin	General Purpose SSD (gp3)	1 x50
Shared	General Purpose SSD (gp3)	1 x1,024

File System	EBS Volume Type	Size (GB)
Backup	Throughput-Optimized HDD (st1)	1 x 8,192

Figure 5.15 SAP HANA Storage Configuration Based on x2iedn.32xlarge

SAP HANA Storage Configuration

Each SAP HANA-certified Amazon EC2 instance supports a maximum amount of Amazon EBS throughput. You can stripe multiple Amazon EBS volumes to get more throughput and IOPS; however, you'll be capped by the maximum amount of throughput and IOPS that the Amazon EC2 instance can support. For example, you can stripe 10 volumes to get, let's say, 4000 MB/s, but if the type of Amazon EC2 instance being used can only support maximum Amazon EBS throughput of 2000 MB/s, then you'll be bound by this supported throughput.

SAP HANA Amazon EBS Storage Type

gp3 provides a balance of cost and performance; however, if you're looking for greater IOPS, throughput, and reliability than a single volume can provide, io2 or io2 block express should be considered. As costs significantly increase with io2 and io2 block express, you should consider them only for production and production-like environments. For other non-production environments like sandboxes and project systems, development gp3 can be considered. This strategy can help you lower your overall TCO related to storage.

Figure 5.16 shows SAP HANA scale-up system with Amazon FSx for NetApp ONTAP.

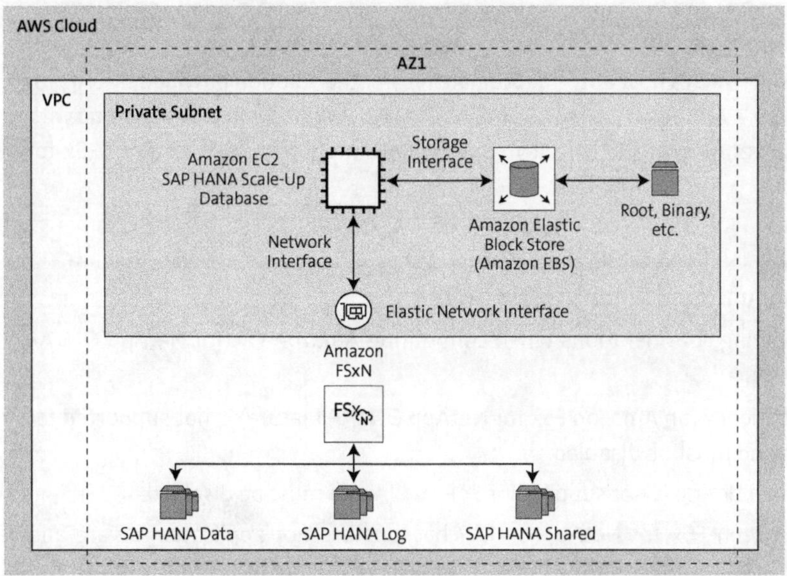

Figure 5.16 SAP HANA Scale-Up Database with Amazon FSx for NetApp ONTAP

Figure 5.17 shows SAP HANA scale-out system with Amazon FSx for NetApp ONTAP.

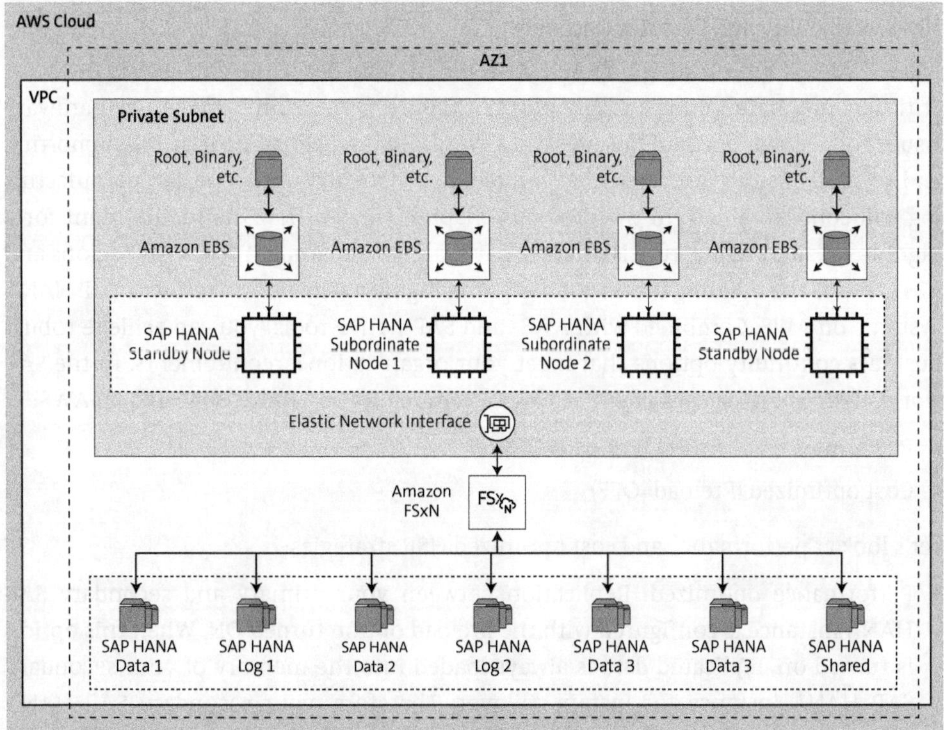

Figure 5.17 SAP HANA Scale-Out Database with Amazon FSx for NetApp ONTAP

> **SAP HANA Scale-Out Support for Host Auto-Failover on AWS**
>
> You can use SAP HANA's host auto-failover, an automated solution provided by SAP, for recovering from a failure on your SAP HANA host. On AWS, SAP HANA scale-out with host auto-failover (i.e., with a standby node) is only supported with Amazon FSx for NetApp ONTAP.

> **SAP HANA on Amazon FSx for NetApp ONTAP**
>
> Note the following considerations when configuring Amazon FSx for NetApp ONTAP for SAP HANA on AWS:
>
> 1. Storage efficiency (an Amazon FSx for NetApp ONTAP feature) is not supported for SAP HANA and must be disabled.
> 2. Capacity pool tiering is not support for SAP HANA and must be disabled.
> 3. Default Amazon FSx for NetApp ONTAP backups are not application-aware and cannot be used to restore SAP HANA to a consistent state. However, you can use storage snapshot feature of SAP HANA to execute a database consistent backup using Amazon FSx for NetApp ONTAP.

High Availability and Disaster Recovery

For an SAP system running on SAP HANA (or any database), the database is the most critical component and is a single point of failure. If there are software or hardware issues for a single point of failure such as the database, global operations supporting end-to-send business processes such as sales or order entry or continuous manufacturing will come to a halt. In the previous chapter, we saw how the requirement for a highly available architecture is driven by availability SLAs and MTTR. AWS Regions and AZs serve as the building blocks for designing highly available and resilient SAP HANA systems on AWS. Combined with AWS- and SAP-native tools, you can achieve robust business continuity options that meet your organization's requirements. In the SAP world, there are two types of SAP HANA system replication fully supported on AWS:

- Performance optimized (Preload=ON)
- Cost optimized (Preload=OFF)

Let's look at performance and cost optimized HSR strategies:

- Performance optimized: Replication between your primary and secondary SAP HANA instance is configured with the preload option turned ON. When this option is turned on, replicated data is always loaded into the memory of your secondary SAP HANA instance for instant failover. Therefore, your secondary SAP HANA instance must be sized like your primary SAP HANA instance. You must ensure that your client traffic is redirected to your secondary SAP HANA instance after the failover. This architecture provides you with the advantage of implementing your SAP HANA solution across multiple AZs with the ability to failover instantly and continue your business operations during an AZ failure, a rare occurrence. We do not recommend hosting your SAP HANA instances across AWS Regions in this setup to avoid latency while replicating in a synchronous mode. You can get the lowest MTTR and can achieve zero data loss with this option. As with this approach, you must have the same-sized compute for your secondary up and running on the second AZ, which makes it a costlier option, but you can achieve the lowest possible MTTR.

Even though SAP solutions on AWS are architected in a highly available, fault tolerant configuration, the failover still needs to be managed at the application layer (only applies to ASCS and database). As mentioned earlier, with this architecture you "can" achieve the lowest possible MTTR. Regarding failover options, you can consider manual, customer HA/failover solution, or a certified clustering software such as SLES for SAP HAE or RHEL for SAP HA Add-Ons.

SAP HANA Cluster Solutions Support

The HA setup on Linux is supported by certified SAP HA partners and not by SAP directly. SAP does not provide any specific cluster guides for the implementation of SAP clusters within Linux HA environments as this task is handled by the ISVs such as Red Hat or SLES in collaboration with hardware platform providers like AWS.

The Pacemaker cluster uses an overlay IP address, similar to what we saw with the ASCS/ERS layer, described in Section 5.3.1, to connect the application servers to the primary SAP HANA instance.

Figure 5.18 shows an SAP HANA performance-optimized scenario for a production SAP HANA scale-up system running on AWS.

Additionally, SAP HANA scale-out performance optimized scenario for a production SAP HANA scale-out system is also supported on AWS.

- Cost optimized: Replication between the primary and secondary SAP HANA instance is configured with the preload option turned off. When this option is turned off, replicated data is NOT loaded into the memory of your secondary SAP HANA instance for instant failover. Therefore, your secondary SAP HANA instance can be sized smaller than the primary SAP HANA instance to save on costs. Another option is to size the secondary same as primary and use it to share a non-prod SAP HANA database in a *multiple components one system (MCOS) configuration.*

Figure 5.19 shows a cost-optimized SAP HANA configuration using a smaller secondary across multiple AZs in a single region.

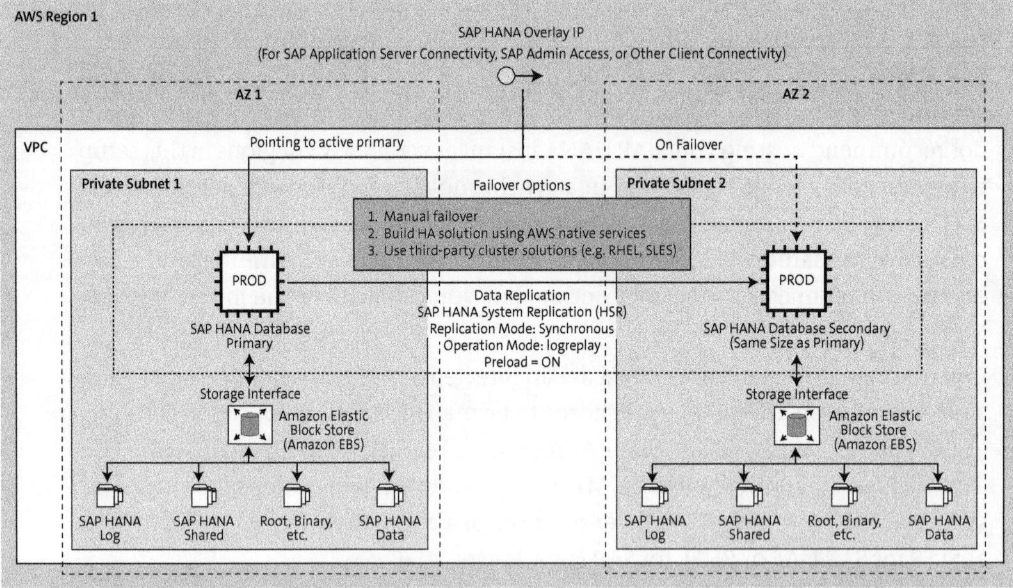

Figure 5.18 SAP HANA Scale-Up Performance Optimized

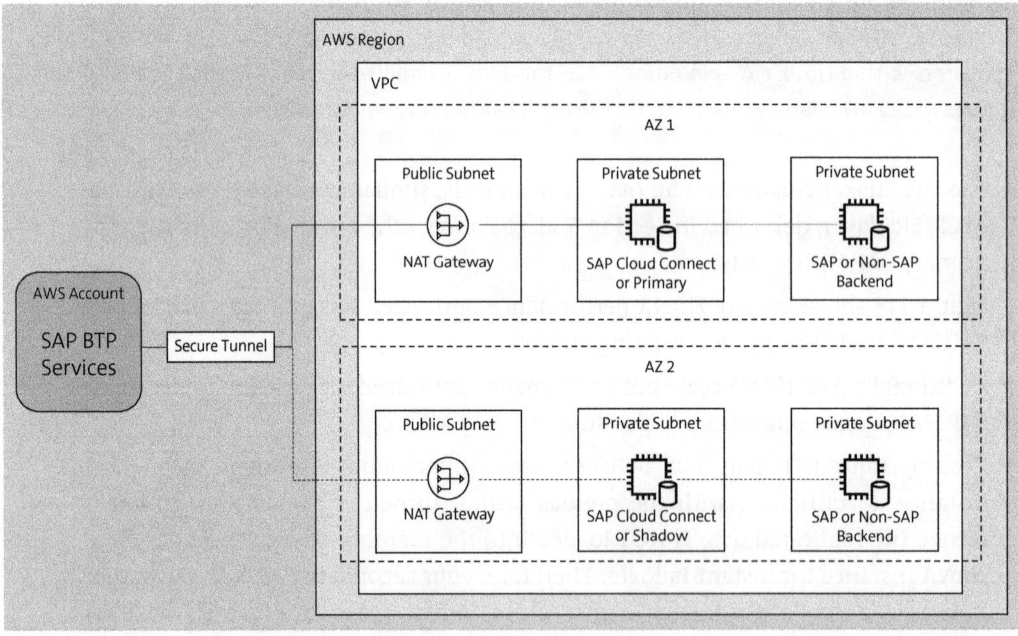

Figure 5.19 SAP HANA Cost Optimized Using Smaller Secondary

- The architecture shown in Figure 5.19 can also be deployed across multiple regions using asynchronous replication which is a typical multi-region DR scenario.

> **Cost-Optimized SAP HANA Using a Smaller Instance**
>
> When you resize the secondary system before a takeover, there is no reserved capacity. The requirement for a production-sized instance is subject to the current availability of the Amazon EC2 instances in your AZ.

- Figure 5.20 shows a cost-optimized SAP HANA configuration using a shared secondary across multiple AZs in a single region.

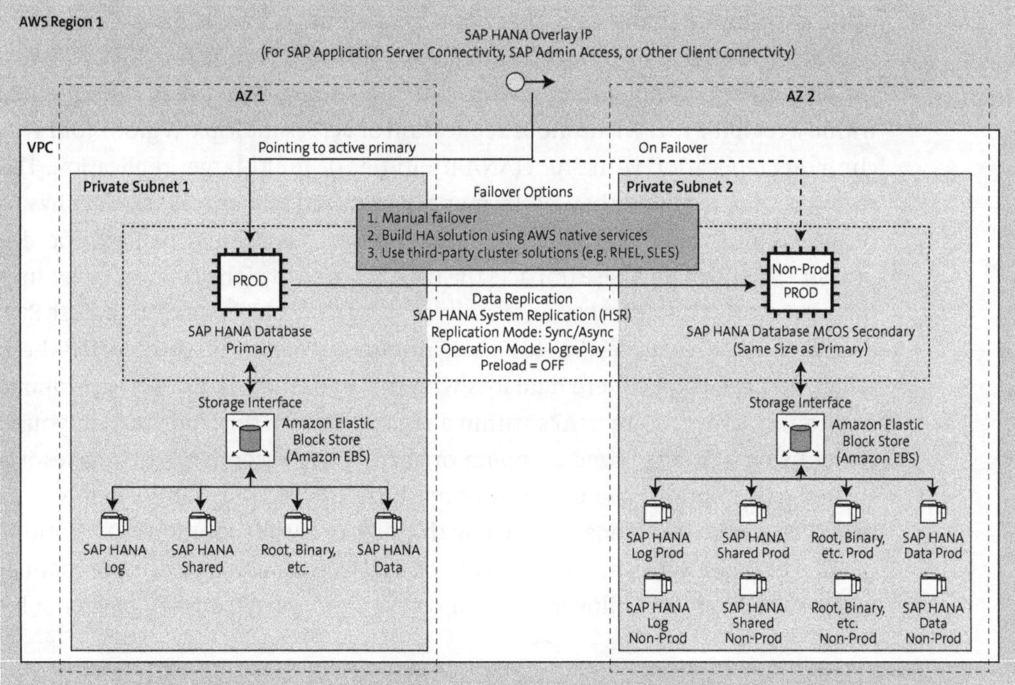

Figure 5.20 SAP HANA Cost Optimized with Shared Secondary

- As we are going with an MCOS approach, this setup requires additional storage to operate the additional non-production instances. During a takeover, the instance with lower priority can be shutdown to make the underlying host resources available for production workloads. The process of shutting down the non-production instance can be automated using HA/DR hooks or using third-party cluster solutions. The key parameter to take note of is `global_allocation_limit` for all systems running on the secondary nodes. This ensures that no one instance with the `global_allocation_limit` set to 0 occupies the entire memory available on the host. This architecture can also be deployed across multiple regions using asynchronous replication which is a typical multi-region DR scenario.

[»]

Cost-Optimized SAP HANA Using a Shared Secondary Instance

Before going with this approach, make sure you have the proper approvals from your internal security team to host production and non-production on the same instance using different storage volumes to store data. In our experience, some organizations don't' allow this option due to data compliance requirements.

- Before you redirect your client traffic to your secondary SAP HANA instance after the failover, you need to ensure that the instance is re-sized to the same sized as primary or in the shared secondary option, stop the non-prod (sacrificial) database before bringing up production database. This architecture provides you with the advantage of implementing your SAP HANA solution across multiple AZs within a region with synchronous replication or across multiple regions for asynchronous replication with SAP HANA multi-tier or multi-target replication. The recovery time is higher than performance optimized scenario as the data has to be loaded into memory after failover since preload has been turned off. The size of your SAP HANA database impacts the time taken to load the column tables into main memory. This affects your overall recovery time objective. You can use SQL scripts to get a rough estimate of the minimum memory required for these tables. You can achieve zero data loss with this option when using synchronous replication across multiple AZs within a region. As with this approach, as you're either using a smaller sized compute or sharing the secondary, this makes it a cost-effective option. You can further optimize the recovery time when you're using cost optimized scenario by using third-party cluster solutions. On Options of failover, as we saw with performance optimized scenario you can use manual failover, build an HA/failover solution or use third-party certified cluster solutions.

- The most important aspect that is often overlooked is the sizing of the secondary instance when using cost optimized strategies. The short answer on sizing the secondary instance is—it depends on the operation mode you're using. You can disable the preloading of column tables into memory, but if you're using logreplay, you must consider a few additional steps when it comes to sizing of the secondary. With logreplay, redo logs (i.e., changes on the primary database) are constantly replayed on the secondary database, and for that reason, you need to look at size of column store tables with data modified in the previous 30 days from the current date of evaluation. For sizing related to the secondary instance, refer to SAP Note 1999880 available at *http://s-prs.co/v577617*.

[»]

SAP HANA Cost-Optimized Secondary Sizing

The logreplay operation mode may not be able to provide true cost optimization. The potential cost savings are higher with larger database instances. The delta_datashipping operation mode can be an alternative. However, this mode has limitations, including higher recovery times and an increase demand for network bandwidth between the replication sites. If your business requirements can afford higher network bandwidth and relaxed recovery times, the delta_datashipping mode can be a viable option.

Because of the criticality of SAP HANA systems, its typical for organizations to look at both HA and DR. SAP HANA multi-tier replication provides a chained replication model where a primary system can replicate to only one secondary system at any given point of time. In this scenario, there can be a mix of performance and cost deployment options. The primary and secondary system can be deployed in an HA setup using a Pacemaker cluster. The tertiary or DR system can be a cost optimized deployment. An active non-production instance can run on the same node, in the MCOS installation model. This setup is shown in Figure 5.21 with three SAP HANA databases, two configured in HA across multiple AZs in a region and third provisioned in another region. You can also consider HA for the tertiary DsR instance by provisioning another instance in the second AZ in the DR region.

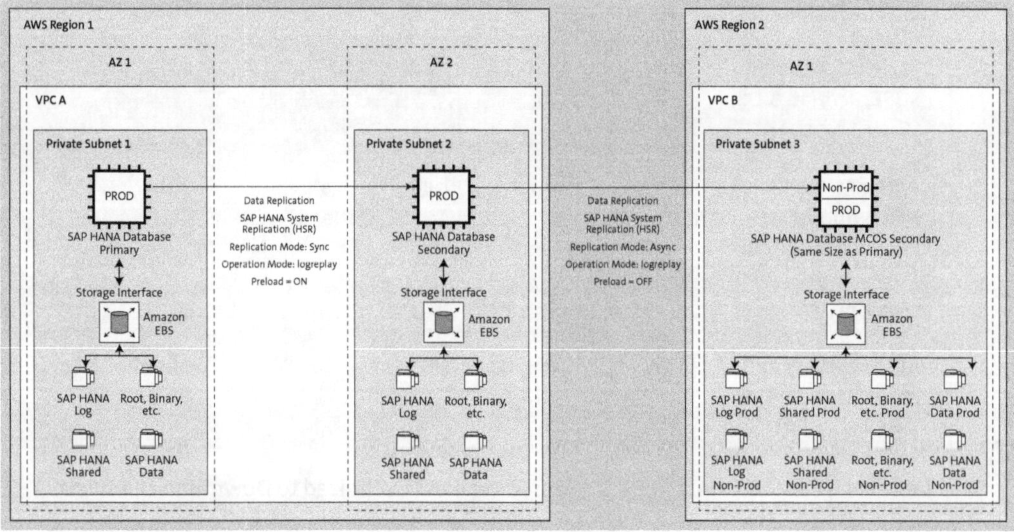

Figure 5.21 SAP HANA Multi-Tier Replication across AWS Regions

In an SAP HANA multi-tier scenario, replication happens sequentially, from the primary to the secondary system, and then from the secondary to the tertiary system. Starting with SAP HANA 2.0 SPS 03, SAP HANA provides multi-target system replication configuration for a single primary system to replicate to multiple secondary systems, which is fully supported on AWS.

Now that we've covered HA and DR in depth, Table 5.4 shows the availability mechanisms for SAP HANA and their related RPO, RTO, and costs.

Availability Mechanism (Solutions)	High Availability (Multi-AZ)			Disaster Recovery (Multi-Region)		
	RPO	RTO	Annual Cost	RPO	RTO	Cost
Auto Recovery and SAP HANA Backup/Restore	Medium	High	Low	-	-	-
Auto Recovery and HSR without Data Preload (Warm Standby)	Low	Medium	Medium	-	-	-
Auto Recovery and HSR without Data Preload (Warm Standby and Dev/QA)	Low	Medium	Medium-High	-	-	-
Auto Recovery and HSR with Data Preload (Hot Standby)	Low	Medium	High	-	-	-
Partner Solution – SUSE, RedHat, SIOS, Veritas, NEC etc.	Low	Low	High	-	-	-
Auto Recovery and HSR with S3 Cross-Region Replication (Hot Standby and OOR DR)	Low	Low	High	Medium	High	Low
Auto Recovery and Multi Tier HSR (Hot Standby and OOR DR)	Low	Low	High	Low	Medium	High

Table 5.4 SAP HANA HA and DR Options: From Cost Optimized to Downtime Optimized

5.3.4   RISE with SAP Architecture

RISE with SAP is a complete offering of the SAP ERP software, industry practices, and outcome-driven services for migrating your SAP ERP to the cloud. In practical terms, RISE with SAP includes a cloud ERP in the form of SAP S/4HANA Cloud Private Edition, SAP SaaS solutions, technical managed services, and infrastructure as a service (IaaS) from cloud service providers. SAP delivers and manages the software, support, infrastructure, and technical services for RISE with SAP under one SLA. SAP offers RISE with SAP on your choice of cloud service provider, including the ability to run on AWS.

Two product types available: a multi-tenant edition and a single-tenant edition. The multi-tenant SaaS service is currently offered on SAP's Converged Cloud (SAP data centers) and Google Cloud Platform only. More than 90% of SAP customers focus on the single-tenant edition. The multi-tenant edition is ideal for new, high-growth companies willing to accept "fit-to-standard" best practices as well as subsidiaries and other smaller business units of larger companies that need a fast way to stand up core SAP S/4HANA capabilities. What's still missing is feature parity with the single-tenant (on-premise) version of SAP S/4HANA. The single-tenant edition is a managed service hosted in SAP's data center, Microsoft Azure, Google Cloud Platform, or AWS. The single-tenant edition of RISE with SAP is available on AWS in two flavors:

- RISE with SAP and SAP S/4HANA Cloud Private Edition (also known as PCE standard)
- RISE with SAP and SAP S/4HANA Cloud Private Edition tailored option

In this section, we'll cover AWS architecture related to AWS region and AZ selection, SLAs, HA/DR, and network and connectivity under RISE with SAP.

RISE with SAP: AWS Regions, AZs, SLAs, and DR

In the ever-evolving world of RISE with SAP on AWS, refer to the latest documentation for RISE with SAP, available publicly at *https://www.sap.com/products/erp/rise.html*.

AWS Regions

Amazon EFS is a mandatory service for RISE with SAP and needs to be available in the region before SAP can start the enablement. This coordination of required services with service teams is supported by the SAP account team. At the time of writing, 31 out of 32 AWS regions are enabled for RISE with SAP. The only region that is *not* enabled for RISE with SAP is SFO (us-west-1). You can select from among the enabled AWS regions to get the SAP service.

Availability Zone Selection and Definition

Ideally, you would want to have the RISE with SAP service in the same AZ as other non-SAP workloads running on AWS because of lower latency, private connectivity, and lower egress costs. In this case, you need to make sure that the same AZ IDs are used for RISE with SAP and for the non-SAP deployments in the customer AWS accounts. To achieve this goal, you can request the zone ID information from SAP via a ticket to the server management team. The team will furnish the information on the AZs of the RISE with SAP landscape and share it with you, and then you can adapt your non-SAP landscape accordingly.

For workloads already existing on AWS, where the RISE with SAP AZ selection needs to be influenced, you must inform SAP, specifically the Cloud Architect & Advisor (CAA), about your preferred AZ ID. During the "Prepare" customer phase, SAP can manually adjust the AZ selection.

Service-Level Agreements for RISE with SAP

Let's look at the production, non-production, and overall SLAs available under the RISE with SAP construct:

- Production SLA: For SAP production systems, SAP now defaults to 99.7% with the option for 99.9%, which would include a (Pacemaker) cluster implementation. You can pivot to higher SLAs at an additional cost depending on the criticality of your SAP system.

- Non-production SLA: SAP offers 95.0% system availability percentage for non-PRD SAP systems unless a higher system availability percentage is identified in a supplement or an order form. You can pivot to higher SLAs at an additional cost.

- SLAs across hyperscalers: SAP offers same availability SLAs on all cloud providers and platforms! However, customers/enterprises should do their due diligence in identifying the right cloud platform based on reliability, security, and innovation metrics.

AWS Regions High Availability/Disaster Recovery Options: Short- and Long-Distance Options

During its inception, RISE with SAP used predefined DR pairs for AWS regions. Starting in October 2023, Flex DR was introduced where the customers can select the DR region of choice.

With optional DR services purchase for RISE with SAP for SAP S/4HANA Cloud Private Edition; SAP ERP, private cloud edition; or SAP S/4HANA Cloud, extended edition, the RTO is 12 hours, and the RPO is 30 minutes for long-distance DR. In contrast, for short-distance DR, the RTO is 12 hours, and the RPO is 0 minutes. For Google Cloud and Microsoft Azure, short-distance DR is only available in selected regions. On AWS, all regions can provide short-distance DR. If you purchase the optional SAP S/4HANA Cloud, 4-hour recovery time objective, private edition or SAP ERP, 4-hour recovery time objective, private cloud edition, the RTO is 4 hours (instead of 12 hours), and the RPO is unchanged as indicated in this section. Table 5.5 shows the supported DR services with RTO and RPO.

DR Services Supported	Type	RTO	RPO	AWS Architectural Pattern
Short-distance DR	Standard	12 hrs.	0	Multi-AZ (single region)
Short-distance DR	Enhanced	4 hrs.	0	Multi-AZ (single region)

Table 5.5 DR Services with RISE with SAP on AWS

DR Services Supported	Type	RTO	RPO	AWS Architectural Pattern
Long-distance DR	Standard	12 hrs.	30 mins.	Multi-region
Long-distance DR	Enhanced	12 hrs.	30 mins.	Multi-region

Table 5.5 DR Services with RISE with SAP on AWS (Cont.)

You can combine HA across AZs, with DR across AWS regions. In this case, you would receive an HA/DR solution with HA within the same region and (long-distance) DR across AWS regions. From a commercial perspective, you'll purchase only one DR option. The HA/DR combination is available on AWS in *all* regions and on Azure and Google Cloud Platform in *selected* regions.

Network and Connectivity

You'll need connectivity between the AWS cloud where your RISE with SAP solution is running and your on-premise data centers. You also need a connection for direct data transfer (to avoid routing data via your on-premise locations) and communication between SAP systems and your applications running on AWS cloud.

Some connectivity options to link your on-premise location to an AWS RISE with SAP account include the following:

- Direct network connectivity to the RISE with SAP environment with SAP VPC
  - Amazon Site-to-Site VPN (site-to-site only; no AWS client VPN)
  - AWS Direct Connect
- An AWS Direct Connect Gateway using existing connections in customer account
- AWS Direct Connect or VPN via AWS Transit Gateway in your customer account. At the time of writing, SAP also introduced AWS Transit Gateway cross-region connections using SAP-provided AWS Transit Gateway. SAP provides AWS Transit Gateway in SAP accounts only for AWS Transit Gateway to AWS Transit Gateway peering use cases.

Encryption over AWS Direct Connect

VPN over AWS Direct Connect directly to RISE with SAP account is not possible. If you want to encrypt traffic with AWS Direct Connect, the connection must be established from your on-premise system to your AWS account through the following methods:

- Using AWS Transit Gateway in with VPN over AWS Direct Connect (recommended)
- Using MACsec
- Using public VIF to connect to public interfaces which is not recommended.

Following are your options for connectivity from customer AWS Account to RISE with SAP account:

- Amazon VPC peering
- AWS Transit Gateway must reside in your AWS account and in the same region as RISE with SAP. Cross-connections to a different AWS region can be done via an AWS Transit Gateway-to-AWS Transit Gateway connection.

Our overall recommendation is to use AWS Transit Gateway to connect RISE with SAP account with your customer AWS account. There are no charges for cross-AZ traffic, and you can leverage existing AWS Transit Gateway and AWS Direct Connect implementations for RISE with SAP. AWS Transit Gateway is the preferred option to connect multiple customer AWS accounts RISE with SAP account. For basic/simple setups to connect one AWS account with your RISE with SAP account, Amazon VPC peering can be used.

RISE with SAP Data Egress Costs

SAP has an internal building block of 2 TB/month egress traffic. However, this limit is not part of the customer contract; instead, this is an SAP internal cost block, assigned to each PCE standard customer by default. The customer is paying on a user basis and does not need to care about data traffic charges, unless explicitly listed in the customer contract.

5.3.5   Microsoft SQL Server

Microsoft SQL Server is one of the oldest RDBMS solutions that SAP has been supporting for its suites of products. Up to now, for OLTP SAPS based on sales and distribution benchmarking, SAP internally uses SQL Server running on Windows to calculate SAPS on any given hardware platform. On AWS, Enterprises have been running Microsoft SQL Server for SAP and non-SAP applications for more than 12 years. When SAP started supporting its workloads on AWS, SQL Server was one of the first non-SAP databases to be supported. AWS supports only BYOL or licenses purchased from AWS for running SQL Server for SAP workloads. Often, enterprises are licensing SQL Server with an SAP material code (which Microsoft calls ISVR or in the SAP world—a runtime license), which can be procured directly from SAP or an authorized SAP partner. If the SQL Server license was purchased from any other party, the relevant licensing terms must be verified with the entity that sold you the original license.

For an SQL Server architecture on AWS, let's look at compute, OS, storage, and HA/DR and how they are typically deployed on AWS.

Compute

For any hardware platform including AWS, unlike SAP HANA, SQL Server does not require any additional certifications to run your SAP workloads. However, as described earlier in Section 5.1, you saw how imperative it was to select only the SAP-supported and -certified Amazon EC2 instances for pure database server such as SQL Server. Since AWS gets new Amazon EC2 instances certified regularly from SAP, we highly recommended you refer to SAP Note 1656099 and the AWS documentation online for the latest list of SAP supported and certified Amazon EC2 instances that you can use for SQL Server.

For SAP NetWeaver on SQL Server, Amazon RDS for SQL Server is not supported. However, you can use for SAP BusinessObjects and SAP Data Services audit, messaging, log, and commentary databases.

With respect to sizing, CPU type, and mapping of on-premise SQL servers to Amazon EC2 instances, you should consider the following information:

- SAP EarlyWatch Alert reports generated from SAP Solution Manager for existing SAP systems.

- Using Windows native tool to gather and analyze historical use data for CPU/Memory with Performance Monitor/Windows System Resource Manager to right size your environment.

- Using AWS native discovery tools such as AWS Migration Evaluator as well to get to the information related to vCPU, memory, and storage requirements and use it to map against a supported and certified Amazon EC2 instance.

- Quick Sizer for greenfield SAP implementations. Additionally, account for the Quick Sizer buffer. Quick Sizer tools provide sizing guidance based on assumptions that for 100% load (as per your inputs to tool) system use will not exceed 65%. Therefore, there is a fair amount of buffer already built into Quick Sizer recommendations. See the documentation for guidance.

MS SQL Server for SAP Deployment and Licensing Considerations on AWS

Based on SAP Note: 2139358, Microsoft has imposed strict restrictions on how you can deploy SQL Server for SAP applications on hyperscalers other than Microsoft Azure. This limit is primarily based on the license type, ELA with Software assurance and License Mobility, and the version of SQL Server, as summarized well at *http://s-prs.co/v577618*. In short, in our experience, we often see enterprises ending up using a dedicated host instead of shared tenancy. Dedicated hosts could imply higher costs; however, there are options to optimize costs by deploying multiple Amazon EC2 instances running SQL Server on a single dedicated host. In this scenario, to right-size and price the SAP on SQL environment on AWS, we highly recommend reaching out to your local AWS Account Team or SAP on AWS Solutions Architects via *http://s-prs.co/v577619*.

[+] MS SQL Server Edition for SAP Deployment on AWS

Per SAP Note: 555223, from a technical consistency perspective, SAP prefers you always use Enterprise Edition in all non-productive systems as well as the productive ones. Some customers have been told they can use Developer Edition in non-productive systems due to the way they licensed SQL Server. While this may be allowed by Microsoft, and fully supported by SAP, from a technical perspective, SAP prefers that you keep the same stack all the way through the landscape (OSs, database version, database client versions, SAP kernel, etc.) so that in any support scenario SAP is 100% sure that what you're running is exactly what SAP have tested and released. (SAP tests only with Enterprise Edition.)

[+] MS SQL Server Developer Edition for SAP Deployment on AWS Supportability Tip

Per SAP Note: 555223, if you purchased your SQL Server license from Microsoft or a third party other than SAP, contact that seller, they should assist you with SQL Server licensing questions, not SAP. If the company that sold you the SQL Server license says that you can use Developer Edition in your non-productive SAP systems, then you can use it, and SAP will support it. If you bought your SQL Server license from SAP (known as a Runtime or ISVR license), then you may not use the Developer Edition for any SAP system, whether productive or non-productive. This limitation is explicit in the reseller contract between Microsoft, and SAP and your SAP Sales Executive cannot grant you any different rights.

Operating System

SQL Server for SAP applications can only be deployed on a Windows OS on AWS only on an x86-based CPU architecture. Within the realms of SAP's Product Availability Matrix, refer to the latest documentation from AWS as well as SAP Note 1656099 for OS versions supported for SQL Server for SAP on AWS.

Storage

In this section we'll cover topics related to Amazon EBS, Amazon EFS, and Amazon S3 for SAP on SQL Server on AWS:

- **Amazon EBS**
  On AWS, typically, Amazon EBS persistent storage is used for SQL Server. Within the Amazon EBS service, you have the option to use gp2, gp3, io1, io2, and io2 block express, which all are solid state drives (SSDs) and certified and supported for SAP workloads on AWS. Additionally, each Amazon EBS volume is automatically replicated within its AWS AZ to protect you from failure, offering HA and durability.

Thus, you can configure a RAID 0 array at the OS level for maximum performance and not have to worry about additional protection (RAID 10 or RAID 5) for your volumes. The cost-optimized/high-throughput magnetic Amazon EBS volume type (st1) is an ideal candidate for storing SQL Server backups.

General purpose SSD gp3 volumes balance price and performance IOPS as the throughput can be increased from a high baseline value on running instance without any downtime. io2 Block Express volumes provide the highest performance consistently and resiliency for mission-critical applications, with the lowest (submillisecond) latency specifically for SAP HANA production systems.

- **Amazon EFS**
  Amazon EFS must not be used for DB data files or log files due to serious throughput and latency constraints.

- **Amazon S3**
  Similar to SAP HANA, you have the option to use Amazon EBS as a backup file system for SQL Server. To save on backup costs you can automate copying of the old backups from Amazon EBS to Amazon S3 using custom scripts.

MS SQL Server for SAP on AWS Storage Supportability

Amazon FSx for NetApp ONTAP and Amazon FSx for Windows File Server are not supported for data and log files for SQL Server for SAP on AWS.

High Availability and Disaster Recovery

For an SAP system running on SQL Server, the database is the most critical component and is a single point of failure. For SQL Server, following are some of the solutions related to HA and the architectures associated with it:

- **SQL Server AlwaysOn Availability Groups**
  AWS Regions and AZs serve as the building blocks for designing highly available and resilient SAP HANA systems on AWS. Combined with AWS and SAP-native tools, you can achieve robust business continuity options that meet your organization's requirements. SQL Server AlwaysOn Availability is fully supported on AWS and it's typical to see enterprises using it to deploy business critical database such as SQL Server on AWS in HA using multiple AZs within a region. It's important to understand SQL Server AlwaysOn provides data replication from primary to the secondary. For failover and automated orchestration, you have the option to use Windows Server Failover Clustering (WSFC) feature which is included in Windows OS. As shown in Figure 5.22, three nodes—primary, secondary, and a witness server spread across three AZs within an AWS region. It's imperative to consider an active directory connection with the same domain. We had mentioned earlier that Amazon FSx

for Windows File Server is *not* supported for SQL Server data and log only however you can use as a shared file system to be used by the witness node to keep information about the database node's status. Unlike Pacemaker solution for Linux, WSFC works with DNS redirection of availability group listener IP address. Each node in the cluster is assigned three IP addresses—the primary private IP, WSFC cluster IP, and availability group listener IP.

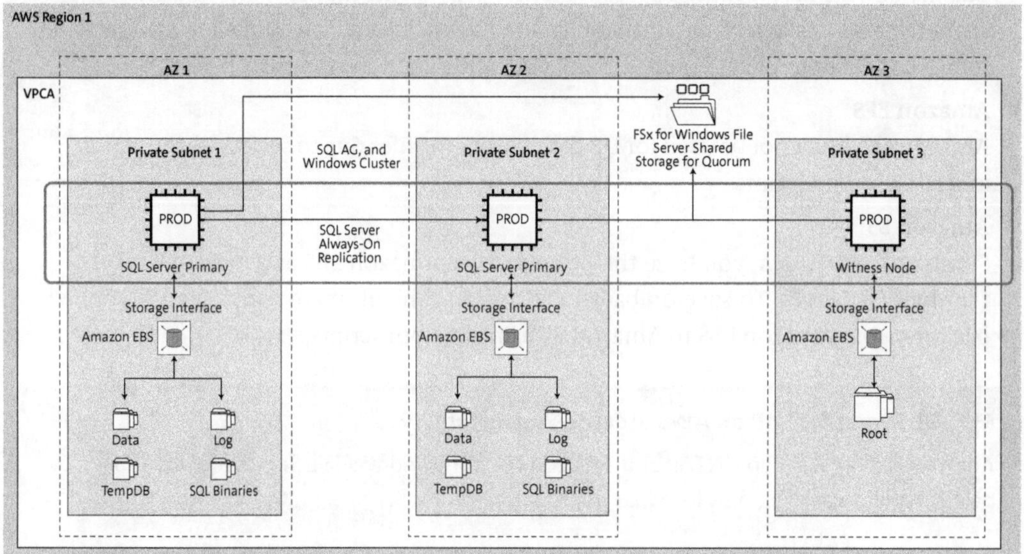

Figure 5.22 SQL Server Always-On Availability Group HA for SAP on AWS

Keeping in mind AWS region and AZ design you must evaluate whether a separate DR strategy is required in addition to the HA design offered by AWS protection. If you do need a DR strategy, then the next step would be to decide on an in-region or a multi-region DR approach. This is typically dictated by data sovereignty and minimum distance requirements. Depending on your specific RTO/RPO, you can implement cold, pilot light, or hot DR architecture.

[+]

SAP on MS SQL Server on AWS HA/DR Tip!

Based on the disaster recovery requirements, additional AZ within the same region or another region can be used to deploy a third database DR node. For data replication, native database replication or storage-based technologies can be used.

Table 5.6 provides a comparison between cold and pilot light DR for achievable RTO/RPO.

DR Architecture	Strategy	RPO/RTO
Cold	SQL Server backup/restore	High/high
Cold	Amazon AMI/snapshots	High/low
Cold	Amazon AMI with frequent database volume snapshots (both data and log)	Low/low
Pilot light	Sync replication (within primary region)	Near-zero/low
Pilot light	Async replication (across regions)	A few minutes/low
Hot	Async replication (across regions)	A few minutes/a few minutes

Table 5.6 SQL Server for SAP on AWS DR Options with RPO and RTO

- **Third-party solutions**
  You can also use third-party tools like SIOS Data Protection Suite, NEC ExpressCluster, or Veritas InfoScale to provide HA for SQL Server. These solutions use WSFC and replicate data from primary to secondary with block level replication of the Amazon EBS volumes.

5.3.6   Oracle

Starting with SAP R/3, Oracle is one of the oldest RDBMS solutions that SAP has been supporting for its suites of products. AWS supports BYOL for running Oracle for SAP workloads and it's common to see enterprises licensing Oracle with an SAP material code (Application Specific Full Use [ASFU] license or Oracle runtime or OEM license), directly from SAP or an authorized SAP channel partner. If the Oracle license was purchased from any other party, the relevant licensing terms must be verified with the entity who sold the original license.

For Oracle architecture on AWS, let's look at compute, OS, storage and HA/DR, and how they are typically deployed on AWS.

Compute

For any hardware platform including AWS, unlike SAP HANA, Oracle does not require any additional certifications to run your SAP workloads. However, as described earlier in Section 5.1, you saw how imperative it was to select only the SAP-supported and certified Amazon EC2 instances for pure database server such as Oracle. Since AWS gets new Amazon EC2 instances certified regularly from SAP, we highly recommend you refer to SAP Note: 1656099 and AWS documentation online to get the latest list of SAP supported and certified Amazon EC2 instances that you can use for an Oracle database.

With respect to sizing, CPU type, and mapping of on-premise SQL servers to Amazon EC2 instances, you should consider the following:

- SAP EarlyWatch Alert reports generated from SAP Solution Manager for existing SAP systems.

- Automatic Workload Repository or AWS Oracle report that can collect, process, and maintain performance statistics for problem detection and self-tuning purposes.

- Using AWS native discovery tools such as AWS Migration Evaluator as well to get to the information related to vCPU, memory, and storage requirements and use it to map against a supported and certified Amazon EC2 instance.

- The Quick Sizer for greenfield SAP implementations. Additionally, account for the Quick Sizer buffer. Quick Sizer tools provide sizing guidance based on assumptions that for 100% load (as per your inputs to tool) system use will not exceed 65%. Therefore, there is a fair amount of buffer already built into Quick Sizer recommendation. See SAP's Quick Sizer guidance for details.

For Oracle for SAP on AWS an important SAP Note to refer is 2358420. Few important takeaways from the note are:

- Oracle Database for SAP applications is supported on Amazon EC2 only. Amazon RDS is not supported with the exception of SAP BusinessObjects BI Platform—check SAP Note 3015613 for details.

- Only single instance configurations of the Oracle Database are supported on Amazon EC2.

- Oracle Real Application Clusters (RAC) is *not* supported on Amazon EC2.

Operating System

Oracle for SAP applications can only be deployed on an Oracle Enterprise Linux (OEL) OS on AWS only on an x86 based CPU architecture. Within the realms of SAP's Product Availability Matrix, refer to the latest documentation from AWS as well as SAP Note 1656099 for OS versions which are supported for SQL Server for SAP on AWS. Oracle requires that the AMI used for Amazon EC2 instances is based on Oracle Linux 6.4 or later, 7.1 or later, and 8.2 or later for both Oracle Database Server and SAP NetWeaver Application Servers with Oracle Instant Client.

Though Oracle Linux has no licensing fee, in order to ensure that your SAP system is fully supported on Oracle Linux you must purchase an Oracle Linux Premier Support subscription from Oracle. To optimize on costs, you can consider Limited Premier (cheaper than Premier) for Amazon EC2 instances less than or equal to 8 vCPUs.

Storage

In this section, we'll cover how Amazon EBS, Amazon FSx for NetApp ONTAP, Amazon EFS, and Amazon S3 is relevant for SAP on Oracle on AWS:

- **Amazon EBS**
  On AWS, it's typical to use Amazon EBS for persistent storage for Oracle. Within the Amazon EBS service, you have the option to use gp2, gp3, io1, io2, and io2 block express, which all are SSDs and certified and supported for SAP workloads on AWS. Additionally, each Amazon EBS volume is automatically replicated within its AWS AZ to protect you from failure, offering HA and durability. Thus, you can configure a RAID 0 array at the OS level for maximum performance and not have to worry about additional protection (RAID 10 or RAID 5) for your volumes. The cost-optimized/high-throughput magnetic Amazon EBS volume type (st1) is an ideal candidate for storing SQL Server backups.

General purpose SSD gp3 volumes balance price and performance IOPS as the throughput can be increased from a high baseline value on running instance without any downtime. io2 block express volumes provide the highest performance consistently and resiliency for mission-critical applications, with the lowest (submillisecond) latency specifically for production Oracle databases. Table 5.7 shows a typical file layout of Oracle database for SAP on AWS.

Amazon EBS Volume	Number of Volumes	Volume Group	Logical Volume	File System	FS Type
gp3	One	vgSwap	Lv_swap	Swap	Swap
gp3	One	vgSAP	lv_usrsap lv_DAA lv_usrsapSID lv_sapmnt lv_ascs lv_hostctrl	/usr/sap /usr/sap/DAA /usr/sap/SID /sapmnt /usr/sap/SID/ASCSnn /usr/sap/hostctrl	ext4
gp3	One	vgOracle	lv_oracle lv_oraclient lv_oracleSID lv_sapreorg lv_oraarch	/oracle /oracle/client /oracle/<SID> /oracle/${DB_SID}/sapreorg /oracle/${DB_SID}/oraarch	ext4
gp3	Two volumes in RAID 0	vgMirrLog	lv_mirrlogA lv_mirrlogB	/oracle/SID/mirrlogA /oracle/SID/mirrlogB	ext4

Table 5.7 Oracle for SAP File System Layout on AWS

Amazon EBS Volume	Number of Volumes	Volume Group	Logical Volume	File System	FS Type
gp3	Two volumes in RAID 0	vgOrigLog	lv_origlogA lv_origlogB	/oracle/SID/ origlogA /oracle/SID/ origlogB	ext4
gp3	Two volumes in RAID 0	vgSAPdata	lv_sapdata1 lv_sapdata2 lv_sapdata3 lv_sapdatan	/oracle/${DB_ SID}/sapdata1 /oracle/${DB_ SID}/sapdata2 /oracle/${DB_ SID}/sapdata3 /oracle/${DB_ SID}/sapdatan	ext4
ST1	One	vgSIDbackup	Lv_backup	/backup	ext4

Table 5.7 Oracle for SAP File System Layout on AWS (Cont.)

Additionally, Oracle databases are supported on the following storage management solutions on Amazon EC2 instances:

- Oracle Automatic Storage Management (ASM): Figure 5.23 shows an Oracle database layout using Oracle ASM. There are two ASM disk groups—DATA and RECO, with five and three ASM disks (Amazon EBS volumes), respectively.

- Amazon FSx for NetApp ONTAP using Direct NFS with NFSv3, NFSv4, and NFSv4.1.

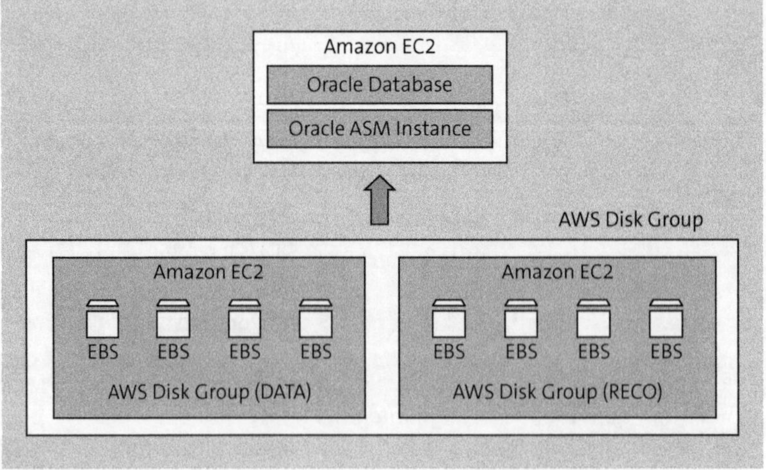

Figure 5.23 Oracle Disk Layout Using Oracle ASM for SAP on Oracle on AWS

– Amazon FSx for NetApp ONTAP is certified for Oracle databases on SAP NetWeaver and supported for SAP applications using Oracle databases in a single or multi-AZ deployment. You can use Amazon FSx for NetApp ONTAP as the primary storage solution for Oracle database data, log, and backup volumes with supported Amazon EC2 instances. Use separate storage virtual machines (SVMs) for Oracle data and log volumes to ensure that your I/O traffic flows through different IP addresses and TCP sessions. Each of the file systems related to sapdata, origog, mirrlog, and control files should have their own Amazon FSx for NetApp ONTAP volume.

- **Amazon EFS**

  Amazon EFS or any other NFS mounted file system must not be used for Oracle data files, log files, and control files due to serious throughput and latency constraints.

- **Amazon S3**

  As mentioned earlier, you have the option to use Amazon EBS as a backup file system for SQL Server. To save on backup costs, you can automate copying of the old backups to Amazon S3 using custom scripts or use the OSB cloud module, which might require additional licenses.

High Availability and Disaster Recovery

For an SAP system running on Oracle, the database is the most critical component and is a single point of failure. For Oracle for SAP, some solutions related to HA and the architectures associated with it include the following:

- **Oracle Data Guard**

  AWS Regions and AZs serve as the building blocks for designing highly available and resilient Oracle database systems on AWS. Combined with AWS and Oracle-native tools, you can achieve robust business continuity options that meet your organization's requirements. Oracle Data Guard is fully supported on AWS, and typically, enterprises use it to deploy business critical database such as Oracle on AWS in HA using multiple AZs within a region. It's important to understand Oracle Data Guard provides data replication from primary to the secondary. For failover and automated orchestration, you have the option to use Oracle Data Guard Fast Start Failover (FSFO).

  Oracle Data Guard maintains these standby databases as transactional consistent copies of the production database. Then, if the production database becomes unavailable because of a planned or an unplanned outage, Oracle Data Guard can switch the role of standby database to the primary role, minimizing the downtime associated with the outage. Fast-Start Failover enables an automated failover to standby database, in case the primary database goes down by incident or network loss. An observer process is used to monitor the network connectivity and availability of the databases. The observer is a separate OCI client-side component that runs on a different server from the primary and standby databases and monitors the availability of the primary database. We recommended you place the observer node in another AZ (not with the primary and standby Oracle node) in case of outages for both AZs. As shown in Figure 5.24, three nodes—a primary server, a secondary server, and a witness server—are spread across three AZs within an AWS region. Thus, you must consider an active direct connection with the same domain. The third node configured with an AWS auto-scaling group (minimum 1, maximum 1, desired capacity 1, across all AZ in region), will work as an observer node to maintain and centralize the creation, maintenance, and monitoring of Oracle Data Guard configurations for HA cluster.

Oracle database provides three protection modes:

- Maximum performance: The primary database is not affected by any delays in writing redo log data to the standby database.

- Maximum availability: A commit occurs when all the redo data needed to recover transactions has been written to the online redo log and to at least one synchronized standby database. If Oracle Data Guard is not able to write to the standby database, the behavior will be similar to the maximum performance protection mode.

- Maximum protection: Changes must be written to both, the online redo log and to the standby database for every transaction. If Oracle Data Guard is unable to write the redo stream to at least one standby, it will shut down the primary database.

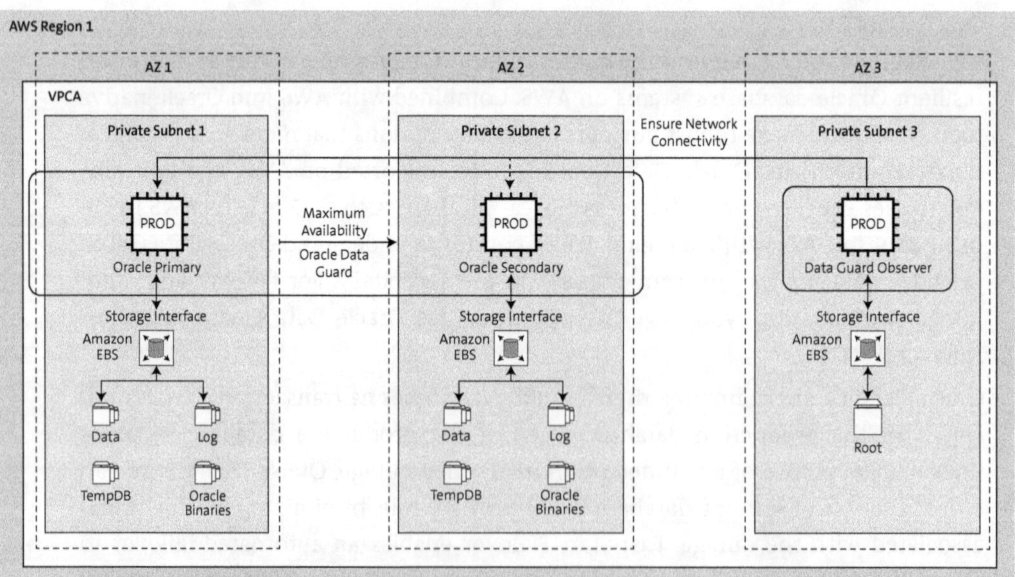

Figure 5.24 SAP on Oracle on AWS HA with Fast-Start Failover (FSFO)

As mentioned earlier, Amazon EFS is not supported for Oracle data and log; however, you can use it as a shared file system for the witness node server to monitor a database node's status.

SAP on Oracle on AWS HA for SAP Application Servers

SAP NetWeaver (7.0x to 7.5) products requiring Oracle Database 18c (min 18.5.0) or 19c (min 19.5.0) must run on Oracle Enterprise Linux (OEL) 6.4 or higher. This applies to both the Database and the Application Servers that require the Oracle Client as per SAP Note 105047. HA for Application servers can be done using horizontal scaling as mentioned in Application Architecture section. For ASCS/ERS, as it does not connect to the database and does not require Oracle Client, you can consider them to run on OSs like SUSE/RHEL and use the Pacemaker solution available within the SAP flavor of their OS. Or another option is to consider a third-party HA solution like Veritas, SIOS etc. across the board for both database and application servers including ASCS/ERS components.

Keeping in mind AWS region and AZ design you must evaluate whether a separate DR strategy is required in addition to the HA design offered by AWS protection. If you do need a DR strategy, then the next step would be to decide on an in-region or a multi-region DR approach. This is typically dictated by data residency and minimum distance requirements. Depending on your specific RTO/RPO, you can implement cold, pilot light, or hot DR architecture.

SAP on Oracle on AWS HA/DR

Based on the DR requirements, additional AZ within the same region or another region can be used to deploy a third database DR node. For data replication, native database replication or storage-based technologies can be used.

Table 5.8 provides a comparison between cold and pilot light DR for achievable RTO/RPO.

DR Architecture	Strategy	RPO/RTO
Cold	Oracle Server backup/restore	High/high
Cold	Amazon AMI/snapshots	High/low
Cold	Amazon AMI with frequent database volume snapshots (both data and log)	Low/low

Table 5.8 Oracle on AWS for SAP DR Options

DR Architecture	Strategy	RPO/RTO
Pilot light (maximum availability—recommended)	Sync replication (within primary region)	Near-zero/low
Pilot light (only maximum performance)	Async replication (across regions)	A few minutes/low
Hot (only maximum performance)	Async replication (across regions)	A few minutes/a few minutes

Table 5.8 Oracle on AWS for SAP DR Options (Cont.)

- **Third-party solutions**
  You can also use third-party tools like SIOS Data Protection Suite, or Veritas InfoScale to provide HA for Oracle database for SAP on AWS. These solutions typically will require additional licenses however do provide a single pane of glass HA capabilities for managing and operating HA across SAP application servers and Oracle database for the entire SAP estate.

5.3.7   SAP ASE Service

SAP acquired Sybase in 2010 including Sybase ASE database. A single standalone installation of SAP ASE service typically comprises one "dataserver" and one corresponding "backup server." In multi-server installation many dataservers can share one backup server. A dataserver consists of system databases and user databases.

For SAP ASE service architecture on AWS, let's look at compute, OS, storage and HA/DR, and how they are typically deployed on AWS.

Compute

For any hardware platform including AWS, unlike SAP HANA, Sybase ASE does not require any additional certifications to run your SAP workloads. However, as described earlier in Section 5.1, you saw how imperative it was to select only the SAP-supported and -certified Amazon EC2 instances for pure database server such as Sybase ASE. Since AWS gets new Amazon EC2 instances certified regularly from SAP, we highly recommend you refer to SAP Note: 1656099 and AWS documentation online to get the latest list of SAP supported and certified Amazon EC2 instances that you can use for SAP ASE service. Within the boundaries of release combinations listed in the Product Availability Matrix, SAP ASE service 15.7.0.051 or higher is supported for NetWeaver based SAP applications.

With respect to sizing, CPU type, and mapping of on-premise SQL servers to Amazon EC2 instances, you should consider the following:

- SAP EarlyWatch Alert reports generated from SAP Solution Manager for existing SAP systems.

- Using AWS native discovery tools such as AWS Migration Evaluator as well to get to the information related to vCPU, memory, and storage requirements and use it to map against a supported and certified Amazon EC2 instance.

- The Quick Sizer for greenfield SAP implementations. Additionally, account for the Quick Sizer buffer. Quick Sizer tools provide sizing guidance based on assumptions that for 100% load (as per your inputs to tool) system use will not exceed 65%. Therefore, there is a fair amount of buffer already built into Quick Sizer recommendation. See SAP's Quick Sizer guidance for details.

Operating System

SAP ASE service for SAP applications can be deployed on a Linux (SLES/RHEL) or Windows OS on AWS only on an x86 based CPU architecture. Within the realms of SAP's Product Availability Matrix, refer to the latest documentation from AWS as well as SAP Note 1656099 for OS versions which are supported for SAP ASE service for SAP on AWS.

Storage

In this section we'll cover Amazon EBS, Amazon FSx for NetApp ONTAP, Amazon EFS, and Amazon S3 for SAP on SAP ASE service on AWS:

- **Amazon EBS**
  On AWS it's typical to use Amazon EBS for persistent storage for Oracle. Within the Amazon EBS service you have the option to use gp2, gp3, io1, io2, and io2 block express, which all SSDs and certified and supported for SAP workloads on AWS. Additionally, each Amazon EBS volume is automatically replicated within its AWS AZ to protect you from failure, offering HA and durability. Thus, you can configure a RAID 0 array at the OS level for maximum performance and not have to worry about additional protection (RAID 10 or RAID 5) for your volumes. The cost-optimized/high-throughput magnetic Amazon EBS volume type (st1) is an ideal candidate for storing SQL Server backups.

  General purpose SSD gp3 volumes balance price and performance IOPS as the throughput can be increased from a high baseline value on *running instance without any downtime*. io2 block express volumes provide the highest performance consistently and resiliency for mission-critical applications, with the lowest (sub-millisecond) latency specifically for production Oracle databases. We highly recommend you have separate Amazon EBS volumes for /sybase/<SID>/sapdata_1, /sybase/<SID>/sapdata_n, /sybase/<SID>/saplog_1 /sybase/<SID>/saptmp, /sybase/<SID>/sapdiag, /sybasebackup, and /usr/sap.

- **Amazon FSx for NetApp ONTAP**
  Amazon FSx for NetApp ONTAP is certified for SAP ASE service for SAP NetWeaver.

Amazon FSx for NetApp ONTAP is supported for SAP applications using SAP ASE service database in a single or multi-AZ deployment. You can use Amazon FSx for NetApp ONTAP as the primary storage solution for SAP ASE service database data, log, and backup volumes with supported Amazon EC2 instances. Use separate SVMs for database data and log volumes. This ensures that your I/O traffic flows through different IP addresses and TCP sessions. Each of the file systems related to sapdata, saptmp, sapdiag, and /usr/sap files should have their own Amazon FSx for NetApp ONTAP volume.

- **Amazon EFS**
  Amazon EFS or any other NFS mounted file system must not be used for Oracle data files, log files, and control files due to serious throughput and latency constraints.

- **Amazon S3**
  As we saw earlier, you have the option to use Amazon EBS as a backup file system for SQL Server. To save on backup costs you can automate copying of the old backups to Amazon S3 using custom scripts or use third-party tools such Syntax's Emory solution which might require additional licenses.

High Availability and Disaster Recovery

For an SAP system running on SAP ASE service, the database is the most critical component and is a single point of failure. For SAP ASE service for SAP, following are some of the solutions related to HA and the architectures associated with it:

- **SAP ASE HADR**
  AWS Regions and AZs serve as the building blocks for designing highly available and resilient SAP ASE service database systems on AWS. Combined with AWS and SAP ASE service-native tools, you can achieve robust business continuity options that meet your organization's requirements. The SAP ASE service database has a built-in HA feature called "always-on" to replicate data between a primary and a companion database, which is actually quite similar to what many of you may know from SAP HANA. The always-on option, which provides the HADR solution, requires the ASE_ALWAYS_ON license.

[»] **SAP ASE Service, High Availability Option**

The cluster edition of SAP ASE service allows a single shared installation or multiple private installations to operate on multiple nodes as a shared-disk cluster environment with a single-system view. The cluster edition of SAP ASE service is not supported in conjunction with SAP NetWeaver and SAP Business Suite applications.

This HA feature within SAP ASE service is fully supported on AWS and it's typical to see enterprises using it to deploy business critical database such as SAP ASE service on AWS in HA using multiple AZs within a region. The always-on option is a HA and

DR (HADR) system that consists of two SAP ASE service servers: one designated as the primary on which all transaction processing takes place, the other acts as a warm standby (referred to as a "standby server" in DR mode, and as a "companion" in HA mode) for the primary server, and which contains copies of designated databases from the primary server. The HADR feature included with SAP ASE service supports only a single-companion server up to SAP ASE service 16.0 SP03. It's important to understand the built-in HA feature provides data replication from primary to the secondary. As shown in Figure 5.25, the HADR system includes an *embedded* SAP Replication Server, which synchronizes the databases between the primary and companion servers. SAP ASE service uses the Replication Management Agent (RMA) to communicate with Replication Server, and SAP Replication Server uses Open Client connectivity to communicate with the companion SAP ASE service. The Replication Agent detects any data changes made on the primary server and sends them to the SAP Replication Server. The HADR system supports synchronous replication between the primary and standby servers for HA so the two servers can keep in sync with zero data loss (ZDL). The HADR system supports asynchronous replication between the primary and standby servers for DR. For failover and automated orchestration, you have the option to use Fault Manager (FM). SAP ASE service replication uses the Fault Manager (FM) to monitor the replication between the primary and companion database and execute a take-over, if necessary (e.g., a primary database failure). Because the FM is monitoring the database replication, it must run somewhere outside of the database replication nodes to ensure the HA. Therefore, a third node is required.

But who monitors the FM itself? What happens if the FM fails? Well, if the FM isn't available then there is no monitoring and no replication take-over. So, what if we could use the existing cluster for ASCS and ERS and implement the FM into that solution too? This is exactly what AWS did in collaboration with SUSE, and SAP, which turned out to be a great idea since this not only fulfilled the requirement to install the FM outside of the SAP ASE service database hosts. it also meant the SLES for SAP HA cluster could monitor the FM, which makes the FM highly available. SLES for SAP OS can be configured to use Pacemaker solution with the sap-suse-cluster-connector, which provides the ability to add FM as a new service. This also leads to a cost optimized solution with no requirement to deploy an additional third node and could remove management and operational overhead of maintaining an additional AZ within the region just for the FM node. Further, FM now runs and is monitored in a similar way to ASCS and ERS processes in a highly available configuration.

Keeping in mind AWS region and AZ design you must evaluate whether a separate DR strategy is required in addition to the HA design offered by AWS protection. If you do need a DR strategy, then the next step would be to decide on an in-region or a multi-region DR approach, which is typically dictated by data residency and minimum distance requirements. Depending on your specific RTO/RPO, you can implement cold, pilot light, or hot DR architecture. HADR with SAP ASE service 16.0 SP03

PLO3 and later supports a three-node architecture, with two companion SAP ASE service servers and a DR node in the SAP Business Suite on SAP ASE service implementation. It supports a third external node DR node as the DR server added to an existing HADR cluster. The DR Node option is not certified for non-SAP Business Suite systems.

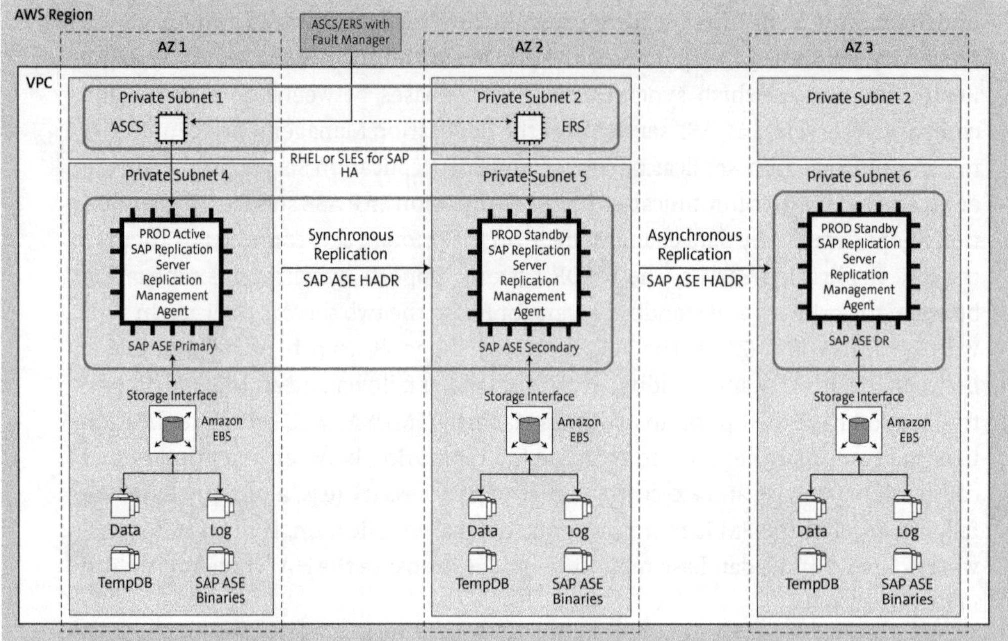

Figure 5.25 SAP ASE Service HADR Always On with HA and DR within an AWS Region Leveraging Multiple AZs

[+]   SAP on ASE on AWS HA/DR

Based on the DR requirements, an additional AZ within the same region or another region can be used to deploy a third database DR node. For data replication, native database replication or storage-based technologies can be used.

Table 5.9 provides a comparison between cold and pilot light DR for achievable RTO/RPO.

DR Architecture	Strategy	RPO/RTO
Cold	SAP ASE service server backup/restore	High/high
Cold	Amazon AMI/snapshots	High/low

Table 5.9 SAP ASE Service on AWS for SAP DR Options

DR Architecture	Strategy	RPO/RTO
Cold	Amazon AMI with frequent database volume snapshots (both data and log)	Low/low
Pilot Light (HADR always on)	Sync replication (within primary region)	Near-zero/low
Pilot Light (HADR always on)	Async replication (across regions)	A few minutes/low
Hot (HADR always on)	Async replication (across regions)	A few minutes/a few minutes

Table 5.9 SAP ASE Service on AWS for SAP DR Options (Cont.)

- **Third-party solutions**
  You can also use third-party tools like RHEL HA Add-On, SLES for SAP HAE, SIOS Data Protection Suite, Windows Server Failover Clustering (for Windows only), or Symantec Veritas to provide HA for Oracle database for SAP on AWS. These solutions typically will require additional licenses; however, they do provide a single pane of glass HA capabilities for managing and operating HA across SAP application servers and Oracle database for the mission critical SAP production systems.

5.3.8   SAP MaxDB

Today, about 20,000 SAP customer installations run on SAP MaxDB and SAP liveCache technology proving SAP MaxDB's reliability, scalability, and availability for configurations of all sizes. SAP MaxDB is also widely used within SAP, for example for the Service Marketplace, SAP Content Server, and also SCN runs on SAP MaxDB. SCN currently is one of the largest SAP NetWeaver Portal implementations with far more than 1,000,000 named users, and it runs on SAP MaxDB and Linux x64.

SAP liveCache technology is an object-based enhancement of the SAP MaxDB database system and was developed to manage complex objects (e.g., in logistic solutions like SAP SCM/APO). For solutions of this type, large volumes of data must be permanently available. Unlike SAP MaxDB, an SAP liveCache instance keeps all data that the system needs to access in memory. It provides data structures and methods for high-speed processing of networks, as for example available-to-promise (ATP) and Production-Planning-and-Control (PPS). SAP liveCache is patented SAP technology that can only be used with SAP applications. Additionally, SAP MaxDB is primarily used for SAP Content Server.

> **[»]** **SAP MaxDB for SAP Application Future and Roadmap**
>
> The core engine of the SAP HANA database is built from a combination of database technologies including SAP MaxDB and in-memory row store. Thus, SAP HANA is the natural born successor of today's SAP MaxDB. Several years from now, SAP MaxDB will seamlessly move to the SAP HANA database. SAP will not force this change but provide guidance only. It is up to you, the customer, to decide when to make the move. The support for SAP MaxDB will continue without any doubt first because of all SAP service and support agreements and second to design and develop a perfect and smooth transition path. So, there is no immediate need to change decisions regarding SAP MaxDB and no need to migrate to other databases supported by SAP. Version 7.9 may not be the last one for SAP MaxDB; there is just no roadmap beyond it yet. According to SAP, SAP MaxDB is still being worked on and will be continuously updated and improved.

For an SAP MaxDB architecture on AWS, let's look at the compute, OS, storage, and HA/DR and how they are typically deployed on AWS.

Compute

For any hardware platform including AWS, unlike SAP HANA, SAP MaxDB does not require any additional certifications to run your SAP workloads. However, in Section 5.1, you learned that it is imperative to select only the SAP-supported and -certified Amazon EC2 instances for pure database servers such as SAP MaxDB. Since AWS gets new Amazon EC2 instances certified regularly from SAP, we highly recommend you refer to SAP Note: 1656099 and the AWS documentation online for the latest list of SAP-supported and -certified Amazon EC2 instances that you can use for SAP MaxDB. Within the boundaries of release combinations listed in the Product Availability Matrix, SAP MaxDB 7.8 or higher is supported for SAP NetWeaver-based SAP applications.

With respect to sizing, CPU type, and mapping of on-premise SQL servers to Amazon EC2 instances, you should consider the following tools:

- SAP EarlyWatch Alert reports generated from SAP Solution Manager for existing SAP systems.

- Using AWS native discovery tools such as AWS Migration Evaluator as well to get to the information related to vCPU, memory, and storage requirements and use it to map against a supported and certified Amazon EC2 instance.

- The Quick Sizer for greenfield SAP implementations. Additionally, account for the Quick Sizer buffer. Quick Sizer tools provide sizing guidance based on assumptions that for 100% load (as per your inputs to tool) system use will not exceed 65%. Therefore, there is a fair amount of buffer already built into Quick Sizer recommendation. See SAP's Quick Sizer guidance for details.

Operating System

SAP MaxDB for SAP applications can be deployed on a Linux (SLES/RHEL) or Windows OS on AWS only on an x86 based CPU architecture. Within the realms of SAP's Product Availability Matrix, refer to the latest documentation from AWS as well as SAP Note 1656099 for OS versions which are supported for SAP MaxDB for SAP applications on AWS.

Storage

In this section we'll cover Amazon EBS, Amazon FSx for NetApp ONTAP, Amazon EFS, and Amazon S3 for SAP MaxDB on AWS:

- **Amazon EBS**
  On AWS it's typical to use Amazon EBS for persistent storage for SAP MaxDB. Within the Amazon EBS service you have the option to use gp2, gp3, io1, io2, and io2 block express, which all are SSDs and certified and supported for SAP workloads on AWS. Additionally, each Amazon EBS volume is automatically replicated within its AWS AZ to protect you from failure, offering HA and durability. Thus, you can configure a RAID 0 array at the OS level for maximum performance and not have to worry about additional protection (RAID 10 or RAID 5) for your volumes. The cost-optimized/ high-throughput magnetic Amazon EBS volume type (st1) is an ideal candidate for storing SAP MaxDB backups.

  General purpose SSD gp3 volumes balance price and performance IOPS as the throughput can be increased from a high baseline value on *running instance without any downtime*. io2 Block Express volumes provide the highest performance consistently and resiliency for mission-critical applications, with the lowest (sub-millisecond) latency specifically for production Oracle databases. we highly recommend you have separate Amazon EBS volumes for /sapdb/<SID>/sapdata, /sapdb/ <SID>/saplog, /sapdb/<SID>/backup, and /usr/sap.

- **Amazon FSx for NetApp ONTAP**
  This solution is certified and supported for SAP MaxDB for SAP NetWeaver. Amazon FSx for NetApp ONTAP is supported for SAP MaxDB version 7.9.10 or higher (including SAP liveCache) with SAP applications in a Single or Multi-AZ deployment. You can use Amazon FSx for NetApp ONTAP as the primary storage solution for SAP ASE service database data, log, and backup volumes with supported Amazon EC2 instances. Use separate SVMs for database data and log volumes. This ensures that your I/O traffic flows through different IP addresses and TCP sessions. Each of the file systems related to sapdata, saplog, sapbackup, and /usr/sap files should have their own Amazon FSx for NetApp ONTAP volume.

- **Amazon EFS**
  Amazon EFS or any other NFS mounted file system must not be used for SAP MaxDB data files, log files, and control files due to serious throughput and latency

constraints. The only exception is SAP Content Server for file-based and SAP MaxDB-based storage.

- **Amazon S3**
  As we saw earlier, you have the option to use Amazon EBS as a backup file system for SAP MaxDB. To save on backup costs you can automate copying of the old backups to Amazon S3 using custom scripts or use third-party tools such Syntax's Emory solution which might require additional licenses.

High Availability and Disaster Recovery

For an SAP system running on SAP MaxDB, the database is the most critical component and is a single point of failure. For SAP MaxDB for SAP applications, following are some of the solutions related to HA and the architectures associated with it:

- **SAP MaxDB HA and Disaster Recovery**
  AWS Regions and AZs serve as the building blocks for designing highly available and resilient SAP ASE service database systems on AWS. Combined with AWS and SAP MaxDB-native tools, you can achieve robust business continuity options that meet your organization's requirements. As shown in Figure 5.26, SAP MaxDB supports hot standby HA solution which is a native feature to replicate logs from primary to one or multiple standby database(s) deployed across two AZs. The standby database is in a continuous restart mode (STANDBY operational state). To apply the changes that are made in the primary database, the standby database performs redo operations of all changes in committed transactions using the redo log entries from the shared log area. The data and log areas of the primary and standby databases, on either Windows or Linux, are located in a storage system such as Amazon FSx for NetApp ONTAP. All databases can access the shared log area simultaneously. Every data volume of each database has its own physical storage area in the storage system. The primary and standby computers are part of a cluster. The cluster software is responsible for monitoring the primary database and initiating the takeover of the primary role by the standby database if the primary database fails. For SAP MaxDB on Windows, it's Windows Server Failover Clustering (WSFC) and for Linux it's (RHEL or SLES for SAP Pacemaker solution).

  Keeping in mind AWS region and AZ design, you must evaluate whether a separate DR strategy is required in addition to the HA design offered by AWS protection. If you do need a DR strategy, then the next step would be to decide on an in-region or a multi-region DR approach. This is typically dictated by data residency and minimum distance requirements. Depending on your specific RTO/RPO, you can implement cold, pilot light, or hot DR architecture.

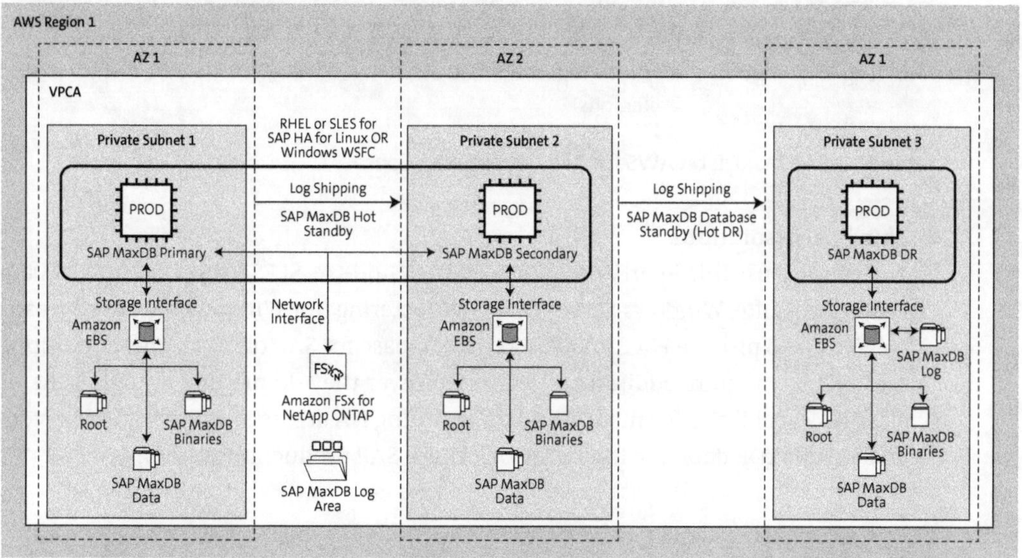

Figure 5.26 SAP MaxDB for SAP on AWS HA Hot Standby and a Database Standby (Warm) Spread across Multiple AZs in a Single AWS Region

SAP MaxDB on AWS HA/DR

Based on your DR requirements, an additional AZ within the same region or another region can be used to deploy an additional database DR node. For data replication, native database replication or storage-based technologies can be used.

Table 5.10 provides a comparison between cold and pilot light DR architectures and achievable RTO/RPO levels.

DR Architecture	Strategy	RPO/RTO
Cold	SAP MaxDB server backup/restore	High/high
Cold	Amazon AMI/snapshots	High/low
Cold	Amazon AMI with frequent database volume snapshots (both data and log)	Low/low
Pilot Light (Log Shipping)	Sync replication (within primary region)	Near-zero/low
Pilot Light (Log Shipping)	Async replication (across regions)	A few minutes/low

Table 5.10 SAP MaxDB on AWS for SAP DR Options

DR Architecture	Strategy	RPO/RTO
Hot (Log Shipping)	Async replication (across regions)	A few minutes/a few minutes

Table 5.10 SAP MaxDB on AWS for SAP DR Options (Cont.)

- **Third-party solutions**
  You can also use third-party tools like RHEL HA Add-On, SLES for SAP HAE, SIOS Data Protection Suite, Windows Server Failover Clustering (for Windows only), or Symantec Veritas to provide HA for SAP MaxDB database for SAP on AWS. These solutions typically will require additional licenses however they do provide a single pane of glass HA capabilities for managing and operating HA across SAP application servers and SAP MaxDB database for a mission critical SAP production system.

5.3.9   IBM Db2

Unlike other database vendors, IBM previously produced a platform-specific IBM Db2 product for each of its major OSs. However, in the 1990s, IBM changed track and produced an IBM Db2 common product, designed with a mostly common code base for L-U-W (Linux-Unix-Windows); DB2 for System z and DB2 for IBM i are different. On AWS, IBM Db2 is supported only on x86 CPU architecture. Solutions from AWS partners can emulate DB2 for IBM i or AIX using the base AWS infrastructure specifically around compute. However, running SAP workloads on these partner solutions is not supported.

For IBM Db2 architecture on AWS, let's look at compute, OS, storage and HA/DR and explore how they are typically deployed on AWS.

Compute

For any hardware platform including AWS, unlike SAP HANA, IBM Db2 does not require any additional certifications to run your SAP workloads. However, as discussed earlier in Section 1.1, we saw how it was imperative to select only SAP-supported and -certified Amazon EC2 instances for pure database servers, such as Sybase ASE. Since AWS gets new Amazon EC2 instances certified regularly from SAP, we highly recommend you refer to SAP Note 1656099 and the AWS documentation online for the latest list of SAP-supported and -certified Amazon EC2 instances that you can use for SAP ASE service. Within the boundaries of release combinations listed in the Product Availability Matrix, IBM Db2 LUW version 9.7 or higher is supported for SAP NetWeaver-based SAP applications.

With respect to sizing, CPU type, and mapping of on-premise IBM Db2 servers to Amazon EC2 instances, you should consider the following options:

- Generating SAP EarlyWatch Alert reports for existing SAP systems from SAP Solution Manager.

- Using AWS native discovery tools such as AWS Migration Evaluator as well to get information related to vCPU, memory, and storage requirements and then use to this data map against a supported and certified Amazon EC2 instance.

- Using Quick Sizer for greenfield SAP implementations. You should also account for the Quick Sizer buffer. Quick Sizer tools provide sizing guidance based on assumptions that, for 100% load (as per your inputs to the tool), system use will not exceed 65%. Therefore, a fair amount of buffer is already built into Quick Sizer recommendations. See SAP's Quick Sizer guidance for details.

Operating System

IBM Db2 for SAP applications can be deployed on a Linux (SLES/RHEL) or Windows OS on AWS only on an x86-based CPU architecture. Within the realms of SAP's Product Availability Matrix, refer to the latest documentation from AWS as well as SAP Note 1656099 for OS versions that are supported for IBM Db2 for SAP on AWS.

Storage

In this section, we'll cover Amazon EBS, Amazon FSx for NetApp ONTAP, Amazon EFS, and Amazon S3 for SAP on IBM DB2 on AWS:

- **Amazon EBS**
  On AWS, it's typical to use Amazon EBS for persistent storage for IBM Db2. Within the Amazon EBS service, you have the option of using gp2, gp3, io1, io2, and io2 block express, which all are SSDs and certified and supported for SAP workloads on AWS. Additionally, each Amazon EBS volume is automatically replicated within its AWS AZ to protect you from failure, thus offering HA and durability. As a result, you can configure a RAID 0 array at the OS level for maximum performance and not need to worry about additional protection (RAID 10 or RAID 5) for your volumes. The cost-optimized/high-throughput magnetic Amazon EBS volume type (st1) is an ideal candidate for storing SQL server backups.

  General purpose SSD gp3 volumes balance price and performance IOPS as the throughput can be increased from a high baseline value on *running instance without any downtime*. io2 block express volumes provide the highest performance and resiliency consistently for mission-critical applications, with the lowest (sub-millisecond) latency specifically for production IBM Db2 databases. We highly recommend you have separate Amazon EBS volumes for `/db2/<DBSID>/sapdata1`, `/db2/<DBSID>/sapdata<x>`, `/db2/<DBSID>/log_dir`, `/db2/<DBSID>/saptmp<x>`, `/db2/<DBSID>/log_arch`, and `/usr/sap`.

- **Amazon FSx for NetApp ONTAP**
  Amazon FSx for NetApp ONTAP is certified for SAP ASE service for SAP NetWeaver

and for SAP applications using an SAP ASE service database in a single or multi-AZ deployment. You can use Amazon FSx for NetApp ONTAP as the primary storage solution for IBM Db2 database data, log, and backup volumes with supported Amazon EC2 instances. Use separate SVMs for database data and log volumes to ensure that your I/O traffic flows through different IP addresses and TCP sessions. Each file system (e.g., related to `sapdata`, `saptmp`, `log`, `log_arch`, and `/usr/sap` files) should have their own Amazon FSx for NetApp ONTAP volume.

- **Amazon EFS**
  Amazon EFS or any other NFS mounted file system must not be used for IBM Db2 data files, log files, and control files due to serious throughput and latency constraints.

- **Amazon S3**
  As mentioned earlier, you have can use Amazon EBS as a backup file system for IBM Db2. To save on backup costs, you can automate the copying of the old backups to Amazon S3 using custom scripts or use third-party tools, such Syntax's Emory solution, which might require additional licenses.

High Availability and Disaster Recovery

For an SAP system running on IBM Db2, the database is the most critical component and is often the most important single point of failure. For IBM Db2, some solutions related to HA and the architectures associated with HA include the following:

- **IBM Db2 HADR**
  AWS regions and AZs serve as the building blocks for designing highly available and resilient SAP ASE service database systems on AWS. Combined with AWS- and IBM Db2-native tools, you can achieve robust business continuity options to meet your organization's requirements. The IBM Db2 database has a built-in HA feature called HADR to replicate data between a primary and standby database, which is actually quite similar to what you may know from SAP HANA. HADR supports up to three remote standby servers.

  HADR is fully supported on AWS, and typically, enterprises use it to deploy business-critical databases such as IBM Db2 on AWS in HA using multiple AZs within a region. As shown in Figure 5.27, the HADR feature consists of two IBM Db2 servers across two AZs in a single AWS region: one designated as the primary server on which all transaction processing takes place, while the other acts as a warm standby (referred to as a "standby server") that contains copies of the designated databases from the primary server. The third server, the DR node, can be deployed within the same region in another AZ or in another region based on your requirements. You must understand that the built-in HA feature provides data replication from the primary system to the secondary system. Standby databases are synchronized with the primary database through log data that is generated on the primary database and shipped to the standbys. The standbys constantly roll data forward through the logs. You can choose from four different synchronization modes. From most to least protection, these options are SYNC, NEARSYNC, ASYNC, and SUPERASYNC. For failover and automated orchestration, you have the option of using Pacemaker for Linux or Windows Server Failover Clustering (WSFC) for Windows.

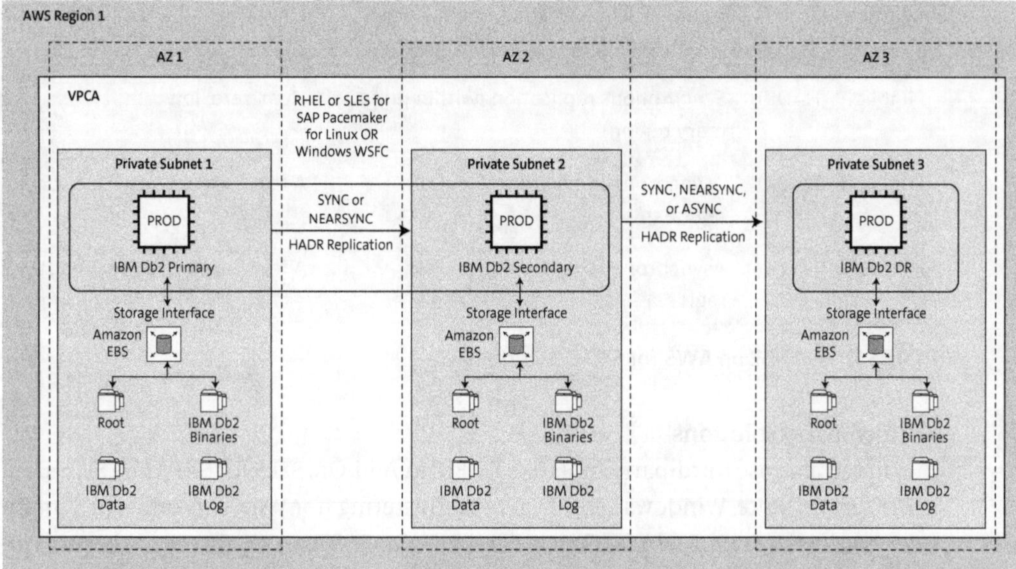

Figure 5.27 Multi-AZ, Single-Region HA/DR for IBM Db2 for SAP on AWS

Keeping in mind AWS region and AZ design, you must evaluate whether a separate DR strategy is required in addition to the HA design offered by AWS protection. If you do need a DR strategy, then the next step is deciding between an in-region or a multi-region DR approach. This choice is typically dictated by data residency and minimum distance requirements. Depending on your specific RTO/RPO requirements, you can implement cold, pilot light, or hot DR architectures. HADR with IBM Db2 10.5 and later supports a three-node architecture, with two standby IBM Db2 servers. It also supports a third external DR node as the DR server added to an existing HADR cluster.

SAP on IBM Db2 on AWS HA/DR

Based on your DR requirements, an additional AZ within the same region or another region can be used to deploy an additional database DR node. For replication, native database replication or storage-based technologies can be used.

Table 5.11 provides a comparison between cold and pilot light DR and achievable RTO/RPO.

DR Architecture	Strategy	RPO/RTO
Cold	IBM Db2 server backup/restore	High/high
Cold	Amazon AMI/snapshots	High/low
Cold	Amazon AMI with frequent database volume snapshots (both data and log)	Low/low
Pilot light (HADR)	Synchronous replication (within primary region)	Near-zero/low
Pilot light (HADR)	Asynchronous replication (across regions)	A few minutes/low
Hot (HADR)	Asynchronous replication (across regions)	A few minutes/a few minutes

Table 5.11 IBM Db2 on AWS for SAP DR Options

- **Third-party solutions**

  You can also use third-party tools like RHEL HA Add-On, SLES for SAP HAE, SIOS Data Protection Suite, Windows Server Failover Clustering (for Windows only), or Symantec Veritas to provide HA for Oracle databases for SAP on AWS. These solutions typically will require additional licenses but often provide a single pane of glass capabilities for managing and operating HA across SAP application servers and Oracle databases for your mission-critical SAP production systems.

5.4 Trade-Offs

In the words of American economist and author Thomas Sowell "There are no solutions; there are only trade-offs." This statement is applicable for the architectural patterns of any software solution including SAP. Simply put, a trade-off means that more of one thing necessitates less of another. Trade-offs related to the architecture for SAP on AWS can be categorized into three areas—cost, availability, and performance, which we'll cover in the following sections.

5.4.1 Cost Based

An optimal design typically does not translate to a low-cost solution. When designing an architecture, we always make trade-offs between costs and features such as reliability, performance efficiency, security, operational excellence:

- **Cost versus reliability**

  When we ask business users or end users about their reliability requirements, the typical answer is zero downtime, 5 9s or "100% availability." As architects, we do acknowledge the intention to get to the lowest downtime—planned or unplanned. And to get to lowest downtime, you might end up designing a full-blown, multi-AZ, multi-region HA and DR solution for SAP on AWS. The question that you need to address is: Is that really required? What does an outage cost to my business, brand, and end-users? Does the cost of an outage exceed the cost to implement and operate an HA and a DR solution annually? Overall reliability SLAs such as MTTR, RTO, and RPO should always be considered while designing the solution. Aggressive or shorter SLAs lead to higher costs. SAP applications can be segregated into categories based on criticality (e.g., high, medium, and low). For example, your core SAP ERP solution, such as SAP S/4HANA, requires a multi-AZ, multi-region HA and DR solution, but SAP Solution Manager does not. Additionally, if the cost of an HA or DR solution exceeds the cost of an outage, then a backup/restore might be a better fit. The graph shown in Figure 5.28 sums it up.

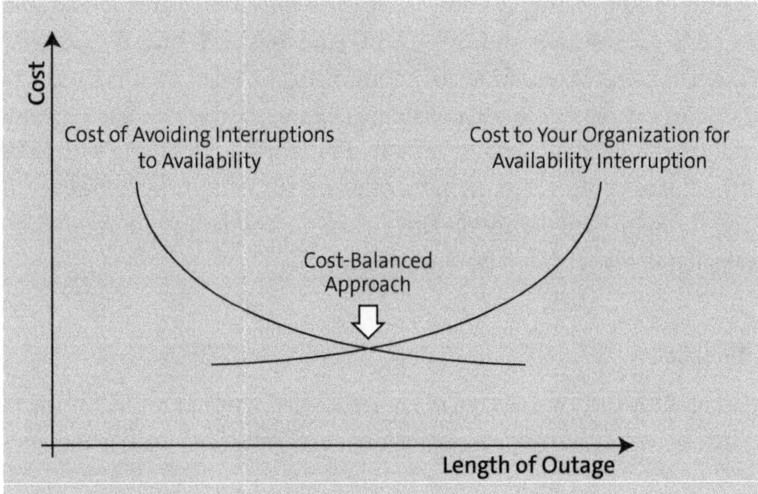

Figure 5.28 Availability Is Not Free

- **Cost versus performance efficiency**
  Increasing the performance of your SAP system beyond what is required would often increase costs. If you design your workloads based on assumed peak capacity, you'll end up with provisioning more infrastructure, thus increasing the costs. Size your systems according to your current performance requirements. Increase system resources such as vCPU or memory as and when required. This model is called "grow as you know" and is highly recommend for SAP HANA databases. In some scenarios, we're just covering what the "sins of our developers," for instance, a non-optimized ABAP report behind a batch job that requires a dedicated application server to run in the same AZ as the primary database. This kind of use of dedicated resources will increase costs.

- **Cost versus security**
  Whether on-premise or in the cloud, you should never compromise on security against costs. You can look into native AWS tools/services that could be more cost effective. For example, instead of investing and managing in next-generation firewalls you could consider a managed firewall like AWS Network Firewall to inspect all the egress and ingress traffic for your SAP workloads.

- **Cost versus operational excellence**
  If and when you start introducing automation related to management and operations for your SAP applications, costs could increase. Generally, however, over time, costs are reduced when considering the benefits (e.g., less human intervention, less downtime etc.). Another area that could be responsible for increased costs is monitoring. Investing in a more sophisticated monitoring tool could increase your costs, but real-time monitoring and anomaly detection could predict events and avoid unplanned downtimes. Application Insights is a feature within Amazon CloudWatch that can monitor SAP applications and SAP HANA databases and could be your single pane of glass cost-effective solution for monitoring. Another area is licensing specifically for OS. For example, you can consider SLES standard versus SLES for SAP for SAP applications and databases wherever you do not need HA. Going with SLES for SAP across the board could get expensive for a large SAP footprint. However, if you go with SLES standard, you'll only get three years of standard support, no SAP-specific features such as sapconf, and no live patching.

5.4.2   Availability Based

When designing an architecture, we always make trade-offs between availability and desired features such as cost optimization, performance efficiency, security, and operational excellence:

- **Availability versus cost optimization**
  When designing an SAP application to be highly available, you must understand your availability requirements by aligning with all the key stakeholders such as your internal IT teams, the C-level suite, and end users (your customers). Some key questions are: How much downtime is acceptable? How much does potential downtime cost your business? How much should you invest in making the application highly available? Is the cost of downtime greater than investing in a highly available system? You might be considering an SAP HANA cost-optimized approach for HA/DR for production systems. However, have you considered Amazon EC2 capacity availability, specifically for a large database, after a failover? To overcome this fear of the unknown on the cloud, you could consider running a QA or non-prod system in an MCOS scenario to ensure capacity but with an increased recovery time or RTO. Based on your budget, the cost of the downtime and reliability SLAs should align with the business before you start necessitating costs over availability.

- **Availability versus performance efficiency**

  AWS typically recommends a multi-AZ approach for HA. However, you could consider a single-AZ approach for HA configuration, which translates to primary and secondary databases and SAP application servers in a single AZ. In this scenario, you're increasing performance by avoiding cross-AZ latency, but you're compromising on availability. What if both the Amazon EC2 and Amazon EBS services within an AZ have failures? What if an AZ goes down? The probability of a complete AZ going down is low; however, individual AWS services like Amazon EC2 and Amazon EBS, which are leveraged to run SAP on AWS, can fail and lead to unplanned downtimes. Along similar lines, if you're considering a single region for DR, you'll get the same/similar performance with respect to end-user connectivity versus a multi-region DR approach, which could introduce increased latency. However, what if a complete region fails? The probability of that is very low and has never happened in AWS's history, but you should always have a contingency plan by, at a minimum, ensuring data backups in another site, another AZ, or another AWS region.

- **Availability versus security**

  Whether on-premise or in the cloud, it is highly recommended not to compromise on security against availability. Security complexity is less directly correlated with resilience. However, there are generally more components to secure for highly available systems. For example, manual certifications or key rotation policies could lead to availability issues. For end-user connectivity for SAP systems configured in HA on Linux, typically, an overlay IP is used, which gets updated to an active node during a failover. For automated failovers, cluster solutions such as Pacemaker from SLES or RHEL, use one of the Amazon EC2 instances in the cluster to initiate an API call to update the route table. However, what if an Amazon EC2 instance gets hacked, and the hacker can update the route table and point the overlay IP to an undesired destination? This vulnerability can be mitigated using the *least privileged access principle* for the Amazon EC2 instance and also introducing an HA/DR hook which updates the route table only based on the route table entry. To summarize, availability, or time it takes to recover a system after a failure, could get compromised and should be aligned with your internal security and compliance protocols.

- **Availability versus operational excellence**

  An HA configuration using a Pacemaker or a third-party proprietary solution is a complex implementation and requires additional operational effort. A misconfigured cluster implementation could lead to higher unplanned downtimes than a system with no cluster configuration. Therefore, depending on the internal skill set or your partner's competency, you should first align on your HA requirements with the business and then based on that, can consider more straightforward availability solutions such as Amazon EC2 auto-recovery, AMI-based recovery, or backup/restore. For example, let's assume an enterprise has thousands of SAP users. They have considered a single-AZ approach with multiple application servers to serve the user load. Little to no operational and management overhead is incurred in maintaining logon user groups and/or in batch job distribution using server groups as compared to application servers split across the AZs. However, what if an AZ or a service within an AZ goes down? You're compromising on availability over easier operations and easier management of your SAP systems.

5.4.3   Performance Based

When designing an architecture, we always make trade-offs between performance efficiency and desired features such as cost optimization, reliability, security, and operational excellence:

- **Performance efficiency versus cost optimization**
  Costs can increase as a result of increasing performance. When you optimize for performance, consider the following factors to manage costs:

  - Avoid estimating SAP workload costs at consistently high utilization. Smooth out the peaks to get a consistent flow of compute and data. Use manual scaling or autoscaling to find the right balance. Scaling up is usually more expensive than scaling out.

  - Costs increase directly with the number of regions. Locating resources in cheaper regions does not negate the cost of network ingress and egress or the cost of degraded application performance because of increased latency.

  - Using dedicated resources to batch process long-running SAP jobs increases costs. You can lower costs by manual scaling or autoscaling for Amazon EC2 instances for SAP application servers.

- **Performance efficiency versus reliability**
  Typically, maintaining reliability lets you support performance efficiency; however, scenarios exist where performance could be compromised, such as the following:

  - A highly available SAP application servers are scaled out and are spread out across multiple AZs; however, when an application server in another AZ communicates with the primary database in another AZ, latency may rise specifically for your critical batch jobs. You can use batch server groups to overcome this latency.

  - For DR, if you're considering another region, you're introducing latency for your end users in the case of a real DR event. You should test these scenarios out by conducting regular DR drills to set expectations with your business users and/or end users.

  - Message queueing can increase reliability but can also reduce performance of the overall end-to-end SAP business process. It's typical to use SAP Process Orchestration to process inbound messages received from a non-SAP system via SAP Web Dispatcher. But what if the network packet gets lost between that non-SAP and the SAP Web Dispatcher layer? You could use services like Amazon Simple Queue

Service (SQS), Kafka, or Confluent, but in doing so, you're increasing the latency between the layers, which could degrade the overall performance.

- **Performance efficiency versus security**

  If access to the data is slow or is so poor that it cannot be accessed, this problem could be considered an unplanned downtime in the following ways:

  - Encryption at rest or in transit can cause performance issues as it introduces latency to encrypt and decrypt the data packets. For SAP databases, volume encryption with Amazon EBS has little to no performance impact; hence, we highly recommend you configure both—Amazon EBS encryption and DB native encryption such as SAP HANA encryption, SQL Server Transparent Data Encryption (TDE), etc.

  - Adding additional network security appliances or next-generation firewalls could cause latency and degrade performance specifically for real-time integration between SAP systems in a VPC against a non-SAP or an SAP BTP service outside of your VPC. Instead of a centralized networking inspection, you can consider a decentralized approach of inspecting network packets with a dedicated AWS Network firewall or a security appliance in the same VPC as your SAP systems.

  - Make sure you integrate critical security alerts and logs into security information and event management (SIEM) systems without introducing high volumes of low-value data. Low-value data can increase SIEM costs, generate false positives, and decrease performance.

  - To optimize performance and maximize availability specifically integrating SAP applications with AWS services such as Amazon AppFlow, application code should first try to get OAuth access tokens from a cache before attempting to acquire a token from the identity provider. OAuth is a technological standard that allows you to securely share information between services without exposing your passwords.

- **Performance efficiency versus operational excellence**

  We highly recommended you invest time and resources in continuously improving the performance of your SAP systems running on AWS, including the following:

  - Plan regular reviews of your SAP workloads with your AWS team, partners, or internal experts at least once a year using the AWS Well-Architected Framework with the SAP Lens as covered in the Chapter 3.

  - Invest in automation related to the deployment, management, and operation of your SAP workloads on AWS.

  - Continuously review Amazon EC2 instance sizing and performance by validating against historical monitoring data. Consider newly released and SAP-certified Amazon EC2 instance types for performance fit and cost optimization. Plan to take advantage of new improvements in your operational backlog.

- Review Amazon EBS sizing and performance by validating against historical monitoring data.

- Monitor SAP on AWS blogs and announcements by subscribing to the SAP on AWS blog feed and AWS's "What's New" feed to stay up to date with newly released service announcements, innovation approaches, and price reductions.

- Plan periodic enhancement work to take advantage of new and improved AWS services by ensuring that your operational budget allows for planned support team effort for the implementation and testing of new AWS services and workload evolution on a periodic basis.

5.5   Summary

In this chapter, we highlighted the importance of leveraging SAP-certified and -supported AWS services and features to guarantee full support from SAP as well as AWS. Within the boundaries of SAP's Product Availability Matrix, we highly recommend referring to the latest SAP Notes and the documentation on AWS for SAP applications and the related OS/DB combinations that are supported on AWS. Once you have identified the AWS services that you can leverage for your specific SAP product, the next step is to design the architecture, which is typically driven by reliability SLAs such as MTTR, RPO, and RTO *keeping in mind the levels of resiliency that is available within a single AWS region*. Depending on the SAP application, its environment, and business uptime requirements, architectural patterns can be either standard, distributed, or highly available.

The chapter also covered reference architectures for SAP application and database tiers. We saw how database HA architectures are unique specifically in terms of data replication and their failover mechanisms to secondary or standby nodes. We also covered some common AWS native features such as Amazon EC2 auto-recovery, AMIs, and Amazon EBS snapshots that can be used as part of your business continuity strategy.

Chapter 6
Migrations

Data center consolidation, digital transformation, and reduction in costs are the most common drivers for migrating SAP applications to Amazon Web Services (AWS). As the most experienced vendor in the public cloud space, AWS and its partners have helped migrate thousands of SAP customers and millions of non-SAP customers to AWS. Based on the lessons learned from these migrations, AWS has developed a proven migration framework and methodology that can be used to migrate mission-critical systems including SAP. By combining SAP native tools and AWS services, you can accelerate your migration to AWS while minimizing risks, costs, and business disruptions.

According to the *Cambridge Dictionary*, a migration is the "process of people travelling to a new place to live, usually in large numbers." The reason to move to a new place could be a new job, better living standards, higher salary, a better place for your kids to grow up, etc. Thus, motives and drivers to consider moving or migrating to a new place always exist. Migrating your SAP systems to AWS is somewhat similar to this analogy. Today, you might be running your SAP landscape in an on-premise data center, a colocation, or a private cloud, maybe even for the past ten years or more, so why move or migrate your SAP applications to AWS? Based on our experience, the answer lies in business drivers—the innovation and modernization of business processes, the adoption of cloud-first/cloud-smart strategy, reductions in costs, the consolidation of systems, and improvements sustainability. Technical drivers might also be in play, such as the 2027 end of support for SAP ERP, reductions in technical debt, improvements in security and reliability, the elimination of data center footprints, and increases in staff productivity and retention. Any combination of technical and business drivers could drive your SAP migration towards AWS.

Additionally, today's SAP customers face some important decisions for the future of their SAP solutions because SAP has introduced SAP S/4HANA and SAP BW/4HANA as the next generation of solutions for existing SAP Business Suite or SAP ERP and SAP Business Warehouse (SAP BW) solutions. SAP has committed to supporting SAP Business Suite until 2027, but many SAP customers are defining their path to adopting SAP HANA now to take advantage of new SAP innovations by considering greenfield or "smart" (selective data transition) greenfield deployments on AWS. Based on our experience, some enterprises start with a disruption-free greenfield SAP S/4HANA Central Finance implementation on AWS to get the journey started with SAP's latest software as well as AWS.

In this chapter, we'll look at AWS's migration framework, application discovery strategy, SAP migration options, and AWS accelerators that can help reduce risks and shorten timelines for your SAP migrations to AWS.

6.1   Migration Frameworks

Three key elements that make up the building blocks of an SAP migration to AWS include the following:

- **Scope**
  What are you migrating?

- **Timeline**
  When do you need to complete the migration?

- **Strategy**
  Why do you want to migrate?

These elements must be aligned and understood from the start of a migration project. Any changes to one of these elements will affect the others. Realignment must be factored into every change, no matter how basic or small a change might seem. Around these building blocks, AWS has developed a framework of migration strategies called the 6 Rs migration strategy and a framework for cloud adoption called the AWS Cloud Adoption Framework (CAF). These frameworks are applicable to any workload, including SAP, that is being considered for migration to AWS. We'll cover these frameworks further in the following sections.

6.1.1   6 Rs Strategy

Built upon 5 Rs that Gartner published in 2011, AWS came up with the 6 Rs migration strategy. A strategy is an approach, and in this case, an approach to a migration of SAP workloads to AWS. Table 6.1 indicates the six most common migration strategies.

"R" Migration Strategy	Methodology	SAP Example
Rehosting	The application is migrated as-is to AWS, also called "lift-and-shift" approach.	SAP ERP EHP 6 on SAP HANA from on-premise to SAP ERP EHP 6 on SAP HANA on AWS

Table 6.1 6 Rs Migration Strategy with SAP Examples

"R" Migration Strategy	Methodology	SAP Example
Replatforming	The application is changed or upgraded or transformed as part of its migration.	SAP ERP on any database from on-premise to SAP ERP on SAP HANA on AWS
Repurchasing	You'll move to a different solution or an application on the cloud.	From SAP Customer Relationship Management (SAP CRM) to Salesforce
Refactoring/ Rearchitecting	The application is redesigned as part of the migration to AWS.	SAP Commerce standard virtual machine (VM) deployment on-premise to Kubernetes on AWS
Retiring	The application will be decommissioned or retired during migration to AWS.	A legacy R/3 or SAP ERP 5.0 system
Retaining	The application is not considered for migration and will stay where it is for now.	A tertiary non-SAP system (e.g., a manufacturing execution system that runs only on IBM Power-based systems)

Table 6.1 6 Rs Migration Strategy with SAP Examples (Cont.)

The two strategies that are specifically applicable for SAP migrations to AWS are rehosting and replatforming. Rehosting is applicable when you want to move your SAP system as is to AWS. This type of migration involves minimal change and might be a natural fit for customers who are already running some sort of SAP system. Replatforming is applicable when you want to migrate from an any database source database (such as IBM Db2, Oracle Database, or SQL server) to an SAP HANA database.

The decision tree diagram shown in Figure 6.1 helps you visualize the end-to-end process, starting from application discovery, and moving through each stage of the 6 Rs strategy.

7th "R" of Migration Strategy

A few years ago, AWS introduced the 7th R—"relocate." However, many enterprises and partners still rely on the 6 key strategies with respect to SAP. Relocate is also known as "hypervisor lift and shift." This approach is similar to rehosting where enterprises can migrate a collection of applications that are hosted in VMWare Cloud or Kubernetes to a cloud version of the same platform. As SAP on VMWare or Kubernetes on AWS is not a certified and supported combination for SAP production workloads, the 7th R is typically disregarded!

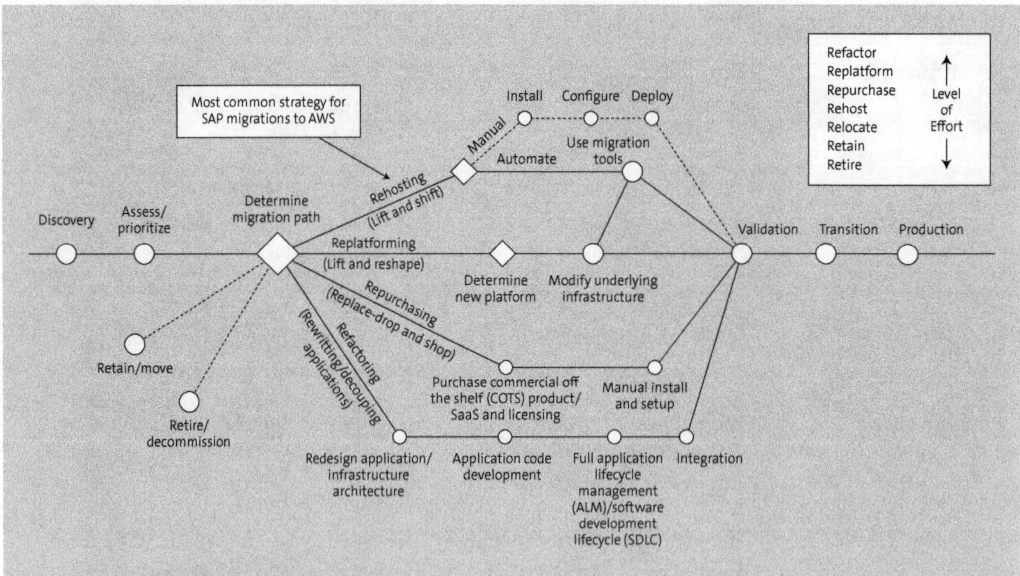

Figure 6.1 The 6 Rs Migration Strategy Decision Tree

6.1.2 Cloud Adoption Framework: People, Processes, Technologies

Building a business that is sustainable for a long time is not easy. Reinvention will be required many times over. Only 50% of businesses who were on the Fortune 500 list in the year 2000 are still there today. Effective leaders rely on data to make informed decisions, look around the corner, and take meaningful action. The introduction of the cloud lowered the cost of storage and compute at scale, and as a result, it set off a generation of reinvention. Now, as new ways to process and analyze large volumes of data become simpler and more cost effective, we see the next wave of reinvention being driven by data.

Similarly, people's evolving expectations and behaviors are putting pressure on organizations to improve digital service delivery. Organizations across the globe are digitally transforming; they are leveraging digital technologies to drive organizational change that allows them to adapt to changing market conditions, delight their customers, and accelerate their business outcomes. For SAP customers, this could be moving to SAP S/4HANA on AWS or leveraging AWS and SAP Business Technology Platform (SAP BTP) services to extend, enhance, and optimize their business processes or a combination of both. Millions of AWS customers, including the fastest-growing startups, largest enterprises, and leading government organizations, are leveraging AWS to migrate and modernize legacy workloads, become data-driven, digitize and optimize business processes, and reinvent operating and business models. Through cloud-powered digital transformation (cloud transformation), they can improve their business outcomes, including lower costs, reduce business risks, improve operational efficiency, become more agile, innovate faster, create new revenue streams, and improve customer and employee experience.

Your ability to effectively leverage the cloud to digitally transform (your cloud readiness) is underpinned by a set of foundational organizational capabilities. The AWS Cloud Application Framework (CAF) identifies these capabilities and has provided prescriptive guidance for thousands of organizations around the world, who have successfully accelerated their cloud transformation journeys. Each organization is unique, and the form in which the cloud is adopted is based on their internal drivers and governance models. AWS CAF provides an adoption model and guidelines that organizations can leverage according to their unique requirements. Think of AWS CAF as an algorithm with variables: You input your values to the variables to come up with a result that highlights gaps or issues that you need to address for a successful migration to AWS. Additionally, AWS and the AWS Partner Network provide tools and services that can help you along each step of the way. AWS Professional Services is a global team of experts with an "SAP on AWS"-specific skillset involving AWS CAF aligned offerings to help you achieve specific outcomes related to your cloud transformation.

As shown in Figure 6.2, AWS CAF comprises of transformation domains that represent a value chain where technological transformation enables process transformation, which enables the organizational transformation that enables product transformation.

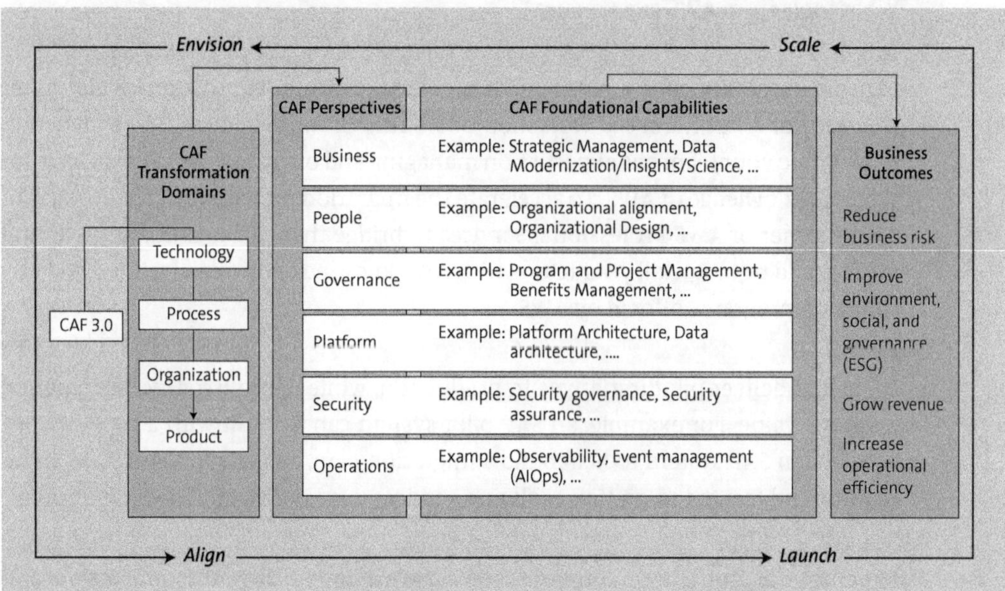

Figure 6.2 AWS CAF 3.0: Transformation Domains, Perspectives, and Capabilities for Driving Business Outcomes Related to Digital Transformation with AWS

Key business outcomes such as reduced business risk, improved environment, social, and governance (ESG) performance, increased revenue, and enhanced operational efficiency are accelerated through cloud-powered transformation domains that are enabled by a set of foundational capabilities. A capability is an organizational ability to leverage processes to deploy resources (people, technology, and any other tangible or intangible assets) to achieve a particular outcome. AWS CAF capabilities provide best practice guidance to improve your cloud readiness (your ability to effectively leverage cloud to digitally transform).

AWS CAF groups its capabilities into six perspectives: Business, People, Governance, Platform, Security, and Operations. Each perspective comprises a set of capabilities that functionally related stakeholders own or manage in your cloud transformation journey.

So how does AWS CAF work? As shown in Figure 6.2, with transformation domains, perspectives, and capabilities as your foundation, a migration is an iterative flywheel rotating around four phases:

1. **Envision**
   Demonstrate how the cloud will accelerate business outcomes. This phase is delivered as a facilitator-led, interactive workshop running 5 hours or more that can be conducted between you and AWS Professional Services, or your partner will help you identify transformation opportunities and create a foundation for your digital transformation. For example, let's say we formulate an end goal of setting up a data lake on AWS using SAP and non-SAP data by a certain quarter of a particular year. What tools from AWS can you use?

2. **Align**
   Identify capability gaps across the foundational capabilities. This phase also takes the form of a facilitator-led workshop of 5 hours or more and results in an action plan. While your team has expertise on managing and operating SAP, it may have little to no knowledge of AWS. To accelerate the migration project, you can leverage an AWS partner or AWS Professional Services to bridge that gap. Eventually, your end goal should be to upskill your teams with AWS knowledge through training and certification programs offered by AWS.

3. **Launch**
   Build and deliver pilot initiatives in production, while demonstrating incremental business value. For example, an SAP pilot system can be built with a copy of your production SAP system to help build a migration plan, envision an end-state architecture, and test the migration tools and services used during the SAP migration.

4. **Scale**
   Expand pilot initiatives to the desired scale while realizing the anticipated and desired business benefits. For example, after a successful pilot, you can start migrating other SAP environments (e.g., Dev, QA, Pre-prod, and Production) using the same/similar methodology that was used for the pilot.

An AWS program underpinned by AWS CAF is AWS Migration Readiness Assessment (MRA). AWS MRA is conducted via workshops between you and AWS (or AWS Professional Services or a partner) to help you gain insights into your organization's cloud journey and its cloud readiness and eventually to build an action plan to identify and narrow the gaps. The end goal is to assess and evaluate your cloud readiness based around the six CAF perspectives spread across 47 different capabilities using the AWS Assessment tool built by AWS Professional Services team. You can create a report showing your cloud adoption and migration readiness scores in each AWS CAF perspective. For example, Figure 6.3 shows your readiness based on your inputs across all six perspectives. Scores in green indicate you meet all the levels of readiness found with successful migrations. Yellow indicates additional prep-work is recommended.

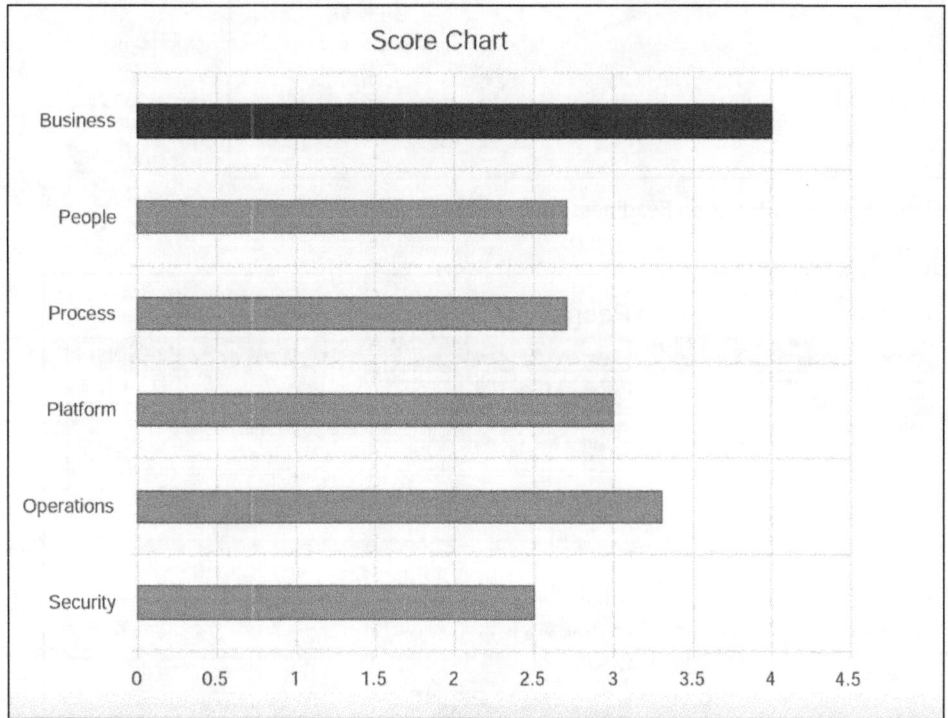

Figure 6.3 AWS Migration Readiness Assessment Report: Score Chart

Figure 6.4 shows the radar chart, which allows you to prioritize your remediation activities.

Figure 6.5 shows a heatmap across the six perspectives. Scores in green indicate you meet all the levels of readiness found with successful migrations. Yellow indicates additional prep-work is recommended.

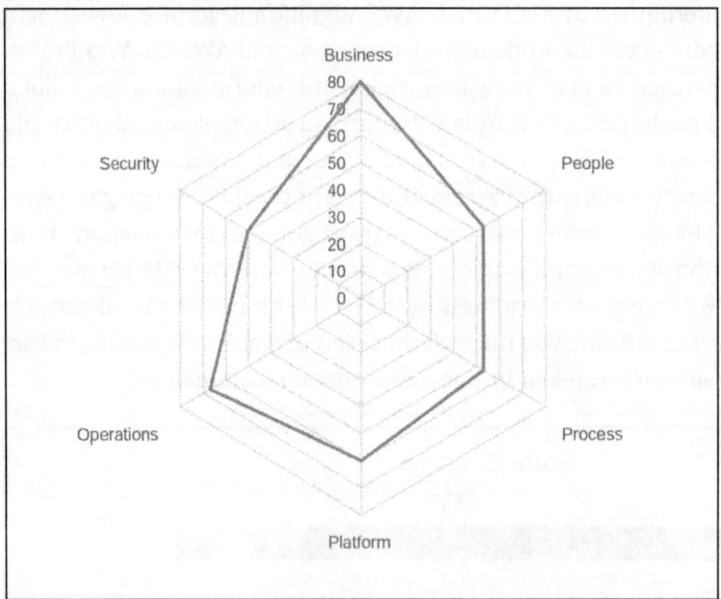

Figure 6.4 AWS Migration Readiness Assessment Report: Radar Chart

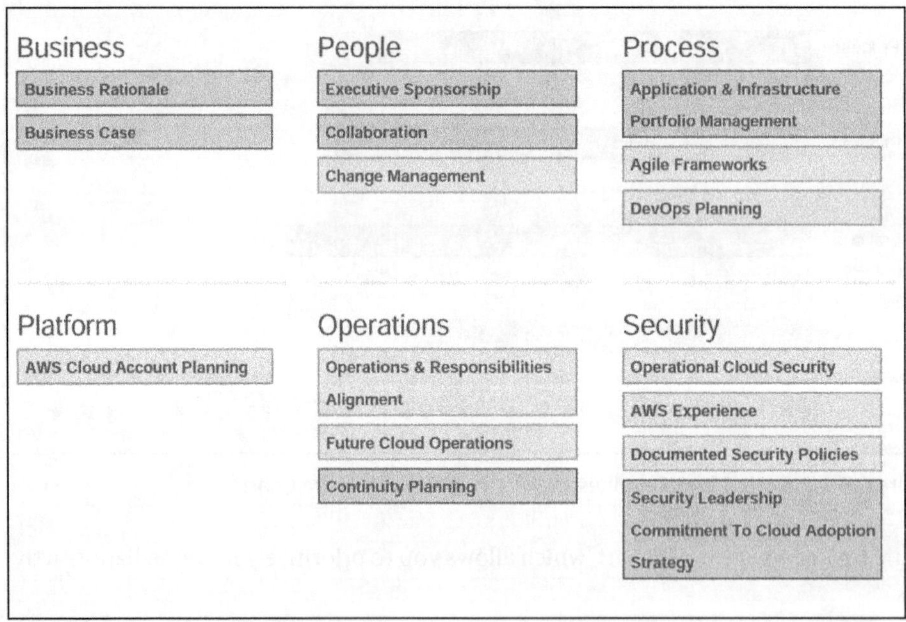

Figure 6.5 AWS Migration Readiness Assessment Report: Heat Map

Figure 6.6 shows a summary and action plan or additional prep-work required for each of the perspective, in this case, the business perspective.

Both AWS CAF and the 6 Rs framework help you understand and plan the broader structure of an AWS migration and what it means to you and your company.

MRA Observations & Actions – Business Case

Readiness Activities	Phase	Summary Observations	Actions & Next Steps
High Level Business Case	Pre-Mobilize	CTO and his office is all in, however the cloud strategy needs to be socialized to the COO, CPO & President. There is a meeting to present the high level business case.	❏ AWS to provide EBC type deck on why cloud/AWS. Need to help provide the business case that can go to the COO, Chief Product Officer & President - Benefits, Credentials, FAQ's (lock-in, security, etc), Anecdotal story (something that goes further than what happened to the power industry) to help business people understand the context. e.g. Cattle not pets like example. ❏ Document the strategic vision of why cloud and why now so others can buy-in and help deliver the vision ❏ State a specific goal around migration to drive alignment and momentum ❏ Select an automated discovery tool to aid in the building of a business case
Key Stakeholder Sign-off	Pre-Mobilize	There is a top level funding commitment for the Research initiatives but the cloud investment still needs to be flushed out.	❏ Customer to define cloud funding estimate in the high level business case ❏ Create a communications plan to include stakeholder buy-in and updates ❏ Build a strategic vision and purpose for the cloud activities to align under
Migration Funding Commitment	Pre-Mobilize	There is also a data center consolidation effort with assumption of lift and shift of workloads to the cloud.	❏ Identify workloads that have a large and near term pay-back to start saving ❏ Do discovery to enable an estimate of migration effort
Specific Migration Workloads Committed	Pre-Mobilize	The core Mobilize work is very mass migration focused. Customer would like to tailor Mobilize for cloud native, net new work. The rest of the framework makes sense for the all-in cloud strategy but need to prioritize product A.	❏ AWS to tailor Mobilize and breakout product A's Mobilize
Detailed Business Case	Mobilize	Need to create a detailed business case	

aws

Figure 6.6 AWS Migration Readiness Assessment Report: Summary and Action Plan

6.2   AWS Migration Methodology

The overarching methodology that is built around the frameworks we just covered uses the knowledge and experience gathered by the AWS Professional Services team from migrating thousands of enterprises to AWS. This migration methodology is called AWS Migration Acceleration Program (MAP). AWS MAP is based on an Agile migration methodology with three phases to help you reduce costs and risk during migrations:

1. **Assess**

   At the start of your journey, you should assess your organization's current readiness for operating in the cloud. In this phase, you leverage free-of-charge workshops to learn about AWS CAF and AWS MRA, as described in the previous section of this chapter. Most importantly, you'll want to identify the desired business outcomes and develop the business case for a migration of SAP and non-SAP workloads or a data center exit strategy.

2. **Mobilize**

   As part of the mobilize phase, you create a migration plan and refine your business case. You address gaps in your organization's readiness that were uncovered in the assess phase, with a focus on building your baseline environment (the "landing zone"), driving operational readiness, and developing cloud skills.

3. **Migrate and modernize**

   During the migrate and modernize phase, each SAP and/or non-SAP application is designed, migrated, and validated.

You can also consider a fourth phase—run or optimize. However, generally, the framework is confined to three phases, and each phase has a set of clear objectives and business or technical outcomes associated with it.

SAP migrations can be complex and thus require coordination and implementation across multiple disciplines and teams. This framework provides prescriptive guidance on accelerating your cloud adoption, building a migration plan, setting up your AWS landing zone, and migrating the first set of your SAP applications to the AWS cloud in weeks, if not months.

Table 6.2 summarizes the AWS migration framework for large-scale SAP migrations representing the three phases and the key activities, associated AWS and third-party services/tools, and the outcomes associated with it. Each phase builds upon the previous one.

	Assess	Mobilize	Migrate
Key Activities	**Discovery Workshops** ■ AWS CAF: Envision and alignment discovery workshops to share information on priorities, SAP landscape, and roadmap. ■ AWS MRA: Assess your readiness in adopting AWS for your SAP workloads ■ Application discovery, dependency, prioritization, and conceptualizing architecture ■ Assess AWS landing zone/AWS Control Tower ■ Learn about SAP on AWS through briefings, online trainings, workshops, and Immersion Days ■ Explore migration and modernization strategy ■ Total cost of ownership (TCO) report	**Validation and Planning (Trial Account, Sandbox System)** ■ Build landing zone/AWS Control Tower ■ Build SAP architecture on AWS ■ Execute pilot migration ■ Upgrade software and database, including Unicode ■ Migrate operating system (OS) ■ Test SAP solution ■ Test backup/restore ■ Test high availability (HA) and disaster recover (DR) solutions ■ Scale storage ■ Evaluate monitoring ■ Identify automation	**Execute Migration (DEV>QA>PRD)** ■ Build SAP target architecture on AWS ■ Migrate SAP applications to AWS architecture ■ Enable monitoring ■ Scale storage ■ Enable HA and DR and validate architecture (in QA and PRD only) ■ Conduct go-live assessment (QA only) ■ Post-go-live hypercare (PRD only)

Table 6.2 AWS MAP for SAP on AWS

	Assess	Mobilize	Migrate
Services/ Tools	■ AWS Migration Evaluator for TCO and business case ■ AWS Migration Hub Orchestrator for visibility into status of migrations across SAP applications ■ AWS prescriptive guidance: Online documentation on migrating SAP workloads to AWS	■ AWS Application Discovery Service (ADS) ■ Migration partner solutions ■ AWS management and governance ■ AWS landing zone ■ AWS Control Tower	■ AWS Application Migration Service (MGN) ■ AWS Marketplace ■ AWS DataSync ■ AWS Transfer Family ■ AWS Snow Family ■ AWS Service Catalog ■ Migration partner and SAP solutions
Outcomes	**Scope and Plan** ■ Business value proposition ■ Architecture concept ■ Migration approach	**Pilot** ■ Architectural design ■ Migration and upgrade plan, including testing ■ Backup/restore processes and procedures ■ HA and DR validation	**SAP on AWS Realization** ■ Migrated SAP landscape on AWS ■ Increased agility/ flexibility with reduced TCO ■ Enable practitioners on AWS ■ Accelerated, repeatable functions through automation ■ Increased performance visibility through monitoring

Table 6.2 AWS MAP for SAP on AWS (Cont.)

AWS MAP for SAP Consideration

The AWS MAP for SAP also includes financial investment programs by AWS to cover your infrastructure costs during the course of an SAP migration to AWS. This program keeps evolving, and we highly recommend engaging with your AWS Account Team to reap the maximum financial benefits if you're considering an SAP migration to AWS.

6.3 Current State Discovery

One common denominator with AWS CAF and the AWS Migration Methodology's assess phase is application portfolio discovery. To derive a TCO assessment related to an AWS infrastructure, you'll need to asses business cases, migrations costs, the to-be architecture on AWS, and scope (number of systems) for any SAP or even non-SAP application that you plan to migrate, all of which requires portfolio discovery and analysis. Once you have your SAP system inventory all nailed down, you can start applying the 6 Rs migration framework. For example, an SAP ERP on Oracle database could be migrated to AWS with a database migration to SAP HANA. This migration would be categorized as "replatforming" because the underlying database is being changed, which has the domino effect of the application being changed too (code changes/remediation). In the next two sections, we'll cover how you can perform application discovery, calculate appropriate sizing, identify dependencies, and work through decision criteria on prioritizing SAP applications to be migrated to AWS.

6.3.1 Application Discovery and Sizing

In this section, we'll cover SAP application discovery and sizing related to SAP workloads on AWS. This information will help you leverage AWS's flexibility to ensure right-sizing, which can be done during or after the migration and can help optimize infrastructure costs for your SAP workloads running on AWS.

Application Discovery

Before any SAP migration, some of the most important questions that you need to address include the following:

- How many SAP applications are being migrated or are in scope?
- Are the planned software and OS versions compatible with AWS?
- What are the key performance indicators (KPIs) of the SAP applications to be migrated? Peak utilization of compute and memory?
- How large is the database that is being migrated?
- What other systems need to integrate with this system, and how are they accessed?
- Will you use the same OS, SAP software, and database software on the source and target?
- Is there an OS platform change?
- Is there an SAP software version change?
- Is there a database platform change?

So, how do we find the answers to these questions? What if SAP is part of a larger data center migration initiative? Who has the metadata with respect to infrastructure (compute, storage, networking) for SAP applications? The first step is to identify your sources and the related metadata of your SAP applications to be migrated. Key metadata may include the application type (e.g., SAP, web app, software as a service [SaaS], mainframe, database), vCPU, memory, storage, application architecture (such as standard, distributed, or HA), disaster recovery (DR) options (such as backup and restore, pilot light, warm standby, multi-site active/active), OS licensing costs, estimated annual costs for infrastructure, interdependencies, and so on.

Some common sources of all this required information include the following:

- **SAP application owner interviews and questionnaires**
  Start by identifying the key SAP stakeholders within your organization that can fulfill the data requirements. These individuals are typically members of the service management, operations, capacity planning, monitoring, and SAP support teams as well as the SAP application owners. Establish working sessions with members of these groups. Communicate data requirements and obtain a list of tools and existing documentation that can provide the data. To summarize, the source data is all based on conducting user and application owner interviews and filling out questionnaires manually. In this way, you can collect what is called institutional or tribal knowledge that can also be documented in a knowledge base. The data fidelity or level of trust with this approach is typically low, with 25% to 40% data accuracy, 65% to 75% assumed data, or data is older than 120 to 150 days old.

- **Configuration management databases**
  Leverage your existing CMDB, and depending on the size of the environment this could be a simple excel sheet or an RDBMS. SAP Solution Manager provides the Landscape Management (LMDB) capability, which is a customized version of CMDB for SAP. Furthermore, LMDB can leverage SAP System Landscape Directory as a data source, and both can provide you with an overview of the SAP systems in your IT landscape. However, you might not get application-to-application dependency information or network dependency information with these SAP tools. Additionally, an SAP application migration could be part of the larger scheme, such as a data center exit migration (including non-SAP workloads) to AWS. In this scenario, SAP tools will not meet your requirement, and you might need to pivot to an enterprise-wide CMDB. The data fidelity or level of trust with using a CMDB approach is typically medium with 50% data accuracy, 50% assumed data, or data is older than 90 to 120 days old.

- **VMWare vCenter exports**
  If you're running majority of your SAP and/or non-SAP applications on VMWare, a vCenter export help you create the system inventory. In this case, the data fidelity is typically medium-high with 75% to 80% data accuracy, 25% to 20% assumed data, or data is older than 60 to 90 days.

- **Application performance monitoring**
  Using a monitoring tool such as SAP Solution Manager, or an enterprise-wide third-party monitoring tool, you can access details about an SAP and/or non-SAP system

inventory including application performance metrics like CPU, memory, and storage, which can help you right-size the environment on AWS as well. The data fidelity is typically high but could be limited to critical production applications. There could be ~5% assumed data or data that is 0 to 60 days old.

All these data source collection methods are traditional approaches that are manual, time consuming and error prone and thus fall into the category of low-to-medium confidence data. Additionally, you still might get incomplete details on the system inventory with application-to-application dependencies or network dependencies.

Some automated options from AWS and the AWS partner ecosystems are available to collect a high-confidence, well rounded data on application sources and the related metadata around peak utilization for compute and storage and application or network dependencies. For SAP applications, we highly recommend running these tools for the timeframe during which your system utilization is likely to be high, for example, during year-end closing or a major sales event.

- **AWS Application Discovery Service**
  This AWS native service helps automate the discovery of servers and their dependencies. This service collects server specification information and performance data, including details about running processes and network connections. This tool can be found in AWS Management Console or AWS Command Line Interface (CLI), which collects server specification information, performance data, and details about running processes and network connections. This discovery can be an agent based, or you can opt for an agentless discovery mode and manual CSV ingestion. Analysis and visualization of data is performed with AWS native analytics services such as Amazon QuickSight and Amazon Athena. No costs are associated with using AWS ADS. The recommended runtime for the collection of data is 4 weeks or during a timeframe when you have peak utilization, for example, during a period-end financial closing.

- **AWS Migration Evaluator (formerly TSO Logic)**
  By engaging with the AWS account team or AWS solutions architect or partner, this service can help you discover what you have on-premise, how it's used, and how much it costs to operate and then determines what the costs would be, with right-sizing, on AWS. AWS Migration Evaluator accounts for physical and virtual servers as well as for storage. It also analyzes Microsoft-specific workloads and makes suggestions for optimized Windows licensing strategies. AWS Migration Evaluator should be your default assessment tool. By understanding your on-premise compute and storage usage and costs, you can identify over-provisioning and where costs could be lowered. This option is highly recommended for workloads based on utilization and costs. Analysis can be conducted using an agentless connector or through manual ingestion in a formatted CSV file. This solution does not identify application dependencies. However, you can input data collected from AWS Migration Evaluator into AWS Migration Hub Orchestrator to visualize, tag, and group servers manually

to build an application dependency mapping. No costs are associated with using this service. The recommended runtime for collection of data is 4 weeks or during a timeframe when you have peak utilization, for example, during a period-end financial closing.

- **AWS Partner Solutions**

  AWS partner network solutions, popular among them Cloudamize and Cloudchomp, can help you automate a collection of system inventories and help identify application-to-application and network dependencies. They can support each phase of the cloud journey, from migration planning to ongoing cost/performance optimization, to maximizing the cloud return on investment (ROI).

> **Application Discovery Tools: Agentless versus Agent-Based**
>
> Agentless based discovery gathers server information regardless of OS. This approach usually takes less time for an initial assessment, and no information is collected about software and dependencies. Agent-based discovery installs an agent on one or more hosts in a data center and typically collects a richer set of data related to system processes and network dependencies.

Let's consider an example. We have two enterprises: Enterprise A is planning to migrate only SAP and non-SAP workloads (supporting the SAP ecosystem), and Enterprise B is planning to migrate all their workloads including SAP, non-SAP, web apps, other commercial off-the-shelf (COTS) applications, non-SAP databases, etc. to AWS.

For Enterprise A with 100 or more SAP servers including databases, the SAP Basis team, and/or their incumbent MSP will typically have an SAP system inventory based off of SAP Solution Manager or a CMDB, which could be a Microsoft Excel sheet. Thus, typically, you won't leverage AWS or third-party application discovery tools. The system inventory usually includes data about hardware type, CPU, memory, storage, networking, and architecture including DR and backup requirement, which should be good starting point to come up with a migration plan. However, if Enterprise A has an SAP ecosystem comprising of hundreds or thousands of SAP servers, then we'd highly recommend you use an automated agent-based application discovery tools like AWS ADS or a third-party tool to realize any unidentified network or application dependencies which can further help you do wave or a phased migration planning.

For Enterprise B, being a data center exit-driven migration, we expect to have hundreds or thousands of SAP and non-SAP servers and databases. Therefore, the recommended approach is to use application discovery tools such as AWS ADS or a third-party tool. Additionally, a well-rounded system inventory can help you identify applications that can be refactored to run in a cloud-native fashion to gain maximum benefits with respect to costs with optimal performance.

Another important aspect to keep in mind is that AWS ADS and most third-party discovery tools do not have the ability to identify whether a physical server or a virtual machine (VM) is running an SAP application or an SAP database. You may ask why is that important? Well, because discovery tools cannot ensure that the recommended compute, storage, or networking AWS service that they recommend is supported by SAP on AWS. And as we mentioned in Chapter 5 Section 5.1, you must ensure that you're using only the certified and supported AWS services for SAP workloads. Application discovery tools do provide capabilities around right-sizing the environment by recommending Amazon Elastic Compute Cloud (EC2) instance types based on the source server's peak utilization, but 'll you need to manually ensure that the Amazon EC2 instance type selected is certified and supported for SAP workloads. For example, for an SAP ERP application server, your discovery tool may recommend an m6g.4xlarge Amazon EC2 instance type. At the time of writing, the m6g Amazon EC2 instance family is not supported and certified for SAP workloads, and thus, manual effort is required to ensure that you select the right Amazon EC2 instance type for SAP applications and databases. If not done correctly, discovery could not only lead to an unsupported (by SAP) configuration on AWS, but it also can lead to unexpected TCO deviation for AWS infrastructure costs specifically for a large-scale SAP migration. One way to tackle and minimize manual effort for this limitation is to ensure proper tagging is done based on the business data at the source to easily identify SAP applications and databases. Some manual effort is still required to ensure the right instance type is being used. Another way is to group the applications based on the system processes that are identified on the source servers by your discovery tool. AWS ADS supports capturing of running processes, and based on *SAP* processes, you can group the servers that belong to your SAP applications. Some manual effort is still required to ensure the right instance type is being used.

SAP Application Discovery: Support

For SAP migrations to AWS-related activities such as sizing, TCO assessment, migration strategy, and high-level architecture design, we highly recommend engaging your local SAP on AWS Solutions Architect or AWS partner with an SAP on AWS competency. They can help ensure that you're following all the recommendations based on best practices put forward by SAP as well as by AWS.

SAP Application Discovery: TCO Assessment Support and AWS MPA

For initial TCO assessment, if you're engaged with an AWS partner with an SAP on AWS competency, they do have access to tools like AWS Migration Portfolio Assessment (MPA). AWS MPA is used for detailed portfolio assessment (server right-sizing, pricing, TCO comparisons, migration cost analysis) as well as migration planning (application data analysis and data collection, application grouping, migration prioritization, and wave planning). CMDB and application portfolio data in varied formats can be imported into AWS MPA with a web-based data ingestion process.

Sizing

You may have an oversized on-premise SAP environment (e.g., to support future growth), so measuring current peak utilization data is a better approach to gain AWS's flexibility than measuring the allocated resources to your SAP servers. By using application discovery tools, you can identify your SAP system inventory with actual peak utilization related to CPU, memory, and storage to right-size your SAP servers. Right-sizing is the process of matching supported and certified Amazon EC2 instance types and sizes to your SAP workload performance and capacity requirements at the lowest possible cost. It's also the process of looking at deployed servers and identifying opportunities to eliminate or downsize without compromising performance. Right-sizing is a key mechanism for optimizing AWS costs, but this task is often ignored by enterprises when they first move to the AWS cloud. They may lift and shift their SAP environments and expect to right-size later. Speed and performance are often prioritized over cost, which results in oversized instances and a lot of wasted spend on unused resources.

AWS's flexibility with instance upsizing and downsizing whenever you need can lower your AWS infrastructure costs, and thus we highly recommend you follow the "know as you grow" model for SAP workloads on AWS. For compute, select the Amazon EC2 instance types that are certified and supported for SAP based on the current peak utilization for CPU and memory analyzed by the discovery tools and then ensure they are a close match to SAP EarlyWatch Alert reports for a timeframe during which your system utilization is likely to be high (e.g., during year-end processing or a major sales event). Similarly, for storage, look at your monitoring reports to ensure that you're sizing your Amazon EBS volumes and shared file systems, for instance, Amazon Elastic File System (EFS) and Amazon FSx, with the necessary throughput and IOPS that are required for optimal performance and in line with the data discovered by the discovery tools. To summarize, whether you're taking a manual approach or an automated approach to right-size, always refer back to SAP native reports and capabilities when right-sizing your SAP environment on AWS. You should always size your SAP systems on AWS based on your *actual* peak utilization and not an *assumed* peak utilization.

For example, using AWS ADS, you can identified a server in the report, which you manually tagged as an SAP application server for your SAP S/4HANA application. The current server configuration for that application server is 8 vCPU and 64 GB RAM. Based on the analysis done by AWS ADS, the peak utilization of this application is 25% of total vCPU and 40% of total RAM. So, instead of selecting r6i.2xlarge (8 vCPU, 64 GB RAM), you can consider r6i.xlarge (4vCPU, 32 GB RAM), which meets your performance KPI requirements and is half the cost of an r6i.2xlarge. In addition, as a best practice, refer to your SAP EarlyWatch Alert reports to validate your sizing. In a large-scale SAP migration, right-sizing can significantly lower your TCO related to AWS infrastructure costs.

In addition to data discovery tools, for SAP migrations and new implementations, other data sources are available. Table 6.3 provides an overview of data sources for sizing SAP workloads on AWS.

"R" Migration Strategy	SAP Application Servers	SAP HANA	Any Database
Rehosting (lift and shift)	■ SAP EarlyWatch Alert reports ■ Source SAP Application Performance Standard (SAPS) rating	■ SAP EarlyWatch Alert reports ■ SQL statements ■ SAP HANA Studio	■ SAP EarlyWatch Alert reports ■ Source SAPS rating ■ Database-specific reports, for instance, Oracle Automatic Workload Repository (AWR) reports
Replatforming (database change)	■ SAP EarlyWatch Alert reports ■ Source SAPS rating	■ Any database to SAP HANA ■ SAP standard ABAP reports ■ SQL statements, ■ SAPS requirement ■ Rule of thumb (memory = 2 × database size)	■ Any database to any database ■ SAP EarlyWatch Alert reports ■ Source SAPS rating
Replatforming (OS changes)	Same as rehosting	Same as rehosting	Same as rehosting
Refactor (greenfield)	SAP Quick Sizer	SAP Quick Sizer	SAP Quick Sizer

Table 6.3 Additional Data Sources for Sizing SAP Workloads on AWS

[»]

SAP Quick Sizer Considerations

You should account for the SAP Quick Sizer buffer. SAP Quick Sizer tools provide sizing guidance based on assumptions that, for 100% load (as per your inputs to the tool), system use will not exceed 65%. Therefore, a fair amount of buffer is already built into SAP Quick Sizer recommendation.

Using a combination of application discovery approaches and right-sizing, and tying it up with SAP- and database-specific performance reports, at this point, you should create a fact-based, comprehensive application inventory with a detailed information about integrations, batch jobs, and satellite systems that are part of the SAP system landscape, including the versions of your OSs, databases, and SAP and non-SAP applications. This inventory should ideally contain all compatibility, support, licensing, and compliance requirements. Additionally, any technical requirements, such as the current size and projected size of databases, network capacity, required compute performance, and HA/DR should be included for each application.

For SAP applications, analyze in the context of the seven common migration strategies (the 7 Rs), for moving applications to the AWS Cloud (i.e., retain, retire, rehost, replatform, repurchase, refactor, and relocate). Add the most applicable "R" to the inventory for each non-cloud-native application. An accurate and complete application inventory will help you identify opportunities for determining application dependencies and prioritizing SAP applications to be migrated, which we'll cover in the next section.

6.3.2   Application Dependencies and Priorities

A common myth about application priority is that the highest priority applications should be migrated first. When you're performing wave planning, there is a high possibility that only a few of the highest priority applications will be in the first set of applications being migrated because the others are not ready. This mismatch could arise for a variety of valid reasons, such as application dependencies, business constraints, or resource availability. Application priority is a critical factor in wave planning, but it shouldn't be the only factor you consider. This section will cover application dependency, prioritization, and wave planning for SAP workloads being migrated to AWS.

Application Dependency

In Chapter 5, Section 5.1, under supported SAP products on AWS, we saw that SAP application servers and its database form a tightly coupled SAP system and thus should always be migrated together, not only for technical reasons (to avoid latency) but for ensuring SAP support as well. On the other hand, two SAP systems, such as SAP ERP or SAP S/4HANA and an SAP BW system connected via remote function call (RFC) can be considered loosely coupled systems. To summarize, in the SAP system inventory, it's imperative to identify the components belonging to the same SAP system along with its application-to-application dependency. If only small to medium number of SAP systems are being migrated (i.e., 100 to 300), then you could consider taking a "big-bang" approach but with a criterion of prioritizing less business-critical SAP environments first. In other words, you would migrate sandbox SAP systems first, followed by DEV,

QA, Pre-Prod, and eventually Production. The "big bang" in this context refers to all the production systems being migrated to AWS in a single wave so all the SAP systems and their application dependencies move together. If you have a large and highly complex SAP application landscape, we highly recommend you split the migration into distinct waves and coordinate with separate migration teams. For instance, perhaps your Americas SAP systems migrate first, followed by your EMEA SAP systems, or an SAP BW application is migrated first in year 1 of an enterprise's cloud journey, followed by SAP S/4HANA in year 2. The goal is to maintain the acceleration, momentum, and consistency of the migration effort while concurrently keeping each wave a manageable size from a resource and complexity perspective.

As shown in Table 6.4, technical dependencies for SAP applications can be categorized into four categories. These four categories can help you identify dependencies to build a high-confidence system inventory with a tangible migration wave planning.

Dependency Category	Description	Example
Application-to-infrastructure	Hard link between software and physical or virtual hardware	Non-SAP satellite or a tertiary warehouse management system that runs only on IBM Power systems
Application-to-component	Interaction between components running in different infrastructure assets	SAP application server running on one VM and an SAP database on a physical hardware
Application-to-application	Interaction between SAP applications or application components with other SAP or non-SAP applications or their components	An on-premise SAP database and AWS data lake integration, SAP ERP integration with a third-party tax compliance solution
Application-to-infrastructure service	They are app-to-app dependency where the given infrastructure service itself is an application	Microsoft Active Directory (AD) and DNS associated with large groups of all SAP applications

Table 6.4 SAP Application Technical Dependency Categories

By conducting SAP application owner interviews, addressing questionnaires, using CMDB, and a discover tool (AWS or an AWS Partner tool) can all help you identify application-level dependencies, which gets you one step closer in building a migration wave plan.

The next step and a key element in migration planning is to establish a well thought through prioritization criteria. The end goal of this exercise is to understand the order in which applications will be migrated. The strategy involves a repetitive and progressive approach to evolve the prioritization model. For enterprises with SAP systems and some satellite systems, the choice could be between migrating a less critical (compared to an SAP ERP) SAP BW system or migrating a satellite system running a legacy application running on end-of-life hardware to a SaaS solution on AWS. Ideally, you would want to start off with applications fall into low-risk and low-complexity categories. In a data center migration or an exit, unless under a time crunch, SAP workloads are and should be considered as the most critical, high risk, and highly complex with respect to migrations and are usually migrated towards the end of a migration timeline. Migrating less-critical, low-risk, and low-complexity applications helps you build momentum and gain confidence on managing and operating applications in AWS before migrating mission-critical systems such as SAP.

Additionally, your initial criteria should prioritize applications with a small number of dependencies, running in cloud-supported infrastructure, and from non-production environments. An example would be an SAP application with 0 to 3 dependencies ready to rehost as-is in a development or test environment. Your SAP analytics applications are typically a good fit in this context—SAP BW and/or SAP BusinessObjects, or any of the third-party visualization tools. These criteria are also valid for defining the pilot applications and potentially the first and second migration waves, again depending on the level of cloud adoption maturity and your confidence levels.

Application Prioritization

For large migrations, determining the prioritization of the applications could become a huge undertaking. So, to overcome this manual labor, scoring is conducted with predefined templates available at public AWS repositories (e.g., GitHub). Define a score, or weight, for each possible value of each attribute in your system inventory. An attribute could be the SAP environment, business criticality, regulatory or compliance requirement, OS support etc. For example, if the "environment" attribute is selected, and the possible values are production, pre-prod, test, and development, each value is assigned a score, a greater number representing higher priority. Although optional, we recommend assigning a multiplying factor for importance or relevance to each data point. This optional step provides a higher-level differentiator to emphasize what is more important, which helps keep the criteria aligned as you iterate on assigning scores to the values. The ideal distribution should look like a standard bell curve, with a few very high-priority workloads and a few very low-priority workloads. The majority of applications will be somewhere in the middle. Based on the strategy to prioritize low-risk, simple applications for the first few migration waves, Table 6.5 shows an example attribute selection matrix and their value assignments.

Attribute or Data Point	Possible Values	Score (0-99)	Importance or Relevance Multiplying Factor
SAP environment	Test	60	High (1x)
	Development	40	
	Pre-prod	20	
	Production	20	
Business criticality	Low	60	High (1x)
	Medium	40	
	High	20	
OS support	Cloud ready	60	High (1x)
	Unsupported in cloud	10	
Number of compute instances	1-3	60	Medium-High (0.8x)
	4-10	40	
	11 or more	20	
Migration strategy	Rehost	70	Medium (0.6x)
	Replatform	30	
	Refactor	10	

Table 6.5 Application Prioritization Scoring in Assess Phase of AWS Migration Methodology

Another key aspect of the initial prioritization is to include internal teams or business units that show interest in being early adopters of the cloud. These individuals could be a considerable lever in obtaining business support to migrate a given SAP application, especially in the early days. If certain teams in your organization show interest, include the business unit attribute in the scoring. Assign a high score to those business units that are willing to come forward with their applications. Using the business unit attribute will help bring those applications to the top of the list.

Identifying application prioritization is an iterative process that should be done in two phases—during the assess phase and the mobilize phase of the AWS migration methodology. In the assess phase, based on our experience, you should start with noncritical applications to refine migration processes and incorporate the lessons learned. However, in the mobilize phase, and to create a long-term migration plan, the order in which applications are migrated should be aligned to your business drivers. Applying the new criteria will generate a new ranking of applications that will be a key input for wave planning. Table 6.6 shows some example prioritization criteria aligned to business drivers for quick cost reduction, which is typically the business driver for migrating SAP workloads to AWS. The higher the score, the higher is the priority to migrate the application to gain quick cost benefits on AWS.

Attribute or Data Point	Possible Values	Score (0-99)	Importance or Relevance Multiplying Factor
Application	SAP ERP	50	High (1x)
	SAP EWM	50	
	SAP BW	70	
	SAP Business Objects	70	
	SAP Cloud Connector	50	
	Other COTS	20	
Database product	SAP HANA	70	High (1x)
	Oracle	70	
	SAP MaxDB	70	
	Other databases	20	
CPU utilization (average)	More than 36%	60	High (1x)
	Less than 36%	40	
Number of compute instances	11 or more	60	Medium-high (0.8x)
	4 to 10	40	
	1 to 3	20	
Migration strategy	Retire	80	Medium (0.6x)
	Rehost	70	
	Replatform	50	
	Refactor	10	

Table 6.6 Application Prioritization Based on Business Drivers in the Mobilize Phase of the AWS Migration Methodology

Application Prioritization

Identifying the prioritization criteria is an iterative process until all stakeholders generally agree with the output. In our experience, it could take three to four iterations to obtain a baseline version.

Third-party discovery tools can automate the use of prioritization criteria using their proprietary formulas to calculate the complexity scores based on your custom input. However, this process could easily take 4 to 12 weeks depending on the number of systems being migrated.

After your prioritization baseline version is identified, the next step is to iterate on the 7 R migration strategy to ensure that migration strategies are allocated to each application component and its associated infrastructure. This information will be a key factor when estimating the effort, capacity, and skills needed and when creating migration wave plans.

Wave Planning

After identifying application dependencies and prioritizing your SAP and non-SAP applications, wave planning in the next key milestone, specifically in a large migration to AWS. As mentioned earlier, for a small to medium-sized migration (e.g., 100-300 servers), a common practice is to migrate all the *production* SAP systems in a single wave or using a "big-bang" approach. In a wave plan, you can group similar applications together, accounting for infrastructure and application dependencies, the priority of the applications, similarity of application architecture, and business functionality. You can then review the wave plan with your application and infrastructure teams to confirm their availability during the specified migration, testing, and cutover window.

Based on real-life deployments for various AWS customers, some best practices for wave planning include the following:

- Plan migration waves at least 4 to 5 waves in advance. This planning horizon ensures that you always have enough servers for the migration workstream.

- You should start with a few low-complexity applications and apply lessons learned to later waves.

- In early waves (i.e., waves 1 through 5), select fewer servers (less than 10), low-complexity applications, and applications in lower environments, such as development or test environments. Gradually introduce more complexity and more servers into your waves as you progress.

- Define a wave plan that can accommodate the production cutover window. If too many applications are being considered, the cutover window can increase. Try narrowing down the number of applications being migrated.

- Include testing of applications as part of wave planning. Testing should include technical tests such as load testing, HA, and DR failover tests 2 to 3 weeks before go-live specifically for production systems. Non-technical tests should include functional testing of critical business processes.

- Wave planning is an ongoing process, not a one-off task. Do not try to plan all waves at once.

- If you're using a portfolio discovery tool with a complexity scoring feature, use it in wave planning. Migrate the applications with the lowest complexity first.

Once you have grouped your applications based on the application dependency category, the applications must be migrated at the same time or on a specific date. For example, an SAP ERP application server running on a VM and an SAP HANA database running in a separate physical machine, where there are low-latency requirements or high-traffic volumes and complex queries, are likely to be migrated together rather than operating one component in the cloud and the other on-premise. Likewise, independent applications, such as SAP EWM and SAP ERP that interact over RFCs with similar low-latency requirements, can also be migrated at the same time. In addition, nontechnical dependencies must be considered. For example, application release schedules, maintenance windows, and key business dates such as month-end or quarter-end processing will influence the wave plan.

Migration waves can span across 4 to 8 weeks, and they can contain one or more migration events. Dependency groups are combined into waves so that a wave can contain one or more dependency groups. The wave also contains other activities that are required for the migration. These activities include AWS infrastructure setup (e.g., setting up landing zones, security, and operations); migration tooling; and migration activities such as data replication, cut-over planning, testing, and post-migration support.

To measure success and track progress, waves should be aligned to outcomes and business drivers. This alignment will also influence wave duration and the dependency groups that a wave contains. The completion of a wave should reflect a measurable achievement. The planning of a wave can also combine other factors, such as technical guiding principles. For example, waves can be defined by the SAP environment (e.g., development, test, or production) or by migration strategy (e.g., rehost wave or replatforming wave).

Let's consider an example of an SAP migration wave planning for an enterprise with a large SAP application footprint comprising of more than 20 different SAP solutions. To conduct wave planning with this complex landscape, follow these steps:

1. **Make sure you meet the prerequisites.**
   In this case, you must have a complete view of the SAP and non-SAP application portfolio; associated infrastructure (e.g., compute, storage, and networks); and migration strategy.

2. **Map the migration wave dependency and prioritization consideration.**
   Table 6.7 shows a migration wave dependency mapping and prioritization consideration. In this example, instead of scoring or defining a weight, we are simply considering low, medium, and high values for prioritization.

Dependency Description	Consider	Priority	Rationale
Application-to-server relationship	Yes	High	Application with persistent bindings cannot function without the server and must be included in the migration wave. Includes application servers, web dispatchers, and database servers.

Table 6.7 Migration Wave Dependency Mapping and Prioritization Consideration

Dependency Description	Consider	Priority	Rationale
Application-to-application communication	Yes	High	Applications that have hard dependencies to other applications (upstream or downstream) must be included in migration wave to avoid potential latency.
Server-to-server communication	Yes	High	Servers that communicate heavily to other servers must be included in migration wave to avoid potential latency.
SAP-to-non-SAP integration points	Yes	High	Non-SAP integration points will be migrated in the same migration waves as their SAP counterparts, when possible.
Server environment	Yes	Medium	Server environments (i.e., sandbox, dev, test, QA, user acceptance testing [UAT], staging, prod) will be segregated in environment-specific migration waves, when possible.
Application SLA/tier	Yes	Medium	Applications that are bronze or silver SLA and tier 3 or 4 will be considered separately from applications that are gold or platinum SLA and tier 1 or 2.
Data classification and compliance requirements	Yes	Medium	Compliance and security controls readiness may impact placement of applications on the migration wave plan.
Input and special factors from application owners	Yes	Medium	Planning interviews with application owners may surface timing considerations based on resource availability, testing, other corporate or application initiatives, licensing, and application dispositions.
Application user count	Yes	Low	Application with higher user counts may be grouped separately from applications with lower user counts.
Application outage windows	Yes	Low	Applications that share common maintenance or outage windows may be grouped together to minimize business impact and downtime.

Table 6.7 Migration Wave Dependency Mapping and Prioritization Consideration (Cont.)

Dependency Description	Consider	Priority	Rationale
At-risk applications	Yes	Medium	Some applications on end-of-life software/hardware (such as Windows 2008) present a significant risk to customer and may need to go in earlier waves.

Table 6.7 Migration Wave Dependency Mapping and Prioritization Consideration (Cont.)

3. **Define migration wave guardrails.**

   Next, we'll define a few guardrails that should be considered to define a baseline for the migration wave planning. Example guardrails include the following:

   - Each wave will be about 8 weeks of planning, design, build, cutover planning, and cutover with post-migration support. Migration waves will overlap.

   - You could have 4 migration teams. The teams will not overlap multiple waves.

   - Each application in its wave will perform its cutover in the final sprint.

   - Each team will have a post-migration sprint after its migration wave for hypercare and post-mortem activities.

   - After each 4th wave (e.g., sprint 8), an open sprint/wave date will be scheduled for contingency and rescheduling needs.

   - SAP and non-SAP applications with core dependencies to SAP ERP could be scheduled to meet go-live needs, and other waves will be planned around that milestone.

   - Waves profiles can be defined by SLA criticality (i.e., bronze, silver, gold, platinum), tiering (tier 1 through 4), and environment (SAP and tier 1 will get priority).

   - Non-production and production systems should not migrate in the same wave, unless an exception is made.

   - Since database servers will be migrated at the host level, they are grouped with the applications they support, even if they are heavily shared database clusters, when possible.

   - Non-persistent connections like AD, domain controllers, and monitoring are not binding for migration waves.

   - Non-SAP integration points will be migrated in the same migration waves as their SAP counterparts when possible. As a result, the SAP migration wave planning will influence the non-SAP migration waves.

   - A migration pilot could be five applications and represent a wide variety of migration patterns from the portfolio. An SAP workload will also be included in the migration.

   - Migration waves will start at 30 to 50 servers each for the first few waves, then slowly grow to 100 servers, and max out at 500 servers per wave if needed.

4. **Create migration wave dependency groupings.**

   Based on earlier steps, Figure 6.7 shows an example migration wave dependency grouping and the MRP for each migration wave.

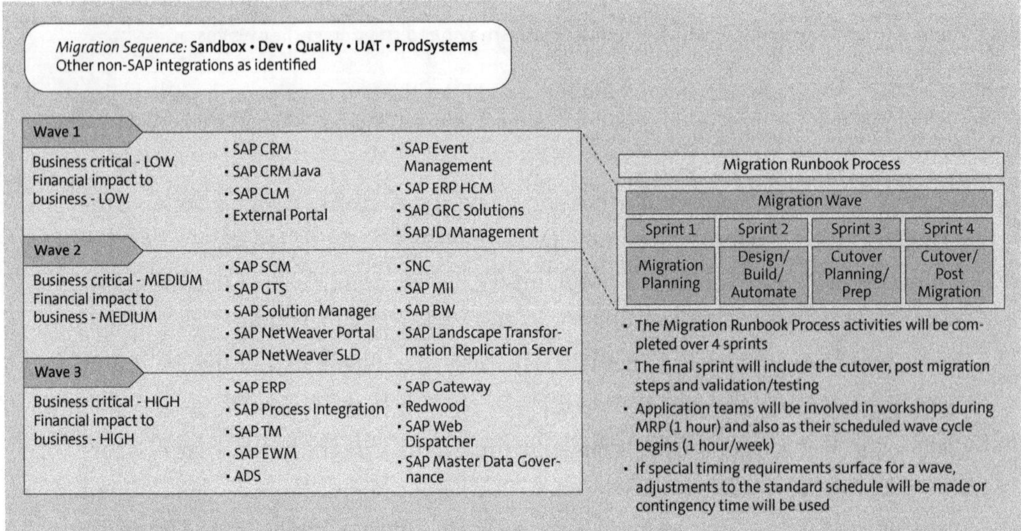

Figure 6.7 Example SAP Migration Wave Dependency Grouping and Migration Runbook Process

5. **Define migration waves**

   Based on earlier steps, the final step is to define the migration waves, which are typically represented on a Gantt chart, as shown in Figure 6.8.

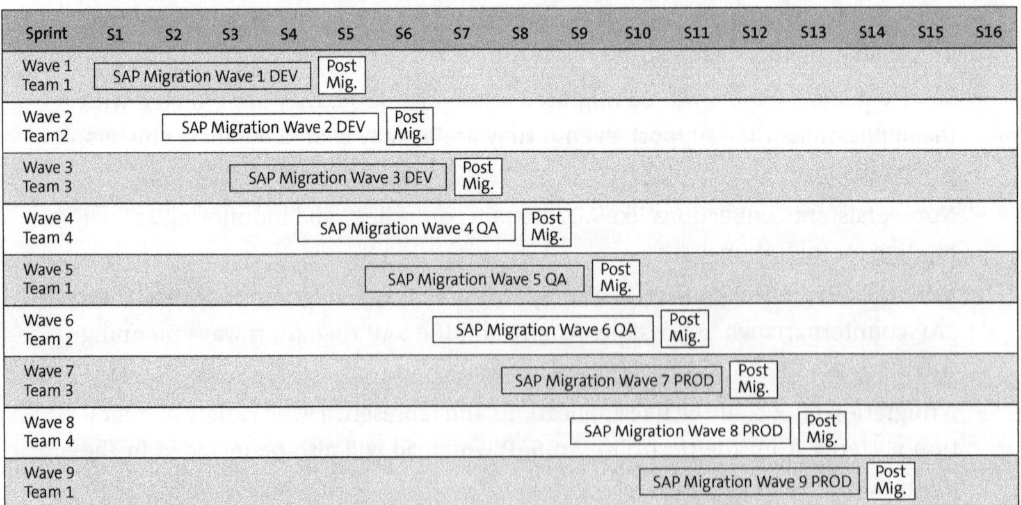

Figure 6.8 SAP Migration Waves on a Gantt Chart

Based on our experience and having been involved in numerous SAP migrations to AWS, following are some of the lessons learned while planning a wave:

- Incorporate flexibility in wave planning for unexpected changes, for instance, a change in design related to technical architecture, an additional set of HA and DR testing before go-live, the introduction of new interfaces integration in the landscape, etc.

- Incorporate automation by leveraging infrastructure as code (IaC) as part of your system builds on AWS to reduce migration timelines.

- Consider cloud-native tools such as AWS Application Migration Service (MGN) to rehost your SAP system on AWS.

- Execute a pre-production migration several times to mimic the production migration.

- Consider third-party interface integration and testing before and after the migration cutover.

- Complexity and duration of testing or mock production cutover is highly dependent on whether you would like to keep the same hostnames or different hostnames for the SAP systems after migration to AWS. So, plan for an extended timeline if you plan to conduct mock failovers if using same hostnames.

- Identify latency sensitive applications and bundle/coupled together in single wave. For example, your core ERP might need third-party tax solution.

- Execute HA and DR drills before production cutover both with and without data.

- Consider downtimes, available human resources, employee attrition, and data transfer link bandwidth while planning.

- Allocate time between waves for hypercare/post-migration support for stabilization and in preparation for the upcoming wave.

- Consider the location of SAP global transport file system while planning waves. If possible, move the transport file system during a production wave.

- Consider migrating an SAP Solution Manager system operating with change request management as part of production/SAP global transport file system wave.

- Understand the difference between technical and business downtime and plan for total downtime in coordination with everyone involved.

- Use SAP tools when possible since using third-party tools may not be supported by SAP.

- Your migration project plan should have a critical path clearly identified so you don't focus on small issues and distract from the important items.

[+] **SAP Migration to AWS: Reducing Migration Timelines**

A majority (i.e., 70 to 80%) of SAP migrations (rehosting) to AWS can take somewhere between 6 to 9 months. The major driver for reducing migration timelines is usually when enterprises are cutting close to the end-of-contract date with the existing infrastructure provider (e.g., a private cloud) or end-of-life with the existing on-premise hardware. A typical migration sequence starts with lower environments such as sandbox, development, QA, and pre-prod before going live with production. To reduce migration timelines, you can consider deployment of a lower environment like QA after the production go-live via system refresh using AWS native backup features such as AMI backup/restore.

6.4   Migration Tools and Services

Part of the portfolio assessment, and along with the 6 Rs migration strategy, a best practice is to include additional information on the migration tools and services that will be used to migrate the SAP applications to AWS. Typically for an SAP application migration, native SAP tools are used; however, AWS does provide accelerators specifically for data transfers from a source environment (on-premise data center, co-location, private cloud, and another hyperscaler) to the target (AWS). By combining both SAP native tools and AWS services, you can achieve your migration goals well within the stipulated timelines. Additionally, the AWS Competency Program has validated that the partners and their tools have demonstrated that they can help enterprise customers migrate their SAP applications and legacy infrastructures to AWS. Let's look at SAP, AWS, and partner tools and services available through AWS migration specialists, with an AWS migration competency partner, or on your own to start the process of moving SAP applications and data to AWS.

6.4.1   SAP Tools

In this section, we'll cover SAP tools such as the Software Provisioning Manager (SWPM), Database Migration Option (DMO) of Software Update Manager (SUM), database-specific tools, SAP Data Management and Landscape Transformation Services, and third-party data transfer tools.

Software Provisioning Manager

SWPM is an SAP native tool built by SAP to streamline provisioning of majority of SAP products, both on the ABAP and Java technology. The tool can be used on on-premise or on cloud and provides a consistent user experience across the platforms. SWPM can be used for the following tasks:

- **System installations**

  SWPM can be used for system installations across all supported SAP platforms and products. The tool provides the same look-and-feel and provisioning procedures that work the same way on all platforms with just minor differences (such as in parameter input phase). It supports standard, distributed, and HA installation setups and has a silent installation feature. In unattended mode, SWPM uses a parameter input file which means it can integrate with your IaC tool such as AWS CloudFormation, Terraform, etc. to automate SAP system provisioning and thus reduce migration timelines. For rehosting or lift and shift of your SAP applications to AWS, SWPM is used to provision SAP systems and its components—the Primary Application Server (PAS), the Additional Application Server (AAS), ABAP Central Services, and database on AWS.

- **System copies and migrations**

  You can use SWPM to create copies of your existing SAP systems or to migrate existing SAP systems to AWS. For migrations, SWPM is typically used for scenarios when an OS or database is being changed, called a *heterogenous migration* in the SAP world. SWPM works by performing database-independent copy procedures by building export files on source systems that are transferred and imported onto the target SAP system on AWS. SWPM is recommended for small to medium-sized databases. However, for large databases (e.g., >2 TB), and to reduce the technical downtime, it requires additional fine-tuning and thus requires in-depth knowledge of the migration procedures within the tool. These procedures include parallel export/import, table splitting, order by, and using distribution monitor. Additionally, you may experience longer downtimes for large databases due to limited downtime optimization options.

- **System renaming**

  SWPM can change a physical hostname to a virtual hostname, for instance, for changing the SAP system ID or to rename an SAP system created as an exact copy of a source system on AWS after migration.

- **Dual-stack split**

  All dual-stack systems must be migrated to single stack sooner or later. SAP HANA does not support dual stacks, and thus, a dual-stack split is a prerequisite to migrating existing systems to SAP HANA on AWS. SWPM can perform a dual-stack split by separating or removing the Java stack on the source system.

Database Migration Option in the Software Update Manager

SUM is an SAP native system maintenance tool built by SAP for a majority of SAP products, both on the ABAP stack and the Java technology stack. This tool can be used on-premise or in the cloud and provides a same user experience across platforms. This tool is used for the following scenarios:

- Release upgrade (major release changes)
- System update (EHP installations)
- Applying SAP Support Packages (SPs)/Support Package Stacks (SPSs)
- Applying Java patches
- Correction of installed software information
- Combine update and migration to SAP HANA using DMO
- System conversion from SAP ERP to SAP S/4HANA

A database migration to AWS transfers an SAP system from one database type to another. SUM offers DMO, which is a combination of an SAP software update with a database migration. DMO was initially a procedure to migrate to an SAP HANA database only. At this time, only the combination of upgrade and migration was possible since typically the source system was not yet on a software level supported by SAP HANA database. Now, DMO allows for the migration to database types other than to SAP HANA databases, such as SAP ASE.

DMO can be used for the following scenarios:

- DMO without system update: This option can be used for a pure database migration to SAP HANA on AWS, without changing the SAP software level or versions.
- DMO with system move: With this option, you can change not only the database (host), but also the application server for rehosting your SAP systems on AWS.
- DMO move to SAP S/4HANA on a hyperscaler (DMOVE2S4): To move/convert SAP from any database to an SAP S/4HANA system or to an SAP S/4HANA foundation, along with a move to AWS, without using the DMO with system move option.

A system conversion is the transition from an SAP ERP system on any platform into an SAP S/4HANA system on AWS. SUM executes the technical conversion of that procedure. In cases where the SAP ERP source system is not yet running on an SAP HANA database, SUM will cover the migration by DMO. In all cases, SUM will take care of the software update, apply new software components, and update existing ones. SUM will partially trigger the data conversion—the transfer of the table content from the old data model to the new, simplified data model of SAP S/4HANA. The relevant SAP S/4HANA releases are all based on SAP Basis 7.50 or higher, so SUM 2.0 will be the SUM version to use for a system conversion. Unlike SWPM, at the time of writing, SUM does not support updates with in unattended mode yet, which implies it cannot be part of an IaC build.

SUM supports several scenarios for optimizing or minimizing downtimes. Table 6.8 provides an overview of approaches and scenarios applicable for SUM to reduce technical downtimes for a database migration to SAP HANA or a system conversion to SAP S/4HANA on AWS. For any SAP tool such as SWPM, SUM, or DMO, we highly recommend you review the latest SAP Notes. Table 6.8 lists the SAP Notes that serve as the main entry points for more information, restrictions, and prerequisites.

Approach	Abbreviation	Scenario	Availability	SAP Note
Near-zero down-time maintenance (ABAP)	nZDM (ABAP)	Update/upgrade and system conversion	Unrestricted available	1678565
Zero downtime option for SAP S/4HANA	ZDO	Update/upgrade	Available for trained consultants (Training—ADM330e)	2707731
Downtime-optimized DMO	doDMO	Database migration & system conversion	Unrestricted available	2547309
Downtime-optimized conversion	doC	System conversion to SAP S/4HANA	Available for trained consultants (Training—ADM329)	3301507
Near-zero down-time technology	nZDT	System conversion to SAP S/4HANA	Service based	693168

Table 6.8 Important SAP Notes for DMO/SUM

In replatforming scenarios (e.g., migrating from any database to SAP HANA or a system conversion to SAP S/4HANA), doDMO is a common approach for large and medium-sized databases as well to meet the production downtime requirements, which are usually <48 hours (or in other words, over a weekend). doDMO is a trigger-based replication of a certain amount of application table data during uptime. While the system is still up and running, the content of some application tables is already being migrated to the SAP HANA database.

As a result, a reduced amount of data in application tables must be transferred afterwards during downtime. This approach ultimately leads to less downtime in total. The amount of downtime that can be saved depends on the size of the tables handled by the downtime-optimized DMO approach, but also on the amount of data that is changed during the procedure ("delta"). Using a reference run, you can find out how long it will take to migrate these tables. Note that the time saved is not the sum of time saved per table. This result first must be divided by the number of parallel R3load export/import processes, and a delta transfer might be required during the downtime. Since these tables will be transferred initially by R3load during the uptime, consider that the uptime phase of the procedure will be prolonged. Also, ideally, the delta changes recorded on the tables handled by downtime-optimized DMO is completed 99% prior to entering into actual downtime. Allow for sufficient time during the procedure for the delta replication to reach this goal.

Some scenarios to consider for doDMO include the following:

- DMO without system update: Can be combined with doDMO.
- DMO with System Move: This option does not allow to use doDMO.
- DMO move to SAP S/4HANA on a hyperscaler (DMOVE2S4): This option can be combined with doDMO.

SAP HANA to SAP HANA Migration to AWS: Considering doDMO for DMOVE2S4

DMOVE2S4 also enables homogeneous migrations (SAP S/4HANA to SAP S/4HANA on AWS). At the time of writing, for this scenario, downtime-optimized techniques, such as doDMO or doC are not supported. As a best practice, always check the latest SAP Notes for any SAP tools that you plan to use for your migration.

SWPM and DMO/SUM: R3load Transfer Modes Tip

SWPM uses either file mode or socket mode. Using *file mode*, the data is exported and written into a file using compression. Using *socket mode*, the data is exported and transferred over the network. With DMO/SUM, both R3load processes are running on the same host or different host, and they communicate using the main memory of that host, using pipe mode by default. Pipe mode is faster and leaner, but you should ensure that your network bandwidth meets the SAP KPIs as described in SAP Note 3296427.

Database-Specific Migration Tools

An imperative step is to consider your migration strategy based on your downtime window for a production migration. A common technique is to leverage database-specific migration tools, such as backup/restore, for small/medium-sized databases and to use database-native replication technologies for large and very large databases. In this section, we'll cover some database-specific migration tools such as backup/restore as well as database-native replication technologies.

Databases: Backup/Restore

The backup/restore method involves the creation of a backup of the database on the source SAP system and then restoring it on the target database in the AWS environment. This approach is typically used for small to medium-sized databases where the downtime requirement is usually not that aggressive as compared to an SAP ERP or SAP S/4HANA. For example, a Tier 3 (not critical) SAP Solution Manager system running on an SAP ASE database (~200 GB in size) is migrated using, first by deploying a vanilla SAP system manually or using automation (IaC) on AWS, then hydrating the database by transferring the backups to the target on AWS, and finally performing restore proce-dures manually. As the restore process can easily take few hours for a large database, this method may not be ideal for complex SAP landscapes or large datasets. Your tolerance for production system downtime should be the driver for selecting the database/restore approach to migration. Figure 6.9 shows the architecture of migrating an on-premise SAP system to AWS using full database backup and restore.

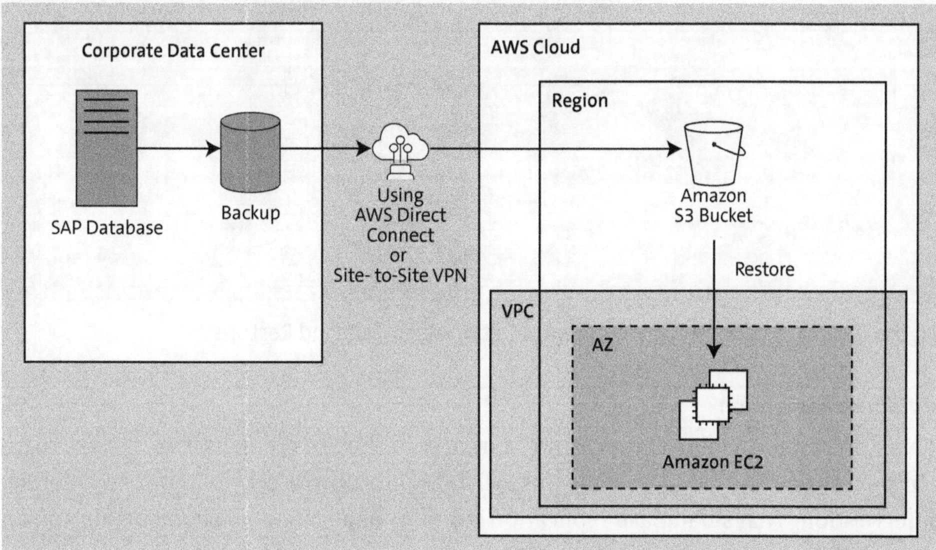

Figure 6.9 Migration Using Full Database Backup and Restore

For moving large backup files, you have the following options:

- AWS Storage Gateway or AWS DataSync or other third-party solutions such as rsync, robocopy, etc.
- AWS Snowball, depending on your downtime tolerance

To reduce the technical downtimes for large databases or where the network band-width between the source and AWS environment is low, a variant of traditional backup/restore is called *incremental backup/restore*. In this approach, a full backup is restored on the target database a few days (e.g., every, 5 days) before cutover. The data-base on the target AWS environment is not started or kept in an "open" state. Subse-quent delta backups—whether incremental or differential backups—are restored on a daily basis until the day of cutover. Eventually, on cutover day, the last delta backup is used to restore the system to its latest state. Effective planning can reduce downtime to a few hours by using log backups only on the day of the cutover. Figure 6.10 shows how the incremental backup/restore process can significantly reduce business downtimes even for large databases.

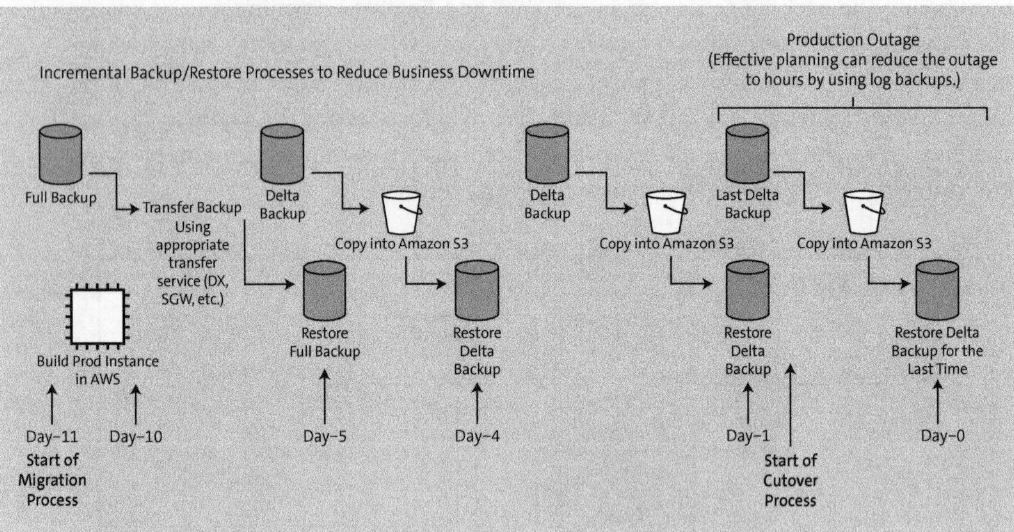

Figure 6.10 Migration Using Incremental Database Backup and Restore

Database Replication

Think of "incremental backup/restore" but now instead of doing backup/restore manually every day, the database does this task for you automatically without any manual intervention. That's database replication for you! Replication could be set up to run every few days (e.g., every 5 days), and then the database handles the delta or log replication from source to target in real time. This strategy is typically used for large and even medium-sized databases where you have stringent downtime requirements. Database-specific data replication technologies include SAP HANA System Replication, Oracle DataGuard for Oracle DB Replication, SAP ASE Replication Server, MSSQL Server Replication, and IBM DB2 Replication. Some capabilities are already part of the database license, while specialized replication methodologies for database migration (e.g., Oracle GoldenGate, O2O, Triple O, etc.) may require additional licenses. Figure 6.11 shows the step-by-step procedure for migrating an SAP system on SAP HANA or any database using database-native replication technologies. This strategy is highly recommended for large databases as well small/medium-sized databases:

1. A new SAP system on SAP HANA or any database is provisioned in your AWS account manually or using IaC (e.g., AWS Launch Wizard for SAP and/or SAP HANA).

2. After provisioning, the database instance can be set up as the target of replication from the on-premise database system.

3. At cutover, all activity on the on-premise system will be halted, and after the remaining synchronization is complete, the SAP HANA-based or any database-based system will be ready for use.

4. Set up the other SAP application servers manually or use the AWS Launch Wizard for SAP (if for an SAP HANA database) to set this up together with the database.

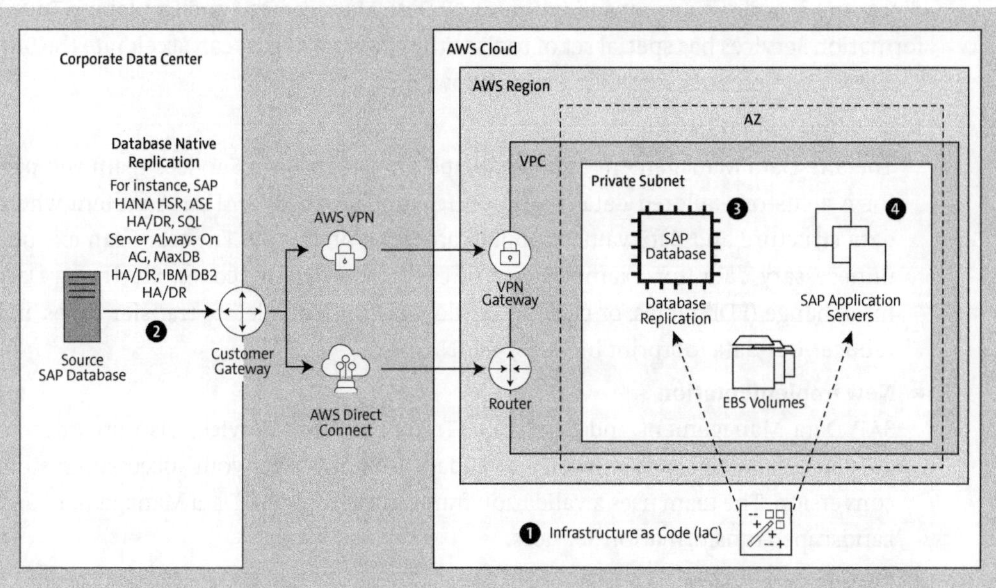

Figure 6.11 Migration Using Native Database Replication Technologies

---

SAP HANA to SAP HANA Migration: Reducing Downtime

Specifically, for low network bandwidth connections between a source and a target on AWS, you can combine initial backup/restore with database native replication technologies. For example, for SAP HANA, you can consider SAP HANA System Replication with initialization via backup/restore. In this strategy, you need to stop the source SAP HANA database and obtain a copy of the data files (essentially a cold backup). After the files have been saved, you may start up your SAP HANA database again. Transfer the SAP HANA data files to AWS, to the SAP HANA server you provisioned in target on AWS. (For example, you can store the data files in the /backup directory or in Amazon S3 during the transfer process.) Stop the SAP HANA database on the target system in AWS. Replace the SAP HANA data files (on the target server) with the SAP HANA data files you transferred earlier. Start the SAP HANA system on the target system and establish asynchronous SAP HANA system replication from your source system to your target SAP HANA system in AWS.

SAP Data Management and Landscape Transformation Services

SAP Data Management and Landscape Transformation Services is SAP's professional services team of SAP consultants within SAP who specialize in SAP S/4HANA implementations on on-premise and or on cloud. This project could be a new implementation, a system conversion, or a selective data transition/implementation of SAP applications (typically SAP S/4HANA). The SAP Data Management and Landscape Transformation Services team also brings in expertise in mergers and acquisitions, divestitures, and system consolidations. SAP Data Management and Landscape Transformation Services has special set of tools within its arsenal that can accelerate the SAP S/4HANA implementation, such as the following:

- **Selective data transition**
  The SAP Data Management and Landscape Transformation Services team will perform a custom-tailored data migration to your new SAP S/4HANA system where data structure and data value mapping are done on the fly. The team can exclude unnecessary data (for example, data of obsolete company codes, Electronic Data Interchange (EDI) tables, or closed workflow tasks) to accelerate transfer times and reduce your data footprint in SAP S/4HANA.

- **New implementation**
  SAP Data Management and Landscape Transformation Services also provides an automated validation framework to validate your data after your successful system conversion. The team uses a validation framework from SAP Data Management and Landscape Transformation Services.

- **System conversion**
  SAP Data Management and Landscape Transformation Services provides a solution for selective historical data and process migration, such as procurement or plant maintenance. Historical data and processes based on the SAP Data Management and Landscape Transformation Services team's technology and selective historical data migration can only be offered for SAP S/4HANA and SAP S/4HANA Cloud Private Edition.

Third-Party Data Transfer Tools

Additionally, AWS also supports third-party or open-source data transfer tools such as rsync, robocopy, or FileZilla (through FTP) for moving data between a source and a target on AWS. For example, you can use AWS DataSync or rsync to move */sapmnt*, and */usr/sap/trans* files from your on-premise network file servers (NFSs) to Amazon EFS on AWS.

6.4.2   AWS Tools and Services

In this section, we'll cover AWS accelerators/services such as AWS MGN, AWS Launch Wizard for SAP, AWS Migration Hub Orchestrator, AWS Network Services, AWS Storage Services, and AWS Professional Services, all of which can be used for migrating SAP systems to AWS.

AWS Application Migration Service

For large SAP migrations, some major challenges include diverse source infrastructure and OS types, busy and continuously changing workloads, downtime and performance disruptions, and tight project timelines and limited budgets. AWS MGN is a native AWS service you can use to rehost (lift and shift) a large set of SAP applications to AWS as-is. It can migrate from any source—physical, virtual, or cloud-based—and supports multiple OSs (Windows and Linux), SAP or non-SAP applications, and databases. AWS MGN is widely used for large migrations because it is highly automated and minimal skillset is required to operate it. To ensure your migrations are successful, you can perform nondisruptive tests prior to cutover. AWS MGN is an agent-based continuous data protection (CDP) tool that replicates in real time the source storage to the target storage, at the block level, using lightweight replication servers. To summarize, it captures the source system's configuration including OS parameters, replicates data changes, and creates a fully functional target environment in AWS.

AWS MGN is primarily used for the following scenarios:

- Rehosting SAP products and applications running on SAP HANA to AWS
- Rehosting SAP products and applications running on any database to AWS

However, in some scenarios, AWS MGN is typically not preferred or not feasible technically, such as the following scenarios:

- Replatforming, for example, a migration from SAP on any database to SAP on HANA on AWS. AWS MGN does not support out-of-the-box capabilities to convert any database into SAP HANA or if the underlying OS needs to be changed. However, you can consider lift-and-shift of SAP on any database to AWS using AWS MGN and then use SAP tools to migrate to SAP HANA or if there any change in the underlying OS. This approach could be feasible in a scenario where you could be running out of compute or storage in your on-premise data center. So, to summarize, it can support SAP heterogenous migrations as well. Another example is to use AWS MGN to transfer SWPM export/import files during a heterogenous migration.
- Rearchitecting/refactoring, for example, for a new SAP implementation.
- Non-x86 CPU architecture, for example, for SAP systems running on IBM Power.
- Replicating individual files/folders/applications.
- Migrating to/from non-block storage, for example, network-attached storage (NFS, Common Internet File System, or Amazon EFS) to an object (Amazon S3).

Figure 6.12 shows the architecture of migrating an SAP system using AWS MGN, which is an agent-based continuous block-level replication service.

AWS MGN: Test Failover

To avoid unintended impacts on production workloads, we highly recommend you test a failover in a network that's isolated from your production network. Most applications require the presence of a domain controller or a Doman Name Server (DNS). Therefore, before the application fails over, you must create a domain controller in the isolated network to be used for testing failovers. The easiest way to do this is to use AWS MGN to replicate a VM that hosts a domain controller or DNS. Then, run a test failover of the domain controller VM before you run a test failover of the recovery plan for the application. Another option is to create a domain controller that's dedicated for AWS MGN to use for test failover. Make these changes only to that domain controller. Reach out to the AD and DNS team in your organization regarding their preferred approach.

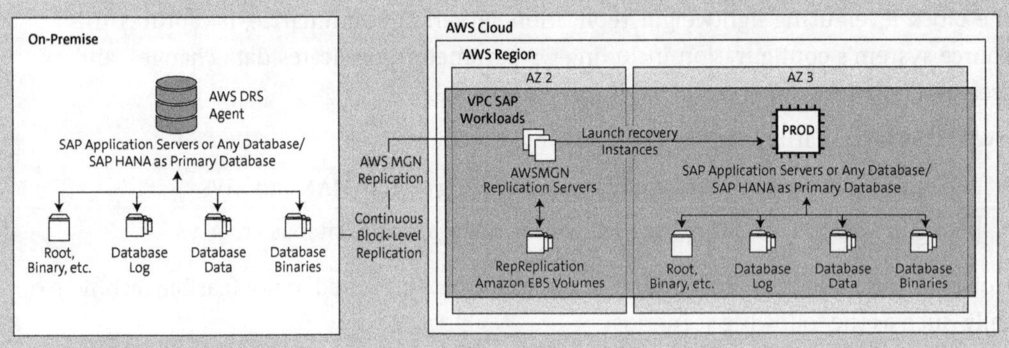

Figure 6.12 Migration Using AWS MGN

AWS Launch Wizard for SAP

AWS Launch Wizard for SAP is a specialized service designed to simplify and expedite the deployment of SAP workloads on AWS infrastructure using a wizard-based user interface (UI) within the AWS management console. Leveraging AWS best practices for SAP, it streamlines the setup process by providing a guided, step-by-step interface for configuring and launching SAP systems. By automating resource provisioning, network settings, and system configurations, AWS Launch Wizard optimizes deployment for SAP applications such as SAP HANA, SAP S/4HANA, SAP NetWeaver on HANA, SAP BW/4HANA, SAP ASE, and SAP Solution Manager on SAP HANA, ensuring they run efficiently and reliably on AWS. This service helps organizations mitigate the complexities involved in setting up SAP environments, enabling quicker deployments while adhering to recommended architectural guidelines for optimal performance and scalability.

During migrations of an existing SAP system or greenfield implementations, this service can be used to deploy vanilla SAP systems running on SAP HANA. Specifically, for migrations, depending on the migration strategy, you can hydrate the database using backup/restore, SWPM export/import, or database native replication technologies. Figure 6.13 shows the power of AWS Launch Wizard for SAP that you can use not only deploy infrastructures but also to perform end-to-end installations of SAP systems including HA using the power of IaC.

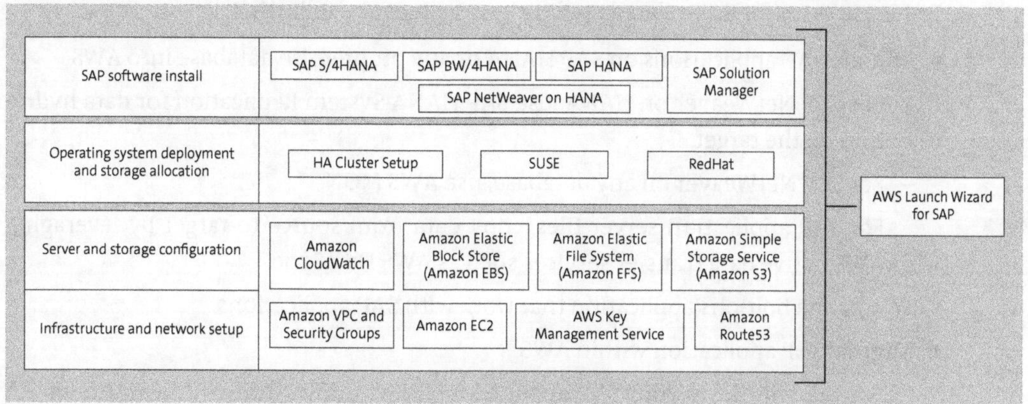

Figure 6.13 AWS Launch Wizard for SAP to Deploy a Vanilla SAP System on AWS

AWS Launch Wizard for SAP brings AWS and SAP best practices together to automate an end-to-end installation for the following scenarios:

- Greenfield implementations
- Migrating existing SAP workloads to AWS

AWS Migration Hub Orchestrator

One of the most common challenges when accelerating a migration to the cloud is the lack of a unified platform to track migration best practices, prescribed tools, and automations. AWS Migration Hub Orchestrator provides a central location to track migration tasks across multiple AWS tools, such as AWS MGN, AWS Launch Wizard for SAP, and even partner solutions. It can also integrate with other native, purpose-built, analytics-related AWS services to analyze server performance and visualize server dependencies using Amazon Athena and Amazon QuickSight. AWS Migration Hub Orchestrator provides key metrics and migration-related progress information for individual applications, regardless of which tools you use to migrate them. Additionally, it supports parallelism by allowing the migration of multiple SAP applications at the same time.

At the time of writing, AWS Migration Hub Orchestrator includes AWS-provided templates for migrating SAP NetWeaver-based applications and SAP HANA databases via AWS Launch Wizard for SAP, AWS HANA Systems Replication, and the automation of many manual tasks. In addition, AWS Migration Hub Orchestrator includes an AWS-provided template for rehosting custom applications using AWS MGN. Migration teams can customize and reuse workflow templates by building on top of baseline recommendations and modifying the steps and dependencies to address the needs of specific workloads and use cases.

Some use cases for SAP-related applications include the following:

- Migrate SAP applications on SAP HANA or any on-premise database into AWS
  - For SAP NetWeaver on HANA, use SAP HANA System Replication for data hydration on the target.
  - For SAP NetWeaver on any database, use AWS MGN.
  - For SAP application server files, copy data from source to target by leveraging AWS native data transfer services such as AWS DataSync.
- To migrate bolt-ons application that work with SAP applications
- Migrate SAP application within AWS
  - For rearchitecting your SAP application, for example, from a non-HA to an HA architecture
  - Change SAP application's OS type/version
  - Change in landing zone architecture, for example, moving to larger subnets
- Operationally create additional sandbox/test/dev systems as copies of existing production/non-production systems that run in parallel for assisting the development/testing cycles.
- Focusing on migrating from on-premise to AWS, before you migrate an application using AWS Migration Hub Orchestrator, you must deploy and configure an AWS Migration Hub Orchestrator plugin on your on-premise VMWare instance and configure. This plugin works along with the service to orchestrate migration activities on source servers. This one-time deployment uses an OVA file that AWS provides and requires connectivity to source and target instances and to the AWS Migration Hub Orchestrator service itself. Once you're done with plugin configuration, and the on-premise servers have been identified, repeat the migration process using migration templates that you have defined for each source application, in migration waves.

To summarize, for an overview of how AWS Migration Hub Orchestrator works, follow these steps:

1. **Discovery and application definition**
   Discover infrastructure and applications with AWS ADS, pick the applications you want to migrate, and configure the AWS Migration Hub Orchestrator plugin.

2. **Select a template and create migration workflow**
   Choose a workflow template to orchestrate your migration process.

3. **Run**
   AWS Migration Hub Orchestrator will guide you to deploy a target system with AWS Launch Wizard and migrate your desired applications for you.

AWS Migration Hub Orchestrator is available to enterprises and to AWS partners at no additional cost. Figure 6.14 shows how to migrate SAP HANA systems to AWS using AWS Migration Hub Orchestrator.

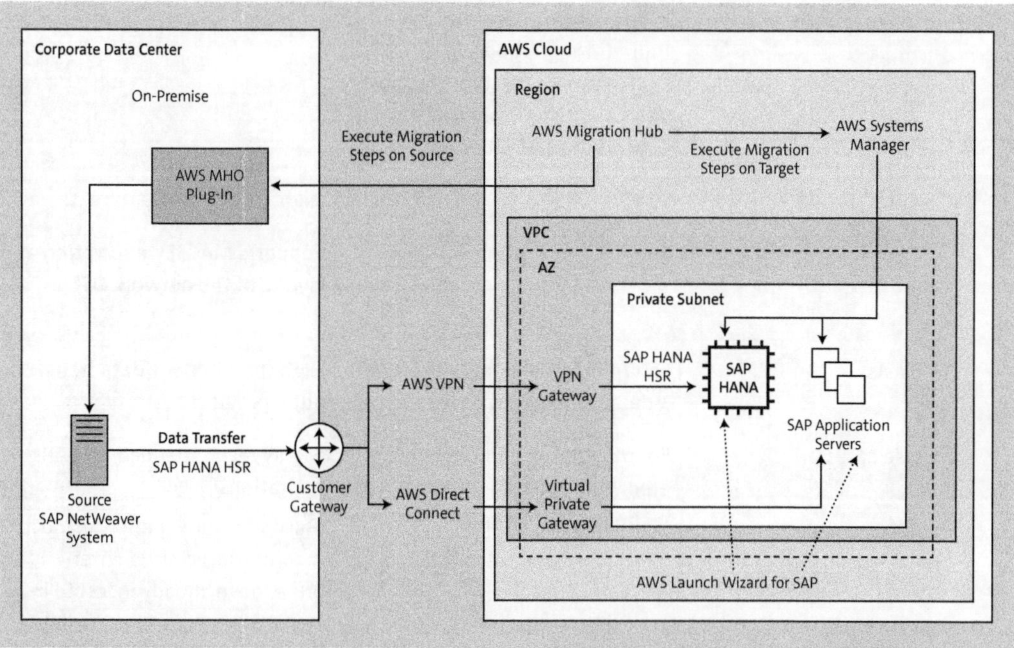

Figure 6.14 AWS Migration Hub Orchestrator: Migrate SAP NetWeaver-Based Applications and SAP HANA Databases to AWS

AWS Network Services: Connectivity Accelerators

As described in Chapter 2, Section 2.2, AWS Direct Connect and AWS Site-to-Site VPN are used to connect a source environment with AWS. AWS Direct Connect can take longer to implement, but in the long run, this choice can significantly reduce the downtime window by providing consistent and stable network connectivity. Table 6.9 provides a comparison between AWS Direct Connect and AWS Site-to-Site VPN.

Tool	AWS Site-to-Site VPN	AWS Direct Connect
Network	Connected over the public and shared networks and thus could be jittery	Dedicated line between source and AWS using a physical connection hence offer consistent bandwidth

Table 6.9 AWS Direct Connect and AWS Site-to-Site VPN

Tool	AWS Site-to-Site VPN	AWS Direct Connect
Maximum throughput	Each AWS Site-to-Site VPN connection has two tunnels, and each tunnel supports a maximum throughput of up to 1.25 Gbps. You can combine multiple VPNs and use ECMP to scale beyond the 1.25 Gbps limitation	Starts from 50 Mbps and scales up to 100 Gbps
Time and effort	Low	High
Security	Supports encryption via IPSec	Supports MACSEC encryption at layer 2 of the network OSI model
Costs	Low/medium (connection/hour + data transfer out costs)	High (port hours + data transfer out + provider costs)
Use cases	For migrations, can be considered with database replication technologies for small/medium-sized databases.	Highly recommended for SAP migrations (Based on our experience, most medium/large-sized enterprises have already invested in an AWS Direct Connect connection.)

Table 6.9 AWS Direct Connect and AWS Site-to-Site VPN (Cont.)

AWS Storage Services (Data Transfer Accelerators)

Whether you're rehosting, replatforming, or refactoring your SAP applications to AWS, some data and files will always need to be transferred from a source environment to the target environment. These files could be related to database backups, SWPM exports, DMO/SUM exports, third-party interfaces, /sapmnt, transports, etc.

AWS offers the several native services for transferring files from a source to a target environment:

- **AWS DataSync**
  An AWS native service that supports discovery and data transfer of data (files) from source to target environment using an AWS DataSync agent. The AWS DataSync agent connects to your on-premise storage system's management interface using a secure port, which then runs a discovery job to collect information about your system. It can work with NAS, SMB, HDFS, and Object storage on-premise storage systems. The agent sends the information that it collects to SAP DataSync Discovery. Using this information, it recommends supported and certified AWS storage services for SAP such as Amazon EFS, Amazon S3, Amazon FSx for Windows File Server,

and Amazon FSx for NetApp ONTAP file systems. Eventually, after selecting the relevant AWS storage service, you can transfer data over secure communication channels. Not only for migrations from on-premise to AWS, AWS DataSync can also be used within AWS in DR scenarios by continuously replicating data from shared file systems such Amazon EFS from one region in one AWS account to another region in another AWS account.

- **AWS Transfer Family**
  An AWS native service that is fully managed and highly available that provides secure file transfer to Amazon S3 and Amazon EFS. It supports the SFTP, FTPS, and FTP protocols to transfer data from on-premise to AWS.

- **AWS Storage Gateway**
  A managed AWS native service that provides on-premise access to cloud data via NFS and SMB protocols. It offers local caching to support low latency access, and asynchronously stores data in Amazon S3 Glacier and Amazon EBS snapshots. The primary use case of AWS storage gateway is to store data backups from on-premise (on tape) to AWS (on Amazon S3). For SAP migrations, it can be used in a backup/restore migration strategy or even in SWPM export/import scenarios as well where downtime requirements are not too stringent.

- **AWS Snow Family**
  Secure physical transport appliances that you can lease to pre-process and move exabytes of data in and out of AWS. The physical devices include AWS Snowcone (8 TB), AWS Snowball Edge (80 TB), AWS Snowball Edge Storage Optimized (210 TB), AWS Snowball Edge Compute Optimized (28 TB), and AWS Snowmobile (100 PB). The AWS Snow family is rarely used in SAP migrations, however, based on our experience; we've seen a few instances where AWS Snowball was used as the source because the private cloud vendor did not allow transfer of data replication or backups over the network. Other reasons include compliance or regulatory requirements. Because shipping time is involved with sending and receiving devices, plus with the time required connect and upload/download data from the devices, you'll need a much larger downtime window. You can consider an incremental backup/restore approach with AWS Snow Family to reduce the downtime window.

Table 6.10 provides a comparison of AWS storage services for SAP migrations.

Tool	AWS DataSync	AWS Transfer Family	AWS Storage Gateway	AWS Snow Family
Transfer method	Online	Online	Online	Offline
Data encryption	TLS	AES	SSL, KMS	SSE-S3, SSE-KMS

Table 6.10 AWS Storage Services for SAP Migrations

Tool	AWS DataSync	AWS Transfer Family	AWS Storage Gateway	AWS Snow Family
Type	Public endpoint, FIPS endpoints, VPC endpoints	SFTP, FTPS, FTP	File Gateway, Tape Gateway, Volume Gateway	AWS Snowball, AWS Snowcone, AWS Snowball Edge, AWS Snowmobile
Target storage	Amazon S3, Amazon EFS	Amazon S3	Amazon S3, Amazon EBS, Amazon S3 Glacier, Amazon FSx for Windows File System	Amazon S3, Amazon S3 Glacier
Transfer through	SAP DataSync agent on a VM	SFTP, FTPS, or FTP client software	Physical appliance or gateway software deployed on VM	Physical devices
What it does	Fast, online data transfer	File transfer through Amazon S3, and Amazon EFS using SFTP, FTPS, or FTP protocols	On-premise, low latency access to cloud storage	Offline migration of data to the cloud
Use cases	■ SAP and database data migration ■ Data protection; archiving old data ■ Data processing for hybrid workloads	■ Sharing and receiving files internally and with third parties ■ Data distribution ■ Data lakes	■ Low latency access for on-premise apps to the cloud ■ On-premise file share backed by AWS ■ Moving backups to the cloud ■ Data protection and DR	■ Data transfers ■ Edge computing ■ Edge storage

Table 6.10 AWS Storage Services for SAP Migrations (Cont.)

AWS Professional Services

We looked at the processes and technologies (tools) used for migrating SAP workloads to AWS. But what about the people? Of course, you can use an AWS SAP competency partner for your migrations. From the people perspective, at the time of writing, AWS is the only cloud provider to have a dedicated technology practice for SAP migrations and implementations. AWS Professional Services include consultants who are highly experienced and knowledgeable on not only SAP but on AWS as well. Often, we've seen enterprises with cloud teams with no SAP experience, or SAP teams with no cloud experience. The SAP on AWS professional services team can bridge this gap by providing prescriptive guidance on migrations as well as on the implementation of SAP workloads on AWS.

Working with the partner of your choice, AWS Professional Services can sub-in/sub-out and lead the migrations as well. Aside from migrations, they offer various other complementary services, such as:

- SAP on AWS Discovery
- SAP HA on AWS Discovery
- DevOps for SAP
- Analytics for SAP
- SAP Extend Discovery, for extending SAP business processes using AWS native services for the Internet of Things (IoT) and AI/machine learning

6.4.3   Partner Tools and Services

Specifically for migrations, and other than SAP offerings, AWS's SAP competency partners have built tools using AWS services and the SAP software logistics toolset including SAP Landscape Transformation framework to reduce business disruptions and shorten migration timelines for enterprises migrating large and complex SAP workloads to AWS.

These tools and services could be an alternative to SAP Data Management and Landscape Transformation Services and are often suitable for the following scenarios:

- For large databases when the SWPM export/import or database replication process does not fit the downtime window due to large amounts of data that must be copied from on-premise to AWS facilities
- SAP enterprises that need near-zero downtime for SAP migrations
- SAP system consolidation (for mergers and acquisitions)
- SAP carve-outs and divestitures
- SAP Human Capital Management (SAP HCM) splits and transformations
- For custom code remediation in highly customized SAP systems when moving to SAP Business Suite on SAP HANA or for an SAP S/4HANA conversion

Based on our experience, Table 6.11 highlights the most notable tools, depending on migration strategy.

Migration Strategy	Migration Scenario	Tool	Partner
Rehosting/replatforming	SAP on any database to any database SAP Suite on SAP HANA to SAP HANA SAP on any database to SAP HANA SAP S/4HANA to SAP S/4HANA	Lemongrass Cloud Platform (LCP) for near-zero downtime migration	Lemongrass
Replatforming	Large and complex SAP S/4HANA conversion (brownfield)	CrystalBridge	SNP
Replatforming	Large and complex SAP S/4HANA conversion (brownfield)	MIG Conversion	MigNow
Replatforming	Large and complex SAP S/4HANA conversion (selective data transition)	SDT Solution	Syniti
Replatforming	Mergers and acquisitions, consolidations, carve-outs/divestitures	DataSync Manager LT	EPI-USE
Replatforming	Large and complex SAP S/4HANA conversion (brownfield/custom code remediation)	Code analysis and transformation	smartShift

Table 6.11 AWS Partner Tools and Services for SAP Migrations to AWS

[+]

AWS Partner Tools and Services Tips

Many other solutions and tools are offered by other AWS SAP partners, so be sure you do your due diligence when identifying and selecting the right tools for your migration scenario.

System integrators and global system integrators (e.g., Accenture, Deloitte, PwC, etc.) have collaborated with partners like SNP, Syniti, and others to create a complete solution for your brownfield or selective data transition migration to SAP S/4HANA or even SAP Business Suite on SAP HANA. This collaboration helps deliver the most comprehensive, fastest, and lowest-risk solution for all your migration requirements related to SAP S/4HANA conversions, SAP Suite on HANA database migrations and/or Unicode conversions.

6.5   SAP Migration Approaches

In this section, we'll cover some proven migration approaches and tool option, both native to SAP and AWS, that you can leverage to accelerate your migrations to AWS. The migration approach in the SAP world can be categorized into two types—homogenous and heterogenous. SAP calls a lift and shift with no change in underlying OS and/or database a homogenous migration, whereas any change in the underlying OS and/or database falls under heterogeneous migration. Another perspective is leveraging the 6 Rs migration approach focusing on rehost (homogenous migration), replatform (heterogenous migration), and refactor (new SAP S/4HANA implementation on AWS).

6.5.1   Rehost (Homogenous Migrations)

Rehosting (also a "lift and shift" or a "homogenous migration") is the process of migrating on-premise (or any other source platform) SAP applications to AWS without modifying the application layer/version, OS, or database. This strategy is used mostly to migrate SAP applications to satisfy specific business goals, such as lowering infrastructure costs or leaving an on-premise data center. SAP applications are rehosted on SAP certified and supported AWS services, such as Amazon Elastic Compute Cloud (Amazon EC2) instances, that meet the requirements of the SAP applications you're migrating. More details on SAP supportability can be found in Chapter 5, Section 5.1.

This migration strategy is useful in any of the following scenarios:

- The SAP application must run, which is majority of the times, on a self-managed or a partner-managed server on AWS (virtual or physical).
- SAP on any database, not on AWS, to SAP on any database to AWS.
- SAP Suite on SAP HANA database to SAP Suite on SAP HANA database migration to AWS.
- SAP S/4HANA to SAP S/4HANA migration to AWS.
- The time and resources to modernize the application are not available. For example, an SAP on Windows to Linux migration could lead to a lower TCO, but your organization may lack the skillset to manage and operate Linux-based SAP workloads.
- Better compliance and security because it uses AWS's infrastructure and security best practices.
- Extensive technical debt that could arise from a heavily customized SAP system, so you prefer to migrate the system as-is without any modifications.

Some advantages of rehosting include the following:

- Minimum effort because little-to-no code or architectural changes are required within the SAP application layer, specifically for SAP business processes.

- Reduced costs as long as your SAP system is designed and implemented based on AWS's best practices.

- Better compliance and security because it uses AWS's infrastructure and security best practices.

- Improved reliability SLAs, specifically around HA and DR.

Some disadvantages of rehosting include the following:

- Additional cost savings might not be realized. For example, when you're migrating SAP running on Windows to AWS, we always recommend you evaluate switching to Linux as part of the migration, which could lead to additional cost benefits.

- Technical debts are carried over.

Figure 6.15 shows the available tools to use in various scenarios for rehosting SAP applications in AWS.

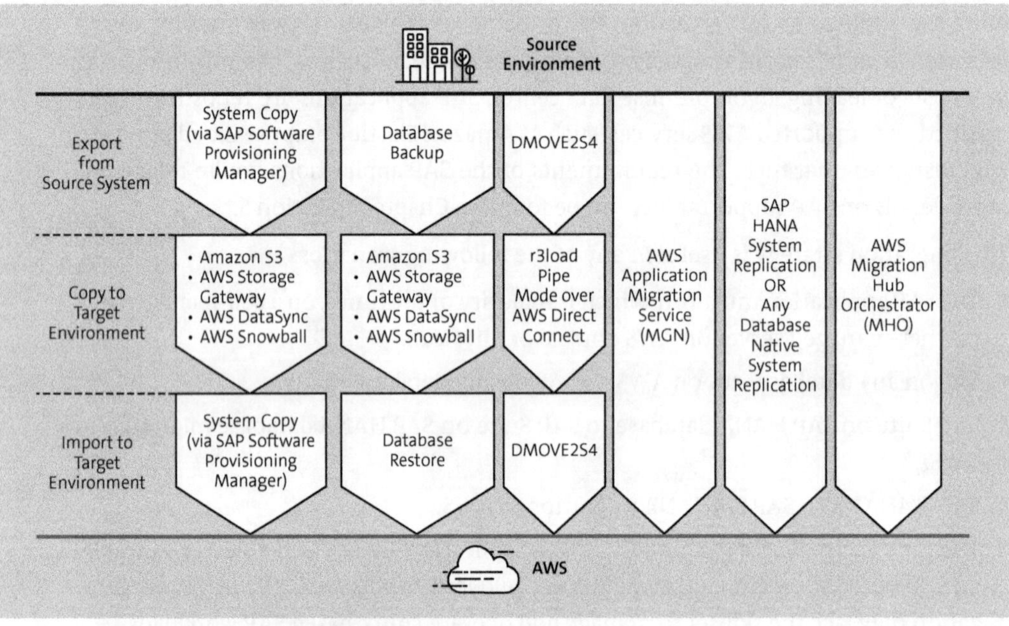

Figure 6.15 AWS Accelerators for Homogeneous SAP Migrations

Storage-Based Replication Migrations Considerations

When using storage-based replication technologies, such as AWS MGN, AWS Migration Hub Orchestrator, or any third-party continuous data protection (CDP) tools for migration, configuration data from the source is copied or moved over as-is on the target side. For instance, OS versions, patches, and third-party security/monitoring agents will get carried over to AWS on the target instance. You could thus require additional cleanup activities or some reconfiguration of infrastructure components. For example, you might need to adjust the storage for SAP HANA to meet SAP's KPIs on throughput and IOPS that are specific for AWS.

6.5.2   Replatform (Heterogenous Migrations)

Re-platforming—lift, tinker, and shift or lift and reshape—is a migration strategy where you move the application to the cloud, and you introduce some level of optimization in order to operate the application efficiently, to reduce costs, or to take advantage of cloud capabilities. The closest correlation to this in the SAP world is a heterogenous migration when the OS/database are changed. For example, moving from SQL Server running on Windows to SAP HANA which runs only on Linux.

This migration strategy is useful in any of the following scenarios:

- Migrating from any database to SAP HANA, for example, from SAP on an Oracle database to SAP on an SAP HANA database

- Migrating from any database to any different database, for example, from SAP on SQL Server to an SAP ASE database to gain cost benefits

- An SAP S/4HANA conversion from SAP Suite (SAP ERP) on any database or an SAP HANA database

- Reduce costs by moving from Microsoft Windows OS to a Linux OS

- Better compliance and security because it uses the AWS infrastructure and security best practices

Some of the advantages of re-platforming include:

- Reduced cost as long as SAP is designed and implemented based on AWS's best practices.

- Additional cost benefits when considering a migration from SAP on Windows to SAP on Linux.

- More options and tools related to compliance and security, because it uses the AWS infrastructure and security best practices.

- Improved reliability SLAs specifically around HA and DR.

Some of the disadvantages of re-platforming include:

- Medium to high amount of effort required in a heterogenous SAP migration specifically around code remediation.

- Potential upskilling of operations team with the new replatformed architecture.

Migration Approach Tip

SAP applications were never designed for the cloud. Hence there will be areas where an SAP deployment does not take full advantage of the performance, reliability, and resiliency options of the cloud. For example, you might be accustomed to running NFS servers on-premise for your SAP file systems such as /sapmnt, /usr/sap/trans, and /interfaces. On AWS, it is highly recommended to leverage Amazon EFS for SAP workloads on Linux to reap the benefits of a fully managed, auto-scalable, and highly available native NFS based shared storage file system. To summarize, as part of your migration, where possible try to design your SAP applications leveraging managed or fully managed AWS native services.

Figure 6.16 shows the available tools to use in various scenarios for replatforming SAP applications in AWS.

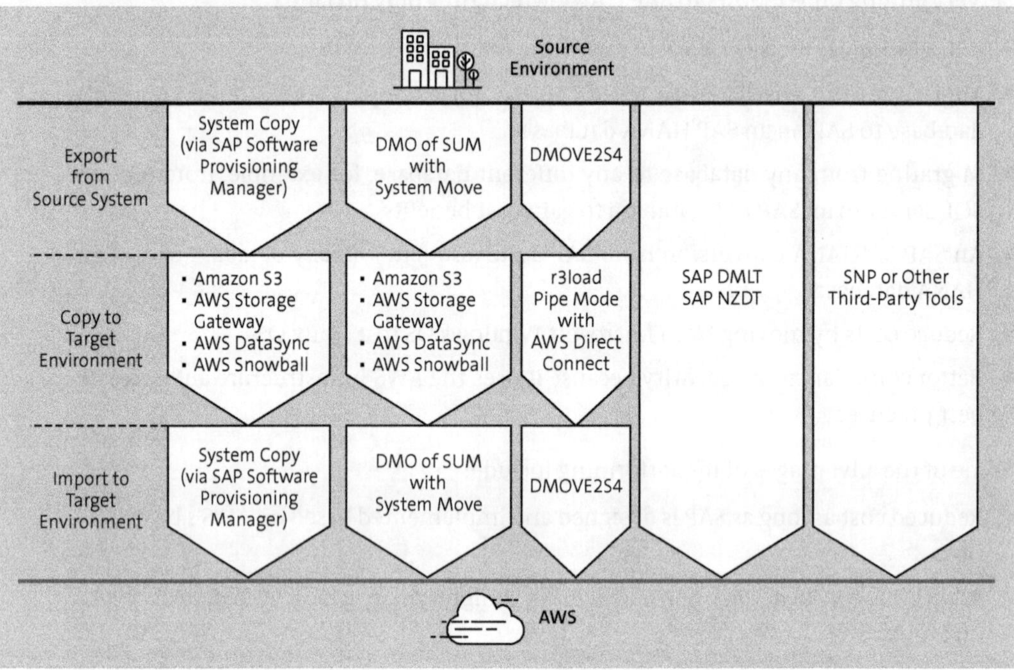

Figure 6.16 AWS Accelerators for Heterogenous SAP Migrations

As described in Chapter 5, for SAP systems running on Oracle or an OS other than Oracle Enterprise Linux (OEL), a mandatory step is to migrate the OS to OEL on AWS. OEL is a spin-off of Red Hat Linux, which implies a similar kernel based as Red Hat Linux. So, if your OS team is already familiar with Red Hat, then you probably can enjoy a smooth transition to OEL from a management and operations perspective. OEL is a free of cost OS; however, when running SAP on Oracle on OEL, you should consider the support costs for OEL, which is offered by Oracle. Table 6.12 shows various migration options specific to SAP systems on Oracle on AWS. RMAN Transportable Tablespaces (TTS),

XTTS (cross-platform TTS), and reduce XTTS are all free of cost tools available for you to migrate to an Oracle database on OEL on AWS. Oracle Golden Gate and Triple O are some license-based options for large database migrations and near-zero downtime requirements.

Source OS	Target: Migrating to Oracle Linux on AWS					
	Backup and Restore?	RMAN Transportable (TTS)?	Traditional RMAN XTTS Transportable?	Reduced Downtime RMAN XTTS Transportable?	SWPM?	Oracle Migration Service (Golden Gate) O2O/ Triple O?
RHEL / SLES	Little-endian	Yes	Yes	Yes	Yes	Yes
OEL	Little-endian	Yes	Yes	Yes	Yes	Yes
Solaris (x86)	Little-endian	No	Yes	Yes	Yes	Yes
AIX, HP-UX, Solaris	Big-endian	No	No	Yes	Yes	Yes
Windows	Little-endian	No	No	Yes	No	Yes

Table 6.12 Migration Options for SAP Systems on Oracle on AWS

Table 6.13 shows the decision matrix covering trade-offs for migrating SAP to Oracle on AWS.

Trade-Offs	Migrating to Oracle Linux on AWS					
	Backup/ Restore	RMAN Transportable TTS	Traditional RMAN XTSS Transportable	Reduced Downtime RMAN XTSS Transportable	SWPM	Oracle Migration Service (Golden Gate)O2O/ TripleO
Cost	Free	Free	Free	Free	Free	Paid
Downtime	Low	Low	Medium	Low, even for very large databases	High	Very low

Table 6.13 SAP on Oracle on AWS Migration: Migration Option Decision Matrix with Trade-Offs

Trade-Offs	Migrating to Oracle Linux on AWS					
	Backup/Restore	RMAN Transportable TTS	Traditional RMAN XTSS Transportable	Reduced Downtime RMAN XTSS Transportable	SWPM	Oracle Migration Service (Golden Gate)O2O/TripleO
Technical constraint	The SAP migration from SUSE Linux or Red Hat Linux to Oracle Linux and vice versa is treated as a homogeneous system copy.	Source and destination platforms must be similar (same endianness, similar OS) and on the same database release	Source must be Oracle Database 10g or higher. Destination has to be on the same or a higher release than source.	Source must be Oracle Database 10g or higher. Destination has to be on the 11.4 or a higher release.	NO	Source and destination need Oracle GoldenGate for data transfer
Special features	Oracle DB tools can be used for backup	Primary database can stay online if used for system copies	Can be performed in iterative steps for very large databases.	Source database is down only for metadata export. Can be performed in iterative steps for very large databases.	SAP Support	Can introduce additional Oracle Database features; includes a reorganization of the database.
Efforts	SAP specific knowledge needed	Oracle RMAN knowledge; no SAP-specific knowledge needed.	Oracle DBA needed	Oracle DBA needed	SAP-specific knowledge needed	Consultant needed who has expert knowledge about database replication with Oracle GoldenGate.

Table 6.13 SAP on Oracle on AWS Migration: Migration Option Decision Matrix with Trade-Offs (Cont.)

6.5.3   Refactor (New Implementation)

In the AWS world, refactoring is migrating an application to the cloud and modifying its architecture by taking full advantage of cloud-native features to improve agility, performance, and scalability. However, from an SAP perspective, refactoring or rearchitecting is more about implementing the latest and the greatest line of product offerings by SAP, which is SAP S/4HANA. Today, most new implementations are on SAP S/4HANA or use a selective data transition (called a *smart greenfield* or *bluefield implementation*). To summarize, you're still deploying SAP S/4HANA on supported and certified AWS services, but at the same time, you're enhancing your business processes by moving to SAP S/4HANA which is typically driven by strong business demands to scale, accelerate product and feature releases, and to reduce costs.

Some common use cases for refactoring a legacy SAP system into a new SAP S/4HANA implementation include the following scenarios:

- The legacy SAP ERP application can no longer address the demands of the business due to its limitations or is expensive to maintain.

- You have an SAP ERP application that is already hindering efforts to deliver products quickly or to address customer needs and demands.

- You have an SAP ERP application that has accumulated many complex processes and code that is heavily customized.

- You want to adopt best practices for process simplification, landscape consolidation, and the rationalization of legacy SAP systems.

6.5.4   Migrating Large Databases

It is common to see enterprises with large or very large databases also migrating their SAP workloads to AWS. Typically, large or very large databases fall into a category of anything which is larger than 10 TB or another way to look at it is a migration which cannot afford large production downtime windows or requires near-zero downtime migration. Many AWS partners - System Integrator (SIs), and Global System Integrator (GSIs) have deep knowledge of the tools involved—DMO, R3load, migmon, third-party tools etc.– as well as manual optimization techniques in order to get even the largest migrations done in least amount of downtime for their clients. What often, however, is not widely known are optimizations possible when migrating SAP NetWeaver systems to AWS to speed things up. Rehosting or Homogenous migrations—same OS/DB platform, no Unicode or endian change—should be done through DB native methods such as SAP HANA System Replication or incremental backup/restore/log-shipping to ensure that the technical downtime is extremely reduced. Re-platforming or a heterogenous migration should ideally be performed by experienced and skilled consultants using the right SAP or third-party tool.

To get the most performance in AWS for your large (or maybe even very large) heterogenous or homogenous SAP migrations, consider the following tips:

- **Know your limits on instances sizes and throughput**
  Every Amazon EC2 instance type provides a certain network and storage throughput. Similarly, every storage type—for example, block storage (Amazon EBS) or shared file system (Amazon EFS or Amazon FSx)—provides a certain IOPS and

throughput. The goal is avoid being capped by your compute resources' storage and network throughput. For large migrations, we always recommend you GO BIG! Upsize the Amazon EC2 instance type to get more storage and network throughput and/or leverage Amazon EBS storage type, such as io2, to get more IOPS, providing more resources to handle CPU or storage-intensive processes such as SWPM export/imports or even SAP HANA System Replication. And as the migration ends, use the flexibility of AWS to downsize these instances and/or this storage to your original/right-sized configuration. From a cost perspective, you should consider the duration for which instances and storage have been upsized.

- **Go big on compute and storage**
  Jumping off from the last point, let's consider an example. Our project is a heterogenous migration of a large Oracle database 10 TB in size. Based on SAP EarlyWatch Alert reports and Oracle Automatic Workload Repository (AWR) reports, you've identified that the right-sized instance for the database is an r6i.16xlarge, with its 64 vCPU, 512 GiB memory, 25 Gbps network bandwidth, 20 Gbps Amazon EBS bandwidth with Amazon EBS gp3 storage configured in RAID 0 for data and log volumes. For the SAP PAS, you have identified r6i.2xlarge, with its 8 vCPU, 64 GiB memory with Amazon EBS gp3 storage type. During SWPM import, you can consider increasing the database instance to r6i.24xlarge, increasing the instance for the SAP PAS to r6i.8xlarge, changing storage type to io2 or io2 block express to get more throughput and IOPS, reducing the timeframe of the import process, and eventually re-sizing it back to the original size after the import process is complete. After migration, related to operations and management of systems on AWS, the same approach can be followed for SAP upgrades as well.

- **Don't forget network optimization**
  In a typical SAP migration utilizing SWPM export/import (r3load) on-premise, the data from the export is either moved through an NFS/SMB share or memory pipes from on-premise to AWS. Due to the distance, and the laws of physics, there will be latency. Therefore, NFS/SMB is not recommended for large migrations in a WAN setup with high latencies. SWPM/migmon offers the native use of FTP, which can be used to saturate the network link. SAP SUM/DMO provides a native rsync script dmosystemmove.sh, which moves data to the target VM. Either rsync or other options like robocopy for Windows are ideal for migration to AWS. AWS DataSync is another option that can be considered. For rsync and DMO, the parameter DMO_SYSTEMMOVE_NPROCS has the default value of 4, which can be increased to have multiple rsync processed, depending on transfer phase to speed things up. You need to decide on the path your data takes when transferred to AWS. AWS Direct Connect or a main VPN connection (one already used for operations) is often the first choice. However, note that these links are often limited in bandwidth and already utilized by other, possibly productive, workloads. You must avoid fully saturating the main AWS link, so as to not impact other workloads. Options available are then rate-limiting the transfer, affecting your migration downtime, or using other links for data

transfers. Dedicated AWS Direct Connect circuits and links or your own VPN gateway (which is easier to set up) would be a first solution, something only used for an SAP migration. Often, firewalls and VPN gateways limit single network links, and thus, you'll want multiple transfer processes to fully utilize the bandwidth. Second, we strongly recommend using the copy method through private connectivity using AWS s3 copy commands manually. The Amazon S3 CLI does provide the best performance most often, can execute parallel threads, and can fully saturate your internet upload speeds. The downside is the need to manually prepare and execute copy operations and the fact that data must be copied twice—once from on-premise to an AWS storage account (Amazon S3 Bucket) and then from this Amazon S3 bucket to actual managed disk (OS) on the target R3load VM(s). However, often the speed advantage makes this approach still the best performing transfer method. Do not be blocked by limitations to your on-premise systems because of firewalls or other potential bottlenecks and instead use many parallel transfers with rsync/robocopy or similar tools. You might also start considering physical devices like AWS Snowball; however, this choice is often not feasible for an SAP migration because AWS does not have an SLA with 24/7 operations to copy data during a downtime window, which is typically over a weekend for production systems.

- **Use instance storage or a temp drive**
  Some supported and certified Amazon EC2 instance types for SAP (like x2idn) provide ephemeral or temporary block-level storage. This storage is located on disks that are physically attached to the physical host. Instance store is ideal for temporary storage of information that changes frequently, such as buffers, caches, scratch data, and other temporary content. An instance store consists of one or more instance store volumes exposed as block devices. The size of an instance store, as well as the number of devices available, varies by instance type and instance size. As this storage is ephemeral, its contents can be wiped if the Amazon EC2 instance is migrated to another host due to maintenance or other reason such as instance shutdown. Therefore, this temporary drive is only used with SAP systems to store the swap area. (swapdisk is created by the Linux waagent.) For some databases like SQL Server, the temporary drive can hold the tempdb to be re-created upon startup. Since the instance store is local to the physical host oi provides faster storage (i.e., higher IOPS and throughput) as compared to Amazon EBS volumes. But can we really use it for migrations? For some scenarios, yes, we can. For most databases, the temp drive can hold the temporary tablespace, where temporary results during index creation are stored. It can also hold online/transactional redo logs if some I/O activity occurs on logs that cannot be removed through the "nologging" typically done during migrations. Particularly for a DB2 migration and split tables, R3load handles split tables sequentially if CLI LOAD is used. Often, a faster approach is to use the R3load SPLITTED_LOAD option for these tables (SAP Note 1058437 - DB6: R3load options for compact installation) so that CLI LOAD can be done in parallel for the table splits. But with SPLITTED_LOAD, you must create another temporary

tablespace, specified by environment variable DB6LOAD_TEMP_TBSPACE, which holds these table fragments. At the end of the migration, this temp tablespace is empty and should be thus created and located on the temp drive. Other often accessed files could be placed here, as appropriate with databases. Finally, if everything for import is running on same instance, including R3load accessing the copied export files, you could think of placing the exported data here. To summarize, you have a fast, local storage that uses another storage path, with another and additional set of quotas that you can work around, depending on your scenario.

- **Fail fast**
Take two-way door decisions in this other methodology that AWS insists on to make high-quality and high-velocity decisions. Failing fast is more of a mental model, which AWS calls one-way and two-way doors. A *one-way door decision* has significant and often irrevocable consequences; an example would be having a single migration strategy for your critical SAP systems. A *two-way door decision*, on the other hand, has limited and reversible consequences: Having a backup migration strategy over your main migration strategy for your SAP systems to AWS is a basic but elegant example of a reversible decision. In case one fails, you can fall back on the other option to complete your migration wtihin the stipulated timeframe.

- **Select the right tools**
Typically, for an SAP migration, you think of R3load and migmon, orchestrated through SWPM. When it comes to SAP HANA migrations, often DMO is used. However, DMO can do a lot more than just use SAP HANA as its target, as described in the SAP Note for DMO v1/2 at *http://s-prs.co/v577620*. Depending on other criteria, you can often use DMO for DB2/Oracle/SAP ASE/MS-SQL as either source or target—but the same database can NOT be both source and target. A database change must occur. Also, migrating non-SAP databases are currently by request only. However, if you do like the benefits DMO offers—integrated package splits/order optimization, learning after each test migration, and SAP updates/upgrades—then see an SAP Note if its available for your migration scenario. For SAP S/4HANA 1909 migrations, don't forget the released DoDMO—downtime-optimized DMO using database triggers to substantially shorten your SAP S/4HANA migration. For more details, see SAP Note 2547309 at *http://s-prs.co/v577621*. DMO uses a different mechanism of transferring files to the target, well documented and also briefly mentioned earlier is the *dmosystemmove.sh* script and possibility to increase parallelism with environment variable DMO_SYSTEMMOVE_NPROCS. For a homogenous system copy, reusing the identical database and OS on-premise as on Azure, you're best served using database replication methods (SAP HANA System Replication, Oracle DataGuard, DB2 HADR, etc.) or simple/incremental backup/restore with log-shipping and aim for extremely short technical downtimes. Keep in mind an OS change but retaining the same database can mean a heterogenous migration if endianness changes and thus backup/restore or replication is not possible.

- **Bring the A-Team**

  The right tool for the job is important, but so is the person handling the tool. Particularly for large system migrations, do not treat test migrations as learning exercise. Instead, leverage some of the many very experienced consultants who can jumpstart your migration project right away with the correct approach from Day 1. Leverage AWS Professional Services, SAP MaxAttention, SAP Data Management and Landscape Transformation Services, and/or NZDT consulting services, and/or third-party professional services from SAP on AWS competency partners.

- **Use SAP monitoring tools in DMO**

  Before you start a complete DMO test run, we highly recommend using the benchmarking tool to evaluate the migration rate for your system, to find the optimal number of R3load processes, and to optimize the table splitting. A benchmarking run as well as a full DMO run will measure and note the real migration times for each table. This information is stored in the duration file. Copy the duration file from one run to the download folder of the next run. SUM will then consider the table migration time (instead of the table size) for calculating a table split that is more appropriate. For a complete DMO run, the migration will only happen after all the preparation steps are done. You may reset the DMO run to repeat the migration with adapted parameters, but it will take a while until the DMO run reaches the migration part. Instead, for a test run, you can directly tell SUM to repeat the migration after it finishes. A benefit of this approach is not only that you'll directly have the next migration run, but SUM will also automatically use the duration files from the previous migration run for the new split calculation. If you follow all these tips and still need to further reduce the migration downtime part, you may consider the downtime-optimized DMO approach (also known as "uptime migration"). This option is available for a migration targeting an SAP HANA database. As a rule of thumb, the standard setup should be able to reach a migration rate of around 300 GB/hour. Additionally, we highly recommend you run multiple mocks to nail down the downtime.

- **Use AWS tools/services for monitoring bottlenecks**

  The Amazon CloudWatch, and Amazon CloudWatch Application Insights dashboards for SAP can provide visibility into system performance during an SWPM import process. Because each Amazon EC2 instance and storage volume has limits, you should constantly monitor for bottlenecks and up-size or resize based on certain metrics (i.e., CPU, memory, volume throughput, volume IOPS, and volume queue length) to ensure the SAP migration goes as quickly as possible. You can set up a quick dashboard in Amazon CloudWatch, which contains metrics of the import VM and its storage, and share this dashboard with other members of the team, as applicable. OS-level tools (e.g., nmon, iostat, iotop, etc.) are good but often are just indicators. A 100% disk utilization in iostat does not mean that's the "fastest she will go"; rather that for 100% of CPU time, the disk was occupied with at least one request. A disk can have 100% utilization and still have lot of room on the IOPS/

throughput side during a massive parallel activity like migration. In addition to disk utilization metrics, you can add the VM network bandwidth ingress/egress, which is not a percentage value but in absolute bytes per second. With this approach, you can create a single pane of glass to monitor both disk performance and VM performance, with filtering on a particular time stretch during migration as needed.

- **Align resources with migration timelines**
  Inform end users, business users, CCoE teams, and other third-party vendors on the migration timelines specifically regarding testing and expected production downtimes.

6.5.5   Migrations with RISE with SAP

As more enterprises look to move to SAP S/4HANA Cloud Private Edition, one option is to consider RISE with SAP. Instead of a software solution, think of RISE with SAP as more of a financial construct under which SAP delivers and manages the software, support, infrastructure, and technical services for SAP systems under one service level agreement (SLA). RISE with SAP is SAP's foray into a subscription-based licensing model. This approach aims to simplify the management of your SAP environment, minimizing the number of separate contracts and support SLAs to deal with. RISE with SAP is offered on your choice of cloud service provider, including the ability to run on AWS. The choice of cloud service provider can play a big role in your ability to realize the true potential of RISE with SAP.

Table 6.14 describes the migration strategies, by implementation pattern, for using RISE with SAP to move to SAP S/4HANA Cloud, Private Edition.

Implementation Pattern	Description	Migration Strategy
Greenfield	New SAP S/4HANA implementation	N/A
Selective data transition/smart greenfield	New SAP S/4HANA with selective data transition from legacy ERP systems	SAP Data Management and Landscape Transformation Services, third-party tools (SNP, EPI-USE, etc.)
Brownfield	Conversion of existing SAP ERP system to SAP S/4HANA Cloud, Private Edition	DMO/SUM
Rehosting	Lift and shift of existing SAP S/4HANA	<ul><li>SWPM export/import</li><li>SAP HANA System Replication; SAP HANA tenant copy/move</li><li>Backup/Restore</li></ul>

Table 6.14 Implementation Patterns and Migration Strategies under RISE with SAP on AWS

Implementation Pattern	Description	Migration Strategy
Rehosting	Rehosting of existing SAP ERP on SAP HANA to SAP ERP on SAP HANA under RISE with SAP with limited duration before converting to SAP S/4HANA dependent on SAP contract. (It can be provided only on exceptional basis for partner-led projects and requires agreement by SAP.)	<ul><li>SWPM export/import</li><li>SAP HANA System Replication (copying and moving tenant databases)</li><li>Backup/restore</li></ul>
Replatforming	Replatforming of existing SAP ERP on any database to SAP ERP on SAP HANA with limited duration before converting to SAP S/4HANA dependent on SAP contract. (It can be provided only on exceptional basis for partner-led projects and requires agreement by SAP.)	<ul><li>SWPM export/import</li><li>DMO/SUM with system move</li><li>DMOVE2S4</li><li>nZDT options by SAP are supported</li></ul>

Table 6.14 Implementation Patterns and Migration Strategies under RISE with SAP on AWS (Cont.)

Migration Approach under RISE with SAP

Data transfer of file-based migration options (e.g., backups, exports, etc.) can be transferred via Amazon S3. At the time of writing, AWS Snowball is still in pilot phase and can be requested to transfer large volumes of data. Note that AWS Snowball is not available in every AWS region.

6.6   Project and Portfolio Management

Imagine teams making independent decisions for every aspect of cloud adoption; while it may enable quick decision-making, in the years to come, you'll end up with what the industry calls "technical debt," which is a set of problems that must be addressed for efficient IT operations. So, in your cloud journey, don't lose sight of the long-term management of both systems and personnel.

This section discusses the financial and expertise management aspects of transformation projects, such as SAP migrations, and discusses to some degree overall IT portfolio management.

6.6.1   Cloud Centers of Expertise

Thanks to the level of automation in the cloud, the traditional boundaries defining various technology towers—infrastructure, SAP Basis, OS, and database—have blurred. Think of it in terms of being "multilingual"—getting comfortable with more than one language (more than one technology in this instance) to drive cloud-led transformation.

A *cloud center of expertise (CCoE)*, in our experience, is a good way for organizations to create multilingual (that is, cross-functional) teams to support transformation projects.

A CCoE accelerates the transformation process in the following ways:

- Taking a consultative role to enable and advocate cloud adoption through your organization's approved processes and tools across all pillars of digital transformation, namely people, process, technology, and *data*

- Setting up the governance and compliance framework for the cloud (or across multiple clouds) across business units

- Creating recommended practices for system deployments, managing training roadmaps, and keeping up with the evolution of technology

When thinking about creating a CCoE, take an outcome-focused approach rather than a technology-focused one. If a team is measured by how quickly and efficiently a technology is adopted, your organization will have the latest and greatest technology without necessarily serving the needs of the organization, thus creating a different version of technical debt.

The focus of a CCoE should be on building capabilities the different pillars shown in Figure 6.17 and on driving these efforts in collaboration with application, infrastructure, and security teams:

- **Strategy and governance**
  Capabilities include creating policies and tools around risk management, setting up cloud financial management, and articulating standards across AWS Well-Architected Framework pillars.

- **Platform and architecture**
  Capabilities include creating reusable artifacts and automation, application-focused architecture (such as for SAP), and a cloud-first approach as the platform for innovation.

- **Community and change management**
  Capabilities include driving organizational change, building communities of practice, and bridging skill and knowledge gaps.

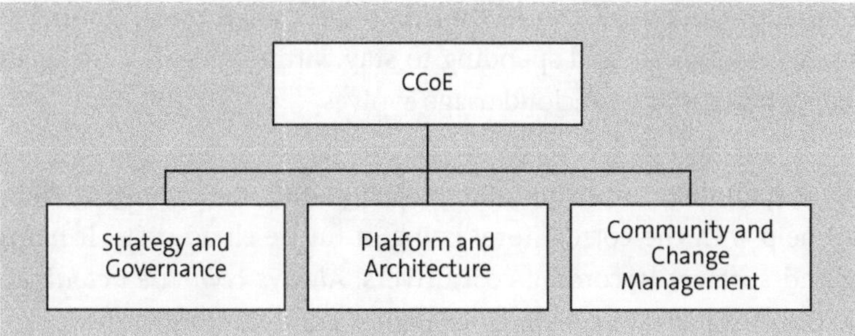

Figure 6.17 Pillars of CCoEs

Cloud Advisory Council

The cloud has brought rapid technological shift, so in a forward-thinking approach, you should also look into a cross-functional advisory council to provide your CCoE with recommendations regarding decision-making, oversight, and feedback on the needs and impacts of the cloud on business.

6.6.2   Cloud Financial Management

The term *cloud financial management*, one of the responsibilities of CCoEs, refers to the policies, processes, and tools for following activities:

- Tracking cloud usage and associated costs
- Spending allocation to business units
- Forecasting future spending
- Optimizing resource usage
- Governing cloud budgets

Figure 6.18 shows the key pillars of cloud financial management:

- **Visibility**
  Cost visibility is the first step to sound financial management; cloud costs can often be complex with charges associated with multiple services, regions, and à la carte options. Therefore, you'll need processes and tools to consolidate billing data, allocate costs to departments, generate reports, etc. Cost visibility planning starts with tagging resources correctly, centralizing data, and connecting billing systems (across cloud providers, in the case of a multicloud setup).

- **Optimization**
  Now that you know where the cost is coming from (visibility), you can look into optimizations, which involves right-sizing workloads (with an application focus such as SAP), eliminating waste, automating shutdown/startup of unused resources, and

leveraging right pricing structures (such as the Compute Savings plan). Optimization also entails governing usage and spending to stay within budget; it's also an iterative and ongoing process as your cloud usage evolves.

- **Forecasting**
  Analyzing historical spending, reviewing new project plans, and leveraging right pricing structures help with the correct forecasting of future cloud costs. It helps budget planning and surfaces upcoming cost drivers. *Always compare actuals to forecasts* while accounting for new initiatives, changing workloads, and growth projections.

- **Governance**
  The right governance aligns usage and costs to business priorities. It also defines roles and responsibilities (who can provision services, what resources they can use, etc.) and ensures compliance with regulations while minimizing TCO.

- **FinOps**
  The FinOps framework promotes collaboration among multiple units of an organizations, such as the finance, IT, engineering, and business teams, to make informed decisions about cloud spend. FinOps teams enable self-service cost analytics, set spending guardrails, implement chargeback mechanisms, offer training, and advise on optimizing workload costs with an application focus (such as SAP). You can also read more about how the FinOps foundation (a project of Linux Foundation) defined its maturity model at *https://www.finops.org/framework/*.

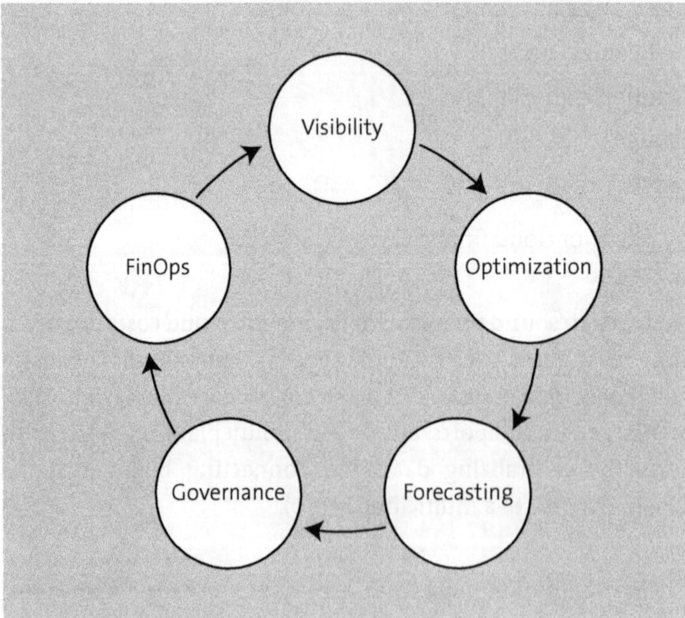

Figure 6.18 Pillars of Cloud Financial Management

> **Personnel Cost**
>
> An important aspect of cost is the time, cost, and effort of personnel; you should always make optimization decisions in that context. For example, if the labor cost to implement an initiative is higher than the savings you would get, you're better off redirecting that effort to a different initiative.

6.7   Summary

The migration of SAP systems to AWS involves a meticulous and phased approach to ensure a seamless transition while harnessing the benefits AWS's cloud infrastructure. In this chapter, we delved into the fundamental strategies, tools, and best practices essential for a successful migration. Starting with an in-depth assessment of your existing SAP landscape, this chapter took you through the process of selecting suitable AWS services, optimizing your architecture for scalability and performance, and devising a robust migration plan tailored to the specific needs of your SAP workloads. We emphasized mitigating risks, such as data integrity and downtime, through meticulous testing, employing AWS services like AWS MGN, and leveraging automation for streamlined execution.

Moreover, in this chapter, we explored key considerations for post-migration optimization, including cost management, security enhancements, and the utilization of AWS monitoring and management tools. The holistic approach we outlined in this chapter aims to empower businesses to execute a smooth, efficient migration, unlocking the full potential of AWS for their SAP ecosystems.

Chapter 7
Operations

Operating SAP on Amazon Web Services (AWS) has significant advantages. However, you must adapt to the AWS cloud operating model to reap the most from the available capabilities. Several components in this operating model differ from how you would approach them on-premise. In this chapter, we'll introduce you to the necessary tools and mechanisms to harness the power of cloud elements while operating SAP on AWS.

Before you start, we must be clear about a fundamental characteristic of the cloud: The cloud is not another data center, and you must not view AWS as merely an infrastructure provider. If you fail to differentiate the cloud in this way, you'll have difficulty grasping AWS's essence and power; the concepts discussed throughout this chapter won't make much sense.

Thus, in this chapter, we'll explore several concepts to ensure the seamless operation of your SAP workloads on AWS:

- What changes occur when operating your SAP workloads on AWS, compared to traditional on-premise data centers?

- What are the essential components or mechanisms required to maximize the benefits of AWS?

- Which AWS-native capabilities or service features can be leveraged to enhance system administration, observability, security, and compliance of your SAP systems?

- How can you adopt cloud-native response mechanisms to automate recovery and remediation activities?

- How can you manage your costs effectively using AWS-native cost control functions?

We'll address each of these concerns in the sections that follow.

7.1   Operating SAP in the Cloud

First and foremost, let's address the elephant in the room: the change in the operating model. Once you identify these changes, you'll gain a fair understanding of how to adapt to them, as well as how to explore new opportunities to enhance your current operating model with cloud-native offerings on AWS.

A non-exhaustive list of changes in the operating model on AWS include the following:

- You'll have visibility into all technology layers, which can lead to information overload. Having information is counterproductive if you don't understand how this new information relates to your decision-making processes, often leading to more false positives.

- Several monitoring and observability channels exist. An essential task is to finalize the tools used to standardize your monitoring framework or integrate these channels with your on-premise monitoring solutions.

- Every team member is empowered to collaborate across technology teams, offering their perspectives. More actors in the game mean more moving parts, contradictory data points, and numerous possible views. Without an organizational culture adaptable to this new working style, you risk creating a chaotic workplace.

- New maintenance mechanisms and change accelerators can reduce system disruptions. However, these capabilities may tempt teams to implement changes more frequently. Therefore, a different governance framework is needed to ensure system reliability, which entails changes in architectural principles, governance models, artifacts, and team collaboration models.

- Not all events require your attention; some can be automated. A beneficial step is to convert some manual runbooks into automation documents and establish mechanisms for notifications in case failures arise. In this way, you reduce your reliance on user complaints to discover problems. Incident management and escalation procedures must be adjusted accordingly.

- With new ways of connectivity and new integration sources, security takes on new meanings. Security now permeates every application stack layer and alters traditional system interactions. Some interactions may involve more steps, others fewer, but either way, your security fabric undergoes significant change.

- Automation can reduce transparency and complicate troubleshooting. Incorporating extensive traceability into your automation efforts can accelerate diagnosis. Alternatively, always have a backup plan to replace automation steps.

- In the cloud, everything is more interconnected than in on-premise environments. Therefore, any change could potentially impact other areas of the landscape, which requires higher levels of collaboration.

- Several project planning activities become obsolete or accelerated. For instance, infrastructure planning, provisioning, setup, and hardening all involve different levels of effort.

- Your IT consumption model must adapt to the cloud's operating expenses model. The visibility of granular cost elements makes the value versus consumption apparent. As the cost structure becomes transparent, team accountability and liability are more clearly defined. Projects now require more business justification than before.

To adapt to these changes and to address operational challenges that may arise in the cloud operating model, SAP cloud architects should have the following skills:

- You must have support channels live and reachable. In addition, you should empower your support personnel with all the paraphernalia of the cloud they need to resolve problems quickly.

- You want to see all of your SAP landscape in a single place. You want to gain a high-level sense of their states, critical metrics, and alarms to gauge their health at a glance.

- You want to ensure the security posture of your SAP workloads and operate in a compliant environment.

- You want to establish suitable tools and mechanisms to identify the performance challenges and quickly troubleshoot the underlying causes.

- You want to control and govern the change through your organizational change management practices. You can establish an overarching administrative function to oversee cloud governance.

- At the end, you want to bring a certain level of autonomy to your operations to reduce your overall operational effort and overhead. You can enable automatic remediation by responding to application and infrastructure events.

- The overall objective is to run a highly available, reliable, secure, performant, yet cost-effective SAP workload on AWS.

Note

In our interactions with SAP architects, we have found that all kinds of mindsets, falling in the spectrum between two extremes. On one end, they expect complete transparency and control over operational processes; thus, they do not generally support automation. On the other end, they want to automate everything but be able to take control if automation fails. Hence, this chapter focuses entirely on addressing the common denominator of how you can carry out SAP operations without resorting to automation. We'll cover the topic of automation in Chapter 8.

7.1.1 AWS Support Model for SAP Workloads

Before you do anything, you must ensure that all your support lines are live and that, if needed, you can reach them as quickly as possible. However, since different parties may handle other technology stacks in your SAP solution, you must establish a clear escalation path to reach the proper support channel with clarity.

As shown in Figure 7.1, you can broadly categorize issues into three different scope areas: infrastructure layer issues, application layer issues, issues in the interdependent layer, where you may have difficulty pinning a root cause to any specific solution layer. These issues in the third scope will require detailed analysis before arriving at any conclusion.

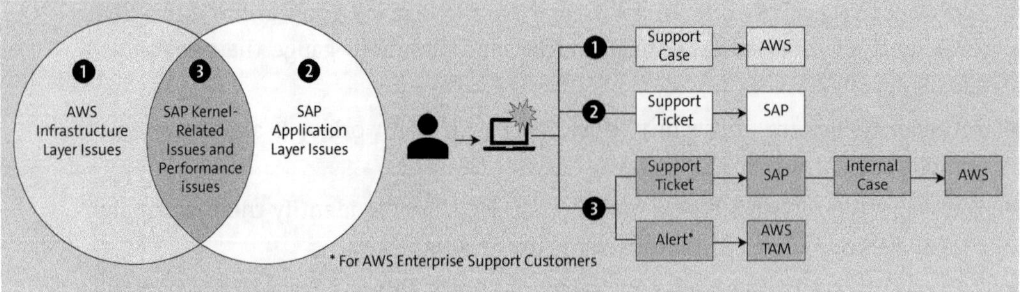

Figure 7.1 Support Paths for Your SAP Issues on AWS

To classify these issues and channel them to the right support channels, let's consider the following issues:

❶ AWS infrastructure layer issues

- Application is up and running, but the system is not reachable.
- Hardware is impaired; health checks are failing.
- Process hung for storage response.
- Infrastructure resource is not available.
- AWS service application programming interfaces (APIs) are not responding.
- Block storage key performance indicators (KPIs) are deteriorating.
- Read/writes to Amazon Simple Storage Service (Amazon S3) are slow.
- Remote user access is slow; response times are high.

❷ SAP application layer issues

- ABAP dumps (Transaction ST22).
- SAP repository object not found errors.
- SAP configuration missing errors.
- SAP setup errors.
- SAP authorization errors.
- Non-availability of free SAP work process.
- Data inconsistency errors.
- Software/SAP component compatibility errors.

❸ SAP kernel-related/performance issues

- SAP database replication is slow.

- SAP transaction response times are high.

- Cluster fencing has failed.

- Cluster takeover has failed.

- Backup is taking too long.

- SAP processes are terminating with operating system (OS) resource errors or infrastructure access failures.

- SAP kernel faults/dumps.

Keep in mind this list is incomplete and only for a loose reference.

The support paths shown in Figure 7.1 are self-explanatory. When encountering an issue related to AWS scope items, you should raise a support case directly with AWS. Conversely, for problems pertaining to an SAP application, open a support ticket with SAP (also known as an *OSS message*). However, in situations involving interdependent scopes, a support ticket should still be raised with SAP. If SAP, upon initial analysis, deduces that the root cause of the problem lies within the infrastructure layer, they will internally forward the SAP support ticket to AWS. Additionally, if you're an AWS enterprise support customer, you can inform your designated Technical Account Manager (TAM) to be on the lookout for potential escalation, especially if SAP indicates that the issue might pertain to the infrastructure layer.

You might question why we advise contacting SAP first, rather than AWS, for issues under the third scope. The reason is simple: It's the most efficient approach. Although AWS possesses dedicated SAP support teams, these teams may not have sufficient access to the application layers to conclusively rule out issues related to application behavior. Thus, it's imperative you have an intermediary who can confirm whether the issue stems from the SAP application layer or provide specific guidance to AWS's teams on where to focus their investigations. Without this clarification, diagnostic efforts at the infrastructure layer may lack precision, potentially prolonging the resolution process.

7.1.2   SAProuter Deployment Options on AWS

Once you raise a support message with SAP, the SAP support team may need to access your system to perform analysis and identify the root cause of your issue. To facilitate this access, you must have SAProuter deployed in your landscape. After moving your workloads to AWS, you can deploy SAProuter in different patterns, as shown in Figure 7.2. Note that all the systems shown in Figure 7.2 are deployed out of a single region. Elastic IP addresses are region-specific only. You cannot move them to another region.

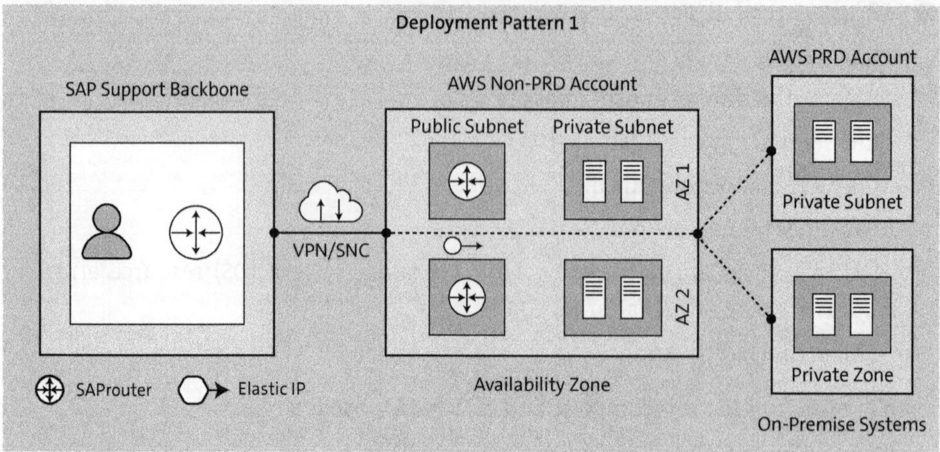

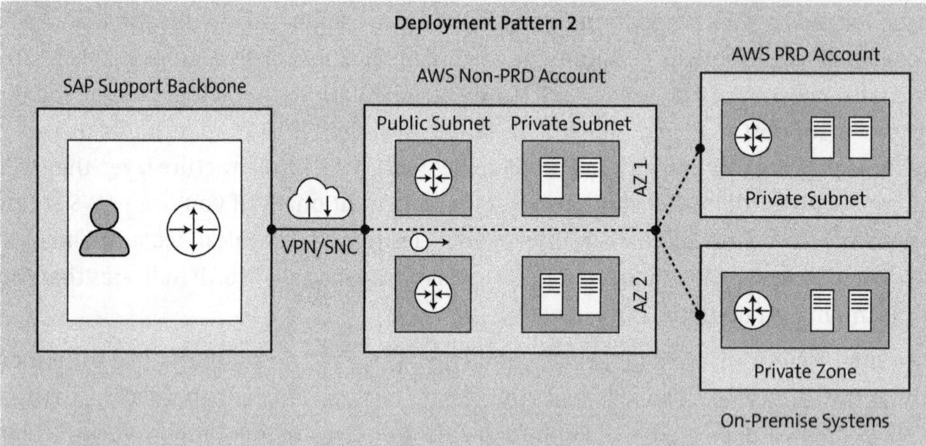

Figure 7.2 SAProuter Deployment Patterns on AWS

Deployment Pattern 1

In this pattern, you have only one SAProuter enterprise-wide, and the rest of the systems, residing in different network zones, are exposed through a permit list of *saprouttab* file entries. You should consider the following points while deploying SAProuter on AWS:

- Deploy the SAProuter instance on a separate Amazon Elastic Compute Cloud (Amazon EC2) instance or co-deploy with another SAP instance (e.g., SAP Web Dispatcher) in a public subnet of the virtual private cloud (VPC). Ensure a connectivity path exists from SAProuter to the rest of your SAP landscape. You can use VPC peering or AWS Transit Gateway to connect systems within AWS and your virtual private network (VPN) or AWS Direct Connect with on-premise systems.

- Create an AWS security group that allows only inbound connection from SAP (side)router IP on the SAP network interface (NI) port (default is 3299). Associate this security group with the Amazon EC2 instance of your(side) SAProuter. Optionally, you can restrict outbound connections to allow only specific connection types. Refer to the default port numbers for each allowed connection type at *http://s-prs.co/v577622*.

- Allocate Elastic IP and associate it with your SAProuter instance.

- Register this Elastic IP with SAP. The procedure described in SAP Note 28976, available at *http://s-prs.co/v577623*.

- Follow the SAP documentation, available at *http://s-prs.co/v577624*, to install and configure the SAProuter.

Configuring saprouttab for Secure Network Communications (SNC) Connections Registered to sapserv9 in Singapore

Listing 7.1 is an example of saprouttab configuration for the first deployment pattern shown in Figure 7.2.

```
# SNC connection to and from SAP
KT "p:CN=sapserv9, OU=SAProuter, O=SAP, C=DE" 169.145.197.110 3299
KT "p:CN=sapserv9, OU=SAProuter, O=SAP, C=DE" <EIP of your SAProuter> 3299
# SNC connection to local WINDOWS system for WTS, if applicable
KP "p:CN=sapserv9, OU=SAProuter, O=SAP, C=DE" <Private IP of WTS Server> 3389
(optional SAProuter password)
# Access from the local Network to SAP
# deny all other connections
D * * *
```

Listing 7.1 Example of a saprouttab Permission Table with SNC for a Central saprouter

In this pattern, you can improve the availability of SAProuter in the following ways:

- Enable Amazon EC2 host auto-recovery to automatically use a new Amazon EC2 instance for SAProuter when the current Amazon EC2 instance fails. You can define an OS run-level script to start SAProuter (either in /etc/init.d/rc3.d or /etc/rc.d/rc3.d based on your OS distribution) so that SAProuter will start immediately as soon as the new Amazon EC2 instance comes up on new hardware.

- To sustain the availability zone (AZ) failure, you can create periodic backups of the Amazon EC2 instance of the active SAProuter using the AWS Backup service. In the event of AZ failure, use this backup image to restore it in a new AZ of the same region. Associate the Elastic IP to the new Amazon EC2 instance of the new SAProuter. Ensure that the saprouttab has the latest configuration to work seamlessly. You can automate this failover through AWS CloudFormation scripting

- For cross-region disaster recovery (DR) scenarios, set up a new SAProuter and register it with a new Elastic IP specific to the DR region. Install and configure SAProuter on an Amazon EC2 instance as you did for your primary SAProuter. To save computing costs, you can back up the Amazon EC2 instance using AWS Backup in the DR region and shut down the instance. However, keep the Elastic IP instance allocated to your account. In the event of a DR failover, you can launch the SAProuter instance from the backup image and assign the registered Elastic IP in the DR region. Note that AWS will charge you for unassociated Elastic IP (refer to *http://s-prs.co/v577625*), but the cost of the Elastic IP is not significant when compared to the savings gained by shutting down the Amazon EC2 instance.

Deployment Pattern 2

This deployment pattern provides a more granular segmentation and more distributed control of an SAP system's access to SAP. You can have a separate internal SAProuter to impose additional access restrictions on the SAP systems that fall within the scope of that SAProuter. Since these instances are only internal, they are neither registered with SAP nor have a public IP to reach them online. You can install them in the relevant private subnets/private zones and configure them to engage in cascading communication between your external SAProuter and the individual internal SAProuters.

To set up additional internal SAProuters, follow these high-level steps:

1. Install the SAProuter application on a separate Amazon EC2 instance or co-deploy in the current Amazon EC2 instance with other applications in a private subnet of AWS or a private zone in the on-premise data center.

2. Allow inbound traffic from the private IP of your external SAProuter on SAP NI port, and outbound traffic to all SAP systems will fall under this SAProuter scope, private IP of external SAProuter, and public IP of SAP (side)router. Use AWS security groups for the SAProuter on Amazon EC2 and use firewall rules for on-premise SAProuter.

3. Your internal SAProuter must have network connectivity between your external SAProuter and all the systems controlled under this SAProuter.

Configuring External saprouttab for SNC Connections Registered to sapserv9 in Singapore

Listing 7.2 is an example `saprouttab` configuration for the second deployment pattern shown in Figure 7.2.

```
# SNC connection to and from SAP
KT "p:CN=sapserv9, OU=SAProuter, O=SAP, C=DE" 169.145.197.110 3299
KT "p:CN=sapserv9, OU=SAProuter, O=SAP, C=DE" <EIP of your External SAP
    router IP> 3299
# SNC connection to internal SAProuter in AWS
KP "p:CN=sapserv9, OU=SAProuter, O=SAP, C=DE" <Internal SAProuter1 IP> 3299
(optional SAProuter password)
# SNC connection to internal SAProuter in on-premise
KP "p:CN=sapserv9, OU=SAProuter, O=SAP, C=DE" <Internal SAProuter2 IP> 3299
```

```
(optional SAProuter password)
# Non-PRD accounts are directly exposed through external SAProuter
# SNC connection of SAP R/3 for a host in non-PRD account
KP "p:CN=sapserv9, OU=SAProuter, O=SAP, C=DE" <Private IP of WTS Server> 3389
(optional SAProuter password)
# Access from the local Network on AWS to SAP
P  <External SAProuter IP >  <Internal SAProuter1 IP>  3299
P  <Internal SAProuter1 IP>  169.145.197.110 3299
# Access from the local Network in On-premise to SAP
P  <External SAProuter IP >  <Internal SAProuter2 IP>  3299
P  <Internal SAProuter2 IP>  169.145.197.110 3299
# deny all other connections
D * * *
```

Listing 7.2 Example of a saprouttab Permission Table with SNC for Internet Facing saprouter

Configuring Internal saprouttab on AWS

Listing 7.3 is an example saprouttab configuration for internal SAProuter deployed in on-premise DC in the second deployment pattern shown in Figure 7.2.

```
# for SSH on host in PRD account
P <External SAPRouter IP> <Private IP of SSH host in PRD account> 23
# From SSH host in PRD account to External SAProuter
P <Private IP of SSH host in PRD account> <External SAProuter Private IP> 3299
# deny all other connections
D * * *
```

Listing 7.3 Example of a saprouttab Permission Table with SNC of an Internal saprouter for Production Workloads

Configuring Internal saprouttab On-Premise

Listing 7.4 is an example saprouttab configuration for internal SAProuter deployed in production VPC in the second deployment pattern shown in Figure 7.2.

```
# for Solution Manger Diagnostics (SMD) connection in on-premise
P <External SAProuter IP> <Private IP of SMD host in on-premise> 443
# From host in PRD account to External SAProuter
P < Private IP of SMD host in on-premise > <External SAProuter Private IP> 3299
# deny all other connections
D * * *
```

Listing 7.4 Example of a saprouttab Permission Table with SNC of an Internal saprouter for On-Premise Workloads

For external SAProuter systems, you can follow the same strategy as the previous pattern. However, for internal SAProuters, since you do not need SAP registration and public IP requirements, you can do the following activities:

- Enable Amazon EC2 host auto-recovery combined with init scripts to start SAProuter upon restart. Amazon EC2 host auto-recovery will retain the same private IP address. Thus, no changes to the saprouttab in the external SAProuter are needed.

- In the event of AZ failure, you can launch a new SAProuter in the new AZ with the backup image of the internal SAProuter. Moving to the new AZ will change the private IP of the internal SAProuter instance. Then, you need to adjust the respective internal SAProuter IP in the external SAProuter configuration and restart the external SAProuter to activate the change.

- You can follow the same approach as AZ failure for the region failover. You can install it from the backup image and adjust IP information in the external SAProuter configuration in the DR region.

- Alternatively, you can also install it from scratch and maintain the new IP address in the external SAProuter configuration.

7.1.3   AWS Data Provider for SAP

SAP support relies on its monitoring infrastructure within the SAP application to analyze performance issues. This monitoring information is available via the OS monitoring transactions, namely, Transactions ST06, OS07, OS07n, or RZ20 for the central monitoring framework (CCMS). SAPOSCOL feeds data to these transactions via the OS interface.

SAP expects SAPOSCOL to behave the same way in virtualized hosts as in non-virtualized hosts. However, due to some nuances in the hypervisor design, specific high-priority functions and infrastructure metrics must be made transparent to the OS level. Thus, SAPOSCOL cannot capture all the required OS metrics through its typical OS interface.

To supplement this deficiency, SAPOSCOL interfaces with AWS Data Provider for SAP (*http://s-prs.co/v5776134*). AWS Data Provider for SAP is an additional agent running at the OS level to interact internally with various AWS services and furnishes all the required host/OS metrics to SAPOSCOL. In addition, AWS Data Provider for SAP also integrates with the SAP common information model provider to expose these metrics to all the tools that exchange data over SAP CIM (e.g., SAP Solution Manager diagnostics, SAP System Landscape Directory (SLD), CCMS, and third-party monitoring tools) For more information, visit *http://s-prs.co/v5776135*.

As shown in Figure 7.3, AWS Data Provider for SAP gathers metrics from Amazon CloudWatch and the OS virtualization layer. In addition, it also extracts host identity details from Amazon EC2 instance metadata services (IMDS) (*http://s-prs.co/v5776136*) and extracts host lifecycle information through Amazon EC2 service APIs (*https:// www.saponaws.org/EC2status*).

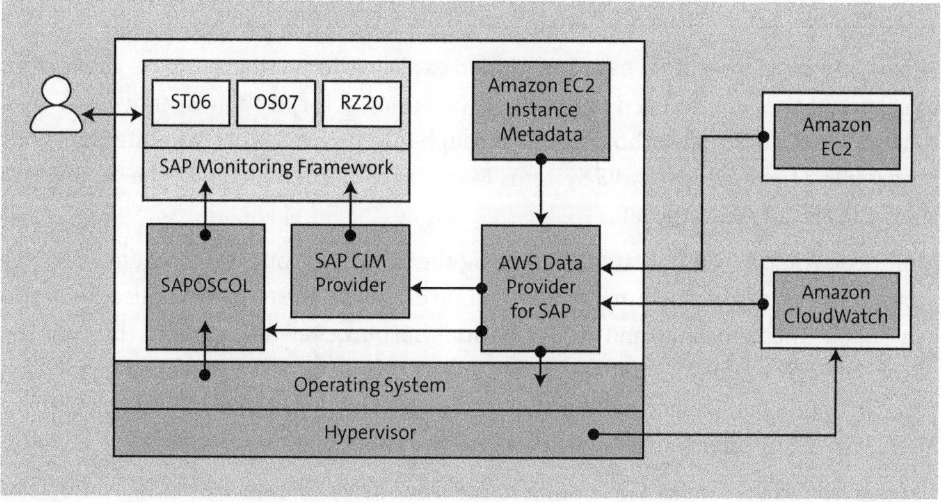

Figure 7.3 AWD Data Provider for SAP

SAP Note 2198693 (*http://s-prs.co/v577626*) describes the critical monitoring metrics that SAPOSCOL extracts from AWS Data Provider for SAP. You can check the SAPOSCOL integration with AWS Data Provider for SAP by running SAPOSCOL in dialog mode (e.g., *saposcol -d*) at the OS level and running the command *dump ccm* at the interactive prompt. You'll see the virtualization metrics if the integration was successful.

> **Important**
>
> AWS Data Provider for SAP is mandatory for running SAP workloads on AWS. Ensure it is active in all the hosts where you've deployed SAP applications. You must enable Amazon CloudWatch detailed monitoring to gather metrics at a more granular level. Refer to SAP Note 1656250 (*http://s-prs.co/v577627*) for more information about SAP support prerequisites.

In the following sections, we'll explain the common operational concepts you must understand to operate SAP workloads effectively on AWS.

7.2   Common Operational Aspects

By the end of this chapter, you'll understand how AWS Systems Manager and Amazon CloudWatch, along with some other AWS native services, can help you manage, monitor, and administer an SAP system in AWS. For operating any solution, including SAP, AWS provides the capabilities shown in Figure 7.4.

First and foremost, to enable these capabilities, your Amazon EC2 instance must be an AWS Systems Manager-managed instance. In addition, these instances must follow an intuitive tagging mechanism to identify or group instances into related entities so that you can target actions to them based on tags/resource groups.

Finally, although we should have mentioned explicitly in each scenario we cover next, to carry out the required actions, each participating actor in the respective activities must possess sufficient authorizations through IAM policies. Once you integrate your Amazon EC2 instance into AWS Systems Manager, you can manage all the capabilities shown in Figure 7.4 on the left.

Moreover, Amazon CloudWatch, leveraging the Amazon CloudWatch Agent, captures metrics and logs from the Amazon EC2 instance, OS, and hosted SAP applications. You can collate these metrics and derive various resource-related insights through the dashboards. By evaluating thresholds and rules emanating out of the resource metrics, you can define and trigger alerts associated with them. Later, you can integrate these alerts into event-driven actions to automate your remediation responses.

Besides, you can leverage AWS Config to continuously evaluate the configuration status of all AWS resources and the custom configurations of your SAP applications. This service will ensure you always follow the best practices/configurations for all the components operating your SAP solutions.

Amazon Inspector's vulnerability scans, IAM access analyzer information, AWS CloudTrail logs, and Amazon VPC flow logs improve the overall security posture of an Amazon EC2 instance and the SAP applications hosted inside it.

We'll discuss these capabilities in detail next.

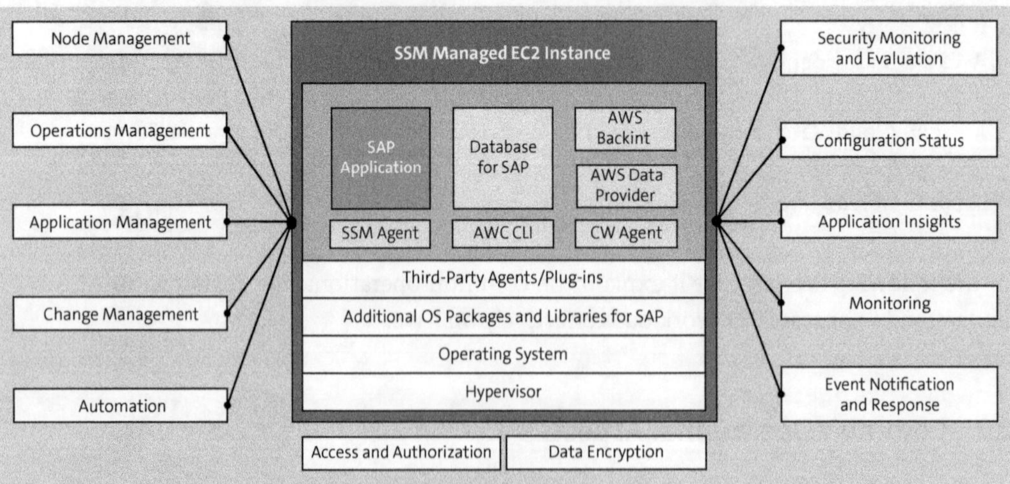

Figure 7.4 Overview of SAP Operations on AWS

7.2.1   Tagging and Resource Groups

Several ways exist to logically reference and group your AWS resources, namely, tags and resource groups, as we'll discuss in this section. You can assign these tags individually while creating the resources or post the creation through AWS Management Console or AWS Command Line Interface (CLI). You can also leverage the **Resource Groups & Tag Editor** to manage these resources in bulk.

Tagging

As mentioned earlier, tagging gives a referenceable identity to an AWS resource. You can use a key-value pair to query, group, and target an action on one or more AWS resources. Tags can be instrumental in managing your AWS resources, but if not governed through organizational standards, they become an unmanageable menace and thus counterproductive. Therefore, your organization should have an intuitive yet consistent tagging policy from the start.

Tagging can also be a powerful tool. You can allow or deny access to critical AWS resources based on tags. You can target critical/disruptive actions on AWS resources based on tags. Also, you can report and manage the cost and the AWS resource consumption using tags. Thus, a helpful strategy is to enforce a robust tagging mechanism to secure critical resources from unintended access. Still, you should keep the ability to delegate escalated access ad hoc to resources when required quickly. We call this access *attribute-based access control (ABAC)*.

Let's start with an example: We only want project team members to start/stop the Amazon EC2 instances belonging to their projects. The ABAC examples shown in Figure 7.5 and Figure 7.6 illustrate how you can enforce and manage access to AWS resources with the help of tags.

```
{
    "Version": "2012 10 17",
    "Statement": [
        {
            "Effect": "Allow",
            "Action":  [
          ❶     "ec2:StartInstances",
                "ec2:StopInstances",
            ],
      ❷    "Resource": "arn:aws:ec2:*:*:instance/*",
      ❸    "Condition": {
                "StringEquals ": {"ec2:ResourceTag/project name": "${ aws:PrincipalTag /project name}"}
            }
        },
        {
      ❹    "Effect": "Allow",
            "Action": "ec2:DescribeInstances",
            "Resource": "*"
        }
    ]
}
```

Figure 7.5 Manage Critical Authorizations Using Resource Tags

You should understand two important aspects from these examples in managing access through tags. You can limit access using IAM condition operators (i.e., `StringEquals`, `StringNotEquals`, `Null`, and `ForAllValues`, discussed next). Second, you can leverage AWS Global Condition Keys to evaluate your tagging conditions. The condition keys most relevant for this occasion are `aws:ResourceTag/tag-key`, `aws:RequestTag/tag-key`, `aws:TagKeys`, and `aws:PrincipalTag`.

As shown in Figure 7.5 ❶, you can only start/stop ❷ an Amazon EC2 instance if the value of the tag "project-name" ❸ matches with the same tag value of the relevant Amazon EC2 instance. Otherwise, they can only read attributes of the Amazon EC2 instance but cannot start/stop them ❹.

However, a fundamental challenge arises with this approach. What if someone else intentionally or accidentally maintains the wrong tag? What if you omit a specific tag entirely? What if someone maliciously manipulates a tag's value to gain additional authorizations?

You can address this challenge by controlling the overall tagging process through service control policies (SCPs), which enforce additional restrictions to set boundaries for the allowed actions within the organizational unit. The example shown in Figure 7.6 controls who can create/delete tags on all the resources, overriding their current authorization even if they acquired through other means.

```
{
    "Version": "2012 10 17",
    "Statement": [
      {
        "Sid":  "DenyTagAddDeleteModification",
❶      "Effect": "Deny",
        "Action": [
            "ec2:CreateTags",
            "ec2:DeleteTags"
        ],
        "Resource": [
❷          "*"
        ],
        "Condition": {
          "StringNotEquals ": {
              "ec2:ResourceTag/project name": "${ aws:PrincipalTag /project name}",
❸            "aws:PrincipalArn ": arn:aws:iam ::111222333444:role/role admins/project admin",
              "aws:RequestTag /project name": [
                  "upgrade",
                  "migration"
              ]
          },
❹        "Null": {
              "ec2:ResourceTag/project name": false
          },
❺        "ForAllValues:StringNotEquals ": aws:TagKeys ": "project name"}
        }
      }
    ]
}
```

Figure 7.6 Tagging Process Governance Using SCPs

In this example, this policy ❶ *denies* the creation/deletion of tags on all the resources ❷, except the principal ❸ performing the action belongs to the same project team and possesses the authorization to do so. In addition, they can only maintain the tag values `upgrade` and `migration` against the tag key `project-name`. Also, they cannot delete the tag `project-name` ❹ if it exists, and they must create the tag if it does not exist. Finally, the SCP ❺ ensures that you maintain the correct tag (`project-name` in this case) without any typos.

> **Note**
>
> More information about controlling access through resource tags can be found at *http://s-prs.co/v577628*.

Resource Groups

A resource group, as the name suggests, is another way of grouping AWS resources to target specific actions on them. As shown in Figure 7.7, you can create resource groups by selecting the **Resource Groups & Tagging Editor** option in the bottom-left corner of the AWS Management Console banner. Alternatively, you can navigate directly to this service using the search bar in AWS Management Console. Resource groups provide even more flexibility and choices to group a resource, using a combination of tags, or AWS resources from an AWS CloudFormation stack. You can leverage resource groups to create ad-hoc combinations and control the scope of the targeted actions.

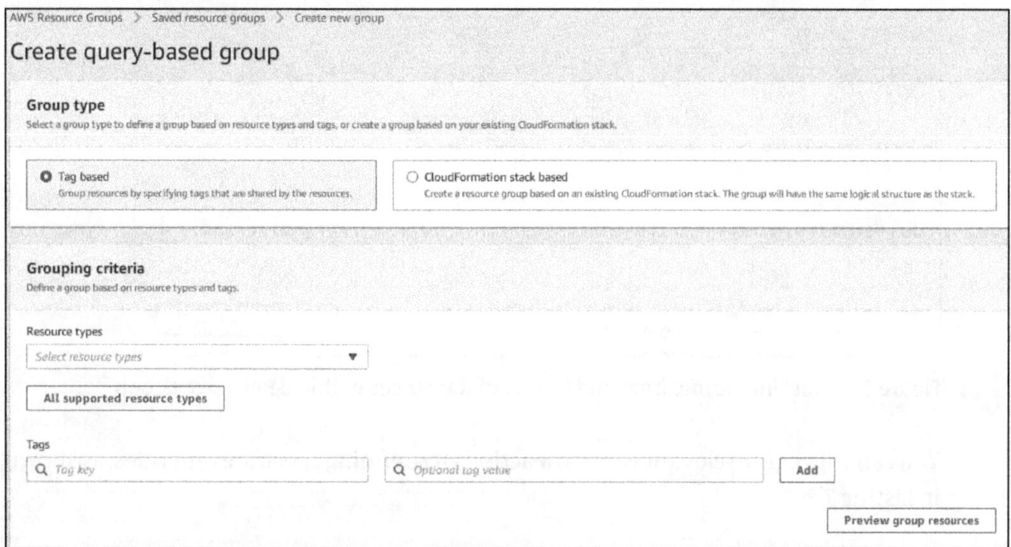

Figure 7.7 Creation of a Resource Group in AWS Management Console

Tags are defined based on a specific characteristic of, or usage context for, those AWS resources. However, resource tags can be subsets (filtering on the resource types) or supersets (combination of multiple tags) of your resources.

For example, let's say you plan to update the SAP kernel of all systems currently on a specific kernel release and patch level. Moreover, it would help if you honored other constraints with the production systems, a subset of them you want to target at a separate maintenance window. Among the rest, some systems with lower OS release levels are unsuitable for the update. The constraints like these are on and on and on. You can unlikely arrive at the proper set of resources using a specific single resource tag to target SAP kernel automation action on them. Often, it is impossible to foresee these situations to formulate a consistent tagging strategy fitting in all potential scenarios. In these situations, where resource groups come in handy.

You can reference these resource groups using AWS Resource Notation (ARN) (e.g., `arn:aws:resource-groups:us-east-2:111122223333:group/Resource-Group`). You can use AWS ARN to control actions and authorizations collectively on all the resources included in those resource groups. This functionality allows IAM administrators to manage task-based permissions and ad hoc role delegations effortlessly. Moreover, you can also enable monitoring over these resource groups using AWS CloudTrail and track their lifecycle events through Amazon EventBridge. Subsequently, you can define follow-up actions using these notifications.

As shown in Figure 7.8, you can enable notifications and track the changes transmitted to Amazon EventBridge through several settings options.

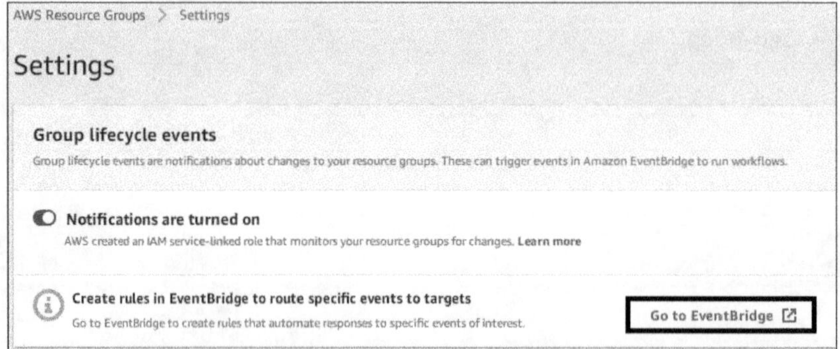

Figure 7.8 Tracking Notification and Events of Resources within a Resource Group

You can trigger the relevant workflow actions by defining specific event rules, as shown in Listing 7.5.

```
aws events put-rule \
    --name "SAPS4ProductionRGEvents" \
    --event-pattern '{"source":["aws.resource-groups"]}'
Output:
```

```
{
    "RuleArn": "arn:aws:events:region:111122223333:rule/SAPS4ProductionRGEvents"
}
```

Listing 7.5 Event Rule to Capture Events from a Specific Resource Group

7.2.2   AWS Systems Manager-Managed Nodes

AWS Systems Manager can perform various host administration activities directly from AWS Console. However, to leverage this functionality, whether a dedicated host, an Amazon EC2 instance, an on-premise system, a virtual machine (VM) on another cloud environment, or an Internet of Things (IoT) device, the node must be an AWS Systems Manager-managed node.

Once a managed node, you can perform the following activities:

- Using **Session Manager**, you can enable users to log on to these nodes securely without sharing passwords and keypairs.

- With **Patch Manager** and **State Manager**, you can evaluate the system's software state and always ensure the most desirable state.

- With **Fleet Manager** and **Inventory**, you can monitor the node's health, compliance status, and state changes and track information on the constituent components.

- Using **SSM Documents**, you can direct the automation activities directly into these nodes to manage the complete lifecycle of all the components in the node.

AWS Systems Manager does not provide this capability to all launched instances by default. To enable this management, you must provide the required IAM authorizations to activate it at the account level or to impart this ability at each node level. As shown in Figure 7.9, using the **Default Host Management Configuration** process in the **Fleet Manager** console, you can turn on the ability of every node to be a managed node at the account level. Then, any Amazon EC2 instance within that account uses Instance Metadata Service Version 2 (IMDSv2) with an AWS Systems Manager Agent version 3.2.582.0 or later installed will automatically become a managed node.

> **Note**
>
> A privileged user must perform this activity, and you must activate this feature in all the regions where you wish to have this automatic capability.

Alternatively, you can enable individual nodes into managed nodes by attaching the IAM policy `AmazonSSMManagedInstanceCore` to the relevant Amazon EC2 instance profile. Then, as shown in Figure 7.10, you can manually associate nodes using the **Quick Setup** option in AWS Console from the **AWS Systems Manager** service page.

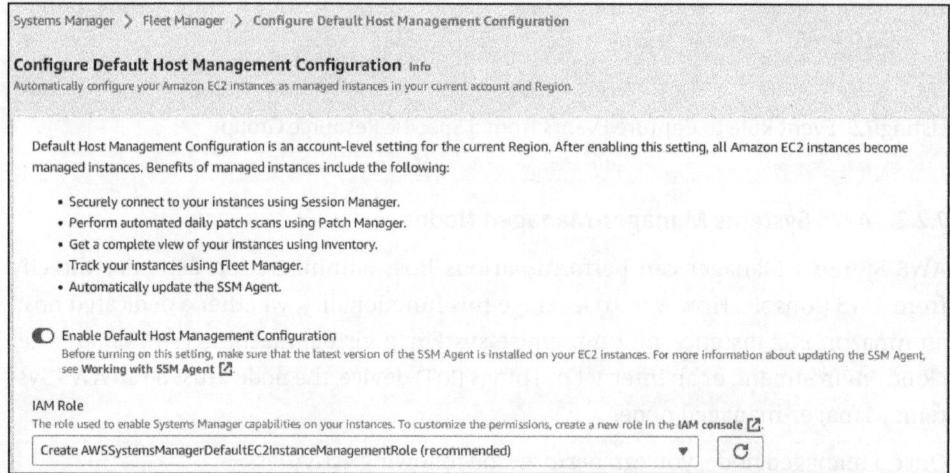

Figure 7.9 Default Host Management Configuration to Convert Nodes into AWS Systems Manager-Managed Nodes

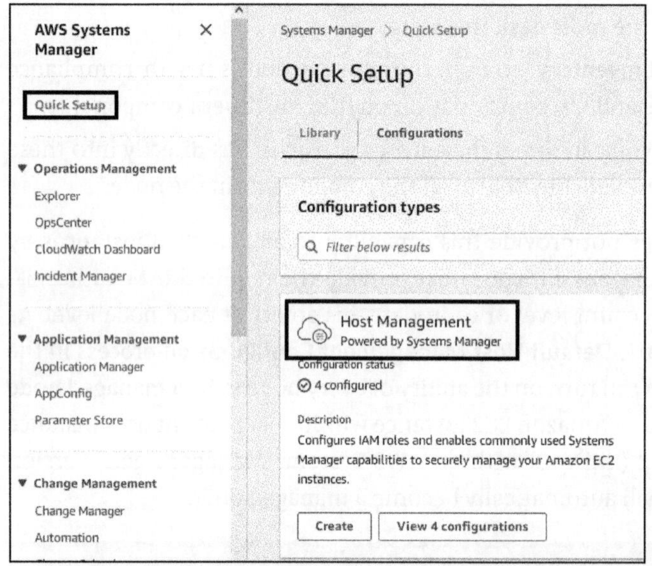

Figure 7.10 Enable AWS Systems Manager Capabilities Using Quick Setup

Note

In either scenario, the presence of AWS Systems Manager agents in the relevant nodes is a mandatory requirement. All AWS Systems Manager actions are integrated through the AWS Systems Manager agent. Specific Amazon Machine Images (AMIs) come with this agent preinstalled; a list of these AMIs is available at *http://s-prs.co/v577629*. If not preinstalled, you can manually install an AMI by following the instructions provided at *http://s-prs.co/v577630*.

7.2.3   Observability

Observability is the ability to assess, infer, and predict the state of your system—in other words, the ability to know how your system is, how it behaved historically, and what potential situation you might expect in the near future based on the collective knowledge of its progress. Under the observability domain, you can achieve these capabilities by harnessing the three essential data streams: metrics, logs, and traces generated by the system. Amazon CloudWatch addresses a significant portion of observability. However, you can augment additional context with the data gathered from AWS Systems Manager capabilities.

As an SAP system administrator, you may have some or all of the following observability requirements:

- The ability to see the high-level health of all the SAP systems on a single pane.

- When required, the ability to dig deep into the metrics, logs, and traces of the system of interest.

- The ability to collect and visualize the complete inventory of system components and release levels.

- The ability to extend the monitoring capabilities beyond what is natively available by capturing custom metrics and logs and deriving meaningful insights from this data.

- The ability to prioritize critical metrics and insights and define alerts/notifications based on certain thresholds.

- The ability to ensure that all the SAP systems are in line with your defined compliance requirements.

- The ability to bind information across multiple data sources within a system to accelerate the troubleshooting process.

- Additionally, you may want to expose your SAP system metrics to other third-party observability tools and solutions.

- Optionally, you may want to collate all your organizational observability data in a single place and discover new usage patterns to establish proactive mechanisms.

We'll guide you through fulfilling these observability requirements in the following sections.

Monitoring versus Observability

In the vernacular, these terms are often used interchangeably. However, both are distinct in relevance and scope. Monitoring is a subset of observability; monitoring limits its scope to leveraging metrics and logs. However, observability extends its scope to include application traces.

Monitoring is more relevant for system administrators, but observability is useful for DevOps engineers. In the context of SAP, we rely heavily on the monitoring side of things, with a few exceptions like dealing with traces in Amazon CloudWatch Application Insights, which we'll discuss later in this section.

Single Pane of Glass: AWS Systems Manager Application Manager

In two ways, you can create a single pane of glass to view all your SAP systems in a single place. One option is through the Application Manager feature in the AWS Systems Manager service. The second option is through the **Resource Health** feature under **Application Monitoring** within the Amazon CloudWatch service.

However, both views have distinct uses and relevance. AWS Systems Manager is useful for managing application inventory and lifecycle activities. This inventory helps you administer systems through capabilities native to AWS Systems Manager, which we'll discuss in Section 7.2.5. On the other hand, AWS CloudWatch application monitoring, as the name suggests, monitors the state and health of the application.

You should first create an application to leverage the Application Manager in AWS Systems Manager, as shown in Figure 7.11. The creation of an application is nothing but grouping a collection of AWS resources under a single name. This process is the same as when we created resource groups earlier in Section 7.2.1. However, as shown in Figure 7.11, you have several ways to define this application via the dropdown list under the **Create Application** tab. Each method has its own distinct purposes and objectives.

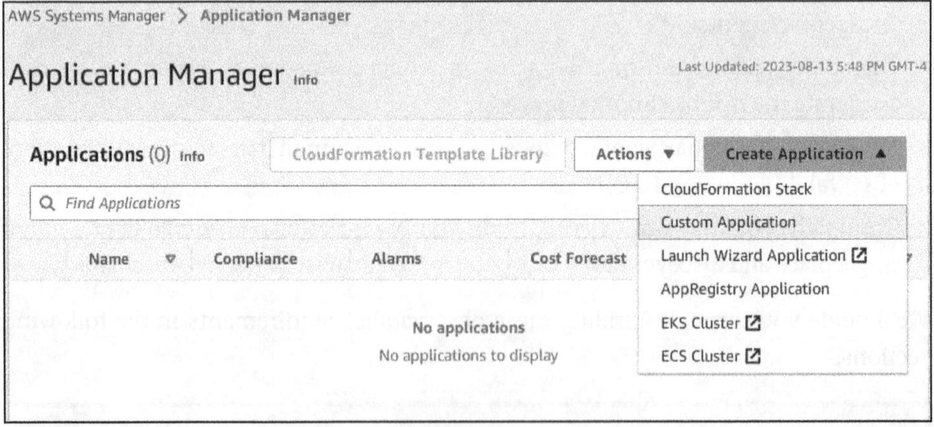

Figure 7.11 Creating an Application to Leverage Application Management in AWS Systems Manager

If you launch your SAP application using a AWS CloudFormation stack or AWS Launch Wizard service, you can use the name of that AWS CloudFormation stack or the name of AWS Launch Wizard deployment to create the application. In addition, you can use any other infrastructure as code (IaC) mechanism available through the AWS Service Catalog service to deploy your applications. You can use the relevant AppRegistry information to create the application. However, to define a custom selection, choose the **Custom Application** option from the dropdown list shown in Figure 7.11.

Custom application options will let you group resources in various ways. As shown in Figure 7.12, you can create an application by including resources based on resource tags, resource groups, and AWS CloudFormation stacks. You can combine resources from one or more options via the checkbox options available on the screen. You can group up to 15 components to create your custom application. Note that the screen shown in Figure 7.12 features other items, like the name and description of the application, that we have not displayed.

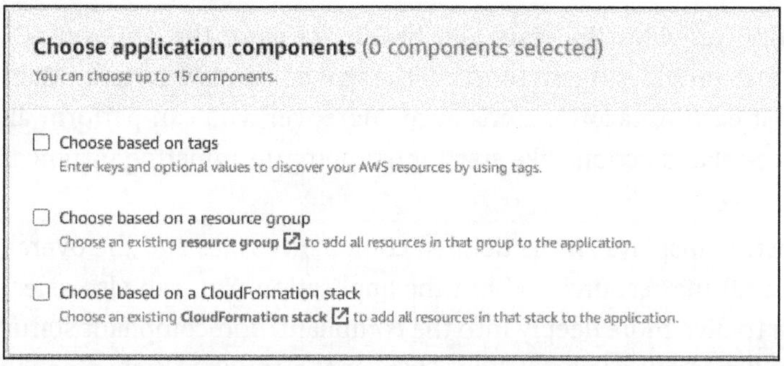

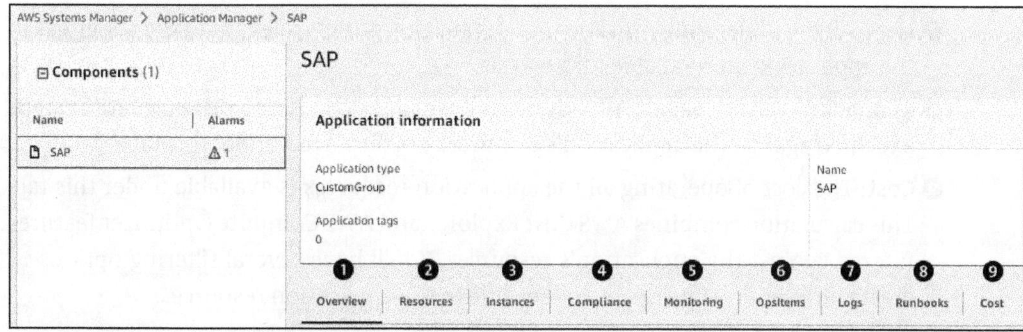

Figure 7.12 Creating a Custom Application for Application Management in AWS Systems Manager

Once you create the application, click on it to see the screen shown in Figure 7.13. The widget on the left under **Components** will show an application's overall status and whether any alarms need your immediate attention. On the right, you'll see several widget tabs that we've numbered from one to nine.

Figure 7.13 Managing SAP Applications Using AWS Systems Manager Application Manager

Each widget tab will have provided information on all the resources within the selected application against different areas, in the following ways:

❶ **Overview**: This tab will provide high-level graphical insights across all the tabular views in the sequence as separate widgets. Click the **View all** button in the top corner of each device to navigate you to the relevant tab for further details.

❷ **Resources**: Under this tab, you can display all the resources included within an application. Select the radio button against each resource and click **Resource Timeline** for details about all the API calls that were targeted against it. From this tab, you'll be taken to the AWS CloudTrail console for a timeline view of all the API action on that specific resource. You can also run automation tasks by selecting the relevant automation documents dropdown list from the **Execute runbook** tab. Through this option, you can perform system administration tasks like extending disks, making backups, resizing Amazon EC2 instances, and starting/stopping Amazon EC2 instances or SAP systems.

❸ **Instances**: This section provides the state and health status of the Amazon EC2 instance. Besides, it also provides the status of all the scheduled AWS Systems Manager activities against each Amazon EC2 instance. Moreover, you can perform all Amazon EC2 instance-related actions like starting/stopping/terminating instances from this screen.

❹ **Compliance**: This section displays all the defined compliance rules and the overall compliance status of all the resources within the application. You can also select individual resources to dive more deeply into the compliant/non-compliant status of each defined compliance rule against it. You also have the option of managing the remediation of failed compliance rules from this screen.

❺ **Monitoring**: Under this tab, you can view Amazon CloudWatch Application Insights, resource alarms, and its management of all the resources within the defined application.

❻ **OpsItems**: This tab displays all the operational incidents triggered through resource alarms or events of the application resources. You can view and manage all the open incidents with their priorities in a single place.

❼ **Logs**: All the underlying Amazon CloudWatch log groups relevant to the resources in the application are available for analysis.

❽ **Runbooks**: All the logs of AWS Systems Manager documents or automation steps executed against the application's resources are displayed under this tab.

❾ **Cost**: The cost of operating all the application resources is available under this tab. This calculation combines AWS Cost Explorer and AWS Compute Optimizer features filtered against the application's resources. You'll have several filtering options to create different cost perspectives specific to the application resources.

Single Pane of Glass: Amazon CloudWatch Resource Health

As discussed earlier, you can create a similar single-pane of glass under the **Application Monitoring** section of **Resource Health** in Amazon CloudWatch. However, this representation is mainly focused on monitoring the Amazon EC2 instances in your account.

Another significant difference of this view from earlier is that there is no grouping of resources based on application. However, you can achieve a similar perspective by leveraging the filters available in the left pane of the screen. You can visualize resources based on their belonging in VPC/AZs or based on their instance types or processor architectures. You can also filter them based on resource tags. You can work backwards to define tagging strategy from the perspective that you wish to see here.

Figure 7.14 shows an example perspective that we defined based on the system type filter viewed in **Map view**. We grouped our resource tags depending on the perceived role of the system. Alternatively, you can also define SAP SID as a resource tag to view all the SAP systems grouped by SAP SID. The options are unlimited to achieve your monitoring needs. Each square block represents a node with a different color coding on it. The coloring is based on the scale defined in the pane on the left against the selected metric (for example, **CPU Utilization**, as shown in Figure 7.14). This view also indicates the running state and the health of the underlying Amazon EC2 instance. If any instance metric enters an **Alert** mode, you can see them here on those instance blocks.

You can drill down into the details by clicking on the required block. A popup window will appear where you can view these details by navigating into the Amazon Cloud-Watch dashboard or switching to the details section of the Amazon EC2 service console.

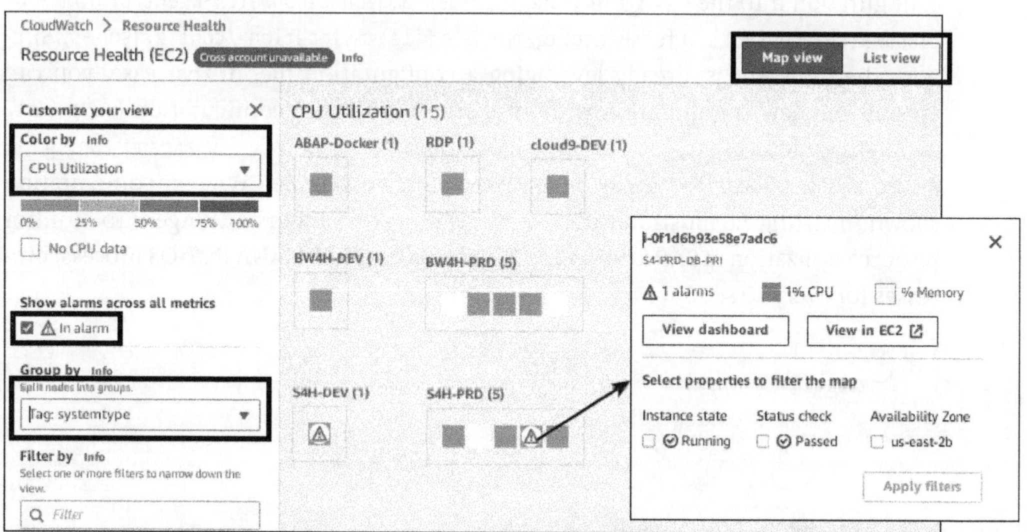

Figure 7.14 Single Pane of Glass Monitoring for SAP in Amazon CloudWatch

Amazon CloudWatch Agent

As you know, Amazon CloudWatch is a metric repository for AWS resources. By default, some AWS services (*http://s-prs.co/v577631*) send a specific list of metrics (*http://s-prs.co/v577632*) to Amazon CloudWatch. Under the **Basic** monitoring, you can send the default metrics at 5-minute intervals to Amazon CloudWatch at no cost. However, for an additional charge, you can also enable detailed monitoring, a mandatory SAP requirement. With **Detailed** monitoring, you can increase the metric resolution to 1-minute intervals.

To extend your monitoring framework beyond what is provided by default, you can gather additional metrics using the Amazon CloudWatch Agent (CWAgent). You can find this list of additional metrics at *http://s-prs.co/v577633*. You can also collect custom metrics, logs, and traces from your applications using CWAgent. Based on your OS, you can install CWAgent by following one of the convenient ways described at *http://s-prs.co/v577634*.

The CWAgent configuration file controls what metrics, logs, and traces CWAgent collects. This file is a JavaScript Object Notation (JSON) file that you can populate invoking configuration wizard at the OS level of the Amazon EC2 instance (`sudo/opt/aws/amazon-cloudwatch-agent/bin/amazon-cloudwatch-agent-config-wizard`) or manually creating or editing the file using your favorite text editor. By default, the file resides at `/opt/aws/amazon-cloudwatch-agent/etc/amazon-cloudwatch-agent.json` on a Linux server or at `$Env: ProgramData\Amazon\AmazonCloudWatchAgent\amazon-cloudwatch-agent.json` on a Windows server. Ideally, you can have any name for the configuration file and place it anywhere in the system; however, you must pass that file on as a parameter to load this configuration into the CWAgent (`sudo /opt/aws/amazon-cloudwatch-agent/bin/amazon-cloudwatch-agent-ctl -a fetch-config -m ec2 -c file:<mylocation>/config.json -s`). Suppose the CWAgent is already leveraging a configuration file. In that case, you can append the new configuration without disrupting the old configuration (`sudo /opt/aws/amazon-cloudwatch-agent/bin/amazon-cloudwatch-agent-ctl -a append-config -m ec2 -c file:<mylocation>/mynewconfig.json -s`). The example CWAgent configuration shown in Listing 7.6 illustrates the use of the `procstat` plugin of CWAgent to monitor resource utilization of SAP (*disp+work* OS process) and SAP HANA (*hdb* OS process) processes for every 10 secs.

```json
{
    "metrics": {
        "metrics_collected": {
            "procstat": [
                {
                    "pattern": "dw.sap",
                    "measurement": [
                        "cpu_time",
                        "cpu_time_system",
                        "cpu_time_user"
                    ],
                    "metrics_collection_interval": 10
                },
```

```
                    {
                "pattern": "hdb.sap",
                "measurement": [
                    "memory_data",
                    "memory_locked",
                    "memory_rss",
                    "memory_stack",
                    "memory_swap"
                ],
                "metrics_collection_interval": 10
            }
        ]
        }
    }
}
```

Listing 7.6 Integrating procstat Plugin into CWAgent Configuration to Capture SAP Process Metrics

Alternatively, you can also push custom metrics programmatically into Amazon CloudWatch using AWS CLI or Amazon CloudWatch APIs in AWS SDK (put-metric-data).

Amazon CloudWatch Application Insights for SAP

So far, we've discussed capturing basic metrics about AWS infrastructure elements using Amazon CloudWatch. We even explored ways to gather enhanced metrics about these resources with the help of CWAgent. With CWAgent custom configurations, AWS CLI, and AWS SDK APIs, you can extend this monitoring framework to populate custom metrics into Amazon CloudWatch.

Besides these capabilities, Amazon CloudWatch Application Insights will expand the observability capability into the application layer. Using this capability, you can gather SAP application-specific metrics. You can extract these metrics from various locations relevant to SAP NetWeaver and SAP HANA.

You can onboard applications from the **Application Insights** option under the **Insights** section of the Amazon CloudWatch console. You can select the application from earlier defined resource groups or AWS CloudFormation stacks. During onboarding, it will install the required components into the respective Amazon EC2 instances, discover the available SAP components, adjust and push the configuration files, and start the necessary services at the instance level to publish the extracted metrics to Amazon CloudWatch.

For this task, perform these steps in the following sequence:

1. AWS Systems Manager Distributor, discussed further in Section 7.2.5, will install and configure the following Prometheus exporters:
   - JMX exporter
   - SAP-HANADB exporter
   - HACluser exporter
   - SAP-SAPHost exporter

> **Note**
>
> Prometheus is a cloud-native, open-source observability and alerting toolkit. It pulls metrics data in time-series format from Prometheus-compatible applications. You can expose these metrics from the application's observability endpoint. Through native integration, the Amazon CloudWatch agent can consume these metrics directly from the same observability endpoint of these applications.
>
> Prometheus exporters are metric aggregators from non-Prometheus application metric sources to port them into a Prometheus-ingestible format. A configured HTTP webserver port exposes these metrics. Then, CWAgent pulls metrics from this endpoint based on the CWAgent configuration (similar to which we customized earlier in this section). As before, exporters consume the following metric sources in SAP NetWeaver and SAP HANA databases to produce enhanced SAP application insights:
>
> - **JMX exporter**
>   Scraps metrics from SAP logs and traces that CWAgent sent to log groups in Embedded Metric Format (EMF); for more information, visit *http://s-prs.co/v577635*.
>
> - **SAP-HANADB exporter**
>   It queries SAP HANA monitoring tables using database interface and expose the results; for more information, *http://s-prs.co/v577636*.
>
> - **HACluster exporter**
>   It calls up Pacemaker commands to periodically gather the HA performance metrics; for more information, *http://s-prs.co/v577637*.
>
> - **SAP-SAPHost exporter**
>   It integrates with SAP Host Control program to gather SAP application metrics; for more information, *http://s-prs.co/v577638*.

2. The AWS Systems Manager Document `AWSEC2-ApplicationInsightsCloudwatchAgentInstallAndConfigure` will install Amazon CloudWatch Agent (if it does not exist) and adjust its configuration to send custom logs and traces relevant to SAP HANA and SAP NetWeaver to the respective Amazon CloudWatch log streams. Besides, it enables Prometheus observability to get metrics published from the exporter services that you configured earlier.

3. The following logs and traces are collected for metrics extraction:
   - SAP NetWeaver:
     - SAP work process developer traces (*/usr/sap/<SID>/*/work/dev_w**)
   - SAP HANA databases:
     - SAP HANA process traces (*/usr/sap/<SID> /HDB<nr>/*/trace/*.trc*)
     - SAP HANA process logs (*/usr/sap/<SID>/HDB<nr>/*/trace/*.log*)
   - HA configuration:
     - */var/log/pacemaker/pacemaker.log*

   For more information about the logs and supported metrics by Amazon Cloud-Watch Application Insights at *http://s-prs.co/v577639*

4. Amazon CloudWatch Application Insights will set up alarms and track these alarm states through Amazon EventBridge. You can respond to these events with notifications or remediation actions.

5. Based on the severity of the events, you can integrate with AWS Systems Manager OpsItems to incorporate them into your organization's incident management process.

AWS Systems Manager Custom Inventories

The AWS Systems Manager feature **Inventory** under **Node Management** provides a holistic view of all the software components within that specific node. This feature not only reports what OS variants or OS packages are present in that node but also provides the ability to capture details about applications running inside the node using the custom inventory option. This feature provides OS administrators and SAP Basis administrators a one-stop shop to manage software components that make up this specific node and its complete list of SAP application inventory.

An AWS Systems Manager custom inventory is an intuitive way to enable system administrators to gather all the required information to operate systems at a scale. However, the custom inventory feature needs a little intervention to shape outcomes as you want.

Generally, an Amazon-owned AWS Systems Manager document called `AWS-GatherSoftwareInventory` gathers the default systems inventory. The default periodicity for invoking this job is every 1 hour. However, you can manually trigger it as many times as you wish to capture changes in inventory. This job has an inherent capability to pull a custom inventory if you place inventory data, as a JSON file, in one of the folders listed in Table 7.1.

Operating System	Path
Linux	*/var/lib/amazon/ssm/node-id/inventory/custom*
Windows	*%SystemDrive%\ProgramData\Amazon\SSM\InstanceData\node-id\ inventory\custom*
macOS	*/opt/aws/ssm/data/node-id/inventory/custom*

Table 7.1 Locations of AWS Systems Manager Custom Inventory JSON Files

This JSON file must have the *.json* extension and follows the template shown in Listing 7.7.

```
{
    "SchemaVersion": "1.0",
    "TypeName": "Custom:SAP",
    "Content": {
        "hostname": "bw4dev",
        "sid": "B4D",
        "sap_instance_details": "ASCS01_D02_HDB00",
        "sap_kernel": "785_400",
        "sap_component_details": "SAPBASIS_756_00",
        "os_vendor": "Redhat",
        "os_level": "8_2",
        "os_cpu_num": "32",
        "os_memory_size": "256",
        "ec2_type": "r6i.8xlarge",
        "instance_id": "i-0972e37bbf71c0cd4",
        "db_type": "SAP HANA",
        "db_level": "2.00_059",
        "db_size" : "240GB"
    }
}
```

Listing 7.7 JSON for Publishing an SAP Inventory into an AWS Systems Manager Custom Inventory

The list of the inventory details that you can construct is limited only by your imagination and your requirements. In the next chapter, we'll discuss how to populate these attributes programmatically through automation. The attributes captured are available for querying and filtering from the AWS Systems Manager Inventory screen. Besides, you can collect this inventory information centrally into an Amazon S3 bucket as "Resource DataSync." Also, by integrating this location with AWS Glue and Amazon Athena, you can query the data to build various visualization models and provide insights into your system's inventory. For more information about "Resource Data-Sync" and building custom inventory queries, refer to *http://s-prs.co/v577640*.

The example shown in Figure 7.15 demonstrates a use case when SAP Basis administrators want to know the current SAP kernel levels of all the systems in the landscape. Similarly, you can leverage the resource data sync feature to extract information from several other application-specific inventories.

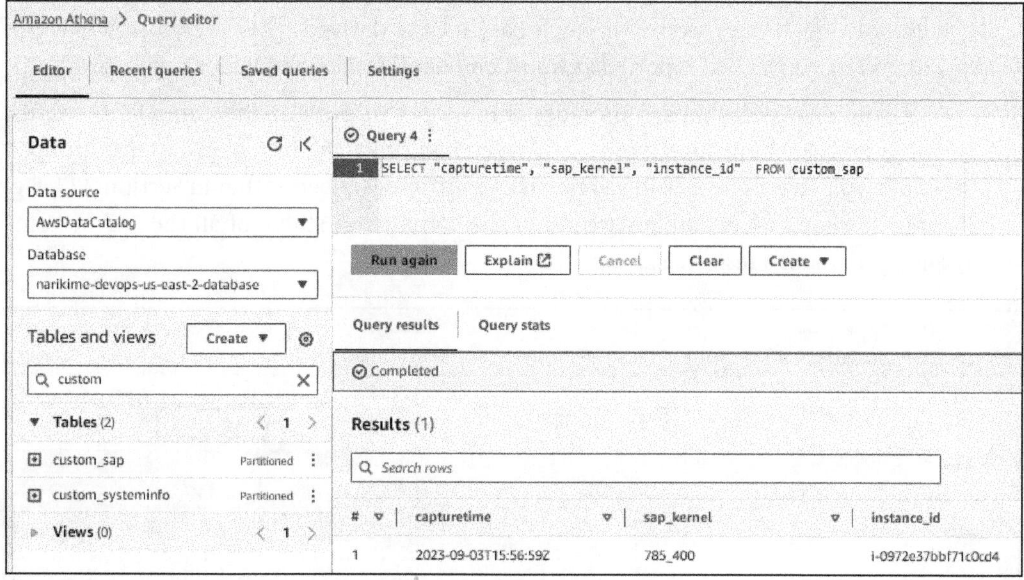

Figure 7.15 Query Custom Inventory Based on Application Attributes

Note

AWS Systems Manager can support up to 20 custom inventory types. You can have multiple JSON files to describe the custom inventory. All the JSON files in the custom inventory folder will be read and integrated into the inventory information. Note that the attribute values of the JSON file only support data of type String. For now, this JSON file does not support all other data types like numbers, dictionary, or arrays.

Compliance Status of Your Systems

As a system administrator, your primary goal is ensuring that all the systems under your control are always in the desired software state as you've defined. In addition, you must also ensure that these software components are up-to-date and free of vulnerabilities.

AWS Systems Manager State Manager uses associations to track and perform actions on managed nodes to confirm or achieve the desired state. Associations are nothing but AWS Systems Manager documents. Some of these AWS Systems Manager documents are predelivered and immediately available for use. However, you can also define a custom association to track your unique requirements. You can report the results of these AWS Systems Manager documents back to the AWS Systems Manager State Manager. The nodes that deviate from the desired state are reported as **Non-compliant** to the **Compliance** dashboard. You can see this status in the **Compliance** feature of AWS Systems Manager, under **Node Management**.

Similarly, deviations from the patch baseline (as derived by OS vendors or custom curated by you) is also reported as **Non-Compliant** in the **Compliance** dashboard.

The compliance dashboard provides three ways to present the compliance status: based on the compliance types (i.e., **Association** or **Patch**), based on systems grouped under a patch group, or using a resource group as described earlier in Section 7.2.1. Figure 7.16 shows an example scenario of the compliance status of all the SAP systems grouped under a resource group.

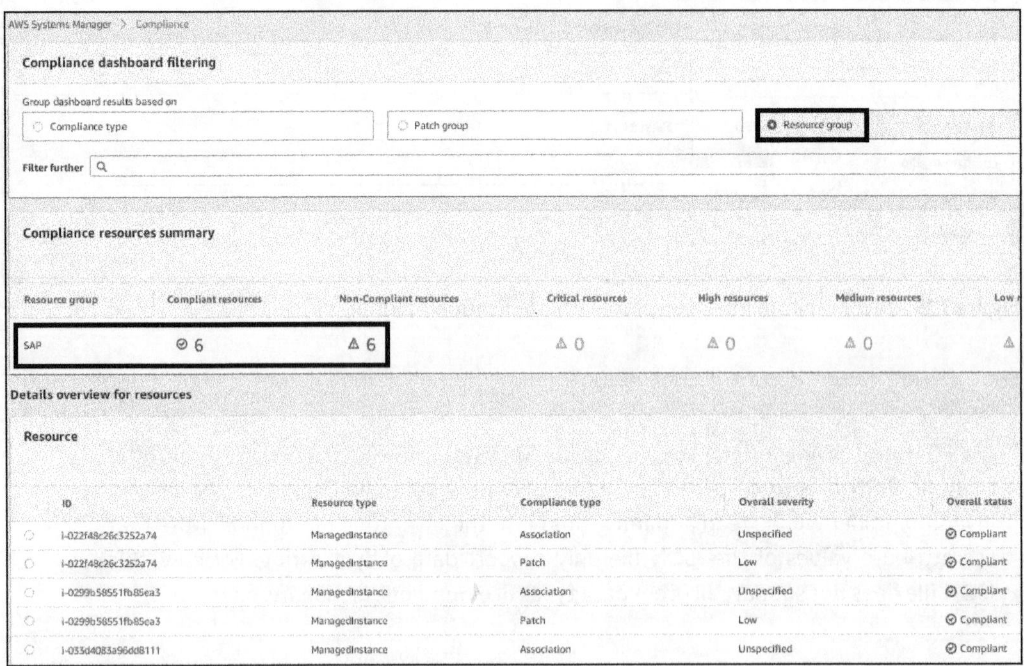

Figure 7.16 SAP Systems Compliance Status Using AWS Systems Manager Compliance Feature

Upon selecting the instance hyperlink, shown in the bottom section of Figure 7.16, you'll see details about associations and patching activities that have drifted from the expected system state.

Amazon CloudWatch Dashboards, Alarms, Events, Logs Insights, and Anomaly Detection

You knew what data sources to tap into to get the data points of your concern. After the required setups are in place, the relevant metrics, logs, and traces flow into the Amazon CloudWatch. This data resides within Amazon CloudWatch until its retention is reached. The resolution of the metrics influences its default retention. Table 7.2 explains the relationship between metric resolution and its default retention periods.

Metric Resolution	Retention Period
1 to 59 seconds (<60 seconds)	3 hours
1 minute	15 days
5 minutes	63 days
1 hour	455 days (15 months)

Table 7.2 Amazon CloudWatch Metric Default Retention Periods by Metric Resolution

However, short-resolution metrics are automatically consolidated into higher-resolution metrics and moved across the designated storage areas. For example, Amazon CloudWatch reduces the metrics to less than 1-minute resolution into 1-minute resolution at the end of 3 hours and keeps it for 15 days. At the end of the 15th day, these 1-minute resolution metrics are consolidated into 5-minute resolution metrics and retained for 63 days. The same cycle continues at the end of the 63rd day to move them to a 1-hour resolution and keep it for 15 months.

If you need to keep the exact resolution for much longer, you can move them to external storage areas like Amazon S3. Later, you can integrate this data with Amazon Athena to derive more granular insights. You can also query and correlate this data with other metrics to create a unified monitoring view.

Once the data resides in Amazon CloudWatch, as shown in Figure 7.17, you can utilize this data in the following ways:

- **Dashboards**

  You can create a customizable metric widget in the Amazon CloudWatch console against selected metrics to track the metrics behavior over some time in the chosen graphical representation (e.g., graphs and charts). Amazon CloudWatch application insights for SAP HANA and SAP NetWeaver come with prebuilt dashboards under custom dashboards view as **ApplicationInsights-SAP**. Besides, you can also create your custom dashboard either directly from the **Dashboards** option in the Amazon CloudWatch console or by using the **Explorer** option under the **Metrics** feature in the same console.

- **Alarms**

  You want Amazon CloudWatch to notify you when one or a collection of metrics hits the threshold or recurring of a specific set of events within a specified duration. You can define Alarms to raise upon meeting the criteria so that you can use these alarms to determine the next course of action. This Amazon CloudWatch feature helps you focus on the application's critical areas. Amazon CloudWatch application insights for SAP have predefined alarm states for each captured metric under this functionality.

- **Events**

  By default, any change in the state of AWS resources gets notified to the default event bus in Amazon EventBridge. Likewise, Amazon EventBridge captures the events raised by Amazon CloudWatch alarms. Amazon CloudWatch Application Insights uses **Alarm State Change** events to transmit the SAP application into the Amazon CloudWatch Application Insights console. Similarly, they can integrate with Amazon Simple Notification Service (SNS) notifications to notify systems administrators about the critical onset of these alarms. Alternatively, these events can invoke AWS Systems Manager Documents, or AWS Lambda functions to automate remediation activities. For example, you can assign a volume resize AWS Systems Manager document to a volume size threshold alarm forwarded to Amazon EventBridge. This way, you can optimize your operational teams' efforts and improve mean time-to-response (MTTR) times to avoid system disruption.

- **Logs insights**

  Using patterns or search strings from the logs residing in the Amazon CloudWatch logs, you can extract metrics or occurrences of events to present them in the desired format. You can use query syntax in combination with regex patterns to sieve through the related log entries from all the log streams of the multiple log groups. By binding them against a timestamp, you can relate all the events co-occurring in all the impacted systems in a single place. You can find the query syntax at *http://s-prs.co/v577641* and the regex pattern at *http://s-prs.co/v577642*. Amazon CloudWatch log insights are convenient during the troubleshooting of issues. They provide a complete picture across systems to understand what happened in the respective application logs and traces.

- **Anomaly detection**

  You can find situations through Amazon CloudWatch alarms when a metric drift from its expected behavior. Anomaly detection employs machine learning algorithms to form a band of possible values of a metric from its historical behaviors. If the metric breaches this band, alarms notify the defined targets and perform associated actions.

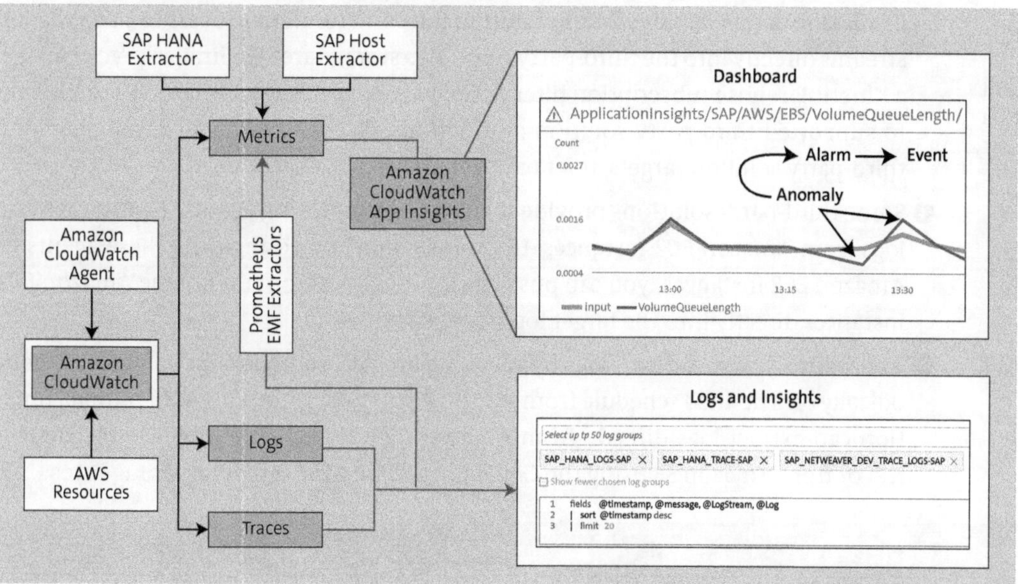

Figure 7.17 Amazon CloudWatch Dashboards, Alarms, Events, Logs Insights, and Anomaly Detection

Amazon CloudWatch Integration with Other Observability Solutions

Ideally, you have an established mechanism for monitoring that is widely accepted across your teams and has become an integral part of your IT operations. In this case, switching to a new AWS monitoring solution by reinventing your operational model is not a rational and cost-effective approach. However, on the flip side, you might lose on the native cloud observability capabilities offered by AWS. To enjoy the best of both worlds, you can integrate Amazon CloudWatch with your current monitoring solution.

Although this integration process varies from vendor to vendor, this process mainly involves the high-level steps shown in Figure 7.18. For vendor-specific configuration, refer to your vendor's documentation.

In general, you would follow these steps:

❶ You need a user with a role granting them all the required API actions for managing Amazon CloudWatch logs. This user must also have a Secret Key (SK) and Secret Access Key (SAK) to allow remote programmatic access.

❷ You may need to install an addon or activate the extension within a third-party solution to enable monitoring functions on AWS. You'll maintain the SK and SAK in its configuration process. This configuration process also involves deploying the required AWS Lambda functions and setting up web hook URLs in AWS.

❸ You can forward the required Amazon CloudWatch log groups to a third-party solution using the Log Group Subscription filter feature, as shown in Figure 7.19. Under the **Subscription filters** tab ❶, create a subscription filter and assign the AWS

Lambda function deployed earlier during the configuration process to push the log streams directly into the third-party solution's storage area ❸. Similarly, you can use a Kinetic Firehose subscription filter to send log data either to Amazon S3 or directly to supported third-party solution destinations ❷. For a complete list of defended third-party solution targets, refer to *http://s-prs.co/v577643*.

❹ Some third-party solutions provide metrics/log forwarding agents to push metrics/logs from Amazon EC2 instances. Upon installing and configuring these agents in Amazon EC2 instances, you can push the metrics/logs from the source Amazon EC2 instances directly into the target log storage area.

❺ Alternatively, you can pull logs by invoking an AWS Lambda function in AWS periodically on a defined schedule from the third-party solution. This AWS Lambda function can extract logs directly from Amazon CloudWatch using the GetMetricData API or from Amazon S3 using the GetObject API into the target log storage area.

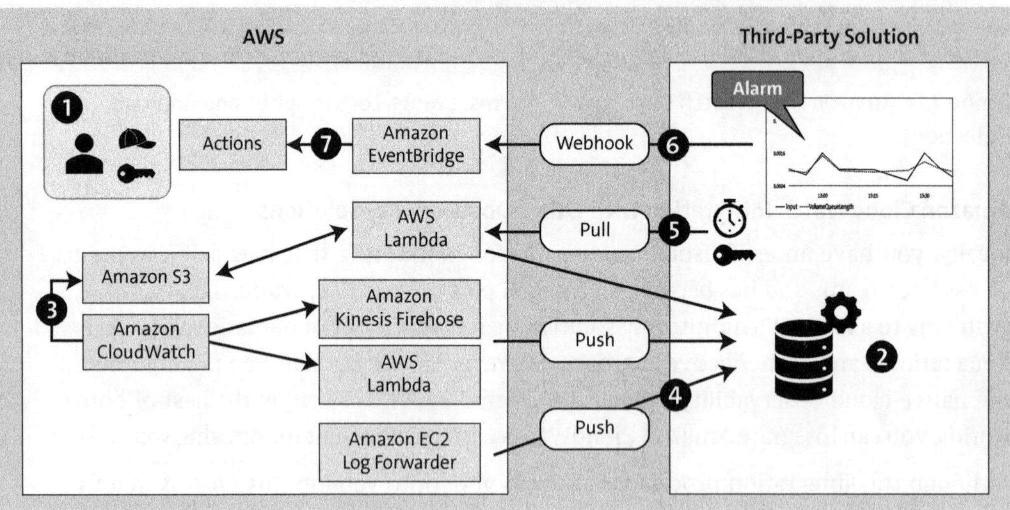

Figure 7.18 Amazon CloudWatch Integration with Third-Party Monitoring Solutions

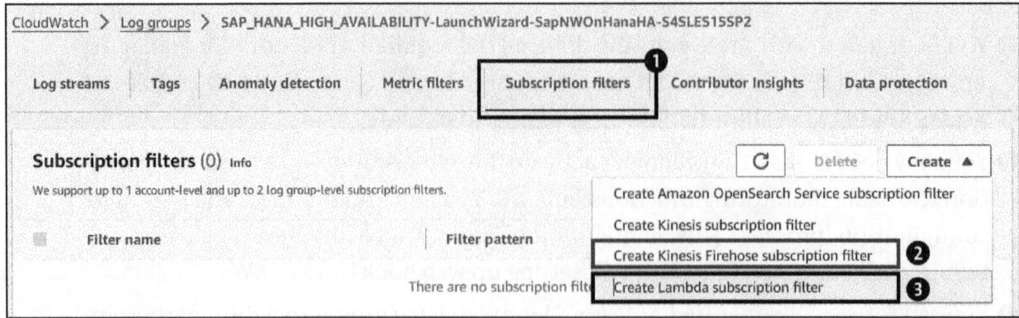

Figure 7.19 Amazon CloudWatch Logs Subscription Filters for Integration with Third-Party Solutions

❻ You can also configure a webhook to Amazon EventBridge to send alarms and events.

❼ You can define the desired remediation actions by integrating them with AWS Systems Manager documents or AWS Lambda functions as event targets.

> **Note**
>
> Be economical on the amount of data that you transmit across these tools. Regex filters can be set up to transfer only the required patterns to target subscribers. Otherwise, your data transmission and storage costs will rise.

Build Your Own Solution: Observability Data Lake

You can consolidate all your observability data into a central place and create your own observability solution in several ways. You can leverage some natively available mechanisms to avoid the heavy lift of building your own observability solution and save on licensing and maintenance overhead through third-party solutions.

Before heading in this direction, you must finalize a few items to consume this data and successfully create actionable insights, such as the following items:

- **The scope**
  The list of namespaces, dimensions, metrics, logs, units, resolutions, thresholds, and alarms.

- **The data sources**
  AWS resources, applications, and service endpoints.

- **The extraction tools**
  AWS SDK, AWS CLI, agents, or log scrappers.

- **Data transfer mechanism**
  Push versus pull, on-demand, scheduled, or in real time.

- **Data storage**
  Amazon S3, Amazon OpenSearch, Amazon Redshift, or Amazon Relational Database Service (RDS).

- **Log format/data schema**
  JSON, CSV, XML, Parquet, or text.

- **Data model**
  Text/search, relational, document, key-value, or graph data models.

The most critical metrics for a typical SAP workload is mainly from infrastructure services like Amazon EC2, Amazon EBS, and Amazon VPC. Since these services tightly integrate with the Amazon EC2 hypervisor, you can capture all these metrics from an Amazon EC2 instance using many tools and agents. Some popular agents/tools, shown in Figure 7.20, operate within an Amazon EC2 instance. As discussed earlier, these agents/tools provide rich infrastructure metrics, clubbing them with SAP application layer metrics. In addition, the AWS Systems Manager agent sends AWS Systems Manager inventory and compliance statuses to the AWS Systems Manager service.

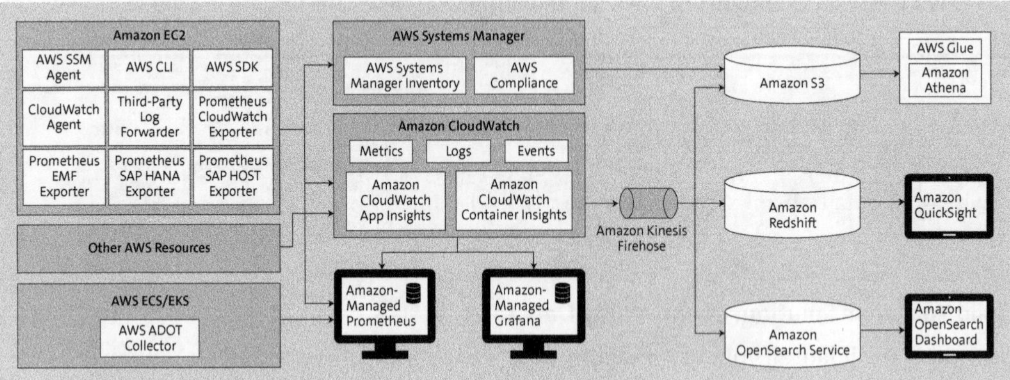

Figure 7.20 Building Your Own Observability Solution

Prometheus agents running in Amazon EC2 instances or the AWS Distro for OpenTelemetry (ADOT) collector running out of container environments can directly send the observability information to the Amazon Managed Prometheus (AMP) service. Moreover, Amazon Managed Grafana (AMG) has native integration with Amazon CloudWatch, allowing you to extract metrics and logs directly into Amazon AMG. However, currently, Amazon AMP has no native integration with Amazon CloudWatch. To bring Amazon CloudWatch logs into Amazon AMP, you must leverage the Prometheus CloudWatch Exporter agent running out of Amazon EC2 instances to extract from Amazon CloudWatch and push them to Amazon AMP. You can leverage Amazon AMP or Amazon AMG to build your own monitoring solutions.

Alternatively, you can send Amazon CloudWatch data in nearly real time to Amazon S3, Amazon RedShift, or Amazon OpenSearch using Amazon CloudWatch log group Kinesis Firehose subscription filters, as shown in Figure 7.19. Later, once the data is ingested into the right targets, you can consume this data using several querying and visualization tools, as shown in Figure 7.20.

Cross-joining the data from AWS Systems Manager inventory and compliance information with Amazon CloudWatch metrics and logs data, you can thus derive a comprehensive view of the operating state of an SAP application—its software state, configuration status, and overall performance situation.

Best Practices

- **Do not create clutter**
  You should stream only the required metrics and logs, define only actionable insights, build only valuable dashboards, and produce only meaningful alarms.

> - **Be frugal on metrics/logs storage**
>   You must follow a rational data retention strategy that fulfills your compliance requirements. Depending on the granularity of the data requirements, you'll need to adjust your metric resolutions. You must decide on Amazon CloudWatch storage and external storage options and its lifecycle options. A good practice is to leverage compression wherever possible to reduce your overall data footprint.
>
> - **Advocate centralized monitoring function**
>   Do not replicate the same data to multiple subscriptions and third-party solution targets. We recommend creating a single pane of glass for all your monitoring needs.

7.2.4   Security and Compliance

As you know, security and compliance are non-negotiable while operating enterprise workloads like SAP. These concerns are amplified when you need a clear line of sight on the physical infrastructure assets that host your workloads. Moreover, when the proper controls are not in place, a constant threat looms when exposed to malicious actors over the internet. Therefore, as an SAP cloud architect, you must find answers to the following questions to run your SAP workloads with confidence on AWS:

- How can I secure my SAP system's perimeter?

- Even if my perimeter is breached, how can I ensure that my SAP systems are not compromised?

- How can I fortify my critical SAP data assets from inside actors? Even from a rogue AWS employee?

- How do I guarantee that every component in my infrastructure adheres to the configuration I defined?

In the following sections, you'll see how these concerns are addressed.

Perimeter Security

We can broadly categorize AWS services in this area into three types of offerings:

- **Infrastructure services**
  You always provision them into a VPC, for instance, Amazon EC2, Amazon EBS, and the Amazon VPC itself.

- **Platform services**
  You must always associate these services with your VPC, though they are physically deployed in AWS-owned VPC. An elastic network interface (ENI) is deployed into your VPC to access these resources., for instance, Amazon Elastic File System (EFS), AWS Transit Gateway, Amazon RDS, and all other services with the tag "Managed" in its service name.

■ **Software services**
These services are accessible only through public endpoint APIs and operate outside of your VPC's boundaries, for instance, Amazon S3, Amazon DynamoDB, Amazon SNS, Amazon Simple Queue Service (SQS), and other abstract services that comprise a significant portion of the AWS service portfolio.

Figure 7.21 shows the possible ways you can integrate AWS resources into an AWS network fabric.

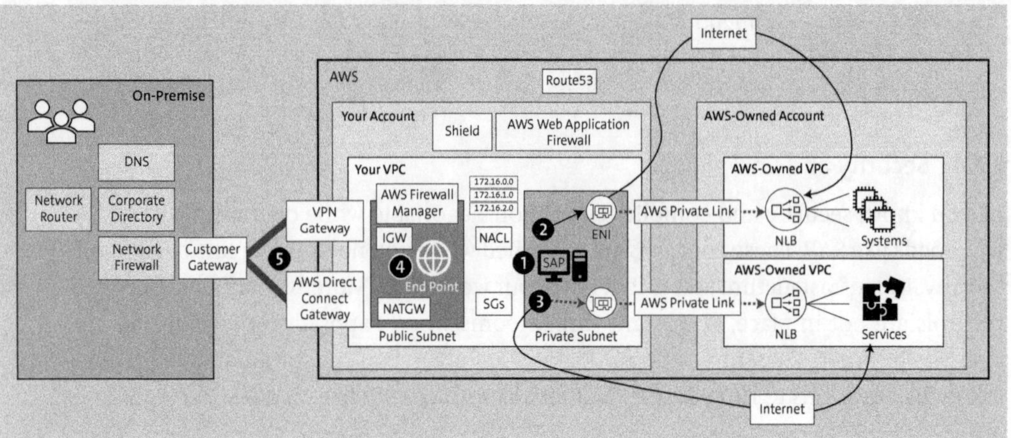

Figure 7.21 Perimeter Scope of AWS Services

You have several options to manage the connectivity among these resources and with the outside world. as shown in Figure 7.21. Let's navigate these options next:

❶ These services are infrastructure services like Amazon EC2 and Amazon EBS. You'll leverage these services to install your SAP workloads. You must deploy them into a VPC. If you deploy them into a private subnet, they are not accessible over the internet. Then, they are deployed with a private IP within the subnet CIDR range. You can further restrict the internal access to these resources using security groups (SGs) and Network Access Control Lists (NACLs). The public internet access to these services is only invoked from the systems inside the private subnet through NAT Gateway (NATGW). No one from outside can reach them as Internet Gateway (IGW) cannot relay the communication to the targets without a public IP. The route table decides if the respective subnet is a private subnet or a public subnet. If the route table has a route to IGW, it is a public subnet. Instead, it may point to NATGW, which is a private subnet. In addition, you can integrate AWS Network Firewall Manager with IGW, NATGW, VPN GW, and DX GW to sniff every packet coming in and out of VPC at the VPC boundary.

❷ These services are AWS-managed services or applications deployed in a physical infrastructure but are managed by AWS up to the runtime of the underlying application/service. These services reside physically in an AWS-owned VPC but get associated with VPC through an Elastic Network Interface (ENI). For all operational purposes, you can treat this ENI as the physical system with its own IP address. Under the hood, all the communication sent to this ENI is tunneled to the target system residing in AWS-managed VPC. As shown in Figure 7.21, the ENI may reach the target system over a public endpoint, but this specific endpoint is not publicly reachable other than the communication from ENI. Using the AWS Private Link feature, you can force the ENI to communicate over a private channel without traversing the public internet. You can find more information about AWS Private Link at *https://aws.amazon.com/privatelink/*.

❸ These services are abstract services. These services do not have any affinity with the VPC. They are hosted in an AWS-owned VPC, but the ownership of these resources is attached to your AWS account. You can reach, access, and manage over a public endpoint from the internet. Even the systems inside the VPC need to traverse through the internet to reach these services. However, as mentioned earlier, you can leverage AWS Private Link to restrict the public access of these services. Besides, it would help if you had robust IAM and resource access policies to ensure the security of these resources.

❹ You want to expose these applications or services to the outside world so your customers can access them online. Since these applications communicate at the application layer, you cannot restrict the access using the combination of IPs and Ports as we did with the help of SGs and NACLs under ❶. You must have more control over the web server communications; thus, you'll need to employ AWS Web Application Firewall (WAF) and AWS Shield to inspect web traffic and restrict malicious actors from access.

❺ This area explains how on-premise entities can communicate with AWS resources. As discussed earlier, AWS VPC is a foundational connectivity fabric with which most AWS resources associate. However, AWS VPC is a virtual network segment allocated with a private IP CIDR range. This network segment is not accessible to the outside world unless it is attached to network edge devices like a VPN gateway or an AWS Direct Connect gateway. These gateways connect with on-premise customer gateways over Internet Protocol security (IPsec) tunnels to establish a dedicated secure communication channel across the data centers. In this case, except for the resources exposed over the internet (as in ❸ and ❹), you can only access them either from an on-premise network or within the VPC. All other connection attempts would fail because these resources are hidden from the rest of the world.

You leverage AWS native tools to ensure the desired connectivity. As shown in Figure 7.22, a network reachability analyzer (NRA) and a network access analyzer (NAA) can help you achieve this goal.

AWS Network Manager service offers both functionalities. At the outset, both trace the network path between two AWS resources. However, NRA helps you troubleshoot the network connectivity between two defined entities. The analysis presents a graphical illustration capturing all the network hops and edge devices on the path, along with their configuration description. This analysis will pinpoint the location of the communication failure (if any).

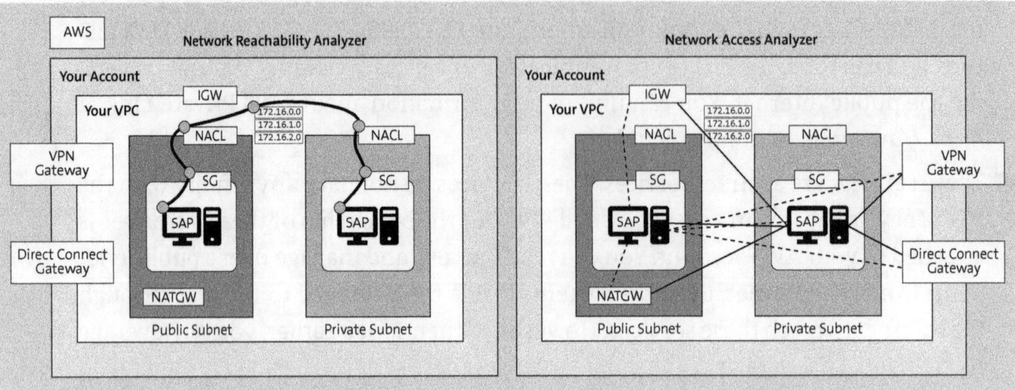

Figure 7.22 Network Reachability Analyzer and Network Access Analyzer

On the other hand, as shown in Figure 7.22, on the right, NAA helps you trace all possible network paths between the entities defined within the scope. This scope can be multiple entities within or across the network segments. If NRA is a one-to-one path, then NAA is a one-to-many, many-to-one, or many-to-many network path. This analysis will provide a holistic picture of who can connect with whom and what network paths are unintended and can pose a potential threat.

Resource Access and Authorization

Network restrictions work well in safeguarding your IT assets from external actors. However, you also need a sufficient defense mechanism against internal actors or someone disguising as an internal actor. You must have a robust access and authorization mechanism to protect systems and applications from these unwelcome actors.

The authorization perspective in the cloud is significantly different from what you saw in the on-premise world. The fundamental distinction in AWS is that every AWS resource can act as a principal. In other words, the actions that one AWS resource can perform on other resources are defined by the IAM or resource-based policies granted to them (*http://s-prs.co/v577644*). In a conventional data center context, you'll always associate a user with authorizations. Managing those users and authorizations is a reasonably straightforward approach.

However, apart from users and their authorizations on AWS, you must also manage the list of resource actions you allow against each resource. To help you contain the action surface area against unintended actors even if users elevate themselves through the resource privileges, consider the following best practices:

- **Multi-account strategy**

  With this approach, you'll segregate your workloads based on their criticality and distribute them into multiple AWS accounts. For example, you could separate your SAP workloads into productive and non-productive resources and deploy them under those accounts. In this way, you can control access sprawl by restricting them to specific account boundaries.

- **Least privilege access**

  Do not create open-ended access policies. Allow only actions that you require for the desired activity. A good idea is targeting these actions on specific resources only. You should further restrict these actions using conditions bloc and ABAC in the policy document.

- **Segregation of duties**

  You should not vest all the critical actions to a single principal. You should distribute essential actions on resources among multiple principals so that even the compromise of a single principal won't cause irreparable damage. For example, you could spread backup creation, encryption, deletion, decryption, backup restore, and encryption key management among three users. Therefore, no single user can destroy or steal the backup assets in a usable form.

- **Service roles**

  Service roles give authorizations to perform actions on your behalf. These roles should be restricted with additional context in the conditions bloc of the policy document to control who can assume these roles. If these conditions are not defined well, the confused deputy problem might arise (*http://s-prs.co/v577645*). You can use "Deny" statements to restrict to the rest of the world and ensure that only specific desired situations are possible. In addition, you can restrict users from launching Amazon EC2 instances with a specific instance role using the `PassRole` IAM action. As a result, users can not indirectly elevate their own privileges beyond what was granted to them through Amazon EC2 instance access. The example IAM policy snippet shown in Listing 7.8 restricts users from launching an Amazon EC2 instance via the Readonly-S3 instance role.

```
{
    "Effect": "Allow",
    "Action": "iam:PassRole",
    "Resource": "arn:aws:iam::111122223333:role/ReadOnly-S3"
}
```

Listing 7.8 Example IAM Policy to Allow Users to Pass Specific Roles to AWS Services

- **Access delegation**

  By using `AssumeRole` in IAM roles and using principals in resource-based policies, you can delegate access to other accounts. However, revoking this delegated access once the delegation request has expired is often forgotten. In those occasions, users may carry elevated privileges beyond their required timeframe. Therefore, we recommend attaching a time duration condition to the policy definition. Once these times are expired, you can make these additional privileges automatically obsolete. The IAM policy snippet shown in Listing 7.9 makes privileges valid only for 3 hours on a given date and time.

```
"Condition": {
            "DateGreaterThan": {"aws:CurrentTime": "2023-09-03T12:00Z"},
            "DateLessThan": {"aws:CurrentTime": "2023-09-03T15:00Z"}
}
```

Listing 7.9 Time-Bound Condition Keys to Restrict Access Delegations to a Specific Timeframe

- **AWS Organizations**
  Set up your organizational structure using AWS Organizations to control enterprise-wide authorizations framework. You can leverage SCPs and permission boundaries to manage IAM authorization boundaries across multiple AWS accounts. You can find more information about its usage in Chapter 4.

Application Secure Login/Single Sign-On

Besides controlling the user and the resource authorizations, enabling secure application logon mechanisms will restrict the resource access to corporate identities only. With this mechanism, you can also remove the threat of password compromises and eliminate the possibility of accessing corporate IT resources outside the corporate network.

You can establish a secure application logon mechanism, called single sign-on (SSO), for SAP on AWS in several ways. If you understand the roles of all the entities involved in this process, you can devise a combination of secure mechanisms to suit your requirements. In SAP's SSO configuration, you mainly deal with four entities: identity providers (IdPs), identity federators (IdFs), service providers (SPs), and service requestors (SRs). The IdF is the central entity that orchestrates the trust relationships among the other entities. However, it can only operate under the agreed access protocols. Based on the combinations in which an SP and an SR want to operate, you would choose the specific IdF accordingly.

In a nutshell, SSO configuration for SAP on AWS is similar to how you achieve this capability on-premise. However, AWS provides additional opportunities to tap into new IdPs and IdFs. You can use them to future-proof your enterprise identity management process. Figure 7.23 shows all the possible ways in which you can set up SSO for SAP on AWS.

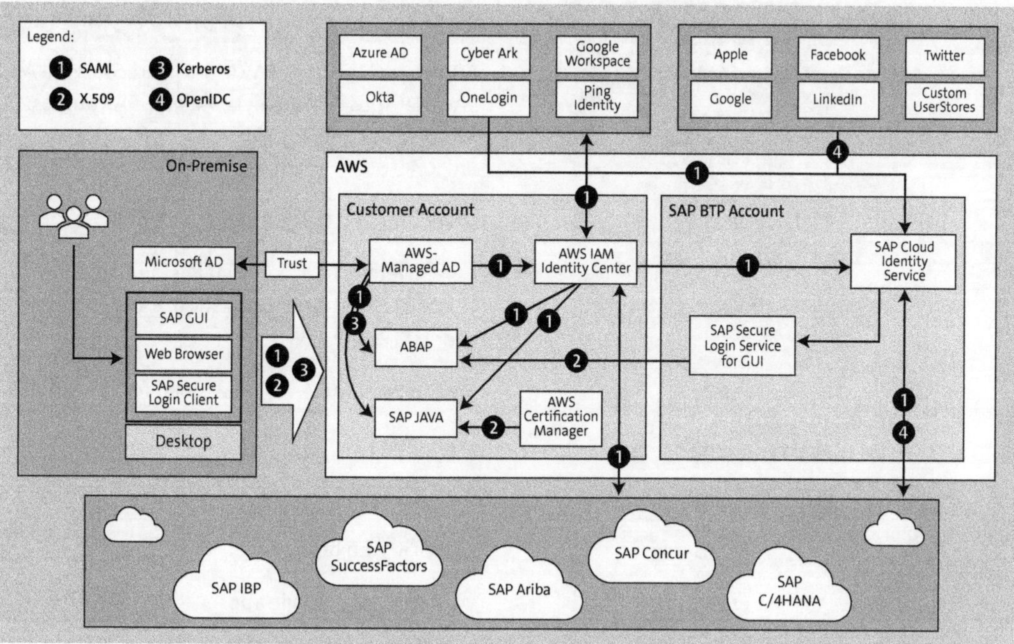

Figure 7.23 SAP SSO Setup Options on AWS

We can classify these entities by the access protocol they support. Table 7.3 described the entities shown in Figure 7.23.

Entity Category	Protocol	Entities
IdP	SAML	Microsoft AD, AWS IAM Identity Center, SAP Cloud Identity Services, AZURE AD, Cyber Ark, Google Workspace, Okta, OneLogin, Ping Identity, SAP IDM (not shown in Figure 7.23)
	OpenID Connect (OIDC)	Apple, Facebook, Google, Twitter, LinkedIn and custom user stores
IdF	SAML	Microsoft Active Directory Federation Service, AWS IAM Identity Center, SAP Cloud Identity Services
	X.509	SAP JAVA, SAP Secure Login Service for Gui
	Kerberos	ADFS
	OIDC	SAP Cloud Identity Services, Amazon Cognito (not shown in Figure 7.23)

Table 7.3 SSO Setup Options Based on Access Protocols

Entity Category	Protocol	Entities
SP	SAML	ABAP over HTTP, SAP JAVA, SAP Fiori, SAP HANA, SAP BTP apps, SAP Analytics Cloud, SAP Success-Factors, SAP Concur, SAP Ariba, SAP Fieldglass, SAP C4C
	X.509	ABAP over SAP GUI, SAP Fiori, SAP JAVA
	Kerberos	ABAP over SAP GUI, SAP HANA extended application services (SAP HANA XSA), SAP Fiori, ABAP over HTTP, SAP JAVA
	OIDC	SAP Concur, SAP Ariba, SAP Cloud for Customer
SR	SAML	Web browser
	X.509	SAP GUI, web browser
	Kerberos	SAP GUI, web browser
	ODIC	Web browser, mobile apps

Table 7.3 SSO Setup Options Based on Access Protocols (Cont.)

Let's explore this table more with an example: Suppose you set up SSO for SAP GUI to connect your ABAP systems. As shown in Table 7.3, you can achieve this in two ways: either using Kerberos or by using X.509-based authentication. Then, you can decide on what IdF you have that supports these protocols. If a Kerberos-based configuration, you can leverage ADFS. For an X.509-based configuration, you'll have two options. If you want to leverage your private root certificate authority (CA) to generate X.509 certification, you may need an SAP JAVA instance to facilitate this. If you do not want to rely on the SAP JAVA instance, you can leverage an SAP BTP service called SAP Secure Login Service for GUI (SLSG) instead. However, currently, SLSG does not support a private root CA certificate chain; it must be an SAP-managed root CA. This support is a roadmap item, and this limitation is expected to disappear in the near future. Similarly, you can arrive at your own combinations depending on your requirements and existing IdF and IdP investments.

Note

Whatever combination you choose may require additional tools or software components to complete the setup. For example, to achieve SSO for SAP GUI as discussed earlier, you'll need SAP Secure Login Client installed on the user desktop, SAP Cryptographic library installed in your backend SAP systems, and the required SAP profile parameters set up. For an X.509 setup, SAP JAVA needs a secure server package installed and a secure client, cryptographic library, and profile parameters. Refer to the relevant configuration guides to perform the setup.

Data Protection

Data is a critical asset in every enterprise. You need to protect your data from unauthorized access, tampering, and destruction. At the same time, the correct set of data should always be available for authorized people all the time without the loss of its integrity. This section will discuss the options to protect data sources from inside actors with access to AWS resources or from the masquerading actors roaming inside your corporate network.

Data at Rest

In the context of SAP, its data resides on Amazon EBS volumes for active access, or you can back up this data in Amazon S3 for future retrieval. You can employ a data encryption process to protect the data at rest. Data encryption ensures that only authorized people can access, view, and manipulate the data that persists in the storage. You can use AWS Key Management Service (KMS) to generate, control, and manage these cryptographic keys. To securely perform encryption and decryption activities, you can integrate these AWS KMS keys with almost all the AWS services (*http://s-prs.co/v577646*).

Moreover, the key control policies enforce an additional authorization requirement to ensure an extra data protection layer. Even if someone has access to data assets, if they are not authorized to use the involved encryption keys, the data is useless to them. In addition, AWS CloudTrail logs all AWS KMS requests and KMS API calls. You can integrate these calls with Amazon EventBridge so that notifications are sent when suspicious KMS access requests arise so you can respond quickly.

For SAP systems on AWS, you can achieve native data encryption in the following ways:

- **Amazon EBS encryption**
  You can use AWS Amazon EBS service-owned keys, AWS KMS-generated keys, or customer keys managed by AWS KMS to encrypt both the boot and data volumes of an Amazon EC2 instance. Once the volume is encrypted, all the data persists on the volume, and the data moving across the volumes is encrypted automatically. AWS also encrypts the snapshots created from encrypted volumes. You can find more information at *http://s-prs.co/v577647*.

  Amazon EBS encryption safeguards against storage access outside of the respective Amazon EC2 instance context. These encrypted volumes will not be accessible if you remove them from the designated Amazon EC2 instance. Accessing this data from a different instance or environment requires the proper encryption context and key authorizations, thus providing an additional security measure controlling who can handle this data.

- **Database native encryption**
  You can use Transparent Data Encryption (TDE) for Oracle and Microsoft SQL, root key encryption for SAP HANA, and symmetric encryption for IBM Db2 and SAP ASE. By using database-native encryption, you'll never expose the data outside of the

database, and only authorized database users can access data through database interface tools.

- **AWS Backint integration with AWS KMS**
  You can integrate the AWS KMS-managed encryption key with AWS Backint to automatically encrypt SAP HANA data and log backups to Amazon S3. You can define the KMS key value against the S3SseKmsArn parameter in the AWS Backint configuration file at `/usr/sap/<SID>/SYS/global/hdb/opt/hdbconfig/aws-backint-agent-config.yaml`.

- **Client-side encryption with AWS KMS wrapper keys**
  You can use the Amazon S3 Encryption client to encrypt files on the client side before you move them to Amazon S3. This client uses all the cryptographic best practices and protects the data keys that encrypt your objects by wrapping them with the provided AWS KMS keys. You can leverage this option for non-SAP HANA backups, where you can take backups locally on Amazon EBS volumes and move them to Amazon S3 after you encrypt them at the source. For more information on Amazon S3 Encryption client, refer to *http://s-prs.co/v577648*.

Data in Transit

Data across the system boundaries may move in plaintext if not transmitted in secure channels. Man-in-the-middle (MITM) attacks can exploit these plaintext data transmissions to eavesdrop on your communications. To remediate this problem, set up SSL/TLS for SAP web-based communications and SNC for the DIAG protocol and remote function call (RFC)-based communications. Enable this setup between all inter-SAP system connections and between user system conversations.

You can use AWS Certificate Manager (ACM) to provision and manage public and private TLS certificates. You can also import certificates issued by third-party CAs. Based on the settings, AWS ACM will automatically take care of key rotations and renewals. You can integrate these certificates natively with AWS Elastic Load Balancer (ELB), Amazon CloudFront, and Amazon API Gateway to enable SSL/TLS communication from users.

To ensure that all your communications are secure, enforce the following policies:

- Allow only secure protocols. For example, configure your security group rules, NACLs, and firewall rules to allow only secure protocols. Close all potentially unsecured protocol ports.
- Set up IPSEC VPN connections between all point-to-point locations.
- Enable secure protocols on AWS ELB.
- Secure database communications by setting up SSL/TLS connections from the database to all database clients. Disable non-secure database connection channels.
- Use only secure data transfer protocols (e.g., SFTP, SCPs, etc.).

Data Privacy

As long as data resides within the SAP system boundaries, SAP's roles and authorization framework will ensure its accessibility and exposure. However, when the data leaves its boundaries (e.g., extracted into external systems), you lose data control. Often, you follow this scenario when you load this data into external data warehouses or data lakes. Though you enforce the data controls on the reporting layer, the raw data residing in Amazon S3 lacks similar access and authorization controls. This situation could open a back door to your data assets. Moreover, these data assets potentially contain personally identifiable information (PII) or intellectual property (IP).

You can use Amazon Macie (*https://aws.amazon.com/macie/*) to automatically discover, classify, and protect sensitive data stored in Amazon S3. This service employs machine learning models to identify patterns that match PII, protected health information (PHI), predefined regulatory documents, API keys, and secret keys.

Amazon Macie provides dashboards and notifications that provide insights into how this data is accessed or transmitted. These security findings can also integrate with AWS Security Hub and Amazon EventBridge. By linking automation to Amazon EventBridge events, you can automatically remediate after generating the findings. You can use this automation to move this data from its insecure place to a more secure place or sanitize it.

AWS Resource Configuration

So far, you've learned how to fortify your network and ensure that applications and their underlying data assets are protected from unauthorized users. While doing so, we proposed several best practices, design principles, and architectural rules. Hundreds of rules exist, if not thousands. This number is constantly growing; thus, keeping track of them is effort intensive, and complying with them is daunting. Therefore, in this section, we provide ways and means to adhere to all of them, often without conscious effort. You're not only assessing these rules for the current workloads; rest assured that you can enforce them during new resource launches and landscape changes.

Well, how can you do that? The answer is AWS Config. In this section, we'll review AWS Config, its conformance packs, and how it can be used to view your overall compliance status.

AWS Config

AWS Config is an AWS service that records the configuration of AWS resources. You can direct these findings into an Amazon S3 bucket to track the change history over some time. In addition, you can also define resource compliance rules for a desired configuration state and evaluate them periodically. AWS Config will send the statuses of these rule evaluations to display in the dashboard and as an Amazon SNS notification on the defined configuration stream. Integrating this Amazon SNS notification with event-driven automation can automatically remediate configuration drift.

AWS, by default, provides a rich set of managed rules you can leverage. In addition, as shown in Figure 7.24, you can define your configuration rule to evaluate any resource compliance state by associating an AWS Lambda function. You can write your evaluation logic and report its status to the AWS Config framework.

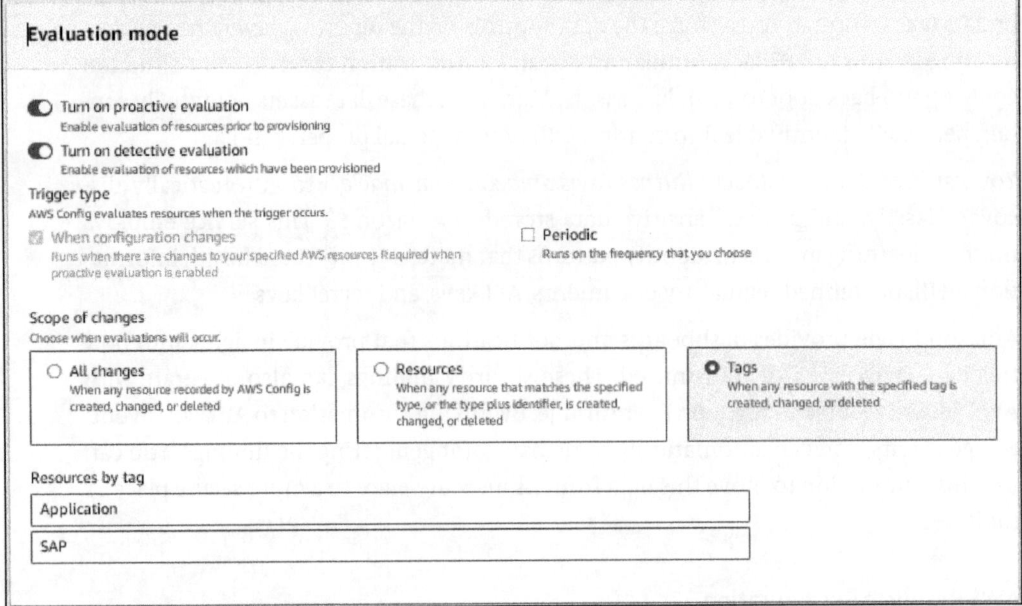

Figure 7.24 Defining a Custom Configuration Rule in AWS Config

To define this custom rule, follow these steps:

1. Go to the **AWS Config** service and click **Rules** in the left navigation panel.

2. On the **Rules** page, click **Add Rule**.

3. Choose the **Create Custom Lambda Rule** option on the **Select Rule Type** page.

4. In the **Configure rule** page, under **Details** input block, give a name to the custom role, and provide the AWS Lambda function ARN where your custom evaluation logic is defined.

You'll see the **Evaluation mode** bloc immediately under the **Details** bloc, as shown in Figure 7.24. You shall see two evaluation modes: proactive and reactive. The proactive option enables you to evaluate a configuration rule while provisioning the resources under this rule scope. The reactive option (also called "detective" on the screen) lets you trigger this rule check whenever a resource configuration changes or on a defined periodicity. Under the scope of changes, you'll include all the resources for which you want to evaluate this rule. As shown in Figure 7.24, tags are a more reliable way of narrowing your scope to the definitive set of resources. We'll discuss the anatomy of the evaluation code of a custom configuration rule using an AWS Lamba in Chapter 8.

Conformance Packs

Since you know you can define custom rules to evaluate the configuration of AWS resources, how can you guard against rule sprawl? How can you systematically govern this function and get a holistic picture of where you stand from the compliance perspective? If you're thinking along those lines, conformance packs are here to the rescue.

A *conformance pack* is a collection of AWS Config rules and associated remediation actions that you can deploy as a single entity. You can create these conformance packs based on the categorization of configuration rules or based on your operational model. For example, you can group these rules under functional areas like operations, monitoring, security, infrastructure, or any way intuitively resembling your governance structure. AWS provides a list of conformance pack templates, which you can immediately deploy or reference to create your own conformance pack. You can access these templates at *http://s-prs.co/v577649*.

After the deployment, as shown in Figure 7.25, you can view the overall compliance score, the history, and the underlying details attributing to this score under the **Conformance packs** section in the AWS Config service.

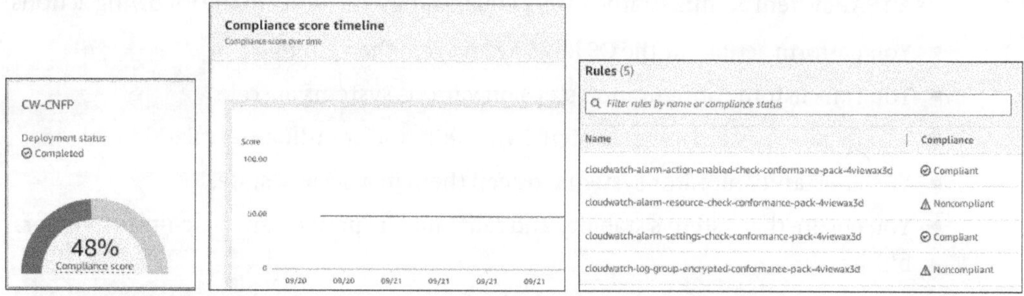

Figure 7.25 Conformance Pack: Overall Compliance Score, History and Details

Overall Compliance Status

The dashboard in the AWS Config service provides a holistic view of the overall compliance status of your AWS resources. Alternatively, you can leverage the **Advanced queries** option in the navigation pane to query the compliance status of your resources. More information about these resource types is available at *http://s-prs.co/v577650*. You can use these resource properties in SQL queries to list the resources with specific criteria. The example shown in Figure 7.26 queries the specific security groups that open for traffic on `tcp` protocol at port 3600 (SAP message server port, for instance number 00).

You can also track the history of resource configuration changes using the **Resource Timeline** feature under the **Resources** option in the navigation panel of the AWS Config service.

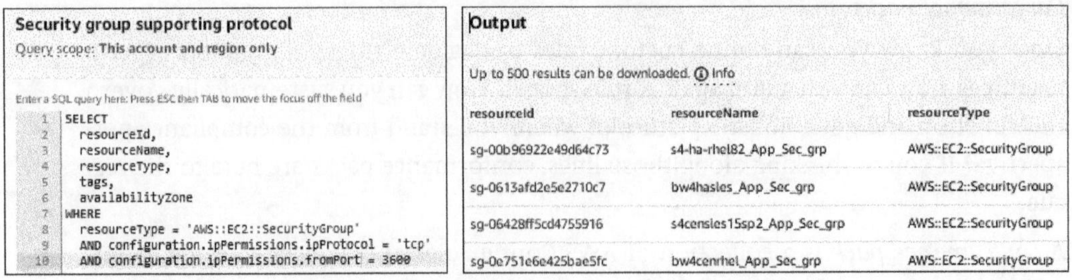

Figure 7.26 AWS Config: Advanced Queries

Note

The recording changes feature in the AWS configuration must be **ON** to track the resource configuration in AWS Config. You can enable this capability in the **Setting** option in the navigation panel.

7.2.5   Systems Administration

As an SAP system administrator, often, you may have to perform the following actions:

- You perform actions at the OS level by logon to the system.
- You run custom configuration steps on various systems at scale.
- You manage system issues, assign ownership, and coordinate escalations.
- You constantly monitor systems to keep them in a desired state.
- You ensure the system's stability and reliability by periodically patching its vulnerabilities.
- You establish a mechanism to honor license agreements and control license costs.
- You set up a quick and robust way to push user packages into multiple systems at scale.
- You build secure system images to create consistency in the system landscape.
- You operate systems securely and ensure their compliance at all times.

In the following sections, we'll explore how the various features of AWS Systems Manager can assist you in performing all these administrative activities.

AWS Systems Manager Session Manager

The Session Manager is a capability of AWS Systems Manager that provides a secure and auditable way to log on to Amazon EC2 instances from the AWS Console. It can initiate sessions to Windows, Linux, and MacOS instances. You can facilitate this logon without opening network ports, maintaining bastion hosts, or managing Secure Shell (SSH) keys. Most importantly, you can start a session without sharing passwords.

In addition, you can govern this access through IAM roles. Thus, you can grant, revoke, and delegate this access as required. This situation is entirely different from what you'd otherwise do by sharing the passwords. Only a password stands between you and the malicious actor to exploit your system. They can employ various brute force mechanisms to guess your password. Moreover, you cannot have complete control over who in your organization can possess this password.

Using Session Manager, you can also restrict which user can start a session, what OS commands a user can execute, and what environment that specific user can log on.

The following prerequisites must be fulfilled to enable this capability:

- You must have a supported OS. The list of OSs supported by AWS Systems Manager is available at *http://s-prs.co/v577651*.

- Install the latest AWS Systems Manager Agent. You can find more information at *http://s-prs.co/v577652*.

- Open HTTPS (port 443) in the instance security group to enable connection to the following endpoints:
    - ec2messages.<region>.amazonaws.com
    - ssm.<region>.amazonaws.com
    - ssmmessages.<region>.amazonaws.com

- Attach AWS-managed IAM policy `AmazonSSMManagedInstanceCore` to the instance profile of the managed node.

AWS Systems Manager Run Command

AWS Systems Manager Run Command enables you to perform administration tasks and configuration changes on managed nodes at scale. This feature avoids the need to log on to the system. You can use the Run Command in various ways. You can run it from the AWS Management Console, the AWS CLI, AWS Tools for Windows PowerShell, or the AWS SDKs.

Like other AWS Systems Manager capabilities, you can control where and what a user can perform. You can package these tasks as AWS Systems Manager documents. You can define IAM policies to restrict actions either based on AWS Systems Manager documents or based on where you can target these actions. You can further limit them to specific tags or resource types using ABAC in the IAM policy document.

Let's consider an example of executing an SAP kernel update on a specific set of SAP systems using AWS CLI. You can restrict this execution to selected instances that have specific tag keys and values. You can define the parallelization of its execution and the error tolerance through the `--max-concurrency` and `--max-errors` options, as shown in Listing 7.10. You can also control the behavior of the document through the `--parameters` option.

```
aws ssm send-command \
   --document-name SAP-kernel-update \
   --document-version "$LATEST"
   --parameters kernelversion=785, patchlevel=400, kernelarchive=mysapmedia
   --max-concurrency 3 \
   --max-errors 1 \
   --targets Key=tag:Application,Values=SAP Key=tag:systemtype,Values=S4H-
DEV, BW4H-DEV \
   [...]
```

Listing 7.10 AWS CLI Command to Execute AWS Systems Manager Documents on Selected Amazon EC2 Instances

In Chapter 8, we'll show you how to create an AWS Systems Manager document.

AWS Systems Manager OpsCenter

The OpsCenter is a capability of AWS Systems Manager that provides a central place where you can manage all the operational items (called *OpsItems*). Operational items are observations or issues raised automatically through resource events or manually created by your teams. Exceptions or the alerts observed in the findings of AWS Config, AWS CloudTrail, Amazon CloudWatch, Amazon CloudWatch Application Insights, and Amazon EventBridge can raise OpsItems. The OpsCenter follows defined criteria (OpsItem rules) to report OpsItems with distinct priorities. OpsItems is a mechanism to prioritize the efforts of your operational teams.

The **Explorer** option under **Operational Management** in the AWS Systems Manager navigation panel provides a holistic view of OpsData reported from different sources. You can control this display by choosing the **Configure Dashboard** option on the screen.

To enable OpsCenter, you must activate the settings outlined at *http://s-prs.co/v577653*. You can view and manage the reported OpsItems from the OpsCenter, which presents the data in different visual formats on widgets under the **Summary** tab. You can navigate to the details by selecting the concerned OpsItem. The **Details** tab provides complete information about what this OpsItems is about, its source, the executed automation in the last 30 days, and the associated operational data for analysis. Also, you can attempt remediation through the runbooks integration available in the OpsItem details section. Besides, you can access the history of related items to determine if you can adopt the same remediation to resolve this item.

Additionally, you can forward critical or concerning OpsItems to the AWS Systems Manager Incident Manager (IM) by choosing the **Start Incident** option on the OpsItem details screen. AWS Systems Manager IM is an incident response mechanism. You can define response plans and engagement models and manage on-call schedules. As shown in Figure 7.27, this solution is a single place to track the entire series of events, metrics, actions, chat conversations, and observations related to an incident. You can also generate post-incident analysis to review and optimize your response action plan for future related events. More information is available at *http://s-prs.co/v577654*.

Alternatively, you can also send OpsItems directly to your IT service management (ITSM) tools like ServiceNow or Atlassian. With the help of the AWS Service Manager connector and OpsItems and incidents, you can also send AWS resource events, AWS Config findings, and AWS Security Hub alerts to these ITSM tools.

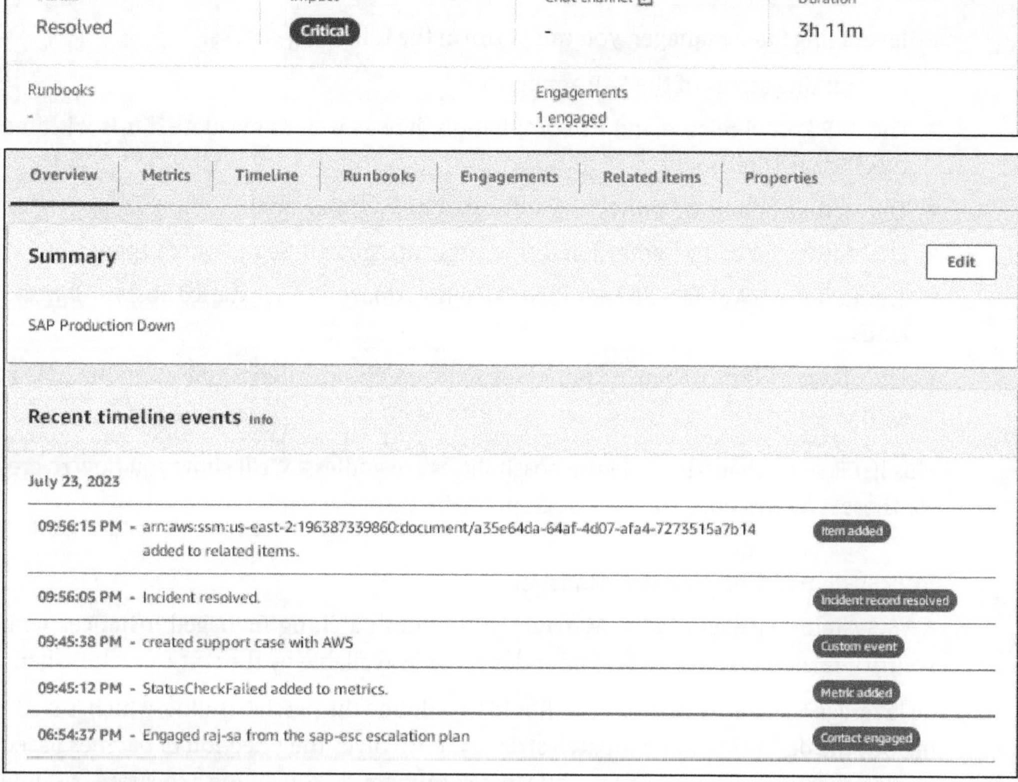

Figure 7.27 AWS Systems Manager Incident Manager

AWS Systems Manager State Manager

The State Manager is a capability of AWS Systems Manager. As the name suggests, this capability ensures the managed node is always in the desired state you defined.

How do you ensure this? With the help of AWS Systems Manager documents! A series of AWS Systems Manager documents, each with a definitive evaluation logic, gets associated with the managed nodes. These associations are triggered on predefined schedules and assess their current state against the expected state in the AWS Systems Manager document. The State Manager captures the results of these associations and reports the overall state of the respective managed node.

From the **State Manager** option, you can create the associations by selecting the AWS Systems Manager documents, their schedule, concurrency, error control, and the passing parameters for the document. You can view the results based on the associations. You can view all the associations running on a specific managed node from the **Inventory** option. In the details page of the managed node, you can find them in the **Associations** under **Properties**. You can view the results based on managed nodes. From the **Compliance** option, you can see how many managed nodes are compliant and how many are not. You can view the results with the perspective of its compliance status here.

By leveraging State Manager, you can perform the following checks:

- The running status of the SAP application
- The SAP kernel release and its compilation date as well as evaluate if it is within a specific age threshold
- The status of malware protection software or IDS/IPS agents
- The status of AWS Systems Manager agents and Amazon CloudWatch agents
- The status of AWS Data Provider for SAP if the Amazon EC2 instance hosts SAP workloads
- The status of Prometheus exporters if Amazon CloudWatch Application Insights is active

This list is not exhaustive, and the possibilities are endless. We'll show you how to create these AWS Systems Manager documents in the next chapter.

AWS Systems Manager Patch Manager

AWS Systems Manager Patch Manager automates patching managed instances with security-related updates and other updates made available by the software publisher.

To leverage this functionality, you should first define the patch baseline, which is nothing but the definition of your patch strategy. It involves the selection of patches based on classification, severity, and the affected components. You define a separate baseline for each supported OS. You'd also decide on the patch approval mechanism based on patch release dates. If your patch baseline is not customized or prepared, the Patch Manager chooses the default patch baseline available for the specific OS.

Once the approved patches are available, the Patch Manager uses the native package manager capabilities within the respective OSs to install the selected patches. You can find more information about this behavior at *http://s-prs.co/v577655*.

The Patch Manager integrates with AWS IAM, AWS CloudTrail, and Amazon EventBridge to provide a secure patching experience that includes event notifications and the ability to audit usage.

You can either scan instances to find the missing patches or scan and automatically install all the missing patches. You can target instances for patching individually or in large groups using Amazon EC2 tags. You can control the parallelization using rate control in the patch baseline.

Patch Manager uses AWS Systems Manager document associations to evaluate the system patch compliance. As discussed earlier in the State Manager section, you can view the outcome of these associations from the State Manager, Inventory, and Compliance options in the AWS Systems Manager navigation panel. You can find the list of AWS Systems Manager documents used in this process at *http://s-prs.co/v577656*.

Note

Patch Manager does not support the upgrade of major OS releases. You should follow a separate process to achieve these upgrades. The respective OS software publisher sets the severity of the patches. They are not validated against other third-party sources such as the Common Vulnerability Scoring System (CVSS) (*https://www.first.org/cvss/*) or from metrics released by the National Vulnerability Database (NVD) (*https://nvd.nist.gov/vuln*). For the Patch Manager to work, it needs access to patch repositories over the internet. The managed node should have a route to the Internet Gateway (IGW) or NAT Gateway (NAT GW).

AWS Systems Manager Distributor

Distributor is yet another capability of AWS Systems Manager under Node Management. As the name suggests, Distributor lets you distribute custom installable units, AWS published agents, and third-party tools into managed nodes at scale.

First, you need to create a Distributor package. The package creation process is nothing more than uploading the installable units into the Distributor and wrapping them into an AWS Systems Manager document. To initiate this process, choose the **Create Package** option on the Distributor service landing page. These installable units are *.zip* files having installation and uninstallation process scripts. You can provide these *.zip* files using an Amazon S3 location.

If AWS or a third-party software publisher provides the package, you can directly upload them into the Distributor as *.msi*, *.rpm*, or *.deb* file extensions. You can combine multiple files—one for each OS, version, and CPU architecture—into a single unit through manifest files. Upon successful creation, you'll see an AWS Systems Manager document created for you.

Second, you should publish the created package. Publishing makes your private packages available for others by editing their permissions. Choose the relevant package from the **Owned by Me** tab on the service landing page and edit the permissions to provide access to others.

Finally, once a package is available for your managed nodes, the Distributor leverages the following AWS Systems Manager capabilities to deploy them into managed nodes. The packages published by AWS, third-party packages, and packages that have been shared with you are already available to deploy.

You have two options, both discussed earlier in this section:

- One-time deployment using AWS Systems Manager Run Command
- On a schedule by using AWS Systems Manager State Manager associations

If your end-user machine is a managed node (e.g., an Amazon workspace or non-Amazon EC2 instance integrated through hybrid AWS Systems Manager activation as described at *http://s-prs.co/v577657*), you can push several SAP client tools centrally into them. For example, you can install or update SAP GUI software, update SAP GUI pad configurations, or deploy SAP SSO clients.

AWS License Manager

AWS offers tremendous flexibility to deploy workloads, including SAP, on Amazon EC2 instances of various compute configurations. However, your enterprise license agreements with your other software vendors may limit your flexibility. Your agreement dictates how many vCPUs or memory entitlements you can leverage and/or how many instances of the software you can deploy. Often, agreements even restrict the CPU core-to-memory ratios.

Besides, you may have to honor other agreements with underlying OSs and database providers hosting SAP workloads. Licensing is more daunting when following a Bring Your Own License (BYOL) model in an AWS environment.

You can use AWS License Manager to comply with these licensing terms. AWS License Manager would track the AWS Marketplace licenses automatically. However, for BYOL licenses, you must complete the following steps:

1. Create a *self-managed license* and define the licensing rules based on your enterprise licensing agreements. You can also define the discovery rules if the product belongs to SQL Server, Windows Server, and Oracle Database. However, for SAP, you can only track based on vCPUs, cores, sockets, and instances as a measure of unit for the licensing terms. You can find an example for self-managed license rule at *http://s-prs.co/v577658*.

2. Associate this license with a designated AMI used exclusively for this workload. AWS License Manager will track the launched AMIs and calculate the provisioned resources against the license rules. However, this option may not be valid for marketplace AMIs if your workload is deployed on them. For example, if SAP HANA in deployed on SLES for SAP marketplace AMI, you cannot associate this AMI for tracking SAP licenses. In that case, you can create a custom Launch template or CloudFormation templates to enforce license tracking through optional parameters defined

in the template. You can restrict users to use only this specific Launch template or CloudFormation template if they want to deploy any SAP workloads. You can find the example IAM policies of this enforcement on Launch templates at *http://s-prs.co/v577659*.

AWS License Manager comes in handy for SAP workloads in the following scenarios:

- **You want to increase your workload without having any licensing implications.**
  Suppose your SAP licensing is based on vCPUs, simply changing the instance generation or processor family, you can gain more SAP Application Performance Standard (SAPS) without changing the instance size. Compare the SAPS configuration for same instance sizes across generations and between AMD and Intel processors from SAP Note 1656099.

- **You want to choose the exact compute configuration for a workload that you're entitled by the licensing terms.**
  Suppose your SQL Server/Windows Server licensing is based on vCPUs. You need a specific vCPU configuration for your workload for which you do not find any exact Amazon EC2 instance configuration. For example, you want an Amazon EC2 instance with 6 vCPUs to honor your licensing terms. However, you either find an Amazon EC2 instance with 4 vCPUs or 8 vCPUs. With lower vCPUs, you're compromising on the performance, but with higher vCPUs, you end up breaching the licensing terms. In that case, you can leverage the vCPU Optimization option to account only required vCPUs over the available vCPUs in a higher vCPU Amazon EC2 instance, as described at *http://s-prs.co/v577660*. You can choose this option while defining self-managed license rules to reflect in your license tracking.

Note

vCPU optimization option will only save you on licensing costs—not the underlying Amazon EC2 instance costs. Meaning, you're still billed for 8 vCPUs though you're leveraging only 6 vCPUs in the scenario mentioned earlier.

Amazon EC2 Image Builder

If you're an OS administrator, your core concerns are to deliver a secure and consistent environment for hosting your enterprise workloads. This includes several steps such as finding a specific OS image, deploying ever changing list of OS and application packages, setting up administration and monitoring agents, applying their latest bug fixes, following application advisories, ensuring security configurations, and enforcing specific infrastructure guidance on where these environments can operate.

In traditional data center setups, this activity is a daunting, repetitive manual process consuming significant human resources. However, in AWS, you can reliably automate this process to ensure that every instance that is deployed has the latest and greatest configuration vetted by your security organization.

Amazon EC2 Image Builder (IB) is a service that enables customers to automate the creation, management, and deployment of customized, secure, and up-to-date server images that are preinstalled and preconfigured with software to meet specific standards. Amazon EC2 IB can provide base artifacts, add/remove software, customize settings/scripts, run tests, and distribute images to multiple AWS Regions. At a high level, this process consists of the following steps:

1. **Create an image recipe**
   Here, you select the base image. This image could be either an AWS curated or marketplace subscription image or a custom image made available. Next, select the add/remove software packages to deploy on top of the base image. Finally, you define storage settings for the target image configuration.

2. **Create infrastructure configuration**
   This is where you define all the launch configuration details, including Amazon EC2 instance type, IAM instance role, VPC and security group settings, and, optionally, key pair input specific to the account and AWS region.

3. **Define distribution settings**
   In this step, we define who can access from where and how an image be made available for others to leverage in their image-building process.

4. **Create components**
   The image Builder component helps define needed scripts and commands that will be applied to the prior Image Recipe. In this step, you'd define these scripts and follow the steps defined in the document definition. These actions will integrate these software packages along with post-deployment configurations. In this definition, you can also incorporate security scans and vulnerability assessments to ensure the required compliance standard of the image.

5. **Create image pipeline**
   This step will combine all the previous steps to achieve an automated process that can be re-used in different combinations. You can run these pipelines on-demand or defined schedules to produce a secure and consistent Amazon EC2 image.

7.2.6   Systems Change Governance

Change is inevitable. However, if handled poorly, every change can potentially disrupt the system's reliability. Therefore, as an SAP system administrator, you must establish a change governance framework in your operation model. In this section, we discuss AWS's strategic capabilities to enforce change controls. In the next chapter, we'll explore tactical mechanisms and tools available in AWS to control the drift in infrastructure as code (IaC) artifacts while you're building automation assets for your operations. SAP application change management is neither impacted nor discussed under the scope of this book. Thus, you can continue to use the same governance framework for SAP application change management.

AWS Systems Manager offers three capabilities to achieve change governance in your organization: Change Manager, Maintenance Windows, and Change Calendar. Let's cover each one next.

AWS Systems Manager Change Manager

This capability under AWS Systems Manager enables your organization to centralize the change governance from a delegated account and define who can perform what through runbook access policies. This process involves several steps:

1. You must have runbooks to perform all the critical activities. Your DevOps engineers create these runbooks and make them available for operational engineers to conduct their activities. AWS provides a rich set of AWS Systems Manager documents for your consumption. You can create your own runbooks if a requirement warrants a custom AWS Systems Manager document.

2. An operational engineer will create a template for their planned action. This template contains details about the activity, targets, and the associated automation runbook to perform this activity. Based on the Change Manager settings, which you can maintain in the Change Manager **Settings** tab, these templates would go through an additional approval mechanism.

3. Upon template approval, the operational engineer will create a change request and associate the approved template against this request. If your template is previously approved, you can skip the earlier steps and create a change request against the approved templates. In this step, the requestor also fills in the runbook input parameters and the required IAM authorizations for successful execution. This request will move in the change approval workflow as defined in the template.

4. This request will wait in the approver's task list if additional approvals are required. An approver approves the request and ensures that the time of its execution is associated with Maintenance Windows (see next section) and Change Calendars (see next section).

5. Approved requests will automatically execute at the designated time. You can see the timelines and execution events on the Change Manager landing page. Alternatively, you can integrate the workflow with Amazon SNS topics in all the above steps to trigger notifications upon completing those actions.

You can also integrate this Change Manager workflow with external Information Technology Infrastructure Library (ITIL) tools like ServiceNow using AWS Service Management Connector. This integration can automate your change management process to trigger changes in AWS from change tickets in ServiceNow.

AWS Systems Manager Maintenance Windows

Maintenance Windows (MW) is a capability of AWS Systems Manager under Change Management. This capability ensures that specific actions on designated targets are controlled to minimize business disruptions. MW is a date and time slot for you to perform the registered tasks in this MW against the registered targets. MW can influence the following task types:

- AWS Systems Manager Run Command
- AWS Systems Manager Automation Workflows
- AWS Lambda Functions
- AWS Step Functions

Suppose you have two categories of activities—disruptive and non-disruptive. You want to control when your operational teams can attempt these activities on the systems based on their availability requirements. For example, you can create two MWs for non-production systems—one for non-disruptive activities and the other for disruptive activities. Non-disruptive activities are allowed to run every 30 minutes, whereas disruptive activities are only allowed overnight. After creating the MW, you can register the non-production systems to both of these MWs, but you register the activities to respective MWs based on the nature of the disruption.

Likewise, you can create another set of MWs for production systems to minimize disruptions. Non-disruptive activities during night hours of each day but disruptive activities for a specific date and time of the month (during downtime window)

These MWs influence the execution windows for the change requests raised by operational engineers. Hence, your changes will automatically find a more reasonable time window irrespective of their change request status.

AWS Systems Manager Change Calendar

Change Calendar (CC) is yet another capability of the AWS Systems Manager to control the change windows for operational activities. CC-defined events supersede the time windows defined under MWs. CC maintains these business events and influences the behavior of planned activities depending on the calendar type.

Let's briefly explore the behaviors of the various calendar types next:

- **Open by default**
  The calendar allows your activities except for the scheduled event defined in the calendar. It is the default behavior if not changed while creating the Change Calendar.

- **Closed by default**
  The calendar blocks all your activities except the scheduled events defined in the calendar. Be careful with this option, as the calendar stops all your activities for the rest of the duration.

Once you create the calendar, it displays all the previously scheduled events by Automation Workflows, Maintenance Windows, and State Manager schedules. You can create calendar events by choosing the **Create Event** tab on the top right corner of the calendar screen. Advisory events are for documentation purposes only; they do not influence the activity window.

Alternatively, you can also upload calendar events from third-party calendar providers. However, they must be in *.ics* format for import. You can also leverage third-party calendar providers, including Google Calendar, Microsoft Outlook, and iCloud Calendar. You can find more information about restrictions on importing third-party calendar providers at *http://s-prs.co/v577661*.

7.2.7 Respond and Recover

So far, we've explored several security and compliance controls to ensure the consistency and reliability of AWS resources. However, in the real world, there are no foolproof mechanisms guaranteeing your resources are safe and secure forever. Hence, it would help if you established an incident response framework to build a defense against potential breaches of controls. In this section, we'll address the following questions you might have:

- How can I get notified of any suspicious activity before it is escalated into a threat?
- How can I protect my AWS resources from evolving threats and vulnerabilities before they are exploited?
- How quickly can I respond to an attack and shrink the blast radius of it?
- How equipped am I to respond to an ongoing security incident effectively?

Amazon GuardDuty

Amazon GuardDuty is a threat detection service fed by threat intelligence from AWS and third-party industry-leading sources. It employs machine learning models to engage in anomaly detection over data captured from various AWS monitoring sources. At the time of writing, Amazon GuardDuty integrates with many data sources to provide a comprehensive access activity across various AWS resources, such as the following:

- **VPC flow logs**
  This feature captures IP traffic coming into and going out of network interfaces in VPC. This information provides clues about who is accessing from where, the paths they traverse, and the ports they target.

- **DNS logs**
  This feature enables the logging of all the public DNS queries that Amazon Route 53 receives and the query forwarding requested from private hosted zones. This information provides traffic patterns on how access to your publicly exposed AWS resources is resolved.

- **Amazon RDS logon logs**
  This feature helps you identify all the logon attempts made to the Amazon RDS database. This information enables you to discover unintended access to database servers.

- **AWS CloudTrail events**
  This feature captures all the API calls fired at AWS resources in your accounts. It captures all control plane and data plane activities aimed at these resources. This information helps you identify all resource access, attempted operations, and the results of its operations through APIs. It also captures the access elevations and the involved identities so that you can pinpoint the source of a resource's access.

- **Amazon S3 server logs**
  This feature enables logging additional data plane activities on top of what AWS CloudTrail can capture. A list comparing Amazon S3 data plane activities with AWS CloudTrail is available at *http://s-prs.co/v577662*.

- **Amazon Elastic Kubernetes Service (EKS) control plane logs and runtime activity**
  This feature provides control plane logs of Amazon EKS operations. These logs make it easy for you to secure and operate your Amazon EKS clusters.

To activate the analysis on these data sources, you must enable them in the settings section of the service. The analysis results are reported to the Amazon GuardDuty console, where you can centrally analyze these findings. These findings include comprehensive details about the involved resources from all the data sources to facilitate further troubleshooting. For the complete list of findings by resource types, severity, and data source, refer to *http://s-prs.co/v577663*.

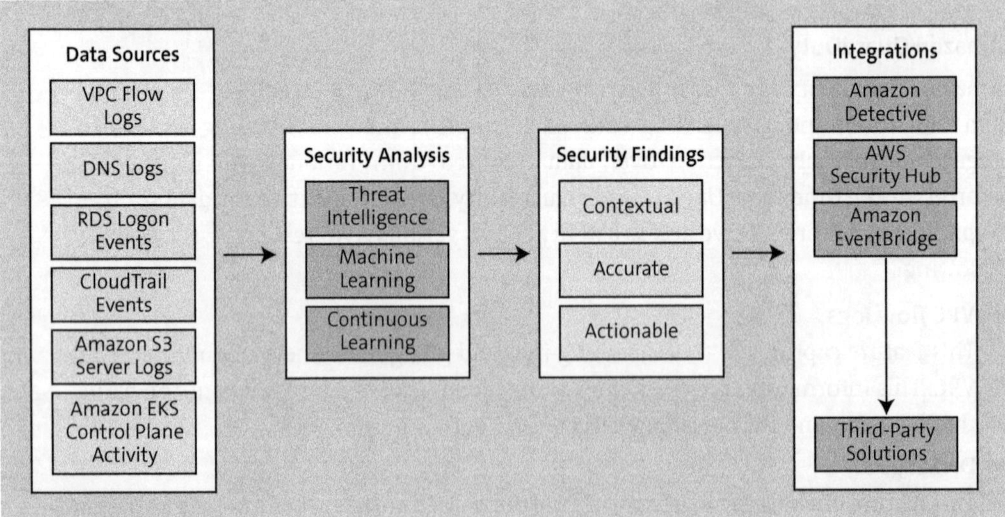

Figure 7.28 End-to-End Integration of Amazon GuardDuty

As shown in Figure 7.28, Amazon GuardDuty can integrate automatically with other AWS security or event notification services. Using these event notifications, you can further forward these findings to other third-party security solutions.

You can also leverage these event notifications to quickly trigger auto-remediation actions to respond to critical security events. You can create AWS Systems Manager Documents to integrate with Amazon EventBridge events and reduce the blast radius of a security event. Some examples of these remediation actions are available at *http://s-prs.co/v577664*.

Amazon Inspector

A vulnerability management solution, Amazon Inspector is an easily enabled, highly scalable AWS-native security solution. This solution covers Amazon EC2, container-based workloads, and AWS Lambda functions. It leverages incumbent AWS Systems Manager agents and scans resources for vulnerabilities. These vulnerabilities include software vulnerabilities and unintended network exposures that may allow malicious actors to compromise workloads, repurpose resources for malicious use, or exfiltrate data.

The core objective of this solution is to avoid *zero-day* vulnerabilities. Amazon Inspector automatically discovers workloads and initiates scans for vulnerabilities. All the discovered workloads are continually rescanned when a new common vulnerabilities and exposures (CVE) document is published or when there are changes in the workloads, including installing new software in an Amazon EC2 instance. However, to leverage this functionality, you must first enable Amazon Inspector in the settings area of the service landing page.

Amazon Inspector leverages several data sources, including its purpose-built scanning engine, which consumes over 50 data feeds to generate findings for CVEs. In addition, it also incorporates vendor security advisories, the National Vulnerability Database (NVD), The MITRE Corporation, open-source feeds, internal research, and licensed data feeds.

Amazon Inspector categorizes these findings into three types: package vulnerabilities, code vulnerabilities, and network reachability-related vulnerabilities. As shown in Figure 7.29, these findings contain several details that help you to find its impacted area, severity, related vulnerabilities, and recommended remediation. It provides CVE information to create contextual scores so that you can prioritize and resolve these vulnerabilities. You can find more information about severity levels and their associated score ranges at *http://s-prs.co/v577665*.

You can repurpose the deep-scanning efforts of this solution to discover the *software bill of materials (SBOM)* of each scanned target. You can direct this collected information to store it in an Amazon S3 bucket. Later, you can integrate this information with Amazon Athena to query and create granular insights from SBOM information.

Affected packages

Name	glib2
Installed version / Fixed version	0:2.56.4-8.el8_2.1.X86_64 / 0:2.56.4-159.el8
Package manager	OS

Remediation

Upgrade your installed software packages to the proposed fixed in version and release.

- dnf update glib2

Vulnerability details

Vulnerability ID	CVE-2022-32792
Vulnerability source	REDHAT_CVE
CWEs	CWE-787
Inspector score	7.8
Inspector scoring vector	CVSS:3.1/AV:N/AC:L/PR:N/UI:R/S:U/C:H/I:H/A:H/MAV:L
Exploit Prediction Scoring System (EPSS)	0.00183
CVSS 3.1	8.8 (Source: NVD)
Scoring vector	CVSS:3.1/AV:N/AC:L/PR:N/UI:R/S:U/C:H/I:H/A:H
CVSS 3.1	8.8 (Source: REDHAT_CVE)
Scoring vector	CVSS:3.1/AV:N/AC:L/PR:N/UI:R/S:U/C:H/I:H/A:H

Finding details | Inspector score and vulnerability intelligence

CVSS v3 (REDHAT_CVE)	Inspector
8.8	7.8

The Inspector score is **lower.** Changed metrics: Attack Vector

CVSS score metrics

Metric	CVSS	Inspector
Attack Vector	Network	Local
Attack Complexity	Low	Low
Privileges Required	None	None
User Interaction	Required	Required
Scope	Unchanged	Unchanged
Confidentiality	High	High
Integrity	High	High
Availability	High	High

Figure 7.29 An Example Finding in Amazon Inspector

You can integrate these findings with AWS Security Hub and Amazon EventBridge. In both integrations, you can trigger a notification to inform the security team or resolve it with an automated response. The event schema of the Amazon Inspector findings includes remediation action that you can extract and attempt accordingly. More details about this event schema and an example Amazon EventBridge rule are available at *http://s-prs.co/v577666*.

AWS Security Hub

AWS Security Hub provides a single pane of the glass of your overall security state at AWS. It helps you assess your AWS environment against security industry standards and best practices, such as AWS Foundational Security Best Practices (FSBP), Center for Internet Security (CIS), the Payment Card Industry-Data Security Standard (PCI DSS), and the National Institute of Standards and Technology (NIST).

As shown in Figure 7.30, AWS Security Hub collects security data across AWS accounts against the security controls defined in the industry standards and best practices framework mentioned earlier. Several third-party security products also feed data into this solution to provide a holistic and comprehensive view of your security posture in a single place. You can find a complete list of internal data providers at *http://s-prs.co/ v577667* and a list of third-party providers at *http://s-prs.co/v577668*. Activating AWS Security Hub is mandatory to leverage this service, which you can activate it in the settings options on the AWS Security Hub service page.

AWS Security Hub offers a response and remediation framework where you can integrate these security findings with automation features to respond to security issues. You can trigger these responses with the help of Amazon EventBridge. To reuse your current investments, you can also send these events beyond AWS's boundaries into third-party security information and event management (SIEM) solutions. You can find the list of integrations under the **Integrations** option on the AWS Security Hub service landing page.

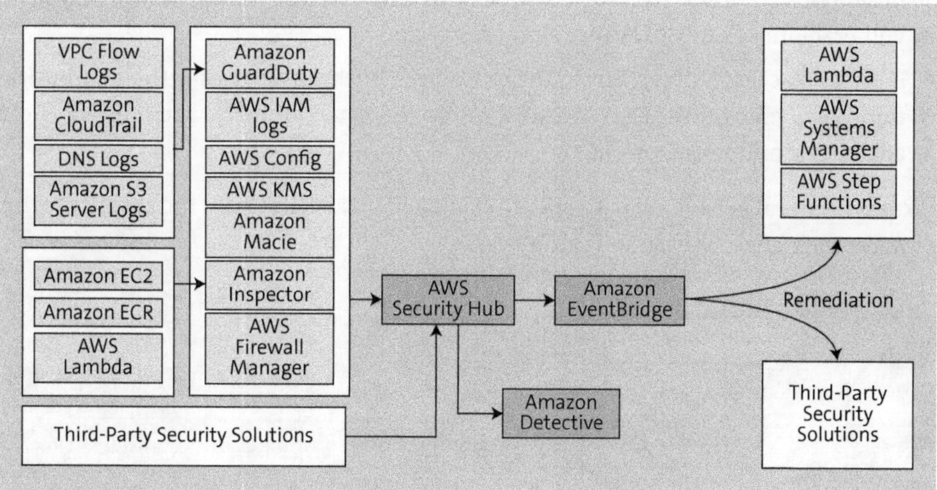

Figure 7.30 AWS Security Hub: End-to-End Integration Scenario

Amazon Detective

Amazon Detective is a service offering that helps security teams intuitively analyze, investigate, and find the root causes of security incidents. This service leverages data from AWS Security Hub, Amazon GuardDuty, and other third-party security providers.

This service uses this data to build a behavior graph involving AWS entities and their relationships. It produces this behavior graph based on a significant period (recommended minimum is 45 days) of telemetry from all your security services. Using machine learning models, this solution can identify how anomalous an action has occurred by leveraging its historical relationship model.

A list of common entity types in the behavior graph data structure is available at *http://s-prs.co/v577669*. Figure 7.31 shows an example of a behavior graph model that Amazon Detective constructed based on information from several sources. These relationship patterns highlight the unusual behavior and associate all the historical information against the involved entities. You can use this information to investigate the threat behavior and its root cause.

The example shown in Figure 7.31 provides answers to the following questions:

- Is there anything unusual about this specific user assuming this IAM role? If not, what is the user's typical activity after taking this role? What personas of users usually need this elevated authorization? Does the current activity fall under the gambit of similar personas?

- If a specific IP address is involved in the action, is this address on the list of previously connected IPs for this Amazon EC2 instance? If yes, is there any unusual magnitude of data transfer involved? Does this Amazon EC2 instance have any open vulnerabilities that could cause similar exploits?

- If the user accesses objects from the Amazon S3 bucket by elevating their authorizations, how many things are accessed within the usual limit? Do these objects hold any critical and sensitive data?

Likewise, several behavioral patterns are evaluated and available for the security teams to make their analysis easy and intuitive.

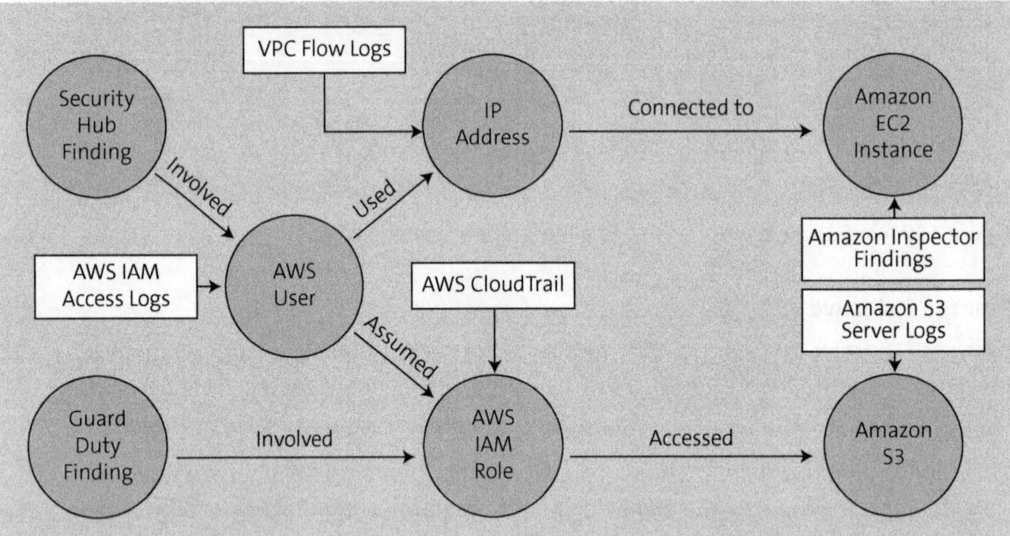

Figure 7.31 Amazon Detective: Example Behavior Graph Data Structure

7.3 Cost Optimization

Cost savings are an essential driver for AWS adoption. You could save up to 50% compared to your on-premise IT costs. However, the temptations of infinite scalability, ubiquitous automation, and quick-and-easy access to on-demand resources may blind you from controlling expenses. In addition, AWS improves your staff's productivity. These productivity gains will improve your ability to do more, so you consume more—the cloud-native models offered by AWS open up new paths for innovation. By embarking on these paths, your traditional cost models become obsolete. Traditional data center cost management has an inherent threshold on cost functions, whereas the pay-as-you model lacks these controls unless you enforce practices to contain them. Hence, an imperative task is to build these controls. The freedom of decentralized cost management in the cloud can easily snowball costs out of bounds if you do not track them constantly.

Conversely, the cloud also provides several avenues to save costs while growing your IT investments on AWS. Cost management best practices are discussed extensively in Chapter 3 under the AWS Well-Architected Framework. In this section, we'll restrict ourselves to the tools and mechanisms available in AWS to track, plan, and optimize your AWS investments.

7.3.1   Cost Visualization

Visualization provides a quick and easy way to gain insights. This section discusses two visualization tools so you can easily track your current costs and control your team's spending—AWS Cost Explorer and AWS Budgets.

AWS Cost Explorer

This cost and usage report can visualize your aggregate costs across all AWS services. As shown in Figure 7.32, with additional filter dimensions, you can analyze your AWS costs and usage at a much more detailed and granular level. Leveraging tags is an intuitive way to track the spending associated with specific applications, including SAP. To determine tag-based cost allocations, you'll need to activate the tags feature in the **Billing** service by choosing the Cost allocation tags option. In the report shown in Figure 7.32, the Cost allocation tag Application has been active since the last week of July. Therefore, this allocation is missing for June and a significant portion in July.

Figure 7.32 AWS Cost Explorer: Tag-Based Cost Visualization

This solution provides the current usage, costs, and historical consumptions of the last 12 months. It also provides an outlook on your future spending by extrapolating insights from your historical spending. You can generate several reports to present your costs from different perspectives for a better understanding.

Moreover, it also generates insights about your Reserved Instances (RIs) and Savings Plans (SPs) utilization coverage and offers recommendations for supplementing your current on-demand resource utilization.

In addition, you can also get suggestions for potential cost savings based on your current usage. You can choose the **Rightsizing Recommendations** options in the **Cost Explorer** navigation panel to view these recommendations. You can further customize your recommendations by selecting additional criteria provided on the screen and, for example, choosing the instance recommendations within the same family so that you can cover this usage with your current RI commitments.

Figure 7.33 shows some recommendations based on the underutilization of several resource metrics, such as CPU, memory, disk, and network utilization, against what you provisioned.

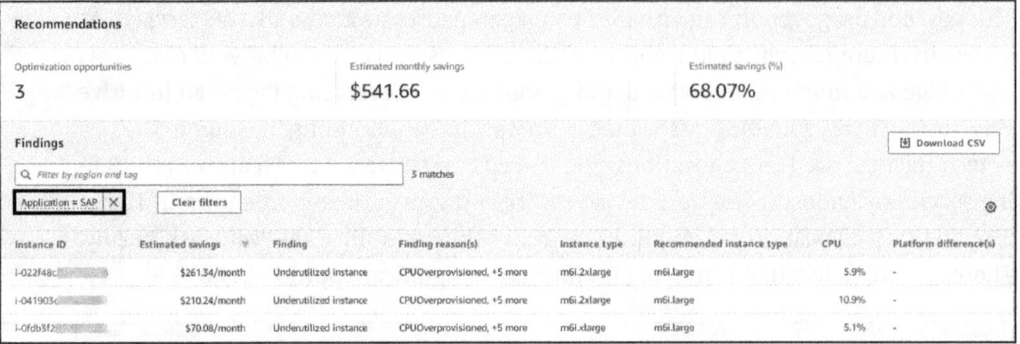

Figure 7.33 AWS Cost Explorer: Rightsizing Recommendations

You can leverage the **Cost Anomaly Detection** option to monitor suspicious spending activities within your AWS account. This option enables you to set up an alert notification mechanism. First, set up a **Cost monitor** to include specific AWS resources for monitoring. Later, subscribe to this monitor for alert notifications. Then, as shown in Figure 7.34, whenever an abnormal activity is detected, it will send messages to the subscribed recipients showing the new service usage ❶, cost monitor name ❷, and abnormal spike in usage ❸.

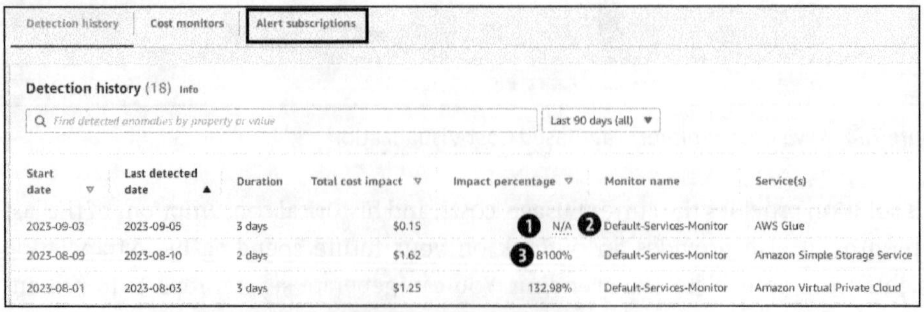

Figure 7.34 AWS Cost Explorer: Cost Anomaly Detection

AWS Budgets

AWS Budgets is a mechanism to set thresholds on your AWS accounts or subaccounts when used under AWS organizations. You can monitor the current and forecasted usage by setting up a budget with the help of available templates. You can also create your customized templates using specific filtering dimensions.

Figure 7.35 shows example of a cost budget you can create using the **Customize (advanced)** template option. You can use **Budget Scope** as the **Filter specific AWS cost dimensions** in this option. In our example, we used the **Tag** dimension to include only SAP resources in the tracking.

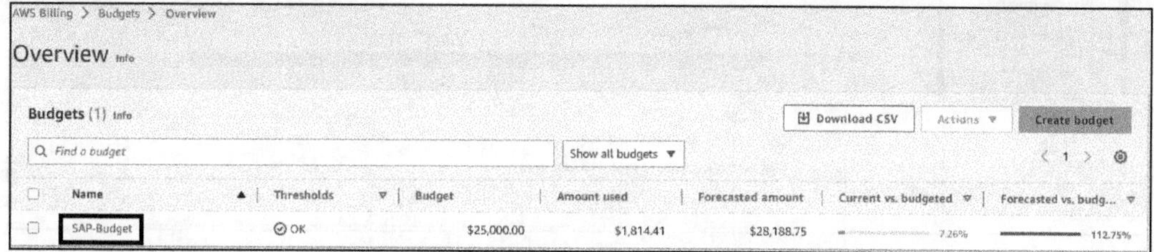

Figure 7.35 AWS Budgets: Tracking SAP Spend Using Tags in Custom Template

7.3.2   Continuous Optimization Opportunities

Along with best practices recommended under the cost optimization pillar of AWS Well-Architected Framework, you can use AWS Compute Optimizer and Amazon S3 Storage Optimization to get continuous optimization recommendations. Let's cover each tool briefly next.

AWS Compute Optimizer

AWS Compute Optimizer is an AWS service that analyzes the configuration of your AWS resources and utilization metrics and incorporates them to propose an optimized resource configuration. It reports whether your resources are optimal and suggests optimization recommendations to reduce costs and improve the performance of your workloads.

This service can also leverage external metrics from partner observability solutions and can analyze resource metrics for an enhanced duration (up to 93 days) to generate optimizations from this data. As shown in Figure 7.36, this tool recommends optimizations for a specific set of resource types; the list is available in navigation panel of the service landing page. You can filter these results by tags to customize your recommendations and find SAP workload optimizations on the subsequent page of the **View recommendations** tab.

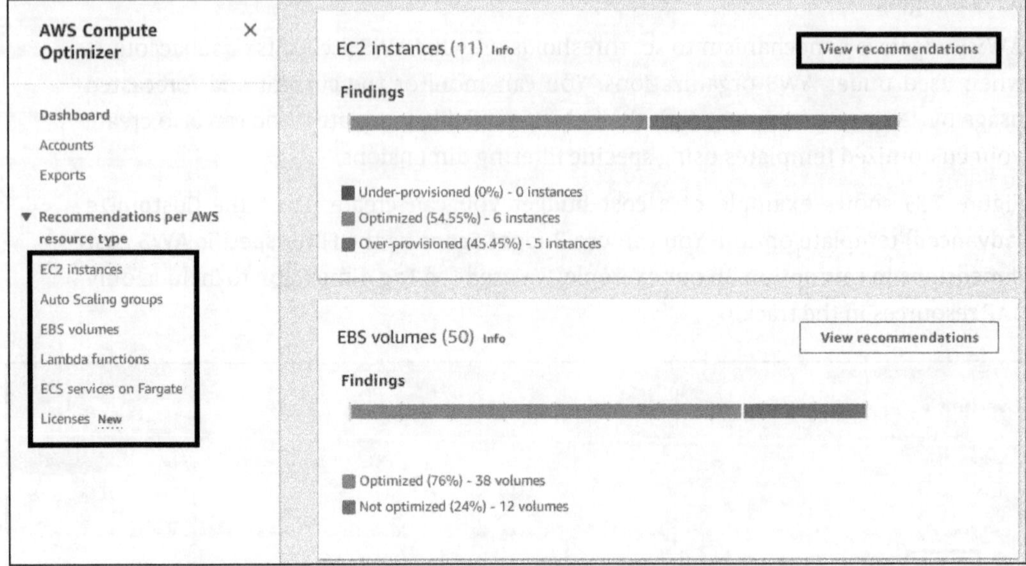

Figure 7.36 AWS Compute Optimizer: Amazon EC2 and Amazon EBS Performance Optimization Recommendations

By optimizing your SAP workloads constantly by adopting the recommendations, you'll save on costs. You've moved from the paradigm of spending on what you want to only on what you need.

AWS Cost Explorer Optimization Recommendations versus AWS Compute Optimizer

Behind the scenes, AWS Compute Optimizer powers the AWS Cost Explorer optimization recommendations, such as rightsizing recommendations and guidance on Reserved Instances (RIs) and Savings Plans (SPs). However, AWS Cost Explorer's perspective aims towards cost optimization, whereas AWS Compute Optimizer focuses more on performance optimization.

Amazon S3 Storage Optimization

After you address compute and block storage, object storage (i.e., Amazon S3) is the next significant cost element. Based on our experience, storage costs often go unnoticed or mismanaged. Due to incorrect lifecycle policies or misunderstood data access patterns, these costs mount slowly in the long term. Though this storage service offers many storage classes to save on costs, identifying the right strategy takes a lot of work.

AWS offers two options in Amazon S3 to give you actionable insights about your current usage and data access patterns. With the help of these insights, you can define more appropriate data management policies:

- **S3 storage lens**

  This option collects Amazon S3 storage metrics about your Amazon S3 storage usage, object attributes, and data growth behavior across your organization. Using this information, you can identify potential cost optimization areas. To start with the Amazon S3 storage lens, you must create a dashboard and wait 48 hours to populate the metrics.

- **Storage Class Analysis**

  This option analyzes storage access patterns to help you decide when to transition the correct data to the proper storage class. Using this analysis, you can define Amazon S3 lifecycle policies to transition less frequently accessed standard storage to other Amazon S3 storage classes. The best approach is to create this analysis for each bucket individually. Choose the **Metrics** tab from the bucket properties and select **Create Analytics configuration** from the Storage Class Analysis widget. You may need to wait sometime to see the information for this analytic configuration.

7.4 Summary

In this chapter, we discussed all the knowledge areas required to run your SAP operations on AWS. We started this chapter by explaining how the operational model changes from on-premise to AWS. Then, we identified what additional prerequisites you should implement to get support from SAP while running on AWS. Then, you learned the importance of tagging and resource groups and the central role of AWS Systems Manager and its capabilities. We dove deeply into observability, security and compliance, systems administration, change management, and the mechanisms to respond to security events. Finally, we touched upon AWS offerings that provide cost optimization avenues to save costs even while you're continuously growing on AWS.

Chapter 8

Automation Tools and Technologies

Amazon Web Services (AWS) provides virtually infinite compute power at your fingertips. Additionally, AWS offers numerous services and tools for deploying, installing, managing, and securing your SAP workloads. These services and tools come with programmatic interfaces to alleviate the challenges of running your SAP workloads and to reduce manual effort. In this chapter, you'll discover how to harness this automation power to operate SAP systems at scale on AWS.

AWS automation streamlines the provisioning and management of cloud resources, enabling businesses to achieve scalability, efficiency, and consistency. Leveraging tools such as AWS CloudFormation, AWS Lambda, and AWS Systems Manager, users can automate repetitive tasks, orchestrate complex workflows, and maintain infrastructure as code (IaC), ultimately allowing for rapid deployments, cost optimization, and enhanced operational resilience in the AWS cloud environment.

Automation on AWS is a game-changer for many organizations for several compelling reasons. AWS automation allows a business to dynamically scale resources based on demand. With tools like auto-scaling, resources can be increased or decreased seamlessly, ensuring optimal performance and cost efficiency. IaC tools like AWS CloudFormation ensure that resources are provisioned consistently across environments, reducing errors that arise from manual configurations. Automated workflows reduce the manual effort in managing and operating resources. This approach not only saves time but also allows IT teams to focus on more value-added activities. Additionally, by automating tasks such as shutting down unused resources or optimizing resource allocations, your organization can make significant cost savings on their AWS expenses. Other benefits include enhanced security, rapid deployment, resilience, high availability (HA), integrated innovation, and improved auditability and compliance.

In this chapter, we'll cover the requirements for starting SAP automation on AWS and establishing DevOps practices to manage changes. We'll review AWS-native tools and other services for developing custom automation solutions. Sample code is provided to demonstrate the integration of the toolset.

8.1   Getting Started

Getting started with AWS automation involves understanding its prerequisites and leveraging AWS services effectively. To embark on this journey, you must possess a basic knowledge of scripting, automation, and programming languages like Python or JavaScript. Installing the AWS Command Line Interface (CLI) and software development kits (SDKs) is crucial for interacting with AWS services programmatically. Moreover, setting up AWS Identity and Access Management (IAM) with the necessary roles and policies is essential for granting the required permissions to automation tasks.

AWS offers a suite of tools for automation, such as AWS Lambda, AWS CloudFormation, AWS Systems Manager, and the AWS Cloud Development Kit (CDK), allowing you to choose the right tools based on your project's needs. Setting up a conducive environment; configuring Amazon CloudWatch for monitoring; ensuring security best practices; managing costs; testing and validating in a non-production environment; and using version control systems like Git are all critical steps to streamline your automation workflow.

To enhance your AWS automation process, utilizing an integrated development environment (IDE) can make the workflow more efficient. Popular IDEs include AWS Cloud9, Visual Studio Code (VS Code), PyCharm, Eclipse, and IntelliJ IDEA, each offering unique features and extensions for AWS development. AWS CloudFormation Designer, although not an IDE, is a visual tool for designing AWS infrastructure, proving useful for those preferring graphical interfaces.

AWS supports a wide array of programming languages for automation, including Python, JavaScript, Java, C#, and more, offering SDKs and libraries to facilitate application development. Refer to *https://aws.amazon.com/developer/* for a complete list of AWS-supported SDKs. For code management, platforms like GitHub, GitLab, Bitbucket, AWS CodeCommit, and AWS CodeStar provide robust repositories for storing and collaborating on AWS automation artifacts.

You can also consider leveraging AWS Code Whisper. This artificial intelligence (AI)-powered service can improve developer productivity by generating code recommendations based on their comments in natural language to create code in an IDE. AWS Code Whisper can also scan your code for security vulnerabilities and provide suggestions to fix them. This solution supports 15 programming languages and can be integrated with various IDEs.

8.2   AWS-Native Tool Set for Automation

The AWS-native toolset for SAP automation encompasses a variety of services designed to support the deployment, management, and operation of SAP environments. In this section, we'll briefly touch upon some key AWS services and features used for SAP automation.

8.2.1 AWS Launch Wizard for SAP

AWS Launch Wizard for SAP offers a streamlined interface for deploying SAP applications on the AWS platform. This service eases the work of planning, sizing, and configuring the necessary AWS resources to support SAP workloads. It guides users through a simplified setup process, automating much of the resource provisioning effort based on the unique requirements of the SAP systems being deployed.

Utilizing this wizard, you can determine the most suitable Amazon Elastic Cloud Compute (Amazon EC2) instance types, Amazon Elastic Block Storage (Amazon EBS) configurations, and Amazon VPC settings for your SAP environment. AWS Launch Wizard for SAP ensures that your setup adheres to AWS's recommended security and operational best practices.

As shown in Figure 8.1, users should perform specific preparatory steps before initiating the AWS Launch Wizard process, either through the AWS Management Console or the AWS CLI. When setting up AWS Launch Wizard, you'll enter the various parameters of your SAP application, such as system IDs and sizes, along with their operating system (OS) specifications. For an AWS CLI deployment, you must store these parameters in a JavaScript Object Notation (JSON) file (the template is available at *http://s-prs.co/v577670*) and provide this file as an input to the command parameter. The required input values will vary depending on the installation scenario. Based on this information, AWS Launch Wizard will choose an optimized AWS environment. Once you confirm this choice, the solution proceeds to automate the rollout of the SAP system, including the provisioning of the underlying AWS infrastructure.

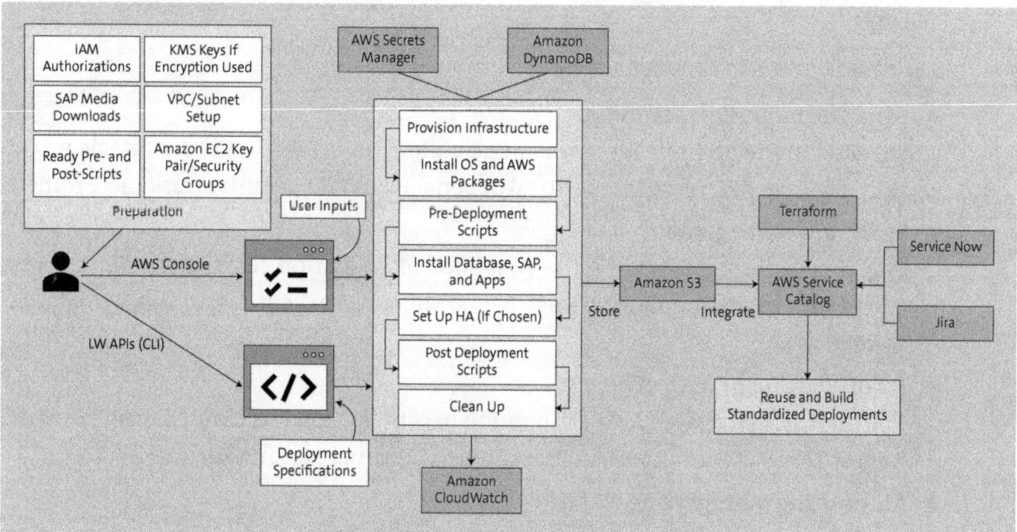

Figure 8.1 AWS Launch Wizard: End-to-End Process for SAP Deployments

You can choose to store the successful deployment template, along with its configuration scripts, in an Amazon Simple Storage Service (Amazon S3) bucket, which can later be integrated into the AWS Service Catalog. Later, this template is available for rebuilding consistent deployments for similar installation scenarios. Additionally, you can invoke this AWS Service Catalog template from Terraform, ServiceNow, and Jira pipelines. Visit *http://s-prs.co/v577671* for more information about the reuse of AWS Launch Wizard artifacts.

The advantage of using AWS Launch Wizard for SAP is that it minimizes the need for deep AWS expertise when deploying SAP systems, allowing SAP technology professionals to concentrate on higher-level application aspects while reducing the time required to get the systems operational on AWS.

> **Note**
>
> You can find additional details about the deployment features supported by AWS Launch Wizard, compatible SAP application versions, OSs, and SAP HANA version combinations, as well as information on SAP installation patterns, preparatory steps, execution sequences, and troubleshooting scenarios at *http://s-prs.co/v577672*.

8.2.2   AWS Migration Hub Orchestrator for SAP Migrations

AWS Migration Hub Orchestrator is a one-stop solution that simplifies the process of migrating SAP systems to an AWS cloud environment. This solution serves as a comprehensive management platform for orchestrating the steps required in a migration journey.

Let's break down its functionalities:

- **Strategic migration planning**
  It assists in the preliminary assessment, aiding in the creation of a strategic migration roadmap that recognizes critical dependencies and organizes tasks accordingly.

- **Workflow management**
  The Orchestrator automates the sequence of migration tasks, streamlining processes such as pre-migration assessments, migration execution, and post-migration validations.

- **Incorporation of AWS expertise**
  The service is built with AWS's migration best practices at its core, ensuring that the migrated SAP environment is configured for optimal performance on AWS.

- **Monitoring of migration activity**
  It offers the capability to monitor the migration progress actively, overseeing resources and task completions, thus providing visibility into the migration lifecycle.

- **Adaptable templates**

  It supplies ready-to-use templates for typical migration scenarios while also allowing for the creation of tailored templates that match unique migration needs.

- **Synergy with AWS solutions**

  AWS Migration Hub Orchestrator works cohesively with other AWS offerings, enhancing its functionality and providing a unified migration experience.

AWS Migration Hub Orchestrator is particularly useful in scenarios where SAP systems must be rehosted, replatformed, or otherwise reconfigured within an AWS cloud environment. Figure 8.2 shows an end-to-end process for an SAP migration using AWS Migration Hub Orchestrator.

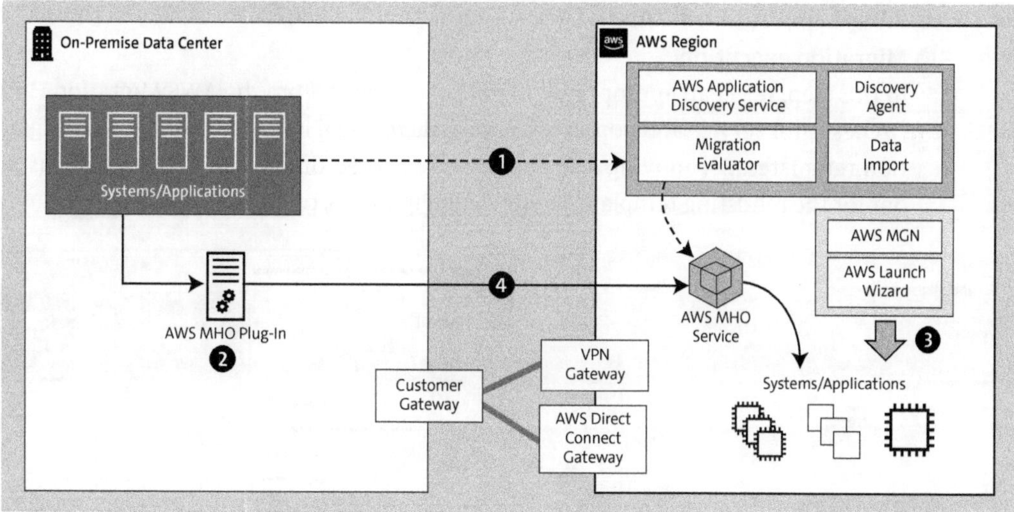

Figure 8.2 AWS Migration Hub Orchestrator: End-to-End Process Flow for SAP Migrations

At a high level, you can perform the whole process in four discrete steps, as shown in Figure 8.2:

❶ Discovery

Initially, an essential task is to compile an inventory of the systems and applications involved in the migration activity. This goal can be accomplished through various methods. The AWS Application Discovery Service (ADS) provides diverse data collection options, both agent-based and agentless, to automate the discovery process and sustain your systems inventory, thereby aiding in migration planning. Furthermore, the results from discovery with AWS Migration Evaluator can be seamlessly imported into this service. As an alternative, you have the option of manually maintaining this information by importing it from a template file into the service. You can find more details about application discovery at *http://s-prs.co/v577673*.

❷ Preparation at source

You'll need to download the AWS Migration Hub Orchestrator plugin and deploy it on a separate virtual machine (VM) at the source. Configure the plugin to register it with the AWS Migration Hub Orchestrator service and establish a trusted connection between the plugin server and the source systems. This step is crucial for enabling secure communications between them. For detailed guidance on configuring the plugin, visit *http://s-prs.co/v577674*.

❸ **Target system setup**

You must set up the target system with the same system information as the source. However, for an SAP migration scenario, AWS Migration Hub Orchestrator offers the flexibility to modify the target system's topology compared to the source. This capability allows for the transformation of a single system installation into a distributed, multi-app server installation or into an HA installation. Similarly, you have the option to convert scale-up systems into scale-out installations.

❹ **Migration execution**

In this step, select an appropriate migration template from the AWS Migration Hub Orchestrator service and populate the workflow. You'll have access to several migration templates. For moving SAP applications hosted on non-SAP HANA databases, opt for the rehosting template, as shown in Figure 8.3, on the left.

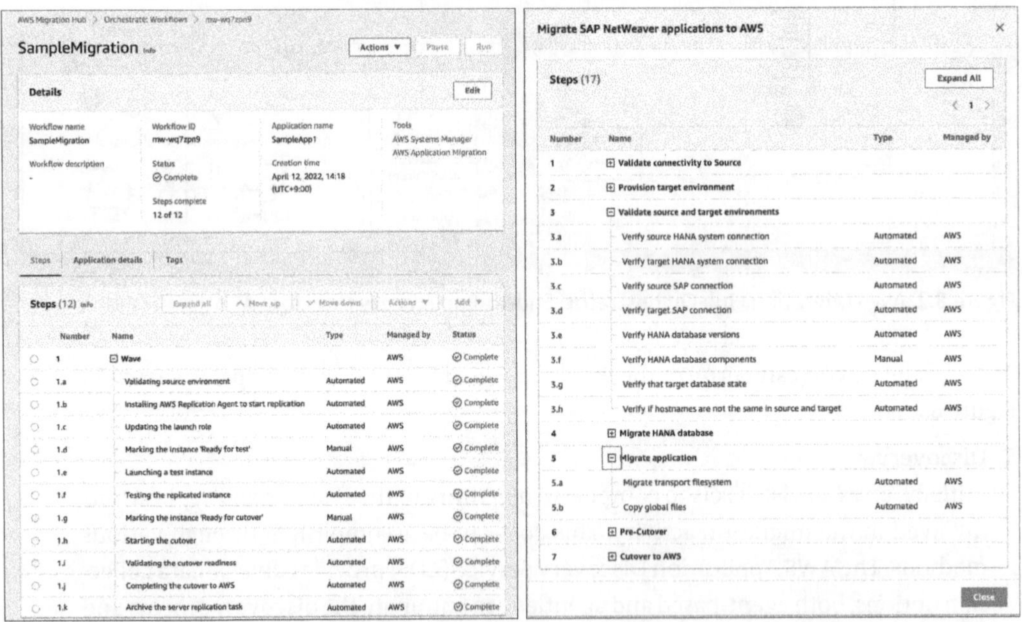

Figure 8.3 Example Workflows of Application Rehosting and SAP NetWeaver Application on SAP HANA

This template utilizes AWS Application Migration Service (MGN) to transfer storage data block-by-block to the target system. Alternatively, for migrating SAP NetWeaver applications hosted on SAP HANA databases, the SAP NetWeaver template is suitable. As shown in Figure 8.3, on the right, this template employs the native SAP HANA System Replication (HSR) mechanism to transfer change data from the source system to the target, along with using a file copy mechanism to replicate other SAP file systems, ensuring the target configuration mirrors the source.

You can customize your migration workflows by adding custom steps and associating them with custom actions through scripts. To add custom steps, follow the process outlined at *http://s-prs.co/v577675*.

By leveraging the AWS Migration Hub Orchestrator for SAP migrations, your organization can expect a more controlled and seamless transition, with minimized risk or disruption to business operations. The AWS Migration Hub Orchestrator ensures that each phase of an SAP system's migration to AWS is executed in alignment with your organizational goals and cloud architecture principles.

> **Note**
>
> You can find additional details about AWS Migration Hub Orchestrator's preparatory steps, types of migrations and corresponding templates, the sequence of the migration workflow, and troubleshooting scenarios at *http://s-prs.co/v577676*.

8.2.3   AWS SDK for ABAP

The AWS SDK for ABAP is a software development kit (SDK) designed to enable ABAP developers to easily integrate AWS services into their SAP applications. This SDK provides a straightforward way for ABAP applications to access various AWS services, thus enhancing their scalability, flexibility, and functionality within the SAP environment.

The typical use cases for AWS SDK for ABAP include the following:

- **Integration with AWS storage services**
  Store and retrieve data from AWS services like Amazon S3, leveraging its virtually infinite data storage as a data integration layer for AWS services or to enrich and build data lakes with SAP data on AWS.

- **Building process extensions into AWS**
  Trigger AWS Lambda functions for serverless computing capabilities. Utilizing a plethora of programming runtimes, you can build SAP process extensions to uphold SAP's clean core philosophy.

- **Enhancing SAP with the power of over 200 AWS services**
  Integrate other AWS services like Amazon DynamoDB or Amazon Simple Queue Service (SQS) to enhance application capabilities. Your imagination is the only limit to extending your application to leverage cloud-native constructs and adopt next-generation computing features.

- **Hybrid SAP application extensions**
  Build hybrid landscapes combining SAP on-premise solutions with AWS cloud services, thus allowing for two-tier application development without disrupting current on-premise SAP workloads.

Setting up AWS SDK for ABAP involves the following steps:

1. **AWS account setup**
   Make sure you have an AWS account with the necessary permissions. Enable AWS credentials (access key ID and secret access key) for an AWS user to configure in the ABAP system for authentication with AWS services.

2. **Source system prerequisites**
   Ensure that your source SAP system fulfills the requirements described at *http://s-prs.co/v577677* to import the AWS SDK package.

3. **Download the SDK**
   Download the AWS SDK for ABAP from the AWS website or GitHub repository using the following command:

   ```
   curl https://sdk-for-sapabap.aws.amazon.com/awsSdkSapabapV1/release/abapsdk-
   LATEST.zip -o "abapsdk-LATEST.zip"
   ```

4. **Import the SDK**
   The zip file contains a list of transports. The complete list of transports and their association with AWS service capabilities can be found at *http://s-prs.co/v577678*. Import the mandatory SDK transports and additional service transports of your choice into the ABAP environment using SAP's ABAP Workbench tools.

5. **Configure AWS SDK**
   Perform SDK configuration steps in Transaction /AWS1/IMG. Refer to *http://s-prs.co/v577679* for more details.

6. **Test connectivity**
   Perform initial tests to ensure that the ABAP system can communicate with AWS services.

For more detailed information, we recommend referring to the official AWS SDK for ABAP documentation at *http://s-prs.co/v577680*, which provides comprehensive guidance. For ABAP code examples refer to *http://s-prs.co/v577681*.

8.2.4   DIY Automation Tool Set from AWS

In addition to the out-of-the-box automation capabilities provided by AWS, you can utilize a variety of other AWS services to develop your own custom automation solutions from scratch.

As shown in Figure 8.4, the AWS DevOps service portfolio is organized into seven functional groups:

- **Change management group**

  Governs the continuous integration/continuous deployment (CI/CD) lifecycle, ensuring continuous integration and delivery of development artifacts. This area will be explored in greater detail in subsequent sections.

- **Invocation group**

  The common service integrations that have the ability to raise events upon breach of specific threshold of the metric or the occurrence of certain critical events that may be of concern for the operational teams. Several other services not mentioned here have this ability either directly or indirectly; however, we only mention the ones that are relevant in operational domains. We already covered these capabilities under Chapter 7 and Chapter 10. Refer to those chapters for more details.

- **Provisioning group**

  The AWS services that help you provision the necessary AWS resources for your applications (i.e., setting servers, VMs, and the other AWS infrastructure components). These IaC tools can model your environment in YAML Ain't Markup Language (YAML) or JSON.

- **Configuration group**

  Involves customizing resources to reach the desired state, such as configuring the OS, database, and applications and implementing governance standards through scripts and tools post-provisioning.

- **Collaboration group**

  These services enable coordination among various AWS tools or services by maintaining state or staging information, acting as an integration fabric where no direct interface exists.

- **Notification group**

  Consists of services that alert on the outcomes of automated actions or state changes. These notifications can be integrated into the automation pipeline to further automate response mechanisms. As shown in Figure 8.4, the notification group is a subset of the invocation group. We covered these services in Chapter 10.

- **Observability group**

  These tools allow teams to observe the health of their AWS environment, identify issues, understand system behaviors, and ensure that performance objectives are met. They play a crucial role in maintaining system reliability and performance optimization. We covered these services in Chapter 7.

Each group plays a specific role within the AWS DevOps ecosystem, from change management to observability, ensuring a cohesive and automated workflow.

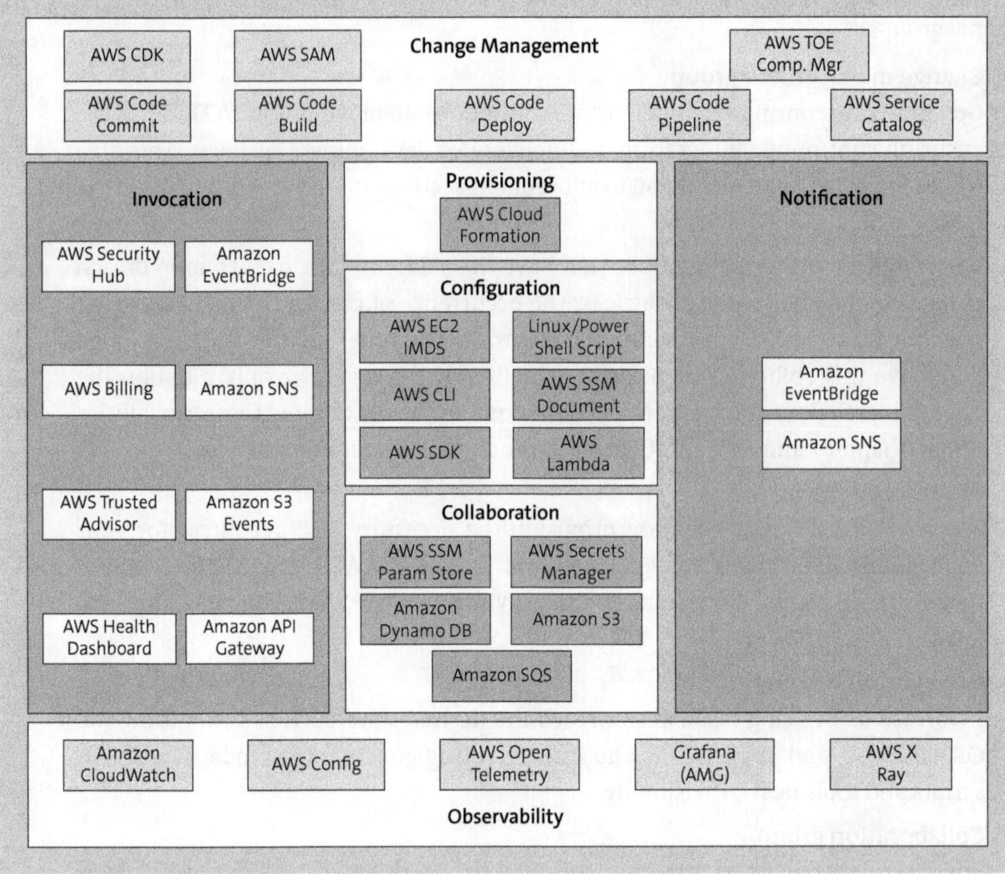

Figure 8.4 DevOps Service Portfolio for Non-Containerized Workloads on AWS

Note

As these capabilities are constantly expanding, this list is not exhaustive. Moreover, with the rapid growth of DevOps practices for SAP, our discussion in this section merely scratches the surface. We've focused on introducing the most prevalent patterns and tools, rather than engaging in an exhaustive exploration of all the possibilities. The automation services we mention in this chapter hold considerable potential. We encourage you to delve into the full feature set of these services for a more comprehensive understanding that enables you to address more complex challenges than those outlined here.

In this section, we explore the services that fall under the provisioning group and the configuration group in detail and show how collaboration groups hold data across automation steps to enable integration across these services.

Infrastructure Provisioning Services

This kind of service enables users to define an IaC that can be versioned, reused, and shared. Let's consider some of the key infrastructure provisioning services available on AWS.

AWS CloudFormation

AWS CloudFormation is like a construction blueprint for a building. Just as a blueprint defines the design, materials, and layout needed to construct a building, AWS Cloud-Formation templates define the AWS resources and settings required to build an application's infrastructure. You create a template in JSON or YAML format that specifies various AWS services like Amazon EC2 instances, Amazon S3 buckets, or Amazon EBS volumes, just as a blueprint specifies rooms, doors, and plumbing.

When you deploy an AWS CloudFormation template, you're handing the blueprint over to a team of builders who then construct the building. AWS CloudFormation automatically provisions and configures the resources as defined, creating a stack. If you need to change the infrastructure, you can simply update the template, much like revising a blueprint, and re-deploy it, with AWS CloudFormation handling the updates accordingly.

What makes AWS CloudFormation stand out for automation includes the following:

- **Idempotency**
  Running the same template multiple times will not create additional resources by mistake. It ensures that the end state always matches the described state in the template, which is key for reliable automation.

- **Atomic operations**
  When deploying stacks, if an error is encountered, AWS CloudFormation rolls back changes to bring your environment back to the last known good state, reducing any manual clean-up tasks.

- **Dependency management**
  AWS CloudFormation automatically handles the order of resource creation based on their dependencies. It waits for a resource to be fully available before moving on to the next, eliminating the need for scripts that manually poll for the resource status.

- **Parameterization**
  You can pass parameters into your AWS CloudFormation templates at runtime, which allows for dynamic input and reduces the need for hard-coded values, making your templates reusable and adaptable.

- **Stack updates and drift detection**
  AWS CloudFormation allows for controlled updates to your infrastructure with minimal downtime. Drift detection identifies when the actual state of a stack drifts from the expected state, which can be crucial for maintaining compliance and ensuring that manual changes do not go unchecked.

- **Nested stacks**
  Complex environments can be managed by nesting stacks within a master stack, allowing you to segment and modularize the architecture for better organization and reuse.

- **Custom resources**
  For resources not natively supported by AWS CloudFormation, you can create custom resources that allow your templates to provision and manage third-party application resources.

- **Change sets**
  Before applying changes, AWS CloudFormation can provide a summary of the proposed changes with change sets, allowing you to review the impact of the changes before execution.

Figure 8.5 shows a skeleton for an example AWS CloudFormation template for an SAP workload.

```
AWSTemplateFormatVersion: '2010 09 09'
Description:>-
  This template deploys the necessary infrastructure for an SAP environment on AWS.

Parameters:
  InstanceType
    Description: EC2 instance type for the SAP application server
    Type: String
    Default: m6i.xlarge

Resources:
  SAPVPC:
    Type: AWS::EC2::VPC
    Properties:
      ...
  SAPSubnet
    Type: AWS::EC2::Subnet
    Properties:
      ...
  SapEbsVolume
    Type: AWS::EC2::Volume
    Properties::
      ...
  SAPApplicationServer
    Type: AWS::EC2::Instance
    Properties:
      InstanceType : !Ref InstanceType
      BlockDeviceMappings
      ...
      ...
Outputs:
  SAPApplicationServerId
    Description: The Instance ID of the SAP Application Server
    Value: ...
Conditions:
  CreateProdResources
    !Equals
      -!Ref EnvironmentType
      -prod
```

Figure 8.5 Example AWS CloudFormation Template in YAML for Provisioning an SAP Application Server

In the example shown in Figure 8.5, the following elements are described:

- **Parameters**
  Allows the user to specify the instance type for the SAP application server.

- **Resources**

  Defines the core AWS resources such as a virtual private cloud (VPC) (SAPVPC), a subnet (SAPSubnet), an Amazon EBS volume for the database (SAPEbsVolume), and an Amazon EC2 instance for the application server (SAPApplicationServer).

- **Outputs**

  Provides useful information upon stack creation, like the application server instance ID.

- **Conditions**

  Sets a condition to create additional resources if the environment type is production.

Besides, several other components exist in the AWS CloudFormation template. A complete list of components is available at *http://s-prs.co/v577682*. For more resource types, visit *http://s-prs.co/v577683*.

The simplified representation in Figure 8.5 illustrates the components of an AWS CloudFormation template. A real-world SAP workload template would be more complex, including additional resources, configurations, and possibly nested stacks. To quick start your journey, review the community templates made available at *http://s-prs.co/v577684*.

AWS Resource Configuration Services

After the infrastructure has been provisioned, you may need to adjust its state to the desired configuration for running your applications. For SAP, once the Amazon EC2 instances for the application and database servers are provisioned, a necessary step may be to install additional OS patches, agents, and OS parameters, as well as the SAP software and configurations, to integrate these systems into your application landscape.

Components of the configuration group, shown earlier in Figure 8.4, can be integrated with each other, as shown in Figure 8.6.

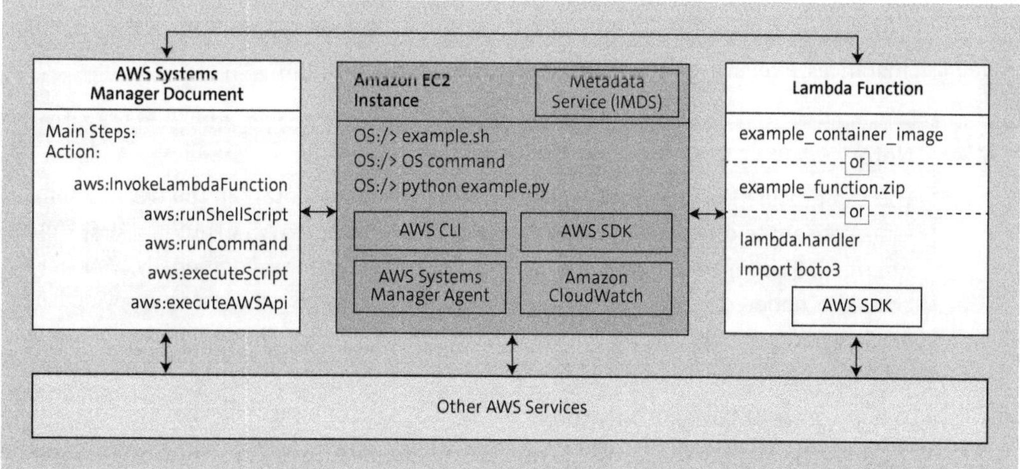

Figure 8.6 The Integration of Configuration Management Services / Tools within AWS

AWS Systems Manager Remote Execution

AWS Systems Manager offers several capabilities for triggering remote executions on AWS resources at scale. This feature enables the performance of configuration actions in various ways, among which are AWS Systems Manager Run Command, AWS Systems Manager Documents, and AWS Systems Manager Automation.

Let's explore how each capability serves a different purpose next:

- **AWS Systems Manager Run Command**
  Allows you to remotely and securely manage the configuration of your managed instances. For example, as shown in Figure 8.7, you can use the Run Command to execute a script or update software on multiple instances without logging into each one.

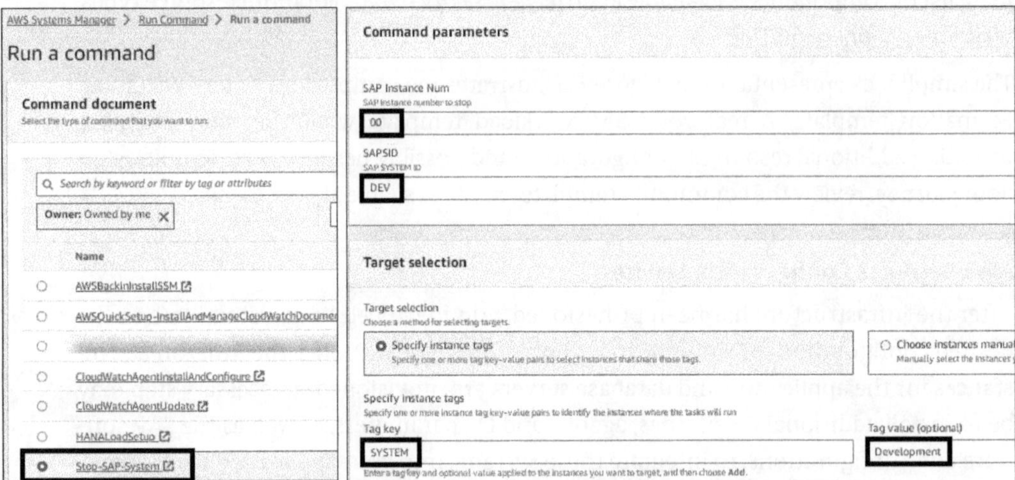

Figure 8.7 AWS Systems Manager: Run Command to Stop SAP System Using Tags and Parameters

This example command will send the **Stop-SAP-System** document to the specified instances, executing the actions defined in that document.

Note

You can execute this document from the Run Command feature in the AWS Systems Manager service console (shown in Figure 8.7), or leverage AWS CLI to invoke it from Amazon EC2 command line (refer to the following section). Additionally, you have access to hundreds of pre-existing AWS Systems Manager Documents provided by AWS for various purposes.

- **AWS Systems Manager Document**
  Defines the actions that AWS Systems Manager performs. These documents can be used by Run Command, State Manager, and Automation. AWS Systems Manager documents can be written in JSON or YAML and include steps for execution.

Example: The Definition of Above AWS Systems Manager Document

Listing 8.1 shows an example of an AWS Systems Manager document.

```
{
  "schemaVersion": "2.2",
  "description": "Stop SAP instance",
  "parameters": {
    "SAPInstanceNum": {
      "type": "String",
      "description": "SAP instance number to stop"
    },
    "SAPSID": {
      "type": "String",
      "description": "SAP SYSTEM ID"
    }
  },
  "mainSteps": [
    {
      "action": "aws:runShellScript",
      "name": "stopSAPInstance",
      "inputs": {
        "runCommand": [
          "#!/bin/bash",
          "lowercase_sid=$(echo '{{ SAPSID }}' | tr '[:upper:]' '[
:lower:]')",
          "runuser -l ${lowercase_sid}adm -c 'sapcontrol -
nr {{ SAPInstanceNum }} -function StopSystem ALL'"
        ]
      }
    }
  ]
}
```

Listing 8.1 Example SSM Document to Stop SAP Instance Running in the Given EC2 Instance

In this example the AWS Systems Manager Document, **SAPSID** is a parameter whose value will be converted to lowercase. The `aws:runShellScript` action executes a command that transforms the input parameter value to lowercase and derives the **sidadm** user. Later, this username is used as a substitute user to run **sapcontrol** command.

Note

AWS Systems Manager supports a variety of actions (as shown earlier in Figure 8.4) that can be specified in AWS Systems Manager Documents to automate tasks and workflows.

Some common AWS Systems Manager actions include the following:

- `aws:runShellScript`: Executes shell scripts on Linux instances.
- `aws:runPowerShellScript`: Runs PowerShell scripts on Windows instances.
- `aws:runCommand`: Executes an AWS Systems Manager Run Command.
- `aws:invokeLambdaFunction`: Invokes an AWS Lambda function.
- `aws:executeAwsApi`: Calls an AWS application programming interface (API) action.
- `aws:executeAutomation`: Executes another AWS Systems Manager automation workflow.
- `aws:changeInstanceState`: Changes the state of an Amazon EC2 instance (e.g., stopping or starting an instance).
- `aws:executeScript`: Executes a script in languages like Python or PowerShell.

These actions enable the automation of various tasks across AWS resources and are central to the functionality of AWS Systems Manager.

- **AWS Systems Manager automation**

  Allows you to automate complex workflows that are repeatable and safe to delegate. These workflows can involve multiple AWS services and operations.

  You can integrate an AWS Systems Manager document into an AWS Systems Manager Automation workflow. First, stop SAP from using the document and then stop the associated Amazon EC2 instance. You would structure the automation document in JSON.

Example

Listing 8.2 shows an example of an automation document.

```
{
  "schemaVersion": "0.3",
  "description": "Automation to stop SAP and then the EC2 instance",
  "assumeRole": "roleArn",
  "parameters": {
    "SAPInstanceNum": {
      "type": "String"
    },
    "SAPSID": {
      "type": "String"
    },
```

```
    "InstanceId": {
      "type": "String"
    }
  },
  "mainSteps": [
    {
      "name": "stopSAP",
      "action": "aws:executeAutomation",
      "inputs": {
        "DocumentName": "Stop-SAP-System",
        "Parameters": {
          "SAPInstanceNum": ["{{ SAPInstanceNum }}"],
          "SAPSID": ["{{ SAPSID }}"]
        },
        "TargetParameterName": "InstanceId",
        "Targets": [
          {
            "Key": "ParameterValues",
            "Values": ["{{ InstanceId }}"]
          }
        ]
      }
    },
    {
      "name": "stopEC2Instance",
      "action": "aws:changeInstanceState",
      "inputs": {
        "InstanceIds": ["{{ InstanceId }}"],
        "DesiredState": "stopped"
      }
    }
  ]
}
```

Listing 8.2 Sample SSM Automation Document to Stop EC2 Instance After Stoping SAP Application

In this automation document, first, the stopSAP step uses aws:executeAutomation to run the Stop-SAP-System document on the specified instance. Then, the stopInstance step uses aws:changeInstanceState to stop the Amazon EC2 instance. Replace roleArn with the appropriate IAM role ARN that the Automation will assume and ensure that Stop-SAP-System matches the name of your previously created AWS Systems Manager document.

In summary, the Run Command is for executing commands on instances, AWS Systems Manager Documents define what actions are performed, and Automation is for orchestrating complex workflows.

AWS Command Line Interface

AWS CLI is a robust tool enabling direct interaction with AWS services via the command line. AWS CLI is widely used for scripting and automating various AWS services, notably in resource management, application deployment, and data querying. AWS CLI is particularly convenient for integration into shell scripts, thus allowing you to create automation scripts at the OS level.

Example Use Case for SAP Workloads

Let's say you need to start an SAP system hosted on an Amazon EC2 instance. You could use AWS CLI to start the Amazon EC2 instance and then execute an AWS Systems Manager document that starts the SAP system with the code shown in Listing 8.3.

```bash
#!/bin/bash
# Set the EC2 instance ID and SAP System ID
EC2_INSTANCE_ID="i-1234567890abcdef0"
SAPSID="DEV"
SAP_INSTANCE_NUM="00"


# Start the EC2 instance
aws ec2 start-instances --instance-ids $EC2_INSTANCE_ID


# Wait for the instance to be in a running state
echo "Waiting for EC2 instance to be in running state..."
aws ec2 wait instance-running --instance-ids $EC2_INSTANCE_ID


# Execute the SSM Document to start SAP
aws ssm send-command \
    --document-name "Start-SAP-System" \
    --targets "Key=instanceids,Values=$EC2_INSTANCE_ID" \
    --parameters "{\"SAPSID\":[\"$SAPSID\"],\"SAPInstanceNum\":[\"$SAP_
INSTANCE_NUM\"]}"
echo "SSM command to start SAP system has been sent."
```

Listing 8.3 Sample Command Sequence to Start SAP Along with EC2 Instance

In this example, the first command starts the Amazon EC2 instance, and the second command triggers an AWS Systems Manager document to start the SAP system on that instance.

> **Note**
>
> To configure the AWS CLI, you need to fulfill the following prerequisites:
>
> - **AWS Account**
>   An active AWS account is essential.
>
> - **IAM User**
>   An IAM user with appropriate permissions.
>
> - **Access key ID and secret access key**
>   Generated from the AWS Management Console under your IAM user's security credentials.
>
> - **AWS CLI installed**
>   The AWS CLI must be installed on your local machine.
>
> - **Configuration command**
>   Run `aws configure` in your terminal and input your access key, secret key, default region, and output format when prompted.
>
> Ensure your IAM user has the necessary permissions for the tasks you intend to perform with AWS CLI.

Amazon EC2 Instance Metadata Services

Amazon EC2 Instance Metadata Service (IMDS) is a service offered by AWS that allows Amazon EC2 instances to access instance-specific data. This capability is particularly useful for managing and configuring your instances automatically without the need to hard-code or manually enter data about these instances. IMDS is accessed from within an instance itself.

Two versions of IMDS exist:

- IMDSv1: Uses a request/response method where each request is independent.

- IMDSv2: Uses a session-oriented method where a token must be created and provided in the header of HTTP GET requests.

IMDS can be used to retrieve information such as the following:

- Instance IDs

- Amazon Machine Image (AMI) IDs

- Instance types

- Local IP addresses

- Public IP addresses

- Security groups

- IAM roles and credentials, and more

To retrieve the instance ID with IMDSv1, you can simply send a GET request to the metadata URL:

```
curl http://169.254.169.254/latest/meta-data/instance-id
```

Example: Bash Shell at Command Prompt

IMDSv2 requires a session token that is used in subsequent requests. Listing 8.4 shows an example of how to use IMDSv2 to retrieve the instance ID:

```
# Step 1: Fetch the session token
TOKEN=`curl -X PUT "http://169.254.169.254/latest/api/token" -H "X-aws-ec2-
metadata-token-ttl-seconds: 21600"`
# Step 2: Use the token to make a request for the instance ID
INSTANCE_ID=`curl -H "X-aws-ec2-metadata-token: $TOKEN" \
http://169.254.169.254/latest/meta-data/instance-id`
```

Listing 8.4 Example of IMDSv2 Commands to Fetch EC2 Instance ID

With IMDS, you can automate various tasks by leveraging this service to extract Amazon EC2 instance self-identities. You can incorporate this information into your automation logic when making AWS CLI or API requests. This scenario is applicable both in local shell scripts and for remote execution using AWS Systems Manager documents or AWS Lambda functions.

Note

You must manage the use of IMDS properly, especially with regard to the credentials it provides. Using IMDSv1 is no longer recommended due to security considerations, but it is still supported. Always use IMDSv2 for enhanced security and follow AWS best practices to secure your Amazon EC2 instances.

AWS Software Development Kit

The AWS SDK is a set of tools and libraries for building applications that interact with AWS services. Developers can access AWS services programmatically using various programming languages. As discussed in Section 8.2.3, AWS provides SDKs in several programming languages. You can choose the right SDK for your organization's skills and your developers' convenience.

AWS SDK for Python (boto3)

Let's look at some example Python code utilizing the **boto3** AWS SDK in Listing 8.5. This code retrieves an instance ID using IMDS and then pushes a custom Amazon CloudWatch metric.

```python
import boto3
import requests
def get_instance_id():
    # URL for the EC2 instance metadata service
    url = "http://169.254.169.254/latest/meta-data/instance-id"

    # Fetch the instance ID
    response = requests.get(url)
    if response.status_code == 200:
        return response.text
    else:
        return "Unable to retrieve instance ID"

# Initialize the CloudWatch client
cloudwatch = boto3.client('cloudwatch')

# Write logic to evaluate the below values programmatically (Hint: sapcon-
trol)
num_sap_active_wp = 10 # find current active work processes
sap_sid = "S4H" # Your SAP SID here
sap_inst_num = "03" # You SAP Application Inst Number

# Get current instance ID
instance_id = get_instance_id()

# Push the custom metric
cloudwatch.put_metric_data(
    Namespace='SAP/Monitoring',
    MetricData=[
        {
            'MetricName': 'WorkProcesses',
            'Dimensions': [
                {
                    'Name': 'SID',
                    'Value': sap_sid
                },
                    {
                        'Name': 'SAPInstNum',
                        'Value': sap_inst_num
                    },
                {
                    'Name': 'InstanceId',
                    'Value': instance_id
                }
            ],
```

```
          'Value': num_sap_active_wp,
          'Unit': 'Count'
      },
   ]
)
```
Listing 8.5 Example of boto3 AWS SDK Python Code

AWS Lambda

AWS Lambda is a serverless compute service that allows you to run code without provisioning or managing servers. It executes your code only when needed and scales automatically, from a few requests per day to thousands per second. You pay only for the compute time you consume, making it a cost-effective and scalable solution.

Let's break down the anatomy of AWS Lambda:

- **AWS Lambda function**

  The core component of AWS Lambda is the Lambda function itself. It consists of your code and any associated dependencies. Functions are stateless, and AWS Lambda can instantiate as many copies of the function as needed to scale to the rate of incoming events.

- **Event source/trigger**

  AWS Lambda functions can be triggered by AWS services or called directly over HTTPS. Event sources are AWS services or custom applications that publish events, and AWS Lambda can be configured to execute a function in response to various event types. Common event sources include Amazon S3, Amazon DynamoDB, Amazon Kinesis, Amazon Simple Notification Service (SNS), and Amazon CloudWatch.

- **Runtime**

  AWS Lambda supports multiple runtimes (Node.js, Python, Ruby, Java, Go, .NET Core, and custom runtimes). When you create an AWS Lambda function, you specify the runtime that you want to use.

- **Execution role**

  An IAM role (execution role) grants your AWS Lambda function permissions to AWS resources. This role is assumed by the function when it's invoked to carry out its tasks.

- **Context object**

  When AWS Lambda runs your function, it passes a context object to the handler. This object provides methods and properties that provide information about the invocation, function, and execution environment.

- **Environment variables**

  AWS Lambda functions can use environment variables for operational parameters or to pass runtime configurations to your application.

- **Layers**

  AWS Lambda layers are a distribution mechanism for libraries, custom runtimes, and other function dependencies. You can use layers to manage your function's dependencies separately, keeping your deployment packages small.

- **Concurrency**

  Concurrency controls in AWS Lambda define the number of instances that can be simultaneously run. You can set reserved concurrency limits on a function to reserve a portion of your account level concurrency limit.

- **Versioning and aliases**

  AWS Lambda supports versioning of functions. You can publish one or more versions of your Lambda function and then use aliases to route traffic to a particular version.

- **Dead letter queues (DLQs)**

  AWS Lambda supports DLQs for handling function invocation errors. If the function fails to process an event, the event can be sent to an Amazon SQS queue or Amazon SNS topic.

- **Monitoring and logging**

  AWS Lambda automatically monitors functions on your behalf, reporting metrics through Amazon CloudWatch. You can use Amazon CloudWatch to collect and track metrics, collect and monitor log files, and set alarms.

- **Extensions**

  Extensions allow you to augment your AWS Lambda functions. They use the AWS Lambda Runtime API to integrate deeply into the lifecycle of an AWS Lambda invocation.

- **Deployment packages**

  The deployment package is a .zip file or a container image that includes your function code and dependencies.

When referring to AWS Lambda, there are typically two modes of operation: You can build your functions directly from your code in AWS Lambda. The function code can be supplied either through the AWS Console's function editor or by bundling the code with all dependencies and uploading it as a .zip file or an Amazon S3 object.

Wondering when to use each method? The direct method in the AWS Console is straightforward. AWS Lambda environments are equipped with preinstalled language packages based on the runtime's type and version. For instance, for the Python 3.6 and Python 3.7 runtimes, it includes popular libraries such as `boto3` for AWS SDK operations, `botocore`, `docutils`, `jmespath`, `dateutil`, `s3transfer`, `six`, and `urllib3`, among others.

If your function code is written entirely within the scope of standard language packages, you can directly paste your code in the editor and create AWS Lambda function. If not the case, if your code requires non-standard libraries or packages, you can address this situation in two ways:

- **Using AWS Lambda layers**

  Layers are used to manage and deploy libraries, custom runtimes, and other dependencies separately from your AWS Lambda function code. For more information about AWS Lambda layers and the process to package them, refer to *http://s-prs.co/v577685*. You can attach these layers to your AWS Lambda functions to provide runtime of additional libraries and packages. Ensure that the version of Python used matches the runtime of your AWS Lambda function. Also, note that the size of the layer and function package together should not exceed AWS Lambda's limits.

  Figure 8.8 shows the process of creating an AWS Lambda layer for the **hdbcli** Python package and then attaching it to an AWS Lambda function that connects at the SAP HANA database layer to query database data: the deployment package ❶, the hdbcli layer ❷, importing the hdbcli layer ❸, connecting to SAP HANA ❹, and making the SQL query ❺.

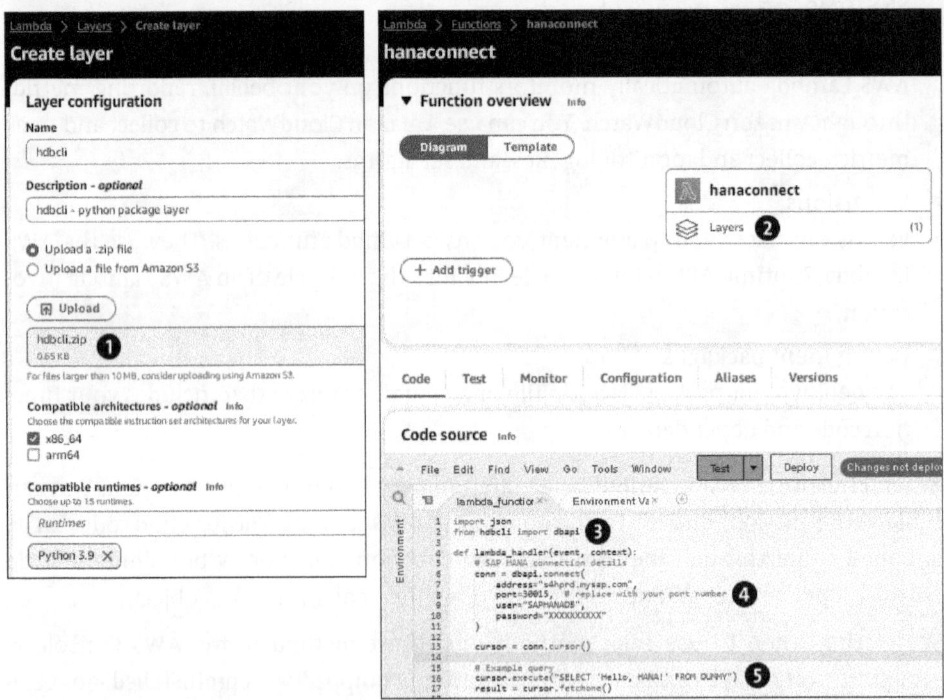

Figure 8.8 Connecting to an SAP HANA Database Using the hdbcli Python Package as an AWS Lambda Layer

- **Using a deployment package**

  Instead of using AWS Lambda layers, you can bundle all language package dependencies into a deployment package and upload this .zip file to your AWS Lambda function. If the .zip file exceeds 50 MB, you can upload it to an Amazon S3 bucket and provide the location as input. Be mindful of the AWS Lambda function's access to the Amazon S3 bucket.

The process of creating a deployment package is similar to the process for an AWS Lambda layer package, but in this case, you'll include all dependencies along with the AWS Lambda function code. For more information on this packaging process, refer to *http://s-prs.co/v577686*.

AWS Lambda Layers versus Deployment Packages

AWS Lambda layers are used to manage and deploy libraries, custom runtimes, and other dependencies separately from your AWS Lambda function code. This approach is particularly useful for the following goals:

- **Reusability**
  Share common components across multiple AWS Lambda functions without duplicating code.

- **Separation of concerns**
  Keep your AWS Lambda deployment packages small and focused on the function code itself.

- **Easier deployment**
  Update common components without redeploying all your AWS Lambda functions.

Containers in AWS Lambda are utilized when a custom runtime is required or when large dependencies or libraries exceed the limits of a deployment package (50 MB zipped, up to 250 MB unzipped). Containers are advantageous for ensuring a consistent execution environment that encompasses the application and its dependencies, especially when dependencies on specific OS libraries and configuration parameters exist.

For instance, let's say you want connect to SAP at the application layer through the remote function call (RFC) interface using PyRFC APIs in the AWS Lambda environment. Because this connection depends on component `NWRFCSDK` with a custom library path, and depends on other Python packages like Cython, you must deploy all dependencies and software components into a custom container. This runtime can then be brought to AWS Lambda through AWS Lambda containers.

Let's explore how you can build and use a container for AWS Lambda:

1. **Create a Dockerfile**
   Define a Docker image that includes your application code and its dependencies.

Code Example: Dockerfile

You can find a sample file at *http://s-prs.co/v577687*.

Contents of Requirements.txt

```
Cython==0.29.23
boto3==1.17.69
botocore==1.20.69
hdbcli==2.8.20
pyrfc==2.4.0
```

Contents of nwrfcsdk.conf

```
# include nwrfcsdk
/usr/sap/nwrfcsdk/lib
```

2. **Build your Docker image**

   Build your Docker image with the following code:

   ```
   OS:/>docker build -t my-lambda-container
   ```

3. **Push your Docker image to Amazon Elastic Container Registry (ECR)**

   First, create a repository in Amazon ECR, then tag and push your Docker image with the following code:

   ```
   OS:/> aws ecr create-repository --repository-name my-lambda-container
   OS:/>docker tag my-lambda-container:latest  <account_
   id>.dkr.ecr.<region>.amazonaws.com/my-lambda-container:latest
   OS:/>aws ecr get-login --no-include-email
   OS:/>docker push <account_id>.dkr.ecr.<region>.amazonaws.com/my-lambda-
   container:latest
   ```

4. **Create or update the AWS Lambda function**

   Now, create or update an AWS Lambda function to use this new container image with the following code:

   ```
   OS:/> aws lambda create-function --function-name my-container-function \
       --package-type Image \
       --code ImageUri=<account_id>.dkr.ecr.<region>.amazonaws.com/my-lambda-
   container:latest \
       --role arn:aws:iam::<account_id>:role/<Lambda-role>
   ```

Once the container is deployed as an AWS Lambda function, you can trigger it based on events to run the SAP data extraction process, including, for instances, changes in Amazon S3 buckets, scheduled events, or API calls via Amazon API Gateway.

In this scenario, using containers can simplify the process of setting up and maintaining the execution environment for your AWS Lambda functions, especially when dealing with complex dependencies like the NWRFC SDK from SAP.

8.2.5   Third-Party Automation Tools

Several third-party automation tools integrate effectively with AWS, enhancing its capabilities. Some notable tools include Terraform and Ansible, as we'll cover in the following sections.

Terraform

An open-source IaC tool created by HashiCorp, Terraform allows users to define and provision a datacenter infrastructure using a high-level configuration language known as HashiCorp Configuration Language (HCL) or in JSON.

Additional Resources

- Terraforms Getting Started: *http://s-prs.co/v577688*
- SAP on AWS code sample on Terraform Registry: *http://s-prs.co/v577689*
- Terraform tutorials for AWS use cases: *http://s-prs.co/v577690*

Ansible

Ansible is an open-source automation tool, or platform, used for IT tasks such as configuration management, application deployment, intraservice orchestration, and provisioning. Ansible code files are called *playbooks*. They are composed of one or more "plays" and describe the tasks to be performed. Playbooks are written in YAML format.

Before running an Ansible playbook, you must ensure that the necessary AWS credentials are available either as environment variables or in the Ansible configuration. This example also assumes that the required Ansible collections for AWS are installed (*http://s-prs.co/v5776131*).

Additional Resources

- Ansible code sample for SAP on AWS: *http://s-prs.co/v577691*
- The plugins in the Amazon.aws collection: *http://s-prs.co/v577692*
- Getting started with Ansible automation controller: *http://s-prs.co/v577693*

8.3   DevOps for SAP on AWS

To enhance the automation of activities, reduce the lead times for fixes, and achieve faster delivery, you can implement AWS DevOps practices tailored for SAP. AWS offers an extensive array of services and tools for this purpose. Figure 8.9 shows the comprehensive DevOps suite available on AWS, which can govern the speed and quality of developing automation products and services for SAP.

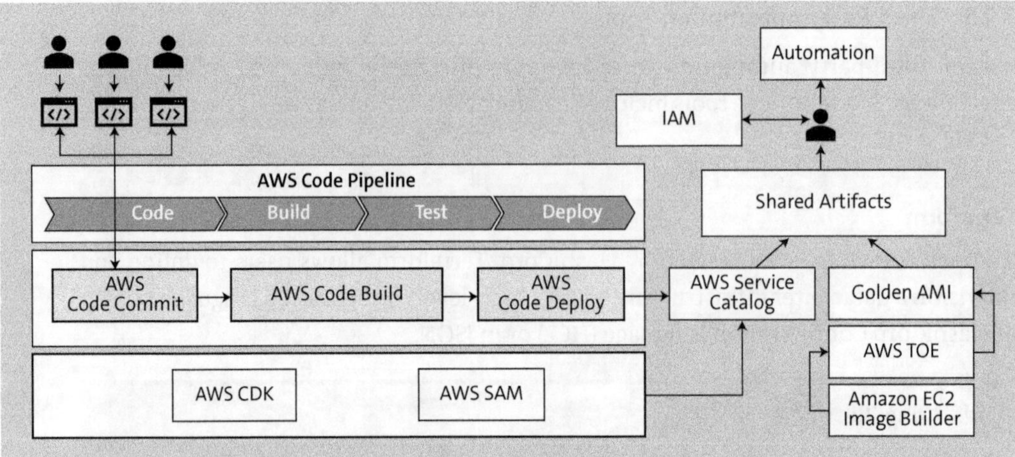

Figure 8.9 AWS DevOps Toolset for Building SAP Automation

DevOps versus Automation

Are you puzzled by the sudden shift in terminology from automation to DevOps? This section delves into the governance aspects of developing automation artifacts. Let's clearly distinguish between the two: DevOps is a comprehensive practice aimed at enhancing collaboration between your development and operations teams. It encompasses various methodologies, including automation, which is a subset of DevOps. Automation specifically uses technology to minimize human intervention in tasks, particularly to expedite operational processes. On the other hand, DevOps encompasses all aspects of development and operations, addressing broader business and technological requirements.

In this section, we examine the elements and components shown in Figure 8.9. We'll focus on establishing CI/CD pipelines with AWS native services and understanding the importance of the AWS Service Catalog for securely managing access to automation artifacts for end users. Additionally, we'll discuss the use of the Amazon EC2 Image Builder service and highlight IAM best practices to ensure secure and robust automation activities.

8.3.1   Continuous Integration/Continuous Deployment Pipeline

As shown in Figure 8.9, on the left, we have a CI/CD pipeline to facilitate the development, testing, and deployment of automation artifacts on AWS. Developers can utilize their preferred IDEs in a distributed environment to craft automation code, subsequently committing their changes to a version control system like AWS CodeCommit. This service enables teams to collaborate on code efficiently and securely in a fully managed environment. To learn more about the features and use cases and how to set up AWS CodeCommit, visit *https://aws.amazon.com/codecommit/*.

Post-development, AWS CodeBuild amalgamates all dependencies and libraries to construct a finished package. This service also offers a controlled testing environment to ensure the product meets its intended objectives. For detailed instructions on using AWS CodeBuild, visit *https://aws.amazon.com/codebuild/*.

Subsequently, AWS CodeDeploy enables deployment in a staging area or the publication of the final product as a community artifact within the AWS Service Catalog. With AWS CodeDeploy, you can automate software deployments to various compute services such as Amazon EC2, AWS Fargate, AWS Lambda, and your on-premise servers. AWS CodeDeploy is designed to make deployments safer and to release new features rapidly. Refer to *https://aws.amazon.com/codedeploy/* for more on AWS CodeDeploy's capabilities and deployment methods and how to integrate it into your CI/CD pipeline.

You can orchestrate the entire CI/CD pipeline from a single place using the AWS Code-Pipeline service. This service automates your release pipelines for fast and reliable application updates. AWS CodePipeline builds, tests, and deploys your code every time a code change occurs, based on the release process models you define. To dive deeper into setting up and managing a pipeline, refer to *https://aws.amazon.com/codepipeline/*.

Additionally, you can employ AWS Cloud Development Kit (CDK) or AWS Serverless Application Model (SAM) to generate deployable AWS CloudFormation templates for inclusion in the AWS Service Catalog. AWS CDK provides you with an open-source software development framework to model and provision your cloud application resources using familiar programming languages. Documentation for AWS CDK can be found at *https://aws.amazon.com/cdk/*. On the other hand, AWS SAM is an open-source framework specifically designed for building serverless applications on AWS. AWS SAM simplifies the process of defining, deploying, and managing serverless applications by providing a shorthand syntax for expressing serverless resources and their event sources. It's built on AWS CloudFormation, which means you can take advantage of AWS CloudFormation's features and benefits for deployment and management while working with serverless architectures.

8.3.2   AWS Service Catalog

With an established catalog of approved resources in the AWS Service Catalog, departments can provision the necessary resources via a self-service portal governed by IAM authorization. This capability allows IT teams to monitor and regulate usage to ensure compliance and to streamline the provisioning process. Through the AWS Service Catalog, organizations can enforce robust governance while enabling departments to promptly and efficiently provision the required AWS resources, thus enhancing autonomy and reducing the administrative burden on central IT.

Some uses for AWS Service Catalog include the following:

- **Standardization**
  Helps in standardizing the AWS infrastructure and services to comply with organizational policies.

- **Self-service portal**
  Provides a self-service portal for users to deploy AWS resources without having in-depth knowledge of AWS services.

- **Governance**
  Allows administrators to apply governance and compliance rules to AWS resources.

- **Cost management**
  Helps in managing costs by controlling the provisioning of resources.

- **Controlled environment**
  Ensures a controlled environment by allowing only approved resources for provisioning.

- **Auditability**
  Provides an audit trail of who requested what service, when, and the configurations they used.

- **Consistency**
  Ensures consistency in the provisioning of AWS resources by enforcing standardized templates.

8.3.3   AWS EC2 Image Builder

Amazon EC2 Image Builder is a service designed to automate the creation, management, and deployment of customized Amazon EC2 machine images (AMIs) for various applications, including SAP systems on AWS. A use case for SAP would involve automating the creation of AMIs with preinstalled SAP software and configurations aligned with SAP best practices. This capability ensures that the underlying compute instances are optimized for SAP workloads, providing a standardized and repeatable process for SAP environment provisioning and maintenance, which enhances scalability and operational efficiency. For more in-depth information, visit *http://s-prs.co/v577694*.

Amazon EC2 Image Builder utilizes AWS Task Orchestrator and Executor (AWS TOE) for executing complex workflows in the image creation process, including software installation, configuration, and testing. Workflows are defined in a YAML document, which directs AWS TOE to execute specified tasks. These tasks can include embedded OS shell commands or calls to external scripts hosted locally or on Amazon S3. Additionally, AWS TOE can integrate with AWS Systems Manager to install packages via the AWS Systems Manager Distributor, and it can also implement Center for Internet Security (CIS) and Security Technical Implementation Guides (STIGs) hardening components into its image building workflow. For more details, visit *http://s-prs.co/v577695*.

8.3.4 Authorization Management

Automation is beneficial when implemented in a controlled and secure manner; otherwise, it can cause havoc in an IT environment. It's essential to have a solid authorization framework before automating operations, including establishing permission controls to handle critical application assets safely. Actions on AWS are managed through IAM and resource policies to ensure access and modification rights are reserved for authorized users and systems. For a comprehensive guide on IAM best practices, visit *http://s-prs.co/v577696*.

In addition to following IAM best practices, consider adopting some additional measures to protect your AWS resources further, such as the following:

- **Segregation of duties**
  Create distinct automation processes for creating, updating, and deleting AWS resources by assigning specific authorization controls to various individuals. Control the deletion process by allocating the tasks of enabling and disabling "deletion protection" to different principals. This approach ensures that only intended resources are deleted during cleanup activities. Additionally, segment the deletion process into three stages to guarantee the phased deletion of critical resources. For instance, delete infrastructure components initially, followed by data and storage assets, and finally assess backup assets to confirm the correct resource deletion. Integrate these stages into additional approval workflows, scheduling them at separate times to allow adequate response time for unforeseen outcomes.

- **Attribute-based access controls (ABAC)**
  Utilize resource tags, maintenance windows, and IAM condition keys to grant specific, time-bound access for principals to execute changes via automation. More details on ABAC can be found in the AWS IAM User Guide at *http://s-prs.co/v577697*.

- **Additional data controls**
  Incorporate extra data controls such as application passwords, process parameter inputs, program status keys, encryption/decryption keys, and vital activity repository objects. Manage these controls using AWS services like AWS Secrets Manager, AWS Systems Manager Parameter Store, Amazon DynamoDB, AWS Key Management Service (KMS) keys, and Amazon S3. These controls should be separate from the main IAM roles or resource policies that authorize specific automation activities to enhance the security of permission administration for automation tasks.

8.4 SAP on AWS Automation Code Examples

In this section, we'll delve into some high-level SAP automation code examples beyond what we've discussed so far in this chapter. These examples serve as foundational code samples, which can be expanded for more varied and complex SAP operations scenarios.

8.4.1   AWS Config: Conformance Pack

As discussed in Chapter 7, a *conformance pack* is a set of AWS Config rules grouped and reported together. Each rule will consist of a definition. AWS Lambda function logic finds the status (and optionally a remediation automation) either through an AWS Systems Manager document or an AWS Lambda function.

To create a conformance pack for SAP configuration checks, you would typically define it in a YAML template and then deploy it using AWS Config. Our YAML templates include a set of AWS Config rules tailored for an SAP environment on AWS.

AWS Config Rules

A sample code block is available at *http://s-prs.co/v577698* to illustrate the structure of a conformance pack for SAP. This example includes configuration rules for verifying whether the provisioned Amazon EC2 instance is SAP certified, whether the underlying Amazon EBS volume has encryption enabled, and whether the assigned security group adheres to the required port configurations.

AWS Lambda Function for SAP

A sample AWS Lambda function is available at *http://s-prs.co/v577699*. This function checks if an Amazon EC2 instance is of the desired instance type, which could be a requirement for SAP workloads.

Remediation AWS Lambda Function for SAP

A sample AWS Lambda function is available at *http://s-prs.co/v5776100*. This function can be triggered if the AWS Config rule finds a non-compliant resource. It also attempts to change the instance type of a non-compliant Amazon EC2 instance to the desired type.

Note

Certain remediation steps might cause service disruption, while delayed action could leave a system vulnerable for an extended period. Therefore, an important step is to carefully select remediation actions. Additionally, not all remediations can be automated. For instance, if an `EBS_ENCRYPTION` check fails, manual steps might be required to achieve the desired encryption status. In such situations, notifying the relevant teams is a practical reactive measure.

8.4.2   AWS Systems Manager: State Management

AWS Systems Manager enables the automation of routine management tasks on AWS resources, including patching, discovering component statuses, and updating software. For instance, if you want to maintain specific SAP kernel release levels across your instances, you can create an AWS Systems Manager document to check the SAP kernel release on Amazon EC2 instances. You can create an AWS Systems Manager association (*http://s-prs.co/v5776101*) using this document to identify the current status of the SAP kernel release on the targeted Amazon EC2 instance.

After the association is in place, the State Manager will execute the document on the specified schedule. You can view the execution results in the console or through the logging method you configured (Amazon S3 or Amazon CloudWatch). Based on this status, you can perform necessary follow-up actions on that specific Amazon EC2 instance.

AWS Systems Manager Document to Check SAP Kernel Release

Visit *http://s-prs.co/v5776102* for a sample AWS Systems Manager document to find the SAP kernel release of an SAP instance running on an associated Amazon EC2 instance. This AWS Systems Manager document defines a shell script that runs on the target Amazon EC2 instances. It uses the **disp+work -version** command, common in SAP environments, to check the kernel version. The script then compares the output against the desired kernel release level passed as a parameter to the document.

Note that the actual command to check the SAP kernel release may differ based on your SAP system setup, and the script should be modified accordingly. Additionally, in the case of non-compliance, the proper remediation steps will depend on your specific requirements and could include notifying an administrator, automatically triggering an update process, or logging the output for further action.

8.4.3   AWS Systems Manager: Custom Inventory

An AWS Systems Manager inventory association job periodically assesses the inventory of managed instances and updates the AWS Systems Manager inventory data. Custom inventory information, stored as a JSON document in the */var/lib/amazon/ssm/node-id/inventory/custom* directory on Linux systems, is integrated into AWS Systems Manager inventory data during this process.

To programmatically update a JSON document for an AWS Systems Manager custom inventory, you can use an OS shell script or a Python script to collect the necessary system data and then populate the JSON structure shown in Listing 8.6 accordingly.

```
{
    "SchemaVersion": "1.0",
    "TypeName": "Custom:SAP",
```

```
"Content": {
        "hostname": "bw4dev",
        "os_vendor": "Redhat",
        "os_level": "8_2",
        "os_cpu_num": "32",
        "os_memory_size": "256",
        "ec2_type": "r6i.8xlarge",
        "instance_id": "i-0972e37bbf71c0cd4",
        "sid": "B4D",
        "sap_instance_details": "ASCS01_D02_HDB00",
        "sap_kernel": "785_200",
        "sap_component_details": "SAP_BASIS_756_00",
        "db_type": "SAP HANA",
        "db_level": "2.00_059",
        "db_size" : "240GB"
    }
}
```

Listing 8.6 JSON Document Representing an AWS Systems Manager Custom Inventory Template

AWS Systems Manager Custom Inventory

Check out the sample code available at *http://s-prs.co/v5776103*. This example Python script retrieves system information and updates the JSON document shown in Listing 8.6 and places the document into the default location of an AWS Systems Manager custom inventory. This script should be adjusted according to your SAP installation type and the distribution of your SAP instances across your multiple Amazon EC2 instances. Ensure that the required Python libraries are in place. Additionally, you can use AWS Secrets Manager to securely manage your SAP HANA database credentials.

8.5 Summary

This chapter focuses on automating SAP systems in AWS, highlighting the benefits of AWS's extensive compute capabilities and tools for deploying, managing, and securing SAP workloads. We outlined the steps to start AWS automation, including basic scripting using AWS CLI and SDKs, setting up IAM roles and policies, and choosing AWS automation tools.

In this chapter, we emphasized using IDEs like AWS Cloud9, VS Code, and PyCharm for efficient AWS workflows. You learned about the various AWS-supported programming environments and code repositories as well as AWS Code Whisper for AI-powered code generation. We also covered AWS-native tools for SAP automation, like AWS Launch

Wizard and AWS Migration Hub Orchestrator, and introduced you other tools like AWS Lambda, AWS Systems Manager, and AWS SDK for ABAP.

The DIY AWS toolset section (Section 8.2.4) includes AWS CLI, Amazon EC2 IMDS, and third-party tools like Terraform. We also explored DevOps for SAP, focusing on CI/CD pipelines and services like Amazon EC2 Image Builder and Service Catalog. Additionally, this chapter provided code samples for AWS Config, AWS Systems Manager's State Manager, and AWS Systems Manager custom inventories to demonstrate the practical application of SAP automation on AWS.

Chapter 9
Multicloud Considerations

The cloud is here to stay, and RISE with SAP is here to stay. Now, with multicloud, the complexities and trade-offs haven't gone away! You must make conscious choices about your organization's direction and think long term when planning for that future.

If you've been asked to look into a multicloud strategy for SAP or for an even broader IT landscape, you're not alone. We've seen many organizations evaluating and implementing multicloud foundations, perhaps for good reason. Figure 9.1 shows how interest in multicloud landscapes over last few years (using searches as a proxy for interest), and the upward movement aligns with what we've seen with large organizations. We realize that multicloud is here to stay, and thus making a conscious decision one way or the other is important. An organization going multicloud to follow trends may learn lessons the hard way!

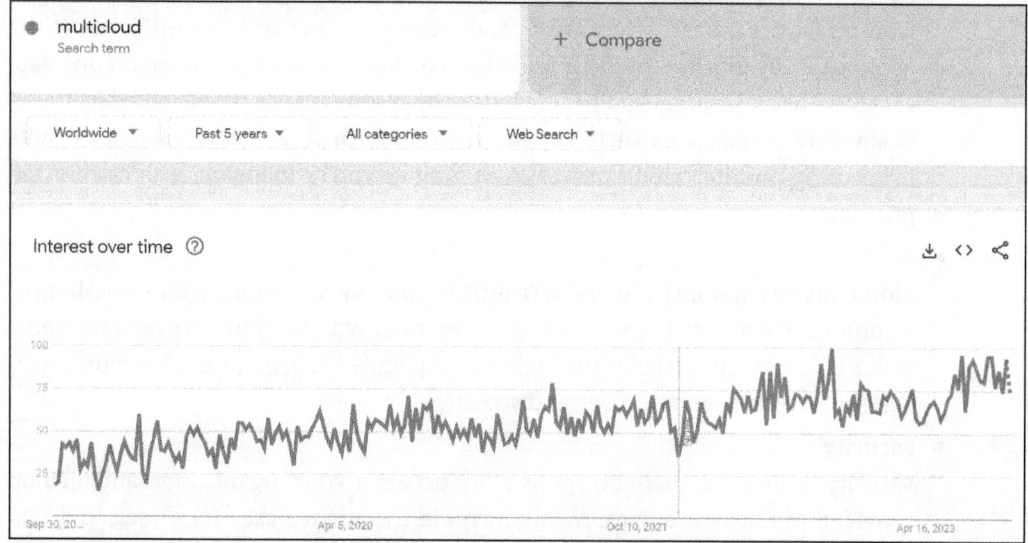

Figure 9.1 Multicloud Search Interest over Time Using Google Trends

In this chapter, we'll discuss how organizations are looking at multicloud, debunk some myths around it, and talk about Amazon Web Services (AWS) tools that you can use in hybrid and multicloud use cases.

[»] **Definition**

For our discussions, we refer to "multicloud" as the intentional use of more than one cloud provider for an organization's technical or business needs. Using different software as a service (SaaS) and infrastructure as a service (IaaS) providers doesn't necessarily constitute multicloud, but using virtual machines (VMs) in multiple cloud providers does, because these approaches follow different service consumption models, and often, you won't get to choose the specific cloud provider for a SaaS solution.

9.1   How Do SAP Customers End Up with Multiple Clouds?

Similar to every other aspect of a cloud strategy, a multicloud evaluation should focus on your organization's business needs and should clearly identify decision points. Before we talk about how organizations end up in a multicloud environment, let's first discuss some common misconceptions.

A few things that do *not* necessarily make the case for going multicloud include the following:

- **Resiliency**
  On the surface, it may seem like using multiple cloud providers equals better resiliency, but that idea ignores the fact that cloud providers have their own resiliency features (at the infrastructure layer), and often, you can configure resiliency at the application layer using the tools provided. Further, you are not relying on one data center with its own single points of failure. Other governance problems aside, statistically, little evidence exists to support the notion that increased resiliency results from using multiple cloud providers as long as you're following well architected principles.

- **Cost**
  Cloud services may differ in pricing a little bit, but if you're looking for something of significance, your organization's negotiated rates with the provider hold the most promise for savings. Usually, the higher your planned consumption, the more leverage you have to negotiate a better discount.

- **Security**
  Security is always a shared responsibility between your organization and a cloud provider. Multicloud environments increase the surface area for attack, and you must implement even more comprehensive measures to safeguard your systems.

In our experience, organizations might end up with a multicloud setup unintentionally, or they might evaluate multiple cloud providers for one or more of following, as shown in Figure 9.2:

- **Acquisitions or independent subsidiaries**

  Acquiring a company that uses a different cloud provider or independent subsidiaries that have their own cloud strategy is the most common way many organizations end up with multiple cloud providers.

- **Compliance or latency requirements**

  If a particular system must meet latency requirements or a specific compliance rule requires systems to be on-premise (or in a certain region for data sovereignty), you could end up with a multicloud setup or a hybrid setup (on-premise and cloud provider together).

- **Software providers**

  If you're an independent software provider that makes the software available over multiple clouds, then you would have a multicloud strategy.

- **Niche services**

  If your organization can benefit from a niche service that one of the cloud providers either released first or has features that your current provider doesn't. In some cases, such as for loosely coupled systems, using multiple cloud providers might make sense.

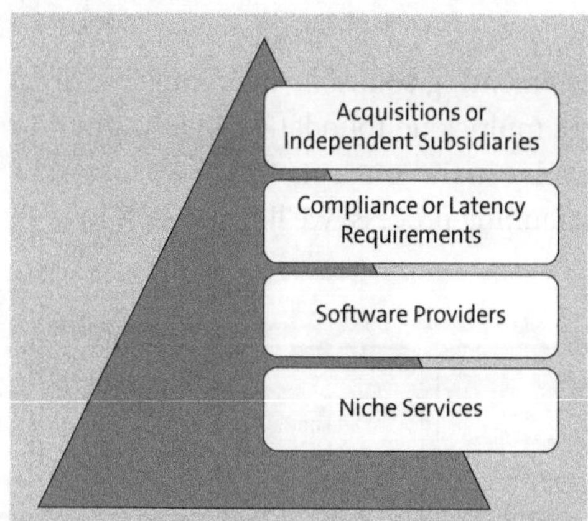

Figure 9.2 Common Reasons for Organizations to Be Multicloud

SAP Business Technology Platform and Software as a Service Products

For SAP SaaS products (such as SAP SuccessFactors, SAP Concur, etc.), usually, you don't have much say over which cloud it resides on, which we exclude from our definition of multicloud. SAP Business Technology Platform (SAP BTP) has a multicloud foundation from SAP's point of view, and the SAP Private Link feature allows traffic to remain on your cloud provider's network backbone. Therefore, choosing the same provider for SAP BTP as your primary cloud provider where possible makes sense so you don't end up navigating multiple cloud provider's configurations.

9.2 Multicloud Planning Aspects

As part of planning for the multicloud, not only must you understand how each cloud service works, but from a bigger picture, you must also understand how your organization will manage its IT landscape efficiently. Multicloud scenarios involve trade-offs among costs, complexity, and risks, as illustrated in the example scenarios described in Table 9.1.

Scenario	Cost Impact	Complexity	Risk
Primary (business-critical) workload in one cloud and secondary in another	Low	Medium	Medium
Workload in one cloud and backup/DR in another cloud	Low	Medium	Medium
Some workloads in one cloud and others in different based on capabilities of cloud	Medium	High	Medium
Cloud-agnostic workload distribution	High	High	High

Table 9.1 Trade-Off by Multicloud Scenario for SAP and SAP-Adjacent Workloads

Before you find yourself too deep into implementing solutions over multiple cloud providers, take a step back and think about a multicloud foundation in terms of how your organization is going to manage and evolve in the long run. Figure 9.3 shows the items you should consider as part of your planning process. We'll cover each item in the following sections.

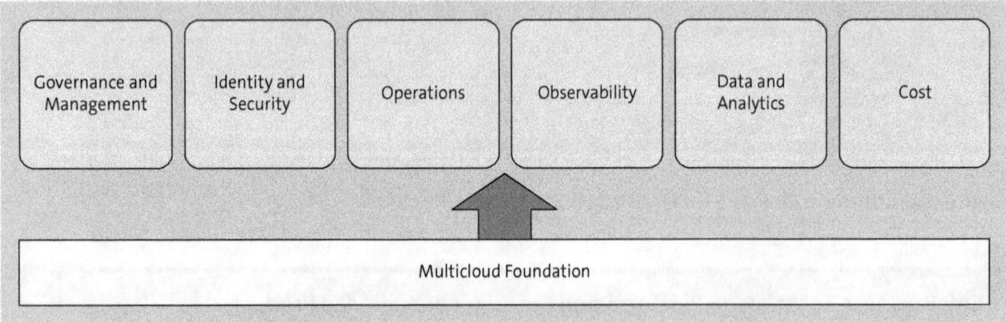

Figure 9.3 Considerations for Multicloud Foundation for an Organization

9.2.1 Governance and Management

Throughout this book, we've discussed various aspects of running SAP workloads on AWS the right way, from design principles to architecture, to operations. When going multicloud, none of these requirements go away. In fact, the considerations we highlight in this chapter are in addition to what we've already discussed. Figure 9.4 shows the management and governance components you should establish when creating your multicloud foundation.

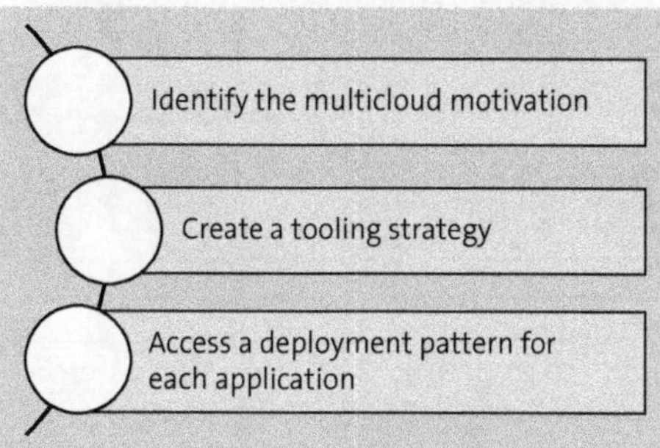

Figure 9.4 Principles of Multicloud Governance Framework

Let's look at each component in a little more detail:

- **Identify the multicloud motivation**
  Every organization has its own journey and knowing why your organization decided to go this route will pave the way to pay attention to certain aspects. For example, if you're motivated by a need for compliance, then you must understand how a different cloud provider addresses it for applications that are in scope.

- **Create a tooling strategy**
  Identifying tools that would work across environments is an important decision with long-term implications. Think about who owns the tools and how they have evolved. You could either look into open-source tools or even cloud provider tools that help with multicloud and hybrid management.

- **Access deployment patterns for each application**
  Not every application supports multicloud deployment and patterns may vary. So, an important step is to think about multicloud in the context of the application. For example, with an SAP architecture, you cannot deploy a database on one cloud and an application on another.

> **Tools Evaluation**
>
> Each function could have its own tools. For example, for deployment, you may choose something like Terraform since its application programming interfaces (APIs) work with multiple cloud providers. Similarly, you could choose tools for monitoring, operations, etc.
>
> In our experience, a more sustainable approach is to identify the right tools and train your employees rather than looking at employee experience and selecting tools based on that information.

> **AWS Tools for Multicloud Governance Examples**
>
> Some AWS tools that you can use for multicloud governance include the following:
>
> - **AWS Systems Manager**
>   Centralized management of resources.
>
> - **AWS Config**
>   Evaluate and access the configuration of resources across cloud providers.

9.2.2   Identity and Security

Security is a shared responsibility between your organization and your cloud providers. So, you should start by reviewing security aspects for each cloud provider and layer your organization's mandates on top of that. Figure 9.5 shows the security and identity principles required in a multicloud foundation.

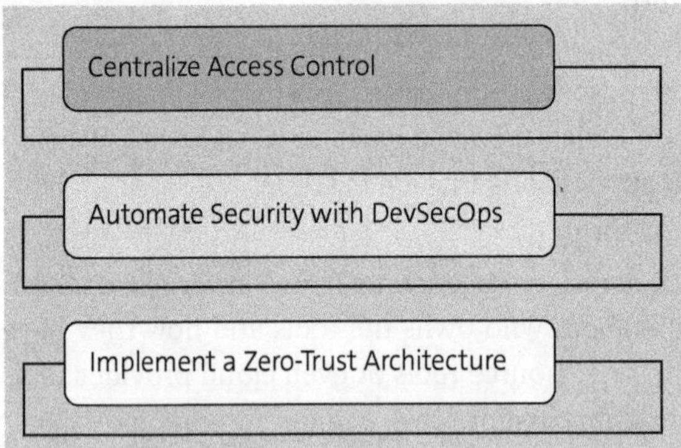

Figure 9.5 Principles of Multicloud Identity and Security Framework

These principles include the following:

- **Centralize access control**
  You would want to manage authentication, authorizations, and policies in a centralized manner while considering factors such as identity federation, multifactor authentication, and audit requirements. Cloud providers may support different authentication models as well, so it makes sense to decouple access control from a particular cloud service.

- **Automate security with DevSecOps**
  DevSecOps combines development and operations with security (or put differently, adds a security approach in DevOps); in practice, every process takes security into account, and automation implements relevant security practices. For example, if your organization requires a security scan for every VM deployed, instead of your

infrastructure team building the VM and then handing it over to the security team for scanning, the code itself could have scanning built in.

- **Implement zero-trust architectures**
In the cloud, no clearly defined perimeters exist; so instead of a conventional network-focused security approach, the zero-trust paradigm eliminates implicit trust and focuses on verifying and validating authorized access at every stage. This approach spans from user identity verification, multifactor authentication, to endpoint compliance, and data protection using continuous risk assessment (i.e., remove implicit trust from users, applications, and infrastructure) Refer to the National Institute of Standards and Technology (NIST) recommendations for implementing a zero-trust architecture, developed in collaboration with several organizations, at *http://s-prs.co/v5776104*.

> **AWS Identity and Security Tools Examples**
>
> Some AWS tools that you can use for multicloud identity and security include the following:
>
> - **AWS IAM Identity Center**
>   Manage identity centrally using sources from different cloud providers.
>
> - **IAM Roles Anywhere**
>   Use this tool to obtain security credentials for workloads running outside of AWS.
>
> - **Security lake**
>   Centralize security data from different sources such as AWS, on-premise, and other cloud providers to get better understanding of your organization's security posture.

9.2.3   Operations

As the saying goes, "Everything leads to operations!" The end result of migration, modernization, and innovation with technology ultimately must be operated efficiently. Thus, an important step is to look at a multicloud foundation from an operational lens. Figure 9.6 shows the principles that you should follow for evaluating the operational aspect of your multicloud foundation.

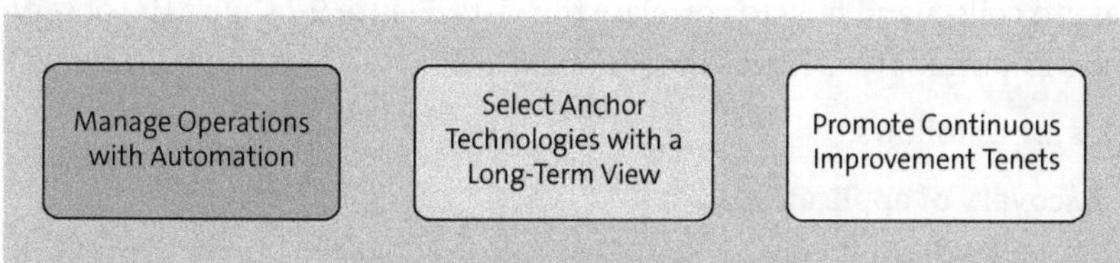

Figure 9.6 Principles of Multicloud Operations Framework

These principles are as follows:

- **Manage operations with automation**
  Similar technological implementations across cloud providers often work differently, with their own feature sets. Automation not only reduces complexity of large operations but also minimizes mistakes, which could have ripple effects.

- **Select anchor technologies with a long-term view**
  To deploy systems across providers, you'll have to select a lot of services of a foundational nature, such as VMs, storage, etc. and expand based on application need. In this context, focus on the end state and think about long-term implications. For example, does using cloud-managed service makes sense for your organization? What direction is SAP headed with the architecture and long-term support of its cloud services?

- **Promote continuous improvement**
  This tenet cannot be overstated! Not only do operational procedures change over time, but also cloud services evolve. Evaluating new opportunities for improvement or even optimizing inefficiencies can go a long way for maintaining business-critical applications such as SAP. This principle ties to your original application architecture as well since you should architect for operational efficiency.

[»]

AWS Operations Tools Examples

Some AWS tools that you can use for multicloud operations include the following:

- **AWS Systems Manager**
  Manage servers on-premise as well as those at other cloud providers using features such as fleet manager, patch manager, etc.

- **AWS Cloud Operations**
  Provides model for operating in cloud efficiently.

9.2.4   Observability

Cloud services can generate more data (in the form of metrics, logs, traces, etc.) than we can effectively process. Thus, to get observability right, an important step is to understand what data to collect and how to correlate this data. Figure 9.7 shows the observability principles important for a multicloud foundation.

These principles are as follows:

- **Automate discovery of applications**
  "Our development system is going live this week, is it onboarded to the observability dashboard?" is a question that shouldn't be part of the equation. Multicloud setups are complex enough, and you don't want to be in the business of using a checklist to onboard new infrastructures and applications. Automated discovery

should be a baseline control implemented in your governance framework. For example, when you create an Amazon Elastic Cloud Compute (EC2) instance for SAP, it should be tagged as such, and from there, SAP-related onboarding processes would kick off.

- **Go beyond monitoring**

  A lot of organizations still rely on reactive monitoring—if a system goes down, you get an alert. While those alerts are absolutely needed, with the amount and type of data you can collect from cloud providers about infrastructures and from SAP application logs, an observability approach should target better understanding of the system (as a whole rather than individual components such as storage, Amazon EC2, application) and include profiling, tracing, etc. Monitoring should not be an afterthought either; rather, you should build monitoring into the design and deployment of the infrastructure and application. For example, SAP deployments require that you enable detailed monitoring for Amazon CloudWatch. Therefore, you should have a configuration that disallows the creation of an Amazon EC2 instance for SAP without enabling detailed monitoring. Another example, if your choice of software is Splunk, data forwarding should be part of the deployment configuration as well.

- **Gather contextual data**

  With a multicloud approach, not only we are dealing with complex applications, but also services that behave differently. So, part of an observability strategy should also include contextual awareness. Creating dependency maps, including all components such as third-party services, load balancers, etc. in the single pane of glass helps avoid blind spots and leads to better mean time to recover (MTTR) by providing actionable insights rather than just saying one among scores of services is down. While all this data is being gathered, transferred, and analyzed, think about the security aspects as well.

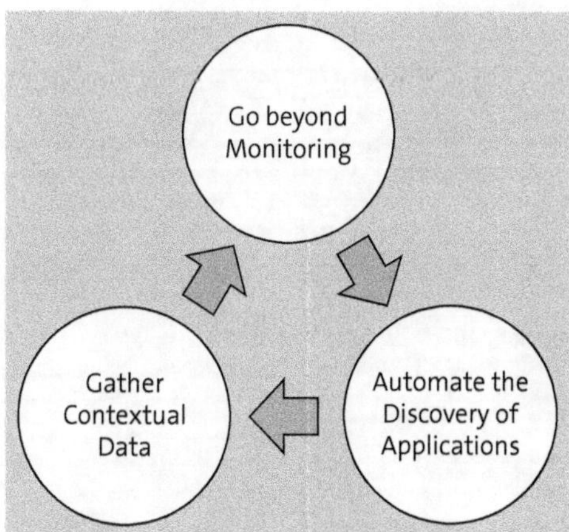

Figure 9.7 Principles of Multicloud Observability Framework

[»] AWS Observability Tool Examples

Some AWS tools that you can use for multicloud observability include the following:

- **Amazon-managed service for Prometheus**
  Prometheus-compatible (open-source) monitoring service.

- **Amazon-managed Grafana**
  Query and visualize using open-source analytics platform.

- **Amazon OpenSearch service**
  Derived from Elasticsearch, it's an open-source search and analytics suite.

9.2.5   Data and Analytics

The saying "data is the new gold" couldn't be more apt in light of generative artificial intelligence (AI) emergence. Whether it's SAP data only or in a larger organization where SAP is just another system, having access to the right data and insights definitely facilitates the type of decisions that an organization needs to stay relevant amidst our fast-changing world. Figure 9.8 shows data and analytics principles for multicloud foundations.

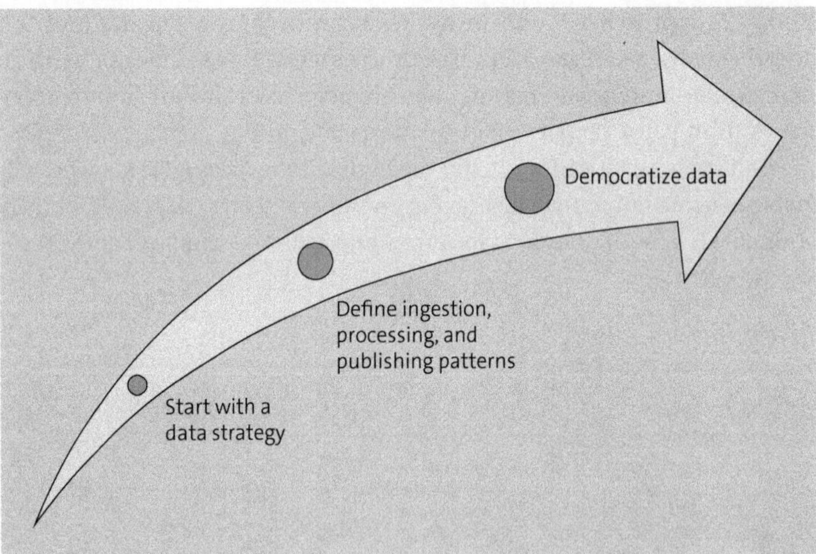

Figure 9.8 Principles of Multicloud Data Framework

These principles are as follows:

- **Start with a data strategy**
  In the beginning, there was data, then big data, and now we have generative AI. To make use of these technical advancements, you still need to start with a data strategy first! Put differently, a data strategy, which is organization's means to collect,

store (for example, to centralize data or not), consume, share, etc. *precedes* data initiatives. You can't train an AI model unless you have the right data.

- **Define ingestion, processing, and publishing patterns**
  In a multicloud environment, data can be siloed, and combining multiple sources often leads to out-of-date data syndrome where the reporting is not using the latest version of data. You need to think about data integration as distinct from data ingestion (whether your organization's SAP data resides within or it would make more sense to extract and combine with other data sources, based on your data strategy). A standard blueprint for collecting, processing, and processing multiple sources of data, from multiple cloud providers, take you to the next step in analytical maturity. This principle also includes access logging and security aspects in the model.

- **Democratize data**
  Data democratization is often associated with access to data, but it's also about empowering people to ask the right questions, to create and validate hypotheses quickly, and to abstract complexity away using low-code or no-code tools. This principle could also mean making the data available to AI training models in a secure way.

AWS Data and Analytics Tools Examples

Some AWS tools that you can use for multicloud data and analytics include the following:

- **AWS Glue**
  Serverless data integration service that supports multiple integrations including SAP.

- **Amazon Athena**
  Built on open-source frameworks, it supports data analysis from multiple sources regardless of where the data resides.

- **AWS DataSync**
  Supports data movement between AWS services, on-premise systems, and other cloud providers.

9.2.6   Costs

Cost is always among the top things being discussed when it comes to moving to the cloud. Using a multicloud setup (for non-SaaS solutions) can exacerbate problems in provisioning and associated fixed cost such as support. (For example, AWS requires a business or higher support plan for supporting SAP deployments.) Multicloud cost management requires a structured approach, as shown in Figure 9.9.

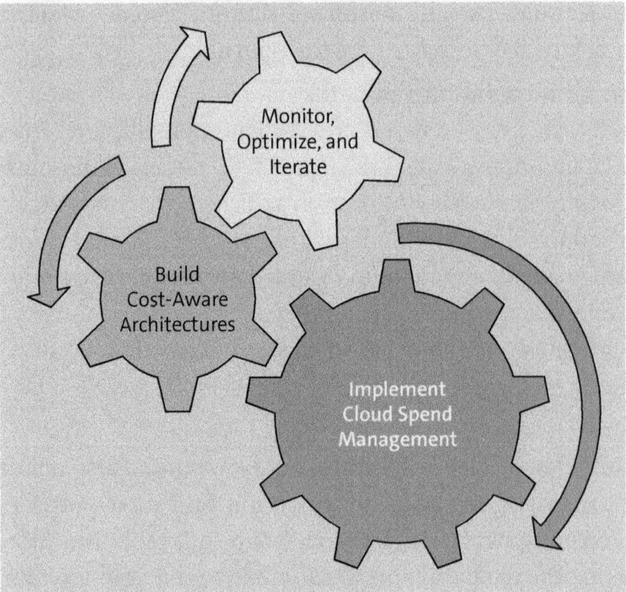

Figure 9.9 Principles of Multicloud Cost Framework

These principles include the following:

- **Implement cloud spend management**
  Spend management in the cloud sets a framework for multicloud cost management and requires implementing the following activities together:

  - Negotiating discounts with cloud provider based on total spend

  - Using the right financial model (for example, a savings plan versus on-demand pricing for workloads that are always on)

  - Budgeting and alerting based on thresholds

  - Implementing chargebacks for subsidiaries (if your organization does that)

  - Aligning business forecasting with cloud resources

- **Build cost-aware architecture**
  Just like security, cost should be part of the discussion throughout the lifecycle, right from design to deployment and operations. In our discussion, a cost-aware architecture means knowing how much business an architecture (or application such as SAP) supports and how incremental business can be supported by lower incremental costs. You should know the costs of subcomponents to make more informed decisions. For example, if you know that SAP training systems are used only a few weeks a year, then you know how much cloud costs to train your business associates on SAP systems. Lots of organizations look at SAP cost in terms of landscape and system cost, such as cost to run an SAP S/4HANA system, which supports organization's financial management, or cost of the Dev/QA landscapes individually as opposed to

cost to run SAP. This change in mindset provides more opportunities to optimize while focusing on the most value to the business.

- **Monitor, optimize, iterate**
  Monitoring and forecasting your cloud spend is a better understood notion than optimizing your spend. Optimization has two aspects: It is an iterative process that should be a part of operational plan similar to DR testing.

Cloud providers launch new features, services, compute instances, often at the same cost or less than previous versions. These events present opportunities to alter your architecture in a way that saves cost; they also use cost as an incentive for you to move to a newer hardware so older hardware can be decommissioned.

Application architectures can change over time, and you can use those changes as opportunities to enhance cost awareness and try to lower the cost of supporting more business.

Business requirements change that affect cloud resources and keeping track of the business value that particular system brings and potentially decommission (or hibernate) resources less frequently used is another way to optimize.

> **AWS Cost Tools**
>
> At the time of writing, AWS doesn't have any official tool for multicloud cost management, but cloud providers let you export cost-related data, which can then be integrated with a visualization tool such as Amazon QuickSight.

9.3   Skilling and Challenges

In Flexera's 2023 State of the Cloud report (*http://s-prs.co/v5776105*), 66% of respondents said managing multicloud is among their organizations' top challenges. This statistic speaks to the fact that cloud patterns are still evolving, and even though more organizations are looking into multicloud approach, its own set of trade-offs exist.

In our experience, multicloud challenges touch all aspects of organizational transformation, namely, *people, process, and technology*, in the following ways:

- **People**
  You learned about creating a cloud center of excellence (CCoE) in Chapter 7, while your organization is adapting and learning about cloud. Involving multiple cloud providers, each with their own uniqueness, can slow down progress and lower efficiency.

- **Process**
  Operational processes, in general, must be tested and updated frequently. Adding multiple cloud providers creates friction for testing and documentation unless your

organization's automation is mature. Consider a DR use case where your primary systems are running on one cloud provider and on standby on another. What would the documentation and process be like when several different services, technologies, and multiple clouds are involved? Data privacy and security processes become more involved as well.

- **Technology**
  Imagine you were thinking of a cloud feature to that improves your process, but it didn't exist and you went through custom development. After several months of effort, it's finally ready and gets you 90% there. Then, you see an announcement that the feature is being released out of the box as a cloud service enhancement! Now, think about this problem across multiple cloud providers—how are you going to keep up with the pace of change when your main focus is supporting a business! Never underestimate the effort required to keep up with changes in cloud technologies, which requires you to adapt for optimal architectures and operations constantly and directly impacts the technology roadmap and potentially impacts business process improvements for your organization.

Any kind of complexity, additional effort, skilling, etc. directly impacts *costs* and your organization's *time to maturity* for cloud capabilities. As part of CCoE, we recommend having a cross-functional team, and as a part of multicloud foundation, it's even more important that all the necessary skills reside in the same team. Another important step is to make decisions based on multicloud foundation requirements and train people where the gap is rather than create processes based on existing skillsets only.

> **Multicloud Skilling Side Effect on Architecture**
>
> Multicloud skilling (or sometimes called *fluency*) has a positive side effect of *understanding cloud design better*. For example, knowing the various points of view of multiple cloud providers regarding availability zones (AZs), namely, how they are built with redundancy and learning how a failover works (or what static stability is), helps you create better application architectures, thus ensuring better application reliability for business operations.

9.4   Vendor Lock-In and Open-Source Tools

In the context of cloud providers, vendor lock-in occurs when the cost of switching providers is high enough (in terms of money or effort or both) to discourage you from thinking about moving your systems somewhere else. For SAP systems, you're not splitting the same application into different cloud providers, so you're probably thinking of other use cases such as DR, niche service, subsidiary SAP systems, etc.

A common notion is that multicloud is a way to avoid vendor lock-in. Let's unpack that idea a little bit:

- **Trade-offs**

  You can easily change vendors only if you use services that provide similar features across the providers such as VMs, storage, etc. At that point, you're not necessarily using any cloud-native service, platform as a service (PaaS) solution that your organization can benefit from. For example, using an Amazon EC2 instance for running SAP GUI would make it easier to move, but you're paying more and creating other artificial restrictions as trade-offs.

- **Appearance of no lock-in**

  From a broader level, if you're using SAP, you're already locked in to SAP suite of software. Organizations that use VMware, for instance, look for a cloud service supporting VMware for easier lift and shift and are thus locked into the VMware technology even though no vendor lock-in is apparent from a cloud provider point of view.

- **Accepting lock-in**

  Your organization is using SAP (and is locked in) because of the value it provides or the capabilities that increase your organization's efficiency. Think of cloud providers the same way, use the best features and capabilities and, if that means being locked in, accept that as the cost of doing business. You should study and manage the risk.

In summary, lock-in is not necessarily a bad thing that must be avoided at all costs; rather, we recommend making more informed decisions directly related to the business benefits. Don't get distracted by the technical details and what only seems like the lack of lock-in.

Along the same lines, if your organization is culturally aligned to using open-source tools, we think that's a good idea and it works. However, using open source as a way to avoid vendor lock-in still has downsides. For instance, Terraform (a popular open-source tool), recently changed their licensing from the Mozilla Public License (MPL) to the Business Source License (BSL). All of these considerations should be part of your organization's tooling strategy, as we discussed earlier in Section 9.2.1.

Portability

Portability is the ability to move workloads from a source (e.g., on the cloud, on-premise) to a different target (i.e., a different cloud provider) *relatively easily*. Consider SAP portability similar to vendor lock-in. As long as you're not moving away from SAP as a software (to a different ERP), you should focus on utilizing cloud services to benefit the business and not make trade-offs for some unforeseen portability needs.

9.5   Multicloud Strategy Recommendations

So far, you've learned about multicloud deployments and how to plan for a multicloud foundation. We would be remiss if we didn't provide some recommendations, as shown in Figure 9.10, for multicloud strategies from a business point of view.

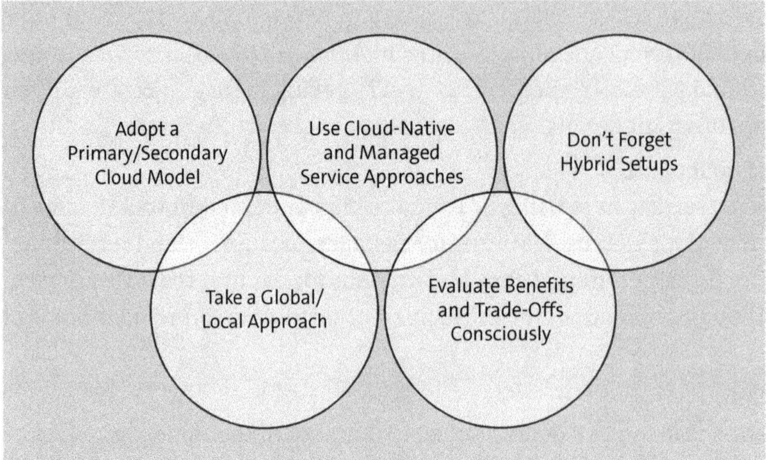

Figure 9.10 Multicloud Strategy Recommendations

Some strategic recommendations include the following:

- **Adopt a primary/secondary cloud model**
  Pick a cloud provider as the primary place where you'll host a majority of your workloads and select a secondary provider for niche use cases. In this way, you're still exposed to each cloud provider, can access different services, and leverage nuances to make more informed decisions, while still taking advantage of an economy of scale from your primary provider.

- **Take a global/local approach**
  Think globally (larger picture) about governance, security, and costs while making use of tools and processes of each cloud provider (locally).

- **Use cloud-native and managed-service approaches**
  Whenever possible, use PaaS solutions (AWS Lambda) and managed services from a cloud provider (such as Amazon-managed Grafana).

- **Evaluate benefits and trade-offs consciously**
  Don't do multicloud on auto-pilot. Think about the business and technical benefits of moving a workload to a different cloud provider (other than your primary). Evaluate these benefits for each application and use case (such as data sovereignty) rather than trying to balance workloads and costs across multiple providers.

- **Remember hybrid setups**
  In the spirit of multicloud, don't forget that sometimes hybrid setup (on-premise

and only one cloud provider) could be the best fit. Starting with a hybrid setup gives you a taste of what it's like to manage SAP/IT landscape at multiple places and allows your teams to learn about the cloud gradually by bridging the skills gap.

> **RISE with SAP and Multicloud**
>
> Your organization can do a multicloud setup even if you use RISE with SAP for your SAP landscape. Some reasons you may consider this option include your organization's data strategy, the desire to use innovative services from cloud providers, and the need to execute other SAP-related projects in a non-RISE with SAP setup.

9.6   When Multicloud Doesn't Work

Just like a relatively straightforward lift-and-shift project can go wrong if not planned properly, a multicloud approach can result in lots of *wasted time, effort, and money* if not done right! Figure 9.11 shows some pitfalls that your organization should consider when going the multicloud route.

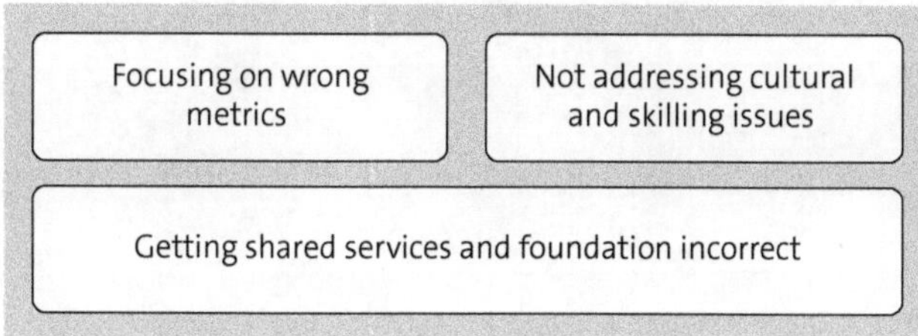

Figure 9.11 Things That Can Derail Multicloud Initiatives

Some potential problems include the following:

- **Getting shared services and even the foundation incorrect**
  We described the details of a multicloud foundation in this chapter. In the context of your organization's IT services for multicloud, shared services for security, cost management, visibility of resources, etc. are important to get right; otherwise, you'll end up creating technical debt in the long run.

- **Focusing on wrong metrics**
  We emphasized looking at the cloud and at IT through a business lens throughout this book, which cannot be overstated. Paying too much attention to vendor lock-in, portability, avoiding cloud-native services, etc., can distract you from business objectives and benefits. Determine your success criteria from a business point of view and align these criteria to technological and multicloud objectives.

- **Not addressing cultural and skilling issues**
  An organization's culture regarding learning, experimenting, and being nimble

matters with initiatives such as going multicloud. A skilling plan, as discussed earlier, is crucial too. Treating multicloud as a mandate without due diligence (and conscious effort) is likely to stump your teams.

9.7   Summary

In this chapter, you learned about multicloud foundations, how to evaluate and plan for it, along with some strategies for success. In the next chapter, we'll discuss, once you're on the cloud, how your organization can take a transformational approach to IT and to business operations using SAP and cloud-native services.

Chapter 10
Transformation

Differentiating yourself in the current competitive environment can be difficult, especially when everybody, including your competitors, has access to the same application capabilities as you do. However, the additional value you can bring to products, processes, and technology can make you the leader of the pack. In this chapter, we'll show you several avenues for extending your SAP applications by augmenting them with Amazon Web Services (AWS) native services and with building transformative architectures.

Transformation enables an organization to bring new ideas to the market faster and reduces the time to respond to customer behaviors and changing market conditions. You can build an adaptive enterprise to quickly cook "secret sauces" and bring that *little extra* to stand apart in the marketplace.

However, the most impending question is "Where to start?" The simple answer is "anywhere in your business value chain." As shown in Figure 10.1, you have multiple options. You can either replace your legacy application components or extend your current solutions with the new generation of digital capabilities (also called *intelligent technologies*) like robotic process automation (RPA), the Internet of Things (IoT), artificial intelligence (AI)/machine learning, and high-performance computing (HPC). Alternatively, you can attempt a complete overhaul of your business model by building new revenue streams, delivery channels, or customer value propositions through cloud-native application models.

In this chapter, we'll present specific foundational tricks with which SAP technical architects can build transformative capabilities to augment the capabilities of AWS from the ground up. However, a successful idea takes a village to adopt and scale in any organization. We hope to lower the barrier to entry and get you started!

We'll follow these themes to kickstart your journey:

- **Act when you want**
  You wait for the proper signal. You must be aware of the situation and act swiftly. You can leverage *event-driven architectures* (Section 10.1).

- **Build magic out of thin air**
  You don't need to own any real estate or "rack-n-stack" to start your journey; you build *serverless architectures* (Section 10.2).

"

- **Delegate the heavy lifting**
  Let the AWS do all the heavy lifting for you. You can weave AWS services through its application programming interfaces (APIs) to extend current capabilities. You can adopt *cloud-native development models* (Section 10.3).

- **Consolidate data**
  You bring all the knowledge sources to a single place. You can infer relationships from insights through this data. You can build a story to inspire action by embracing *modern data architectures* (Section 10.4).

- **Have no boundaries**
  You go beyond the traditional reach of data. You can integrate across vendor boundaries and across multiple clouds (Section 10.5).

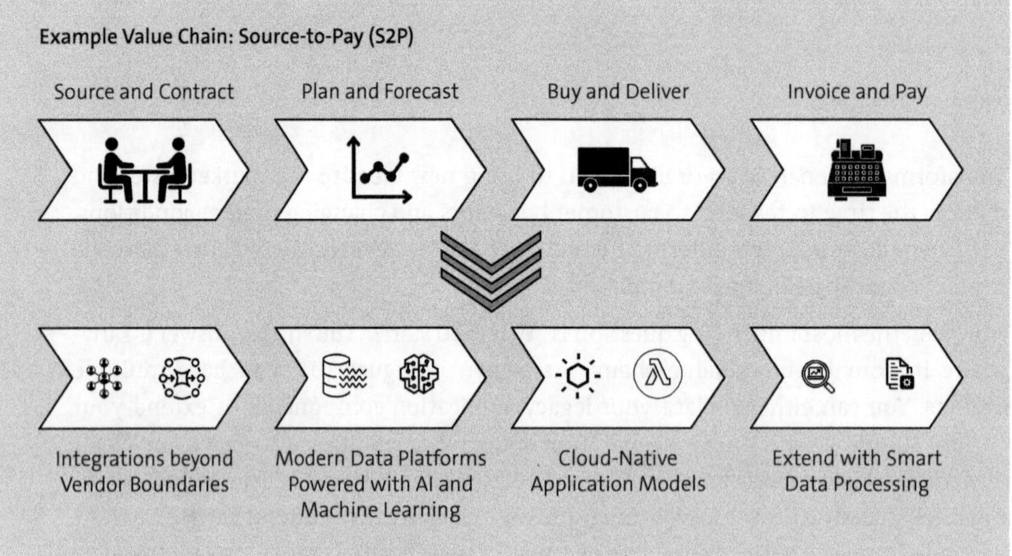

Figure 10.1 Example Value Chain: Transformation Target State

Note

A simple application transformation project is often overloaded with more ambitious overarching vernacular—the digital transformation. Though that transformation is needed for organizations to harness new frontiers at a tremendous scale, we don't want to present the effort as organization-wide heavy lifting. This chapter's motivation and scope are to inspire innovation among SAP technical architects and dare them to leave their comfort zones. You want your architects to dream big, yet start small, without the need to convince the whole universe. They can fail early without inflicting any collateral damage and move on to the next innovation.

10.1   Event-Driven Architectures

Consider *events* as changes in states: Every action performed in an SAP system will change that business object's state. Thus, every SAP system is capable of emitting events. At the time of writing, SAP has exposed nearly 400+ business events for SAP S/4HANA and other SAP solutions (*http://api.sap.com/*). However, earlier releases of SAP systems (e.g., running on the SAP NetWeaver platform) require the additional SAP NetWeaver, add-on for event enablement (*http://s-prs.co/v5776106*) to create and manage events in those systems.

An event-driven architecture enables you to grow your SAP system's functionality incrementally without impacting the current flow of the business processes. In addition, your business systems can communicate beyond vendor boundaries by porting these events. As a result, you can build connected enterprises.

SAP events are observed but not directed, so you need to create the required integration pipeline for these events to reach their intended consumers. Several AWS native services can facilitate event-driven architectures, including Amazon EventBridge, Amazon Simple Notification Service (SNS), Amazon Simple Queue Service (SQS), Amazon MQ, Amazon Kinesis, and Amazon Managed Message Streaming for Apache Kafka (MSK).

Architecture Decision

Though you can use all these services to achieve an event-driven architecture, specific nuances in their capabilities make them the best fit for different use cases. This high-level list can help you make the right design decisions:

- Amazon EventBridge, Amazon SNS, Amazon SQS, and Amazon Kinesis are serverless. You can scale as you grow. These services do not need any minimum commitments or server infrastructures. Conversely, Amazon MQ and Amazon MSK are Amazon-managed services. Amazon manages the underlying computing; thus, you need a minimum commitment to leverage these services. To scale these services, manual intervention is required.

- Amazon Kinesis and Amazon MSK are leveraged for more complex event processing needs where a massive number of events are streamed and persisted for deriving historical outcomes and reporting. They both work on polling, which means the consumer has access to all the events where the consumer decides on the required events for further processing

- Amazon MQ enables compatibility and out-of-the-box integrations with on-premise message broker systems. This service provides an easy lift-and-shift migration option to AWS without the need to refactor your on-premise messaging systems. Otherwise, for new applications, AWS recommends leveraging Amazon EventBridge, Amazon SNS, or Amazon SQS for building cloud-native messaging systems.

> - Amazon SQS is suitable for a polling mechanism to process messages asynchronously in batches.
>
> - Amazon SNS is suitable for a push mechanism to distribute messages to a vast number of consumers via a rigid schema set out by the producer, in what are called *fan-out patterns*.
>
> - Amazon EventBridge is suitable for directing application events from third-party software as a service (SaaS) providers and integration products to several AWS services. Amazon EventBridge has several native event schema discovery and transformation options before events are sent out to the designated targets. You can use this service to build complex event mesh configurations.
>
> This list is not exhaustive: Refer to the relevant product documentation for more details.

This section focuses mainly on Amazon EventBridge, Amazon SNS, and Amazon SQS to build event-driven architectures. In the following sections, we discuss various event-driven architecture patterns. We examine how events in an SAP system can trigger subsequent actions in connected systems, a concept we explore first. We also delve into situations where specific events might lead to complex, decision-based actions in connected systems. Additionally, we look at event fan-out patterns through the lens of *publish-subscribe (pub-sub) notifications*. Finally, at the end of this section, we explore strategies for reducing dependencies between connected systems to enhance the reliability of cross-system business processes.

10.1.1   Event-Based Triggers

In this pattern, you can tap the required events from your SAP systems and relay them to your intended consumers using AWS native services. As shown in Figure 10.2, in any event-driven framework, three components are essential to complete the scenario:

- **Event producers**

  Entities that produce events.

- **Event brokers**

  Entities that stage or store the events to enable scaling and loose-coupling in the architecture.

- **Event consumers**

  Entities that pick the events from the event broker and perform the associated task against the information embedded in that event.

Once the required outbound communication arrangements are maintained for the event in your source SAP system, you can direct these events to the event broker. SAP

Event Mesh is the default native event broker for SAP events. In SAP Event Mesh, these events are registered against the queue to which consumers are subscribed. As shown in Figure 10.3, with SAP Integration Suite and Open Connectors, SAP Event Mesh can transmit these events to SAP SaaS solutions, SAP Business Technology Platform (SAP BTP)/custom applications, and specific AWS services for consumption like Amazon Simple Storage Service (S3) and Amazon SQS. You can use the *webhooks* capability in SAP Event Mesh to extend these events to Amazon EventBridge. Alternatively, you can use the AWS ABAP SDK in SAP systems with an ABAP environment to programmatically transmit events to Amazon EventBridge. Once an event lands in Amazon EventBridge, you can trigger actions into AWS services to extend SAP solution functionalities on AWS.

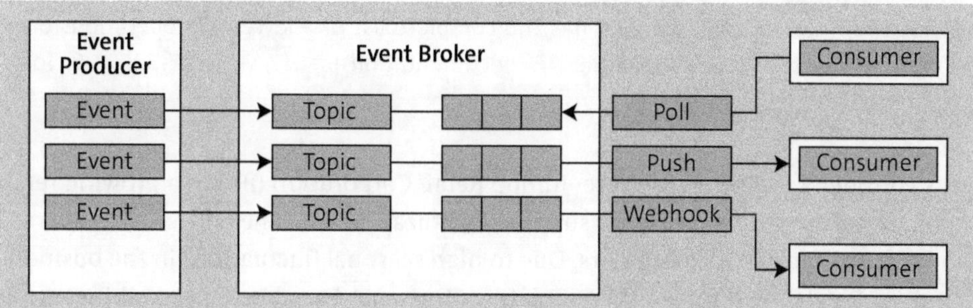

Figure 10.2 The Components of an Event-Driven Architecture

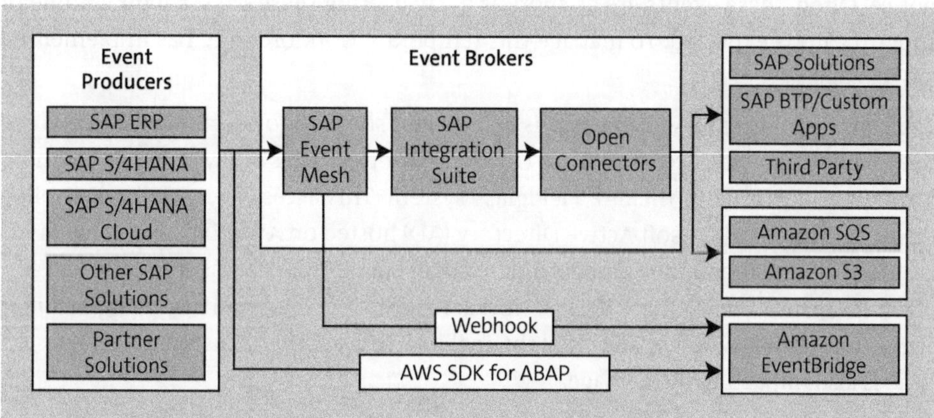

Figure 10.3 SAP Events Integration with Amazon EventBridge

Amazon EventBridge is a serverless event bus with native integration to nearly 20 AWS services, and the list keeps growing. This event bus can ingest events from other AWS services, SaaS providers, custom applications, and third-party event producers. You can learn more about Amazon EventBridge at *http://s-prs.co/v577610T*.

> **Webhook for Amazon EventBridge**
>
> Webhooks require an invoke URL to send events from the SAP Event Mesh. You can create a REST API endpoint using Amazon API Gateway on a webhook resource targeting Amazon EventBridge. Adjust the request headers and the content schema to map to the incoming content model.
>
> We recommend checking out the webinar "Deep Dive on Amazon EventBridge" (*http://s-prs.co/v5776108*; from 23:52 to 32:03) as an example of how to create webhooks for Amazon EventBridge.

> **AWS SDK for ABAP API for Amazon EventBridge**
>
> The ABAP method /AWS1/IF_EVB has the complete list of allowed API operations on Amazon EventBridge. Refer to the API documentation (*http://s-prs.co/v5776109*) for more details.

Let's consider a customer use case: Jumbo Retail Corporation (JRC) is a growing retail company in the US. Its customer support organization relies heavily on a temporary workforce to cater to its customers. Due to high seasonal fluctuations in the business, the need for the workforce is very volatile. Moreover, based on sales and inventory forecasts, JRC runs several sales promotion events intermittently on specific days of the year. Its sales and marketing teams must plan for these events quickly on short notice. Often, these events are as short as a day. During these days, a high volume of support calls is expected. To manage this temporary workforce, JRC has implemented SAP Fieldglass.

Their problem is that SAP Fieldglass is an efficient solution to address JRC's workforce needs, but a worker's access to JRC's IT resources often takes a day after their joining formalities are done in the SAP Fieldglass system. This access is centrally controlled through the JRC's Microsoft Active Directory (AD) hosted on AWS. Creating users in AD is entirely manual, and they are coordinated over email. This process is error-prone and effort intensive. Due to these delays, workers are sitting idle, resulting in high worker acquisition costs (WAC). In addition, the inability to respond to customer requests on time is also impacting JRC's image.

As shown in Figure 10.4, the AD team at JRC team decided to automate this process by integrating its SAP Fieldglass system with the AD service running on AWS.

Some solutions that can be considered include the following:

❶ You can configure an SAP Fieldglass event (*http://s-prs.co/v5776110*) to SAP Event Mesh to trigger an event upon creation of worker profile.

❷ You can send these events to a designated queue in SAP Event Mesh.

❸ Using the webhooks method, you can automatically forward these events from a designated queue in SAP Event Mesh to Amazon EventBridge. You must maintain the event forwarding URL for Amazon EventBridge in the SAP Event Mesh queue configuration.

❹ Based on the event rule defined in Amazon EventBridge, a matched event will call the AWS Systems Manager Run Command with the AD server as a target on AWS (*http://s-prs.co/v5776111*).

❺ The Run Command executes the PowerShell command New-ADUser to create an AD user in the directory server (*http://s-prs.co/v5776112*).

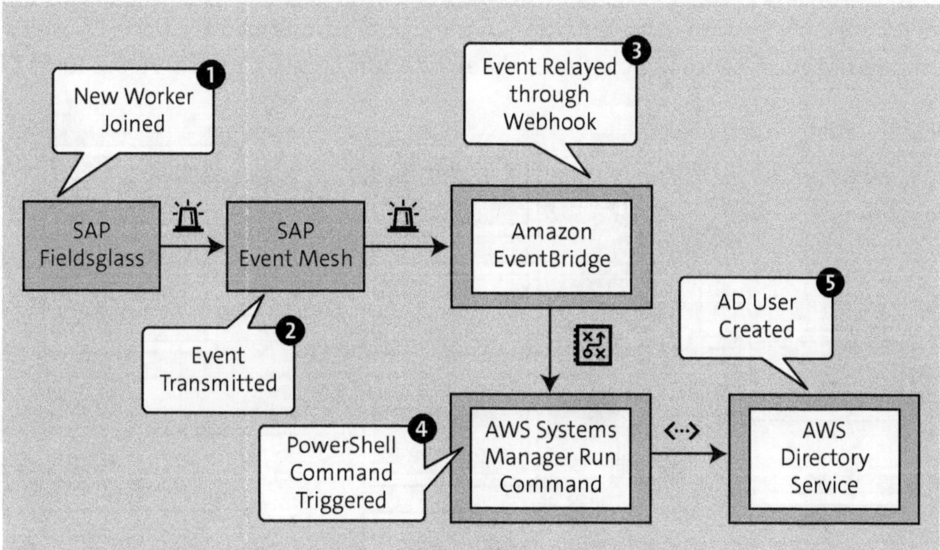

Figure 10.4 Amazon EventBridge Use Case: SAP Fieldglass Integration with AD Service

10.1.2 Rule-Based Orchestration

Sometimes, you may have too many actions to perform based on an event, or you want to run this information through a complex decision structure to arrive at the desired outcome. Thus, a single event from one system can create a ripple effect in your other systems, but all those tasks may need to follow a defined sequence. In addition, you'll want to deal with failures separately by retrying the same thing or trying something different to complete the task. Some tasks in the process are entirely independent, and you can run them in parallel to optimize the overall runtime. The possibilities are endless, but how can we coordinate these tasks so that we don't create confusion in the chain of command? We have two ways to solve this problem. You can design this architecture through a choreography pattern or an orchestration pattern.

In the choreography pattern, every actor coordinates their actions through an independent actor who maintains statuses and controls their actions in the process chain.

You must have a central persistence for every actor to report their statuses. You can use a central queue (e.g., Amazon SQS) or a database (e.g., Amazon DynamoDB). However, you must code the entire coordination for each step in the process. This pattern could lead to monolithic architectures, thus, jeopardizing future extensibility.

To continue our customer example, upon the success of a single step in the onboarding process, JRC would like to bring in more automation to optimize its current onboarding process. Once a user is created in their corporate AD, they must provide a virtual desktop and an Amazon Connect instance for each worker to perform their duties. The current manual process not only creates delays, but these delays are incurring unnecessary IT infrastructure costs by keeping them idle. JRC wanted to automate these steps using a choreography pattern using Amazon EventBridge as an intermediary actor. The architecture shown in Figure 10.5 continues the architecture shown earlier in Figure 10.4.

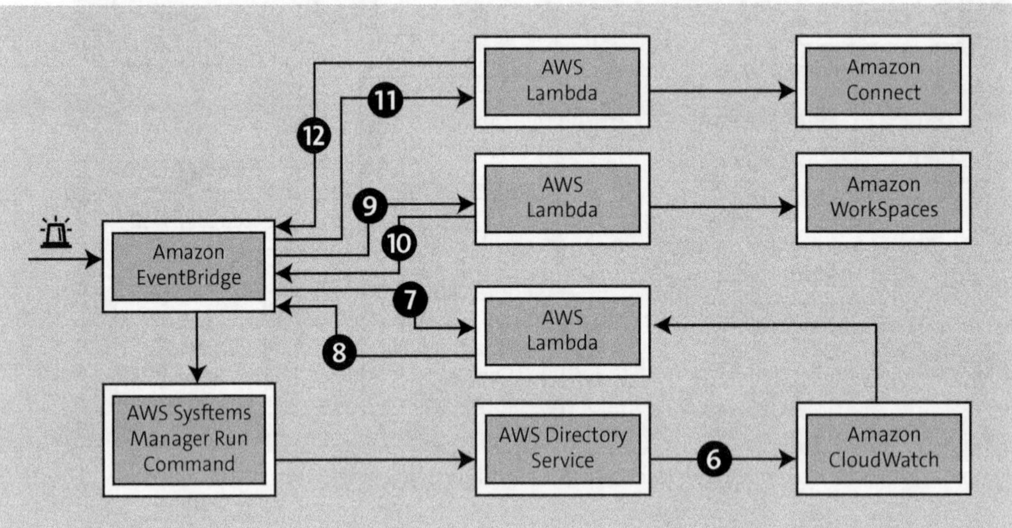

Figure 10.5 Amazon EventBridge Use Case: Multi-Step Choreography Pattern

Once the user is created in AD, the following activities are performed:

❻ AD activity logs are forwarded to Amazon CloudWatch logs. Refer to *http://s-prs.co/v5776113* for more details about AD log forwarding.

❼ After you complete the AWS Systems Manager Run Command, Amazon EventBridge will trigger an AWS Lambda function to analyze Amazon CloudWatch logs. This function will filter through the logs and find the details of the newly created users.

❽ AWS Lambda will return the identified user details to Amazon EventBridge as events.

❾ Once the user creation event reaches the event bus, the associated event rule will trigger the AWS Lambda function to provision an Amazon workspace.

⑩ The workspace completion event is sent back to Amazon EventBridge.

⑪ Amazon EventBridge will trigger the creation of an Amazon Direct Connect instance.

⑫ Upon provisioning the Amazon Direct Connect instance, the completion status is sent back to the Amazon EventBridge.

This architectural pattern works well with a small number of tasks. However, as the number of steps grows, the complexity multiplies because you need to coordinate every step to follow the sequence. Moreover, the troubleshooting of the failed steps becomes more cumbersome because the relevant logs are scattered all over the place. The overall process must be more transparent to track the status of its execution. You must be aware of its design to understand its sequence. Even adding a new step or changing the sequence of any step requires careful planning and intense effort.

We can transform this architecture into an orchestration pattern to address these concerns. The new architecture is shown in Figure 10.6.

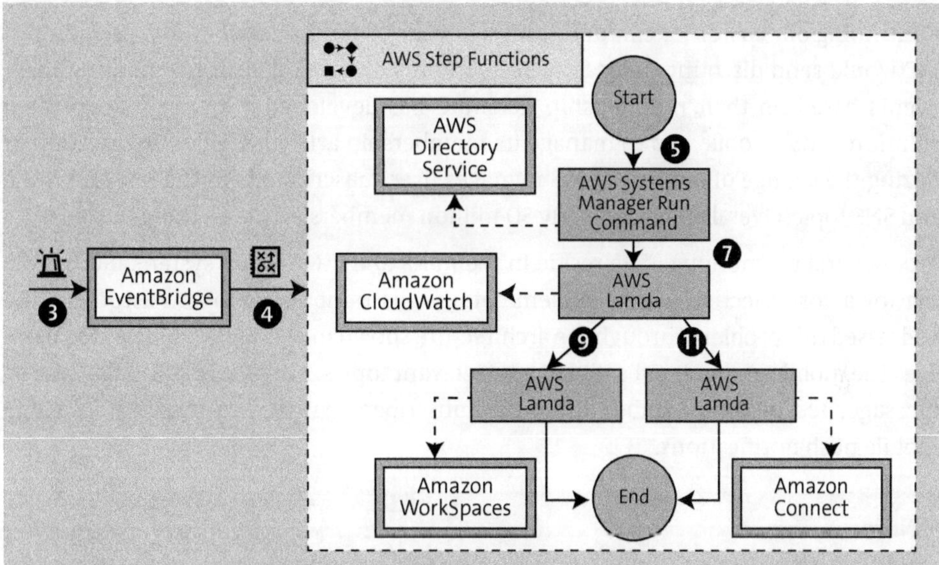

Figure 10.6 Multi-Step Orchestration Using AWS Step Functions

In this architecture, you have only one invocation triggered from the Amazon EventBridge calling AWS Step Functions, but this single function in AWS Step Functions will orchestrate the rest of the process (the corresponding numbered steps are similar to those shown earlier in Figure 10.4 and Figure 10.5). The steps shown in Figure 10.6 are self-explanatory in that they flow in a defined sequence. The main advantage of this architectural pattern is that the status and control of process sequences are visually representative of understanding and troubleshooting any step separately.

Architecture Decision: Amazon Simple Workflow Service versus AWS Step Functions

AWS Step Functions is a serverless service, whereas Amazon Simple Workflow Service (SWF) requires a dedicated infrastructure to run its workflows. AWS Step Functions has integrations with 200+ AWS services, but Amazon SWF has limited integration. Moreover, AWS is not investing anymore in Amazon SWF.

10.1.3   Publisher-Subscriber Notifications

You can use the pub-sub notification pattern to send the same notification to many targets. This mechanism is reliable and affordable to reach your customers in near real-time. You can subscribe your customers to relevant topics on Amazon SNS and publish your messages on those topics. Within no time, Amazon SNS will fan out these messages to all the subscribers of that topic.

Continuing our customer example, JRC runs various promotional events periodically. JRC would send discount codes for selected items to its loyal customers during these events based on their membership tiers. JRC has developed a loyalty management solution (LMS) mobile app to manage its membership activities. Upon registration or during the change of their tier, LMS manages user subscriptions to the relevant Amazon SNS topic. Overall, JRC has nearly 50 million members across all these tiers.

Product and promotional data reside in their SAP S/4HANA Retail system. JRC is looking for a cost-effective way to disseminate its discount codes to its customers. JRC addressed this problem through the architecture shown in Figure 10.7. As the LMS manages the mobile app user subscriptions to relevant topics, Amazon SNS will fan out the messages fed by AWS Lambda to all the subscribers assigned to these topics using mobile push notifications.

Note

Amazon AppFlow is a low-code/no-code serverless service to extract data at scale over the OData protocol from SAP and other SaaS applications. It can perform incremental extracts with the help of the Operation Data Provider (ODP) concept in SAP. For more information, refer to *https://aws.amazon.com/appflow/*.

Note

Amazon SNS can send notifications directly to mobile apps using several supported push notification services. Refer to *http://s-prs.co/v5776114* for more information.

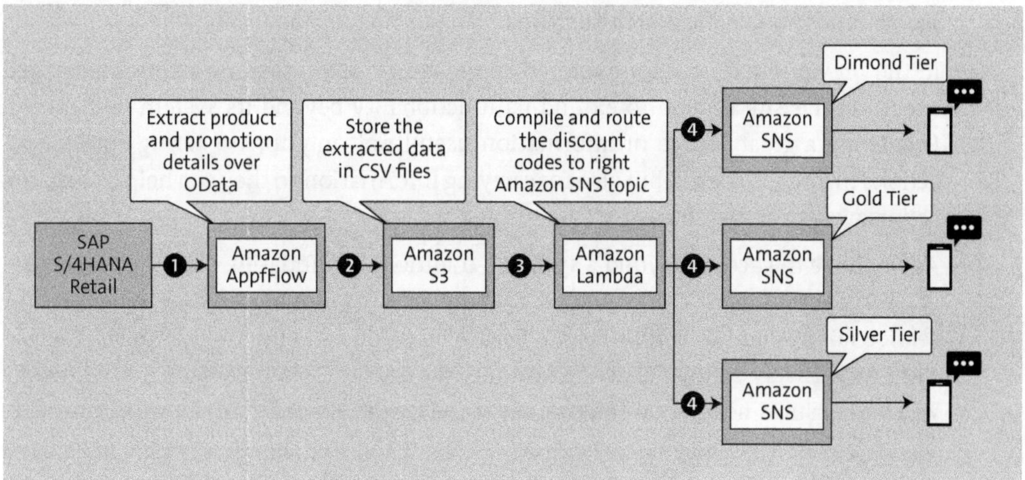

Figure 10.7 Amazon SNS Notification-to-Mobile Push Pattern

Observe the scenario shown in Figure 10.7: You've sent the same message to all the subscribers assigned to the same topic. However, if you want to send a customized message to each user individually, you must address this scenario a bit differently. As shown in Figure 10.8, you can leverage Amazon EventBridge in combination with AWS Lambda to scale this requirement to millions of users.

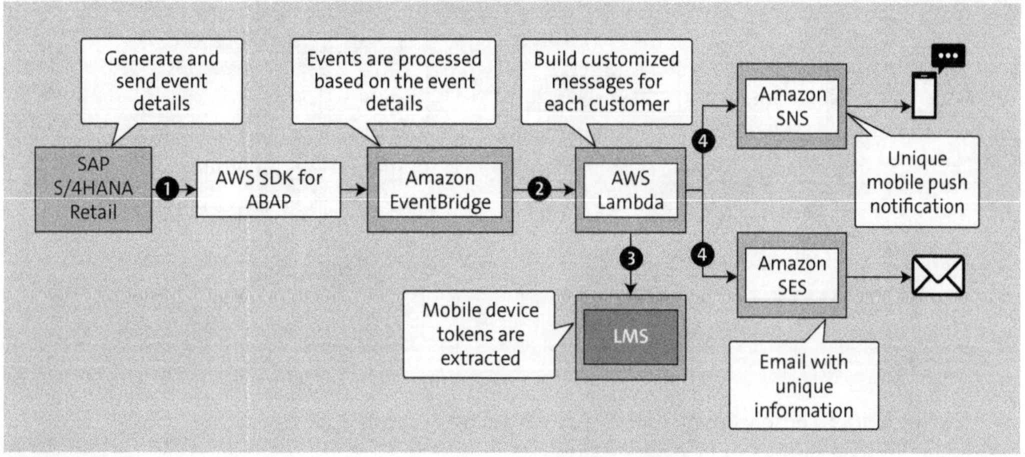

Figure 10.8 Customized Notification Pattern Using Amazon EventBridge and AWS Lambda

In this architecture, AWS Lambda is customizing the target notification utilizing the event details coming from the Amazon EventBridge and the mobile token information coming from the LMS system.

10.1.4   Loosely Coupled Architectures

In the previous sections, we explored some simple use cases in which we leveraged events in the SAP system to extend its functionality beyond its system boundaries. Events are a great source of information useful not only for triggering subsequent actions in target systems, but also for carrying information so they can help offload the action away from the source.

As we move the action beyond a system's traditional boundaries, hopping across several serverless components, you must account for exceptional scenarios to handle, like service failures and communication disruptions. You can employ the "push" mechanism extensively to invoke actions on target systems. This mechanism often works well and ensures nearly real-time action at the target. However, in the absence or failure of any actor in the chain, the process misses the step and ignores the associated actions in the subsequent systems. These errors can impact the reliability of the process.

You can address these challenges by adding error handling into the architectures. In addition, introducing a "poll" mechanism brings resiliency into the process. With these adjustments, our earlier architecture would change, as shown in Figure 10.9.

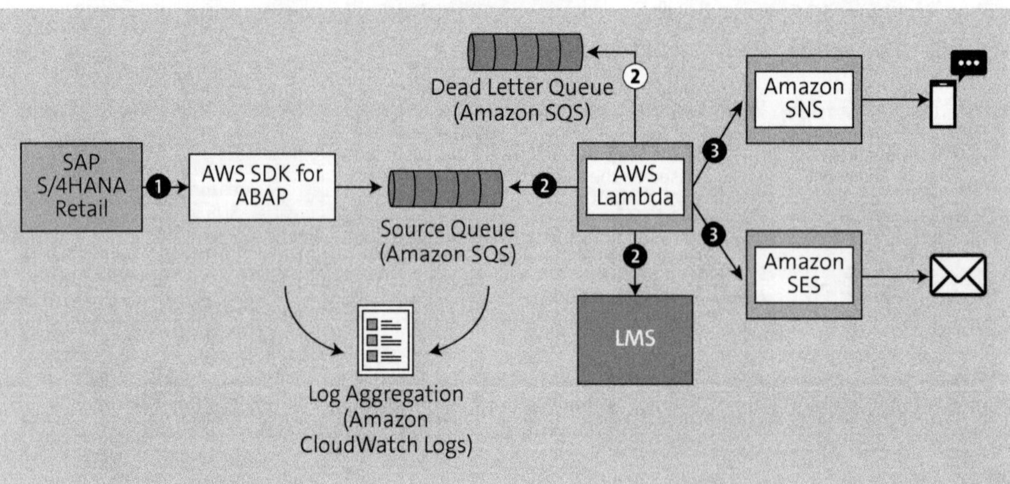

Figure 10.9 Loosely Coupled Architectures Using Amazon SQS

The resiliency of this new architecture is improved by introducing Amazon SQS ❶. The source system sends the events to store them in the source queue ❷. Later, these events are asynchronously processed by AWS Lambda. You can configure Amazon SQS as an event source for AWS Lambda, then AWS Lambda polls continuously for messages and triggers the AWS Lambda function upon message arrival. Upon failure of a message, it will be moved to dead letter queue (DLQ). Based on the configuration settings, you can *redrive* the failed messages in the DLQ back to source queue for later processing ❸. AWS Lambda will scale automatically based on its concurrency settings to consume the messages in parallel.

Moreover, while designing loosely coupled architectures, you should consider the following points:

- You should maintain a certain flexibility to scale each step in the process independently. You should be able to manage, enhance, and troubleshoot these steps separately without impacting other connected steps.

- *Polling* makes the execution asynchronous.

- Often challenging is troubleshooting failures in these architectures. As shown in Figure 10.9, consider *aggregating the logs* of all the components in the process into a single place to improve observability.

- A significant number of AWS services follow an *eventual consistency*. A slight delay could result between the creation of resources and its consumption. Bake-in the correct wait conditions in the process to avoid `resource_unavailable` failures.

- Most importantly, the process should be *idempotent*, which means the target outcome should not impact even if you apply duplicate changes.

10.2   Serverless Application Design

In this section, we'll delve into serverless computing, a paradigm that eliminates the need for you to host and manage computing resources. Let's now explore AWS serverless services, highlighting their advantages and relevance through various architectural scenarios. We hope this discussion provides a comprehensive understanding of how serverless computing can streamline operations, reduce costs, and enhance scalability, particularly in dynamic and evolving technological landscapes.

10.2.1   Advantages and Disadvantages

In the traditional computing world, nearly 80% of the time is spent provisioning, operating, and managing the underlying computing resources. That leaves only 20% of the time to invest in innovation. In addition, serverless computing has several other advantages over the conventional computing model, including the following:

- **Time-to-market**
  You can bring in innovations at an accelerated pace. This model brings lowers the barrier to entry when building new products and services.

- **Scalability**
  The inherent characteristic of the serverless model is its ability to grow at an infinite scale (virtually). Your ideas aren't constrained by server capacity.

- **Built-in resiliency**
  These services are designed to serve with a committed service level agreement (SLA) and are built to withstand local or AZ failures. If you design the architecture well, they can even withstand a complete region failure.

- **Cost savings**
  With the pay-as-you-go cost model, you pay only for your usage. Your expenses and returns are always proportional and predictable.

However, the serverless computing model is not a magic bullet and is not suitable for all types of workloads. Therefore, you should be aware of the following disadvantages when thinking of going serverless:

- Heavyweight, batch-mode, and long-running processing steps are not a good fit for serverless architectures.

- Processes with complex runtimes and configurable environments may not work with the available serverless runtimes.

- Ensuring security across all the moving parts in the serverless model requires a new kind of a security model.

- Due to cold starts and invocation lags, the performance of the serverless components is not always optimal.

- AWS's serverless computing model does not support running monolithic SAP workloads nor ABAP runtime environments. However, you can enhance SAP functionalities by building extensions using AWS serverless services.

10.2.2 AWS Serverless Services

So far, in this chapter, we covered various patterns for extending an SAP functionality by consuming events generated from the SAP. In addition, AWS has numerous other serverless services to build web apps end-to-end with full-stack functionality.

As shown in Figure 10.10, these services can be loosely categorized in the following way:

❶ **Frontends**
  You can create user interfaces (UIs) for web or mobile using Amazon API Gateway and AWS Amplify.

❷ **Runtimes**
  AWS Lambda and AWS Fargate provide computing resources with the required programming environment to run your business logic.

❸ **Data persistence**
  You can store user profiles, interactions, and supporting business data in cloud-native, serverless databases without any management overhead, for example, with Amazon DynamoDB, a serverless variant of Amazon Aurora and Amazon S3.

❹ Integration

You can use these services to integrate data, services, systems, and applications within AWS or externally to build seamless interfaces.

❺ Observability

You can log and trace the actions carried out by these services to monitor the metrics and assist in troubleshooting efforts.

❻ Additional services

These services provide the necessary support functions as well as modern application extensions.

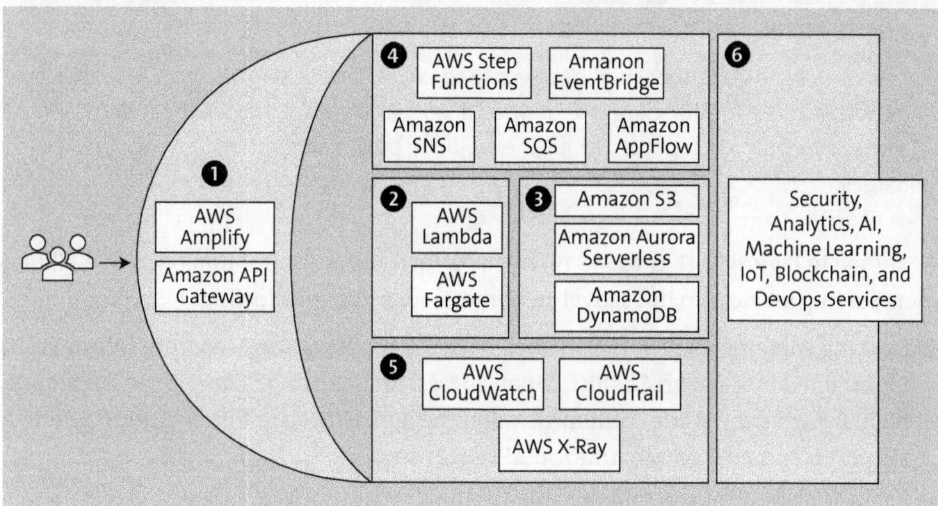

Figure 10.10 AWS Serverless Services

10.2.3   Cloud Application Gateways

In this section, we'll examine a sample architecture to explore the art of possibilities on AWS. You can easily build a mobile application to extend an SAP functionality using AWS serverless services.

Let's return to our earlier customer example from Section 10.1. JRC now wants to start a premium member service to their most loyal customers (as determined by the LMS application shown earlier in Figure 10.8 and Figure 10.9). JRC wants to pilot this feature to a small number of customers before it is scaled for their entire customer base. Customers are categorized into three membership tiers based on how much they spend on JRC's products and services.

To build this mobile app, JRC uses Amazon API Gateway and AWS Amplify. This app is open for existing customers on their e-commerce platform who are business partners on the SAP S/4HANA Retail system as well as for new users who directly sign up through the mobile app.

> **[»]**
>
> **Amazon API Gateway**
>
> A serverless service, developers can create, publish, and manage their APIs at scale in Amazon API Gateway to enable frontend interfaces for customers. Users can interact over RESTful APIs and WebSocket APIs with serverless and containerized workloads deployed on AWS.

> **[»]**
>
> **AWS Amplify**
>
> A serverless, no-code/low-code UI design and development service, AWS Amplify lets you create full-stack interfaces for the web and mobile apps to host static content. This solution has native support for all the popular programming frameworks and languages.

As shown in Figure 10.11, the LMS mobile app carries out several functions to facilitate membership services to JRC's loyal customers, such as the following:

❶ Existing customers are authenticated over OAuth using the user store information maintained in Amazon Cognito. Upon successful logon, AWS Amplify will instantaneously serve the static content through the app. However, for dynamic content, a request is routed through Amazon API Gateway.

❷ For new users, Amazon Cognito will call the user profile management service.

❸ Upon successful profile update, a user store entry is created with a user ID and password. (Optionally, this feature can be extended into a federated identity with external identity providers.)

❹ Amazon API Gateway will call profile management service implemented using AWS Lambda.

❺ For an existing user in the SAP S/4HANA system, the user profile is updated with additional information extraction from LMS. Otherwise, the new user's details will be sent back to create a new business partner on SAP S/4HANA using Amazon AppFlow SDK to update over the SAP OData service endpoint.

❻ The user profile data is stored in Amazon DynamoDB. This data includes a mobile token ID to send mobile push notifications directly via the mobile app (shown earlier in Figure 10.9, step ❷).

❼ The second AWS Lambda function implements the rest of the membership activities, which also occurs at ❽.

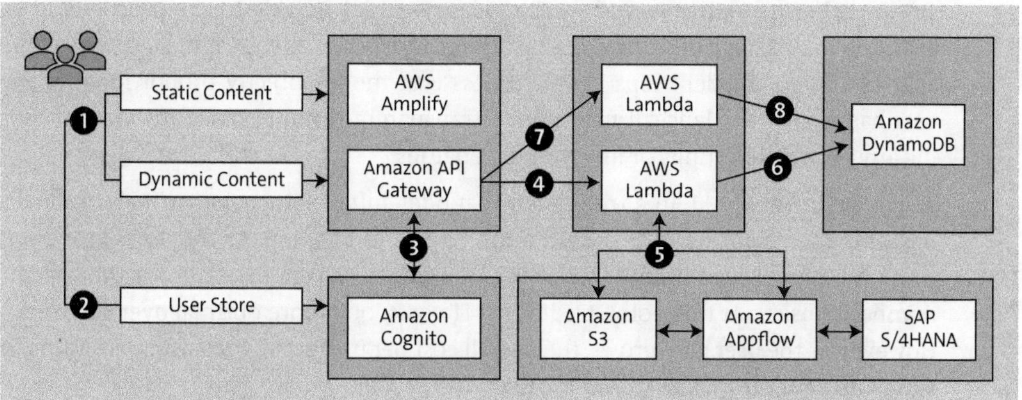

Figure 10.11 Extending SAP Solutions Using Amazon API Gateway and AWS Amplify

10.2.4   Scale on Consumption Model

The total cost of ownership (TCO) in managing monolithic applications has several cost elements. Out of which, computing costs and management overheads amount to nearly 80% of the TCO. In addition, as shown in Figure 10.12, due to the rigid nature of its provisioning behavior, your scaling responses to the user loads are only sometimes optimal. Either they get over-provisioned or under-provisioned. The systems with consistent workloads can predict the workload and scale proactively, but for the systems with un-predictable workload patterns, either you end up wasting resources ("Waste zone" as shown in Figure 10.12) by keeping idle or letting the user suffer ("Suffer zone" as shown in Figure 10.12) as the right amount of resources are not available to carry out their business.

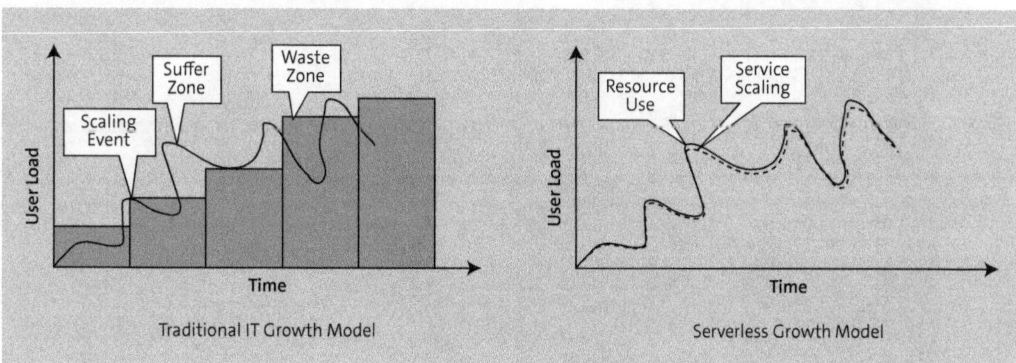

Figure 10.12 System Scaling Patterns: Traditional IT versus Serverless

On the other hand, the serverless IT model responds quickly to the user load, thus avoiding the wastage of computing resources. Moreover, as shown in Figure 10.12, the "pay-as-you-go" model keeps the serverless cost model directly proportional to the user base's growth. Hence, it maintains a transparent revenue structure without having any growth sufferings or innovation penalties.

As shown in Table 10.1, in a traditional IT model, initial setup costs will always arise, meaning there will always be a minimum cost of innovation. As the users grow, you could hit a breakeven point where the setup infrastructure fits right for the specific number of users. At this point, traditional IT costs look more optimal over serverless. However, as the user base grows, the overall cost per user starts increasing, resulting in eating up your returns, as shown in Figure 10.13.

Cost Scenario	Total Cost of Ownership		Cost per User	
	Serverless	Traditional	Serverless	Traditional
"1 User" Cost	$3.00	$100.00	$3.00	$100.00
"100 User" Cost	$300.00	$500.00	$3.00	$5.00
"1000 User" Cost	$3,000.00	$2,400.00	$3.00	$2.40
"10000 User" Cost	$30,000.00	$34,000.00	$3.00	$3.40
"100000 User" Cost	$300,000.00	$390,000.00	$3.00	$3.90

Table 10.1 Hypothetical Cost Example to Compare Cost Models: Traditional versus Serverless

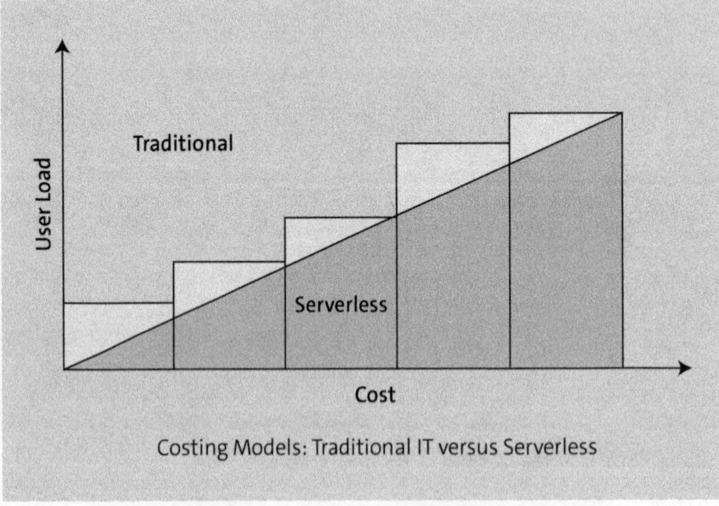

Figure 10.13 Cost Models: Traditional versus Serverless

> **Note**
>
> Volume discounts and the other costs similar in both the models are not considered in the pricing example presented in Table 10.1.

10.2.5   Building without Infrastructures

The serverless deployment model not only helps you start quickly but also enables you to grow incrementally. For example, you can start your journey by building non-differentiating capabilities to experiment and learn in this way. Later, you can slowly augment additional functionalities like automating repeated tasks or removing manual steps in your differentiating capabilities. In addition, you can build these features as side-by-side extensions without disrupting current processes. First, however, you must identify suitable integration patterns among these components, so they do not throttle you as you grow.

As you carve out more and more functionalities, application security, observability, integration, and analytics functions should undergo a complete overhaul to adapt to your new cloud-native paradigm. Thus, consider including these functions in your design from the start, though they are optional in the initial experimenting phases. On the other hand, putting off including these dimensions could result in a complete refactoring of your solution later. The ultimate aim of your journey is to break the SAP monolith into cloud-native microservices, as shown in Figure 10.14.

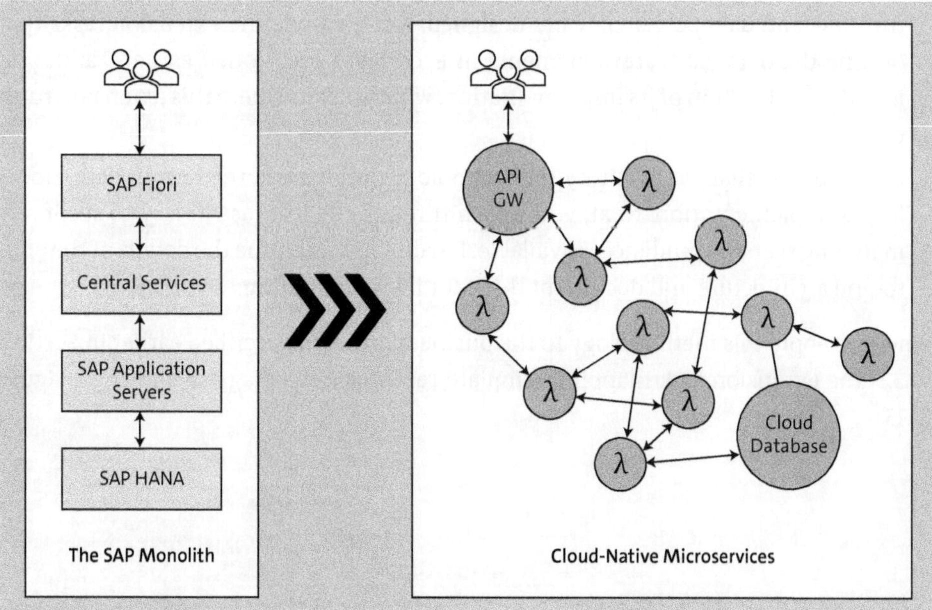

Figure 10.14 Breaking the SAP Monolith into Cloud-Native Microservices

10.3   Application Extensibility with Cloud-Native Services Integration

In earlier sections, you learned how to enhance SAP solutions with AWS services. In addition, we also discussed how to leverage SAP data to create SAP solution extensions outside of your current system boundaries. Moreover, we explored various architectural patterns to build cloud-native capabilities without the constraints of an IT infrastructure. To achieve similar objectives in a more structured way, SAP has provided a technology-agnostic methodology for enterprise architects to formulate organization-specific extension strategies, called SAP Application Extension Methodology (SAP AEM). We'll explore this methodology in detail in this section.

10.3.1   Building SAP Service Extensions as Microservices

SAP AEM has three phases to guide you through the journey of building extensions for SAP solutions:

- **Phase 1**
  In this phase, first, you assess and document the state of your current IT solutions, their lifecycles, and their probable future states. Next, you gather and align your business requirements with your applications lifecycles. Later, you find the gaps and identify the prospective applications for enhancements.

- **Phase 2**
  In this phase, first, you define extension styles, meaning how the application UI, runtime, and data persistency are designed. Next, you define extension tasks that outline the detailed feature elements in each layer mentioned earlier. Later, you identify the location of its implementation, which means where this extension runs.

- **Phase 3**
  In this phase, first, you identify the technology options and the best fit data model for its implementation. Next, you search through the use case library to see if any matching scenario guidance is available. Finally, you fine-tune the details of the end-to-end architecture and document them for implementation.

When we apply this methodology to the business problem described earlier in Section 10.2.3, the extension task mapping template resembles the diagram shown in Figure 10.15.

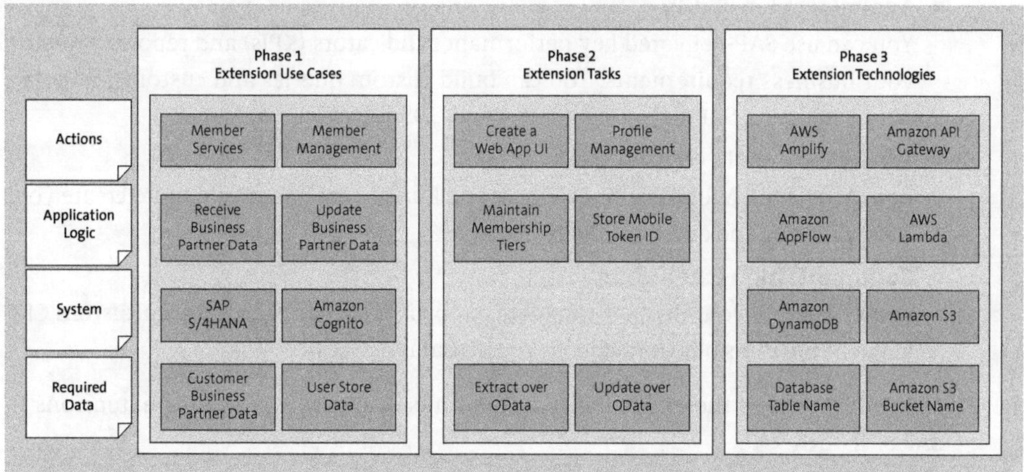

Figure 10.15 SAP Application Extension Methodology: Extension Task Mapping Template

10.3.2   SAP Extension Strategy and Governance

So far, we've explored architectures that leverage the SAP data in some form without making any changes to the existing data model in the source system. However, you may need to adapt the UI, business logic, or data model in the source SAP system in some exceptional situations. Custom changes in SAP systems are highly discouraged (and not even allowed in SAP S/4HANA Cloud Public Edition) to avoid future upgrade challenges. SAP promotes a "clean core" philosophy where changes to the core are streamlined through the SAP S/4HANA extensibility framework, in-app extensibility, or key-user extensibility. Changes made according to this framework won't hinder future lifecycle events. Check the list of all available in-app extensibility usage opportunities in standard SAP S/4HANA via Transaction SCFD_REGISTRY.

This transaction offers the following extensibility tools:

- **UI extensibility**
  You can make UI adaptations by customizing screen elements (including adding external content to your frame).

- **Field extensibility**
  You can extend SAP tables, application core data services (CDS) views, and OData services possible under the given business context.

- **Table extensions**
  You can create or add custom business objects and their corresponding database tables. You can also add/delete fields to the existing business objects.

- **Business logic extensibility**
  You can extend the business logic using Restricted ABAP or RAP (RESTful ABAP Programming). You can only extend or use the repository objects for which APIs have been released.

- **Analytics extensibility**
  You can use SAP-delivered key performance indicators (KPIs) and reports based on your business requirements. You can build custom queries and custom CDS views out of standard, SAP-delivered queries and CDS views.

- **Forms and email extensibility**
  You can use the Adobe Forms Designer and E-mail Template Designer to create your custom forms and email templates.

- **Communication extensibility**
  You can define your own custom data exposure mechanisms to make custom CDS views or business objects available over OData services.

Figure 10.16 shows the extensibility frameworks available for building extensions in SAP S/4HANA.

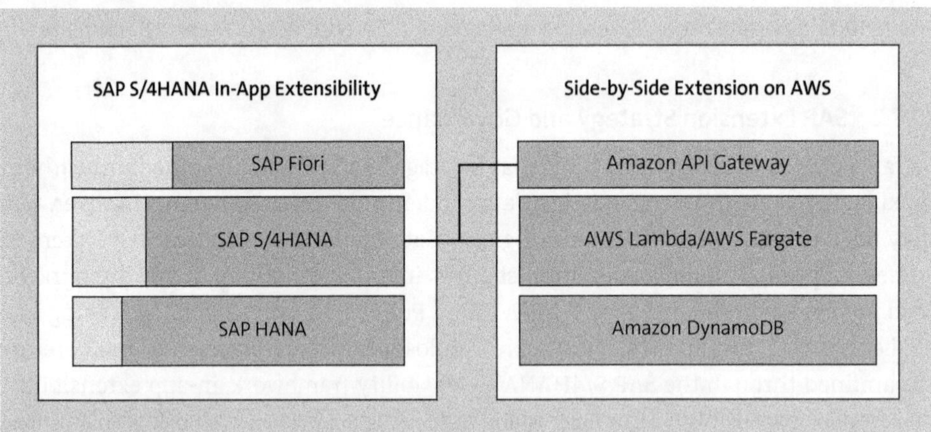

Figure 10.16 SAP S/4HANA Extensibility Framework

10.3.3   Scaling with Container Technology

Container technology created a burst of euphoria in the IT world—with good reason! Containers pave the way for building and managing large-scale applications in a microservice architecture. In addition, a container provides portability over an underlying infrastructure so that you can run your application on any infrastructure without the need for refactoring or rearchitecting. This portability enables organizations to avoid vendor lock-in and operate workloads in multicloud environments. Moreover, you can orchestrate and govern an entire application stack through automation and DevOps practices.

Note

A *container* is a lightweight virtualization construct that holds an application's code, its dependencies, and its runtime together so this application can run independently and reliably on any computing environment. You can package these elements as container images using the *Docker Engine* and store these images centrally in container registries.

Note

Kubernetes is an open-source application to orchestrate containers from their deployment to their management through load balancing, service discovery, automatic scaling, and self-healing. This application provides a reliable and self-managed environment to operate large-scale microservice architectures running in containers.

At the time of writing, only SAP Customer Experience, SAP Data Intelligence, and SAP Datasphere have native support for container technologies. In addition, SAP BTP provides the Kyma runtime (*https://kyma-project.io/*) to build extensions on managed Kubernetes environments. The other monolithic SAP solutions lack support for running container workloads. However, this situation is bound to change in the future.

As shown in Figure 10.17, the SAP BTP, Kyma runtime, environment is built on top of a managed Kubernetes cluster by enhancing its functions through additional addon services. These addon services will simplify cluster management and improve its observability and service discoverability. These added service integration capabilities integrate easily with SAP system events and API frameworks.

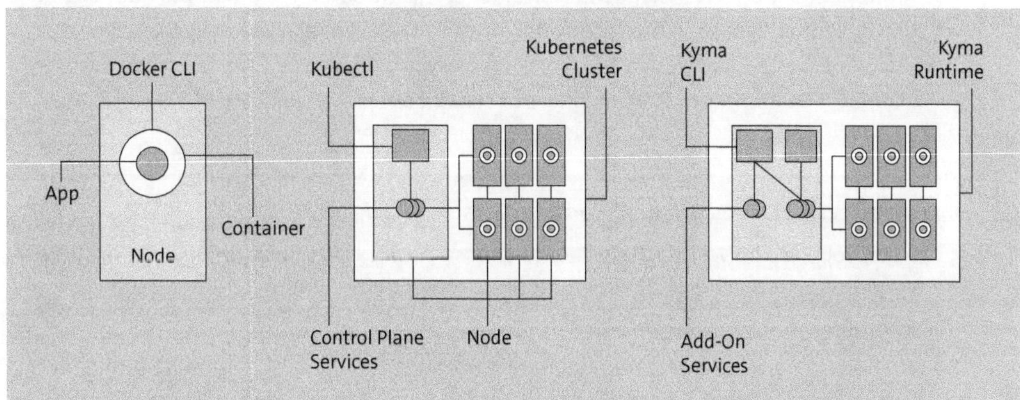

Figure 10.17 SAP BTP, Kyma Runtime, for Developing Extensions on Containers

You can set up a similar environment on AWS using Amazon Elastic Kubernetes Service (EKS) blueprints (*http://s-prs.co/v5776115*). As shown in Figure 10.18, Amazon EKS blueprints are collections of infrastructure as code (IaC) modules that you can package to deploy together with an Amazon EKS cluster. You can use these Amazon EKS blueprints to add Amazon EKS addons and other third-party addons to enhance observability and service networking. In addition, with these Amazon EKS blueprints, you can also integrate other AWS native services using AWS Controllers for Kubernetes (ACK), as described further at *http://s-prs.co/v5776116*.

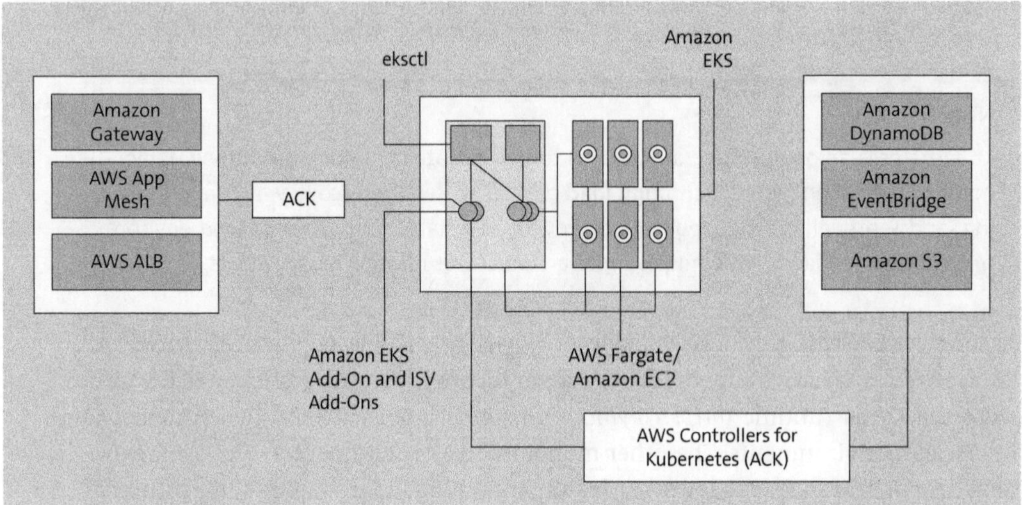

Figure 10.18 Amazon EKS Blueprints Deployed on AWS Fargate with Amazon EKS Addons and AWS ACK

[»]

Note

Amazon EKS is an AWS-managed container orchestration service that provides out-of-the-box functionality to manage scalability and the availability of your container clusters. Amazon EKS can deploy resources on cluster-managed or user-managed Amazon Elastic Cloud Compute (EC2) instances or serverless on AWS Fargate.

[»]

Architectural Decision: AWS Lambda versus AWS Fargate

Now, you may be wondering when to choose AWS Lambda and when to adopt AWS Fargate. The following factors may help you make a clear choice:

- **Runtime environment**
  If you need a custom runtime where you maintain specific OS support through OS packages, settings, and library bindings, AWS Fargate is the right choice.

- **Execution time**
  The execution time of a user interaction or a job completion exceeds the AWS Lambda maximum runtime (i.e., 15 mins), you can shift to AWS Fargate.

- **Service interactions**
  If one or more services/functions require synchronous or interactive communications between them, AWS Fargate is the default choice.

- **Observability**
  AWS Lambda has out-of-the-box observability by pushing its logs and traces to Amazon CloudWatch and AWS X-Ray; however, for AWS Fargate, you'll need additional configuration to achieve this goal.

- **Invocation lag**
  AWS Fargate needs around 60 to 90 seconds to start up, but for AWS Lambda, startup can be as quick as 5 seconds. This lag can impact application performance if numerous invocations are involved.

- **Integration with AWS services**
  AWS Lambda can integrate with almost all the AWS services, interacting with them using AWS SDK. On the other hand, AWS Fargate requires an additional layer (AWS ACK) to facilitate this integration.

10.4   Modern Data Architectures for SAP

All the interactions that an organization carries out with their stakeholders, vendors, partners, and customers leave a trail of information. Even the machines in your shopping floor, robots in your warehouses, and the point of sale terminals in your stores generate more useful events to track and trace your operations. However, this information is collected through different enterprise solutions and stored in different formats within their siloed data stores. Though they capture current business events to direct immediate actions and responses, they often fail to tie relationships to related events occurring at the same time in other connected systems.

Moreover, the solutions and their data stores are designed to fulfill operational requirements with optimal performance. They are not equipped to analyze terabytes of historical data to derive patterns and predict future behaviors. Furthermore, essential information sources in an organization no longer exist in the rigid row-column formats of relational databases. Thus, you must have a comprehensive data consolidation solution in place that integrates all the information source and provides a unified data insight to drive business outcomes.

In this section, we'll delve into some strategies for consolidating data through the creation of cloud data lakes, alongside methods for integrating various data sources dispersed throughout your organization using cloud-native data integration patterns. Additionally, we'll investigate how you can optimize data storage and retrieval by taking advantage of the diverse purpose-built database options available on AWS.

10.4.1   Building Cloud Data Lakes to Improve Business Insights

Data consolidation is not about dumping all the data into a single data store. Consolidation goes much beyond that. When data leaves its boundaries, it moves away from its safety net. You must constantly be reassured of its integrity and security by placing it under new guardrails and avoid unwarranted exposure, yet while still granting seamless access to intended parties.

Several other challenges arise while collating data from different data sources to make it available for diverse data consumers. Therefore, to sustainably address this data requirement in an organization, you must have a visionary strategy to address, not only current data requests, but also to address future growing demands through scalability. As shown in Figure 10.19, a data mesh architecture can help you address all these challenges with seamless integration between data producers and data consumers through transformed data products. In our example, data from various SAP and non-SAP sources is extracted and later transformed into data products in a unified data format available for consumption.

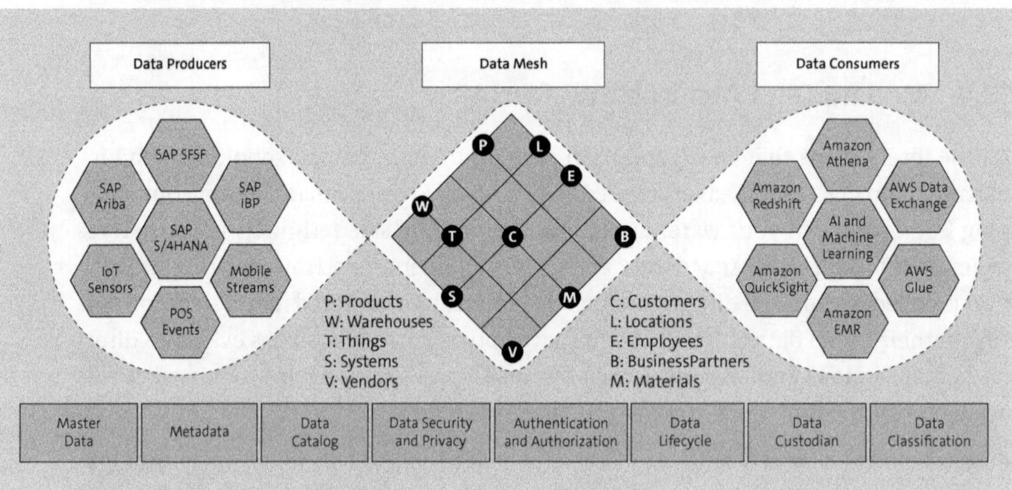

Figure 10.19 Creating a Data Mesh Architecture on AWS

To comprehensively achieve your organizational data goals through a data mesh architecture, you must have a robust data governance framework in place, which we'll call the *Unified Data Governance Framework (UDGF)*. This framework has three dimensions: data domains, data products, and data objects. These dimensions weave together perfectly to create a data mesh architecture.

To define these data elements accordingly, consider the following guidance:

- **Identify data domains**
  You can define a data domain based on your organization's operating model. It should identify the ownership of the data assets underneath it. This segregation should also loosely reflect the distribution of its data sources and their corresponding data consumption patterns to minimize cross-domain data dissemination, including:

- Lines of business (LOBs), for instance, finance, sales, marketing, manufacturing, etc.

- Locations, for instance, the US; Europe, Middle East, and Africa (EMEA); Asia-Pacific, etc.

- Business units, for instance, consumer services, commercial products, devices, healthcare, etc.

- **Create data products**

  These objects are microservices of data. They contain logical subset of curated data objects within the data domain along with its data definitions, classifications, lifecycle attributes, and access attributes. You can have master data as a separate data product, or it is distributed across data products to hold data product-specific master data. Example of data products in the sales data domain include online sales, B2B sales, and indirect sales.

- **Maintain data objects**

  These are consolidated data containers that are extracted and transformed from one or many data sources. These are the technical entities that directly represent the underlying data model. The below are examples of data objects for online sales data products:

  - Master data objects, for instance, customers, vendors, tax codes, and products

  - Transactional data objects, for instance, sales documents, billing documents, customer activities, and customer sentiments

You can build data mesh architecture on AWS using AWS Lake Formation. AWS Lake Formation is an AWS managed service that automates the significant portion of building, managing and safeguarding data lakes on AWS. As shown in Figure 10.20, it is in combination with AWS Glue, it classifies data, identifies master and meta data, and maintains data catalogue against which data lake administrator maintains permission control to grants access to authorized consumers.

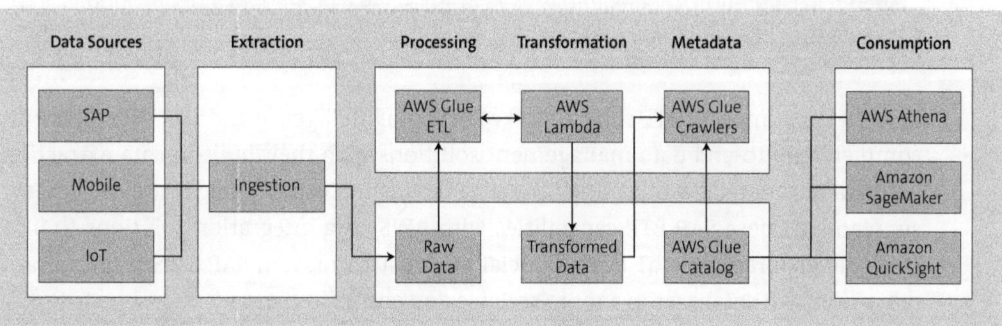

Figure 10.20 Building Data Lakes Using AWS Lake Formation

Administrator can maintain permissions on Lake Formation (LF) either using named resources in data catalogue (i.e., directly against data objects) or LF-tag based access controls (LF-TBAC) associating tags against resources in data catalogue (i.e., indirectly tagging them based on their data domains or data products to percolate those permissions to underlying data objects). You can find more details about AWS Lake Formation at *http://s-prs.co/v5776117*.

10.4.2   Leveraging Cloud-Native Data Integration Patterns

Data sources can have their own data models (relational, columnar, key-pair, document, graph, etc.) Moreover, every data source is structured to serve a defined set of activities optimally (OLTP, OLAP, object stores, event stores, data caches, etc.,). In addition, the applications built on top of these data sources imposes specific schematic rules (column names, data types, record structure) that must be adhere to while dealing with the underlying data. Furthermore, these data sources will only expose their data through limited interfaces to the external world. Then, these interfaces have their own operating limitations in generating the extracts in specific data formats (raw data, CSV, JSON, SML, Parquet, etc.). Thus, the data integration is the core concern that one must address all along the data pipeline. Once the data integration is addressed well, it ensures that the right set of data is available at the right time in the required format at the right place.

As shown in Figure 10.21, some data sources have several extraction methods available (e.g., SAP S/4HANA and SAP BW/4HANA) whereas select data sources may only have exclusive extraction mechanisms (e.g., SAP's SaaS solutions) to integrate their data into SAP analytics data pipelines.

Note

We did not show all the possible options available in each phase of the data pipeline shown in Figure 10.21. For example, AWS offers several other data storage options and AI/machine learning services that are not included.

In addition, some SAP BTP solutions like SAP Data Intelligence and SAP Datasphere are complete end-to-end data management solutions with their built-in data extraction, data enrichment, data modeling, and data cataloging capabilities. However, you can supplement these SAP BTP capabilities with AWS data integration solutions to integrate data sources like IoT events, social media streams, non-SAP SaaS solutions and other non-SAP databases in a more scalable and cost-effective way.

Moreover, you can benefit from the wide range of AWS AI/machine learning services by integrating SAP data sources either through SAP BTP data stores (e.g., SAP HANA Cloud or SAP Datasphere) or through AWS data stores like Amazon S3, Amazon Redshift, and other purpose-built cloud databases. In addition, SAP BTP data stores and

AWS data stores can federate each other's data through defined data connectors; thus, you can leverage the data visualization tools or AI/machine learning services from either place even though the data resides in different data territories. Based on the combinations shown earlier in Figure 10.21, you can loosely pursue several data integration patterns to build your analytics story in AWS.

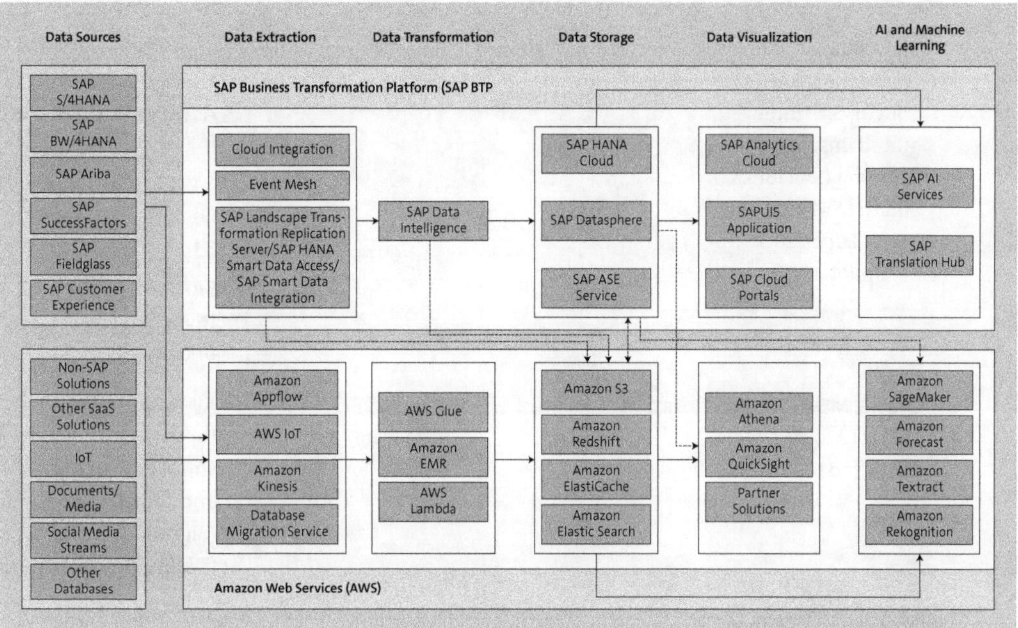

Figure 10.21 SAP Data Integration Patterns with SAP BTP and AWS

The options listed in Table 10.2 are a feeble attempt to present all the possible combinations available at the time of writing. This information is rapidly changing and thus can quickly become obsolete. Refer to the documentation for the relevant service for any new alternative integration options.

Source Systems	Extraction Layer	Extraction Method	Extraction Tool/ Service
SAP S/4HANA, SAP BW/4HANA, SAP ERP, and SAP Net-Weaver (ABAP)	Web services layer	ODP/OData	Amazon AppFlow
	Application layer	SOAP, RFC, REST API, and AMQP	AWS Glue, AWS Lambda, and Amazon EventBridge
	Database layer	ODBC/JDBC	AWS Glue, AWS Lambda, SAP SLT, and SAP HANA smart data access

Table 10.2 Data Integration Options Using AWS Native Services

Source Systems	Extraction Layer	Extraction Method	Extraction Tool/ Service
SAP HANA database	Database layer	ODBC/JDBC	AWS Glue, AWS Lambda, SAP SLT, and SAP HANA smart data access
SAP Ariba, SAP SuccessFactors, SAP Concur, SAP Integrated Business Planning (SAP IBP), and the SAP Customer Experience portfolio*	Web services layer	REST APIs and AMQP	Cloud Integration: Open API, SAP Advance Event Mesh (AEM) SAP CPI (Open Connectors) to Amazon S3 SAP AEM (webhook) to Amazon EventBridge
Oracle, MS-SQL, IBM Db2 LUW, IBM Db2 z/OS, and SAP ASE	Database layer	ODBC/JDBC	AWS Glue, AWS Lambda, SAP HANA smart data access, and AWS Database Migration Service (DMS)
MySQL, PostgreSQL, Amazon Redshift, Amazon Aurora, and Amazon S3	Database layer	ODBC/JDBC	AWS Glue, AWS Lambda, and AWS DMS
MongoDB, Amazon DocumentDB, and MariaDB	Database layer	ODBC/JDBC	AWS DMS
File transfers	Infrastructure layer	SFTP/FTPS/FTP/REST API	AWS DataSync, Amazon File Gateway, and AWS Transfer Family
ServiceNow, Salesforce, Slack Zendesk, Microsoft Teams, Microsoft SharePoint, Instagram, and Infor	Web services layer	OData/REST API	Amazon AppFlow
	Application layer	AMQP	Amazon EventBridge

Table 10.2 Data Integration Options Using AWS Native Services (Cont.)

Source Systems	Extraction Layer	Extraction Method	Extraction Tool/ Service
Event streams, log streams, mobile and app sensors, video streams, Apache Flink, Fluentd, Debezium, Oracle Golden Gate, Kafka Connect, and Striim	Application layer	Data stream APIs, agents, producer libraries	Amazon Kinesis
IoT events	Application layer	MQTT	AWS IoT

* This includes all SAP Customer Experience clouds (SAP Marketing Cloud, SAP Commerce Cloud, SAP Customer Data Cloud, SAP Sales Cloud, and SAP Service Cloud)

Table 10.2 Data Integration Options Using AWS Native Services (Cont.)

One important aspect while choosing the right tool/service for data extraction is whether you need data in batch mode or in (nearly) real time. Although several services can extract data in dual mode, none are efficient enough to serve both scenarios equally. For example, Amazon AppFlow can extract data both in batch mode and in (nearly) real time. However, AWS Glue or AWS Lambda are optimal for getting data in batches.

Similarly, some data sources are not optimally designed to expose data changes in real time. For example, SAP Ariba Open APIs are best suited for asynchronous API requests over synchronous API calls.

10.4.3   Leveraging Cloud Database Models

SAP solutions are certified to run on predefined commercial database platforms. Thus, you can only run them as customer-managed databases deployed on Amazon EC2 instances. Your choices are limited and cannot deviate from the compatibility scope described in SAP's Product Availability Matrix. However, when you build side-by-side extensions for SAP solutions, you have the liberty of adopting modern database models in your microservice architectures.

The needs of modern application design are entirely different than traditional application architectures. Traditional architectures are designed for a fixed number of users and impose rigid constraints on its scalability. As monoliths, they can only scale vertically, and each scaling event incurs downtime. Data persistency is centralized and, therefore, is also a single point of failure from a reliability perspective. In addition, every application has a specific data model upon which the application runs optimally.

However, adopting specific commercial databases for each database model poses several inherent challenges.

Some additional challenges include the following:

- The TCO is exceedingly high.

- They have rigid licensing terms to restrict the freedom of your growth.

- They are built on proprietary mechanisms that limit your application's portability and lead to lock-in.

- Data integration is either controlled or made expensive by upselling the required connectors.

- You need to handle infrastructure setup, capacity planning, performance, scaling and high availability (HA); thus, you'll need an army of experts to take care of these aspects.

AWS offers a rich set of open-source, purpose-built database offerings to address these challenges. All the management aspects of the underlying infrastructure, capacity, scaling, and availability are entirely taken care of by AWS. You should choose the right database based on your application data model and your performance requirements. To provide a high-level understanding of these offerings and where they fit appropriately to reap the maximum benefits, we'll cover cloud database models in this section. You can categorize the AWS offerings into three segments: relational databases, non-relational databases, and graph databases, as we'll discuss in the following sections.

Relational Databases

As the name suggests, the data is distributed into multiple tables, and these tables maintain the relationships among various data points to provide the required view of the data. For example, you can build a billing data model as a collection of tables perhaps called `customers`, `sales`, `items`, `invoices`, and `payments`.

A relational database follows the many-to-one relationship model, meaning many customers are part of a single `customer` table, but each customer resides in a unique row. These databases follow a rigid schema that the application run atop must adhere to. This data model has several advantages apart from reusing the incumbent skill pool. As the connected SAP systems run on relational databases, it becomes easy to integrate that data and maintain the resembling data relationships with them.

However, as mentioned earlier in Section 10.4, relational data models take away from modern application design's scalability and flexibility tenets. Adopt these database models when your data has a long list of related attributes, and you'll have design-time clarity on their data access patterns:

- **Amazon Aurora database**
  AWS-managed and compatible with MySQL and PostgreSQL, this relational cloud database engine is architected to provide higher performance in the tune of 3 times

over PostgreSQL and 5 times over MySQL over similar infrastructures setups. Amazon Aurora leverages a multi-AZ distributed, self-healing storage system that can scale automatically at increments of 10 GB with a maximum of 128 TB per database instance.

You can set up to 15 low-latency read-replicas. Using the Database Backtrack mechanism, you can move back and forth between different time states of the database and recover from accidental data deletions without the need of setting up an entire forensics system in parallel. The Zero Downtime Patching (ZDP) feature of Amazon Aurora significantly reduces the maintenance windows, thus improving the overall application availability. With its Global Database option, you can spread its read-replicas to multiple AWS regions, thus reducing the latency of read requests across user geographies.

Common use cases: You can migrate any application designed to run on a relational data model to Amazon Aurora. You can utilize AWS Database Migration Service (DMS) along with AWS Schema Conversion Tool (SCT) to refactor your high-cost commercial database application to Amazon Aurora. In addition, you can use Amazon Aurora as a database layer for an SAP BTP, Cloud Foundry environment, application you've developed.

- **Amazon Relational Database Service (RDS)**
  Providing AWS-managed, commercial, and open-source relational databases, AWS takes care of provisioning, scaling, availability, monitoring, backup, and maintenance. Thus, you can utilize your database administration personnel in application development efforts instead. MySQL, PostgreSQL, MariaDB, Microsoft SQL Server, and Oracle are databases managed under Amazon RDS.

  Common use cases: You can migrate your legacy, non-SAP application as well as SAP extensions built on these database engines. You can achieve AWS migration with improved reliability but without the need to refactor your underlying applications.

Keep in mind these access patterns warrant navigating through the relationships among the table attributes with complex joins and indexes.

> **Difference between Amazon Aurora and Amazon RDS**
>
> Both are AWS-managed offerings. However, Amazon Aurora is a refined, feature-rich cloud database engine. Due to its cloud-native design, it offers an entirely different set of features that are not available for the comparable database engine under Amazon RDS. Multi-AZ storage replication, fast crash recovery, lower replication lags, and the number of read-replicas possible with Amazon Aurora are significantly better than the comparable database engines in Amazon RDS. However, these features come with an additional cost. For a more detailed comparison, refer to *http://s-prs.co/v5776118*.

Non-Relational Databases

The core data model in this genre of database is to self-contain the whole set of information into a single tuple. We call this tuple a "document." This document has all related attributes packaged together so that an application can consume this information in a single fetch. In this context, note that "non-relational" does not mean that no relations exist across the tables in the database. Instead, the application design relies only minimally on those relations. Moreover, these database engines are not designed to efficiently perform complex joins across the tables although technically achievable. For example, you can build a billing data model into a single table; let's call this table `billing_detail`.

This data model follows the one-to-many relationship model, meaning one document in the `billing_detail` table contains many related attributes devoid of a traditional normalization burden. These databases follow an implicit schema either loosely imposed at the database level or handled in the application code. This data model has several advantages apart from simplifying the database design and lowering application development overhead. These databases can grow virtually infinitely in size without sacrificing on performance KPIs. They can scale horizontally without causing any disruptions in application function.

However, the adoption of these non-relational data models poses significant challenges especially when applications/data are moving from relational data models. They warrant a complete refactoring/transformation into the target data model. Thus, consider adopting these database models when your application has a bounded context, and you want to represent this context with the underlying data model. In addition, this data model is best suited when you do not have design-time clarity on its future growth and predictability over its access patterns.

AWS offers a growing list of fully managed non-relational databases. Some important ones include the following:

- **Amazon DynamoDB**
  This key-value and document database is designed to deliver millisecond performance at any scale. Following a serverless database model with built-in security and management, Amazon DynamoDB can handle more than 20 million requests per second.

  Common use cases: Mobile gaming, advertising technology, shopping carts and inventory tracking in retail industries.

- **Amazon DocumentDB**
  This MongoDB-compatible document database has been redesigned to deliver 2 times the performance over standard MongoDB database. Amazon DocumentDB is designed for 99.99% availability, and it replicates six copies of your data across three AZs.

Common use cases: User profile management, real-time big data analytics over vast operational data, enterprise content management solutions.

- **Amazon ElastiCache**
  A Redis- or Memcached-compatible in-memory caching service, Amazon Elasti-Cache for Redis is a fast in-memory data store that provides sub-millisecond latency to power internet-scale real-time applications. Amazon ElastiCache for Memcached is an in-memory key-value store service that can be used as a cache or a data store to reduce read/write latency of highly performant workloads.

  Common use cases: You can place this cache in front of slow-performing data stores to improve read/write latency. However, keep in mind that Amazon ElastiCache is not ideal for data with a high change rate or a low tolerance for staleness.

- **Amazon MemoryDB for Redis**
  A Redis compatible in-memory database service, you can leverage its in-memory database capabilities to build applications on an SAP HANA database. This service is capable of serving 160 million requests a second and is highly scalable (able to scale more than 100 TB of storage per cluster).

  Common use cases: Fraud detection, e-commerce, enterprise search, mobile apps, and gaming.

- **Amazon Timestream**
  ideal for storing, processing, and analyzing time-series data, Amazon Timestream provides immutable, append-only data storage with advanced capabilities for statistical data analysis techniques (e.g., smoothing, approximation, and interpolation) to compensate for missing and outlier data points.

  Common use cases: Application monitoring, fraud detection, predictive analytics and IoT

- **Amazon Quantum Ledger Database (QLDB)**
  For blockchain applications, this database offers an immutable data store with a cryptographically verifiable log of data changes. Amazon QLDB is designed to scale automatically.

  Common use cases: Banking and financial sector for storing transactions, reconcile supply chain interactions across partners and suppliers, enterprise data management system for critical business documents.

- **Amazon Keyspaces**
  An Apache Cassandra-compatible database service, Amazon Keyspaces is designed for large-scale applications that need fast read and write performance.

  Common use cases: User profiles, device metadata management, event logging.

Graph Databases

This data model follows many-to-many relationships, meaning, every attribute in the database can potentially relates to every other attribute in the database. Thus, this data model follows a different data storing models to facilitate access from one attribute through the reference of another attribute. We call the attribute as edge and the connection between them as vertices in graph model. This data model helps to traverse through the complex connections among all the edges in the database.

Amazon Neptune is a fully-managed graph database service, ideal for applications with complex, highly connected datasets. This service supports diverse access patterns and boasts a high-performance engine optimized for handling billions of relationships with minimal latency. Common uses include recommendation engines, fraud detection, knowledge graphs, drug discovery, and network security.

[»]

Serverless Databases

The following databases are entirely serverless or a serverless variant exists:

- Amazon Aurora Serverless
- Amazon Redshift
- Amazon DynamoDB
- Amazon Timestream
- Amazon QLDB
- Amazon Keyspaces

10.5   Multi-Cloud Integration Patterns

Multi-cloud is imminent if not already happened. SAP is moving all their on-premise solutions towards independent SaaS solutions or SAP BTP extensions. At the same time, all the SAP ISV solutions are also moving either to their proprietary SaaS variants or offered as extensions through SAP BTP channels. Moreover, new enterprises want to tap into new data streams like big data, social media, and IoT devices to derive business outcomes. In addition, you're encouraged to build extensions in a wide-open world where you are at liberty to adopt new development environments and diverse data models. These possibilities for expansion pose new integration challenges that didn't exist in our earlier monolithic, on-premise world. However, SAP is offering a governance framework to address these challenges.

SAP Integration Solution Advisory Methodology is a methodology for enterprise architects to define, document, and govern enterprise-wide integration strategies. This methodology provides a consistent and transparent way to handle integrations across the enterprise entities (people, process, things, data, platforms, and organizations).

As shown in Figure 10.22, you can develop this methodology in three simple steps: First, define the integration domain based on the platform upon which the solution resides and its organizational boundary. Our experience indicates that organizations tend to adopt different sets of governance standards or integration mechanisms when they communicate with external or government entities. Thus, an organizational boundary is added as an additional dimension to the standard SAP Integration Solution Advisory Methodology taxonomy.

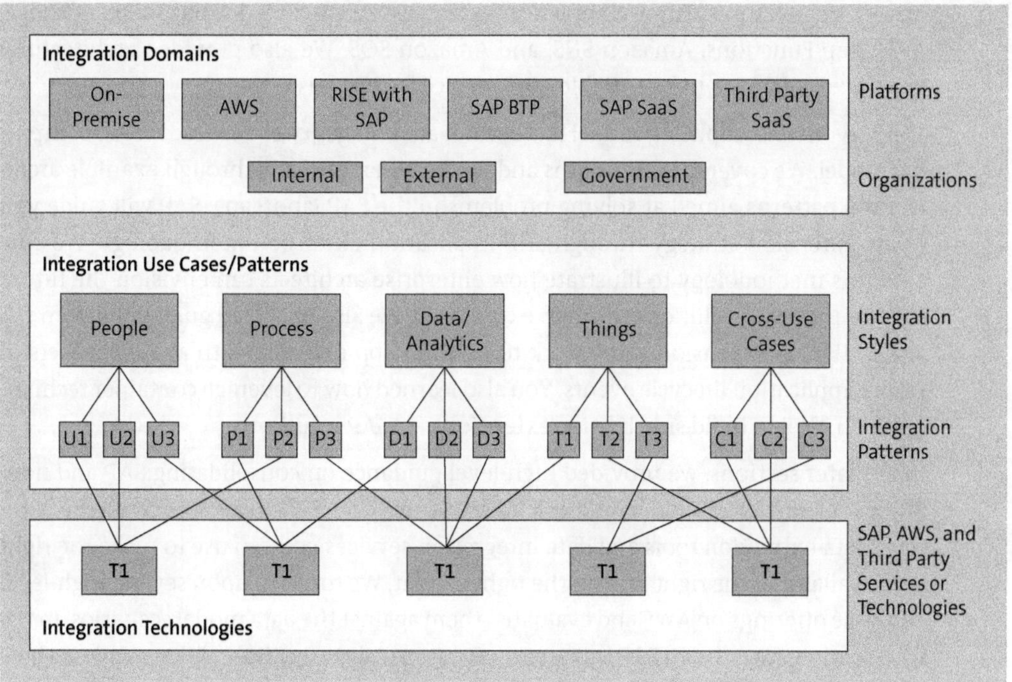

Figure 10.22 SAP Integration Solution Advisory Methodology

Second, you can classify the integration patterns based on their integration styles. You can define these styles based on the participating actors in the communication. These actors are divided into people (employee or customer); processes; data objects; and things (devices, machines, or robots). Integration patterns emerge based on the communication methods across the integration styles. These integration patterns are defined in a technology-agnostic vernacular.

Third, these integration patterns are mapped to the possible technology options to facilitate the required integration outcomes. You can leverage SAP Integration Solution Advisory Methodology templates to maintain these patterns under this methodology. For more information on SAP's latest integration strategy, refer to *http://s-prs.co/ v5776119*.

10.6   Summary

In this chapter, we covered why SAP transformation is required and how you could transform the SAP value chain with promising new-generation technologies. You can use different architectural patterns to create side-by-side extensions on AWS. We covered different ways of integrating SAP events into AWS infrastructures to build event-driven architectures. Through practical examples, we discussed Amazon EventBridge, AWS Step Functions, Amazon SNS, and Amazon SQS. We also provided architectural guidance to decide on the suitable AWS service for your use case.

The serverless computing model has several advantages over the traditional computing model. We covered its strengths and weakness extensively through example architectural patterns aimed at solving problems in the SAP landscape. SAP will guide you on its extension strategy through SAP Application Extension Methodology. We covered this methodology to illustrate how enterprise architects can envision the future state of their SAP solutions through extensions. We also provided guidelines from the SAP S/4HANA extension framework to build in-app extensions to avoid conflicts in future application lifecycle events. You also learned how to leverage container technologies on AWS to build side-by-side extensions for SAP solutions.

In the later sections, we provided high-level guidance on consolidating SAP and non-SAP data sources to build data mesh architectures on AWS. We also covered the common data extraction tools and data integration services you can use to make the right data available at the right time in the right format. We touched upon several high-level database offerings on AWS and evaluated them against the data model scenarios. In the final section, we covered SAP Integration Solution Advisory Methodology and saw how this methodology can help enterprise architects govern organization-wide integration strategies across integration domains.

Chapter 11
What's Next?

Migrating your SAP workloads from on-premise to Amazon Web Service (AWS) marks a pivotal first step in your digital transformation journey. This strategic move grants access to a suite of next-generation IT technologies and services, transforming your value chains into more agile and resilient systems. Migrating to AWS lays the foundation for building an adaptive enterprise equipped to thrive in today's dynamic business landscape. In this chapter, we'll provide a high-level overview on the myriad opportunities available on AWS to further your digital transformation journey.

Now that you've made your move to AWS, the most pertinent question is "What's next?" Where can you go from here? In this chapter, we'll briefly explore a wide array of possibilities that you can consider help your SAP architectures evolve and pave the way for building next-generation SAP solutions. These possibilities are far from exhaustive, and, moreover, the list evolves at a pace faster than anyone can claim to keep up with. Therefore, we hope you consider the following recommendations as guidance rather than concrete next steps.

The upcoming sections will cover SAP modernization patterns, focusing on empowering end users with self-service IT, enhancing SAP processes using artificial intelligence (AI)/machine learning, integrating automation for autonomous operations with less manual effort, and the role of no-code/low-code platforms in enabling business teams to self-manage their development and integration needs.

11.1 SAP Modernization

You can modernize SAP in several ways. We briefly discussed some opportunities, like event-driven architectures, side-by-side extensions, in-app extensions, building data lakes, and developing microservices in Chapter 10. Additionally, we discussed the automation of SAP system operations on AWS and the establishment of DevOps practices with AWS native services in Chapter 8. In this section, we'll cover a few modernization patterns evolving in the SAP community, promising significant impact on future SAP architectures.

11.1.1   Integrate Everything

Innovation is all about breaking barriers and visiting places where you've never gone before. This adventure involves three interesting aspects: First, doing it! Yes, innovation is the first step, not the last step as you would have conceived in a traditional waterfall model. However, you innovate in isolation, kind of like "dipping your toes." The next step is integrating with core capabilities to extend them with this new experience. Then, finally, you'll scale capabilities to your demand. Out of these three steps, you shouldn't worry about the first and final last steps, which are inherent characteristics of cloud computing and are sort of a given on AWS. (We're talking in relative terms in this context. Otherwise, they do have their own set of challenges in handling them.)

We're left then with the more critical step of integration. If not performed correctly, this integration creates a separate silo and ends up being an anti-innovation pattern. Thus, you must be aware of your integration paths with AWS native services before you venture into anything cloud native on AWS. In this section, we'll try to address this challenge and explore the integration paths that exist between SAP applications and AWS services, as shown in Figure 11.1.

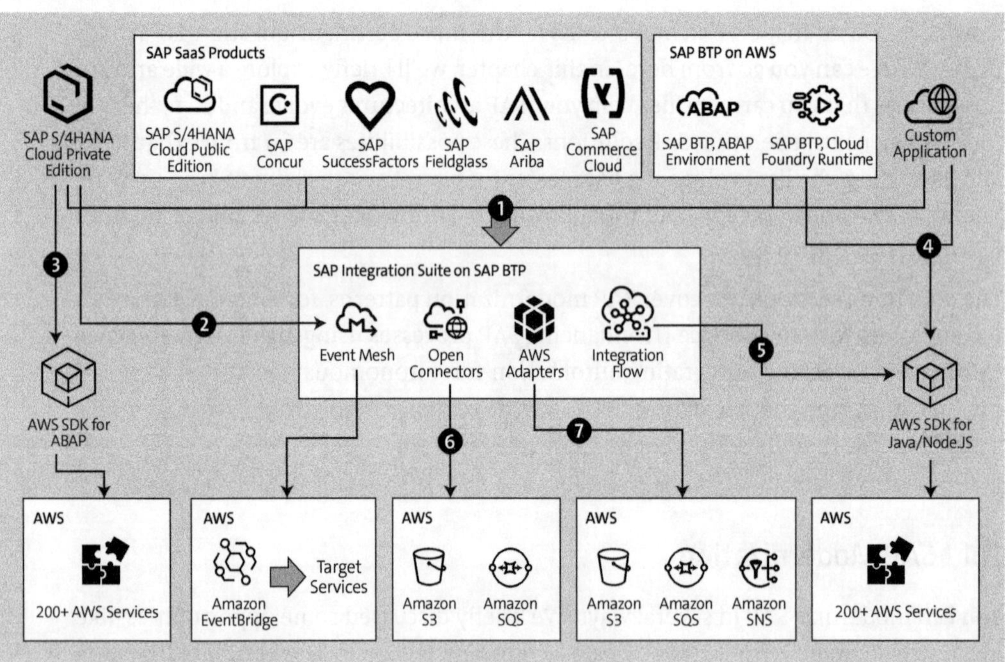

Figure 11.1 Application Integration Options for SAP on AWS

All the integration paths shown in Figure 11.1 can be topics of entire books themselves. However, due to limited space, we'll only discuss some basic premises and provide some related pointers for you to explore further.

Let's get into the integration patterns:

❶ Cloud Integration is a one-stop shop for all integration scenarios in the world of SAP. This solution includes prebuilt integration content, adapters, plugins, and application programming interfaces (APIs) to manage and orchestrate integrations across all SAP applications. Also included are several open-source plugins to let your custom applications make use of it as well. Although all SAP applications can integrate with Cloud Integration, only a few onward integration paths exist for AWS. We discuss them in the last three items of this list.

❷ At the time of writing, some SAP systems, SAP S/4HANA, and a few other software as a service (SaaS) applications can emit their business events to SAP Event Mesh, but this list is growing rapidly. We discussed this option under the event-driven architecture pattern in Chapter 10. Using the webhook mechanism, you can forward these events to Amazon EventBridge. Later, Amazon EventBridge can integrate with other AWS services to consume these business events from the source SAP systems.

❸ Another path is through a software development kit (SDK). The AWS SDK for ABAP is a straightforward way to call AWS services directly from SAP S/4HANA systems programmatically using ABAP. With this option, you can call nearly 200+ AWS services from an SAP S/4HANA system. We discussed this topic in a more detail in Chapter 8.

❹ SAP Business Technology Platform (SAP BTP) applications built on the Cloud Foundry runtime and other custom applications built on programming languages for which an AWS SDK exists can leverage the AWS SDK and interact with 200+ AWS services as SAP S/4HANA does with ABAP. Visit *https://aws.amazon.com/developer/tools/* for more information about AWS SDKs.

❺ SAP BTP integration flows can model communication flows from start to end. In these flows, you can add steps that can be scripted in Apache Groovy and JavaScript. Apache Groovy is a Java-syntax-compatible object-oriented programming language for the Java platform. Thus, you can create steps in Java/JavaScript by incorporating their respective AWS SDK libraries to call AWS services within them. An example scenario is discussed at *http://s-prs.co/v577612O*, which can help you gain a high-level understanding of scripting steps in integration flow.

❻ If not SDKs, you can use prebuilt open connectors to connect with Amazon Simple Storage Service (S3) and Amazon Simple Queue Service (SQS). For now, only these two connectors are natively available; for more information, see *http://s-prs.co/v5776121*. Watch this link because this list may grow in the future. Alternatively, you can build your own Open Connectors using the framework defined at *http://s-prs.co/v5776122*. However, building a custom connector for each service is not a scalable model. This approach becomes unmanageable when you need to ping-pong between numerous AWS services.

❼ Another native option is to leverage AWS adapters to integrate into integration flow. However, the relatively small number of AWS adapters limit our integration to a handful of AWS services. Refer to the list at *http://s-prs.co/v5776123*. More information on using these adapters in the SAP BTP integration context is available at *http://s-prs.co/v5776124*.

Note

These options are popular integration choices at the time of writing. In some edge scenarios, you can achieve integration in different ways that we won't mention, for brevity. As integration continues to gain traction, you might find more integration options in the future.

11.1.2   Consume Data from Everywhere

Another way to break silos is to gain the ability to consume everything being generated in your enterprise. Consuming structured data is SAP's forte, but when it comes to semi-structured and unstructured data sources, its capabilities are not mature as those of AWS. In addition, the virtually infinite scalability and pay-as-you-consume model makes AWS stand apart in this space. Therefore, in this section, we'll explore some data integration options possible for SAP workloads on AWS, as shown in Figure 11.2.

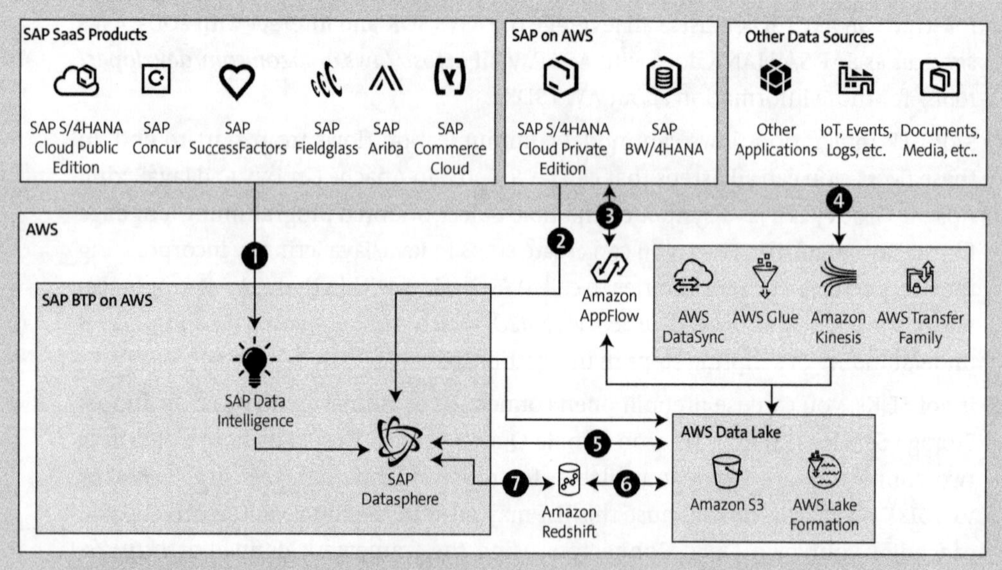

Figure 11.2 Data Integration Options for SAP on AWS

Let's look at the options shown in Figure 11.2 in more detail:

❶ SAP Data Intelligence is a powerful tool capable of extracting business objects data from SAP SaaS products. This solution can push this data to an AWS data lake and to

SAP Datasphere. Some of the capabilities of SAP Data Intelligence are slowly converging with SAP Datasphere. However, at the time of writing, SAP Data Intelligence is still required to bring data from SAP SaaS products. Though not shown, SAP Data Intelligence can extract data from SAP S/4HANA, SAP BW/4HANA, and other structured data from non-SAP systems also.

❷ SAP Datasphere offers native extraction capabilities to directly extract structured data (both business and technical objects) from SAP S/4HANA and SAP BW/4HANA systems.

❸ Amazon AppFlow is a low-code/no-code data integration tool to extract data from SAP S/4HANA and SAP BW/4 HANA systems using the OData fabric. This solution has near-real-time capability to extract data (both full data sets and delta data) using Operational Data Provisioning (ODP) extractors. A write-back feature can also send data into SAP S/4HANA and SAP BW/4HANA systems.

❹ You can consume structured data from non-SAP systems using AWS Glue. This extract, transform, and load (ETL) tool can carry out transformations on your data before sending it to target systems. You can use Amazon Kinesis to capture streaming data from sources like the Internet of Things (IoT), logs, and events. Similarly, you can use AWS Transfer Family and AWS DataSync to transfer semi-structured business documents and emails as well as unstructured objects like images and media files.

❺ A back-and-forth data channel exists between Amazon S3 and SAP Datasphere. Thus, data can transition between these solutions.

❻ and ❼ Similarly, Amazon Redshift can communicate with both SAP Datasphere and AWS data lakes. Additionally, you can leverage the federated query feature to offload query processing to your AWS data lakes.

> **Note**
>
> The data integration space is congested with a plethora of data extraction and management tools. In this chapter, we want to present SAP and AWS native capabilities only. Otherwise, several third-party tools or adapters can perform specific aspects of data integration better than what we mentioned here. Evaluate your options based on your use case, cost, and scalability. Alternatively, you can also build your own data extraction adapter using AWS Lambda.

11.1.3   The Future Big Picture of SAP

Figure 11.3 shows the inevitable future state of an SAP landscape. To manage these disparate systems across different infrastructure domains, having varying governance frameworks for communication, you must have a robust application and data integration regime, as shown in Figure 11.3, on the left. Otherwise, you'll be dealing a nightmare—an unmanageable web of spaghetti integrations, as shown in Figure 11.3, on the right.

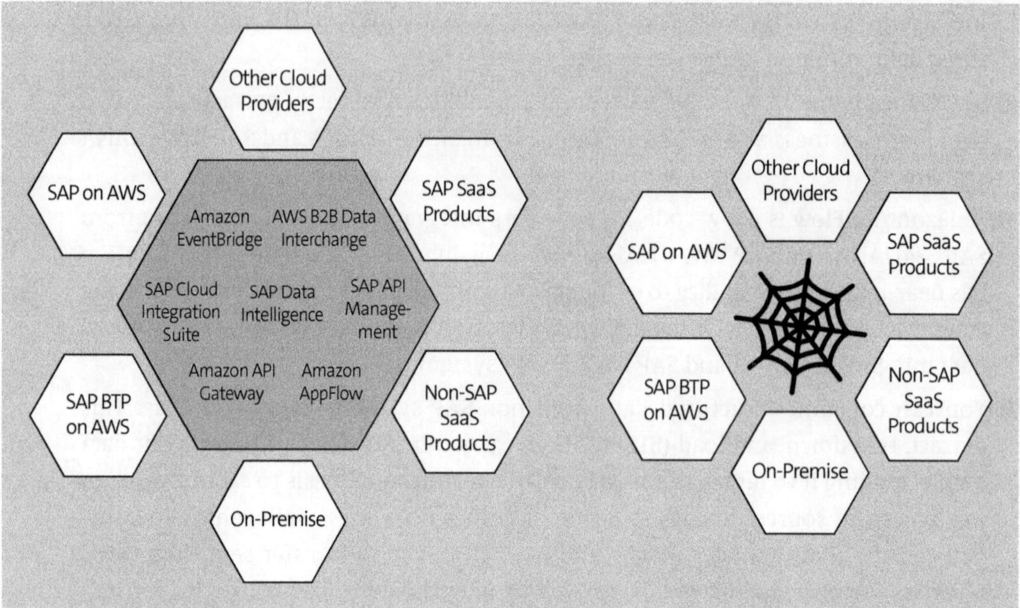

Figure 11.3 The Future Big Picture of SAP Architecture

11.2 Self-Service IT

"The most radical and transformative of inventions are often those that empower others to unleash their creativity—to pursue their dreams." — Jeff Bezos, Amazon Founder

This quote sums it all: The highest level of maturity or the state of perfection in any field is giving consumers/receivers the freedom to choose and reap good outcomes. Thus, we can reflect on a similar philosophy to empower our end users to serve themselves at the time of their choosing.

Among several approaches, the following methods are good starting points:

- **Self-service analytics**
  The democratization of data is easier said than done. Data is the most critical asset in an organization, and if not handled well, bad data can lead to irreparable damage. Thus, to enable self-service activities on your data, you must have a robust data governance and classification mechanism in place, which can be enabled through Amazon DataZone.

  Amazon DataZone is a data management service that provides a quick way to find data inventory, catalog data objects, enable data governance controls, and facilitate data sharing with users in a secure manner. This service comes with a unified data management portal to perform all these activities in a single place, based on the personas of team members. This solution also provides a collaborative space where users can interact with data project teams to gain the required access and control over the data.

Amazon DataZone can work with data assets from Amazon Redshift, AWS Lake Formation, and other data sources, making the data available to users based on established governance controls.

In summary, Amazon DataZone enables users to search, request access, and make this data available to data analysis tools without compromising your organization's data protection rules. A demo of Amazon DataZone that explains an end-to-end use case of enabling self-service analytics is available at *http://s-prs.co/v5776125*.

- **Self-service provisioning**
  You can enable self-service provisioning access for enterprise cloud products using AWS Service Catalog. Once you publish approved cloud products, authorized users can access these products and launch them in their accounts. This service contains the following components:

  - **Portfolio**
    A collection of products that you want to make available to your organization. For SAP, you might have a portfolio specifically designed for SAP-related resources.

  - **Product**
    Represents an AWS resource or a set of resources that you want to make available to users. For SAP, a product could be an SAP HANA instance, an Amazon Elastic Compute Cloud (EC2) instance configured for SAP, or any other SAP-related automation template or script resource. Moreover, this service also supports Terraform open-source products for automation.

  - **Provisioning constraints**
    You can define constraints and rules for each product to ensure compliance and governance. You can also define service actions to associate specific automation actions during product launch. For instance, you can specify that only certain IAM roles or security groups can be associated with SAP instances. You can associate specific AWS Systems Manager documents as service actions to prepare your Amazon EC2 instance for SAP-related approved configuration steps.

  - **Launch constraints**
    Rules that specify how many instances of a product can be launched, in what regions, and with what configuration settings.

  - **IAM roles and permissions**
    Define IAM roles and the permissions necessary for users to launch and manage SAP-related products.

  You can also monitor usage and costs based on product tags. You can even control and manage a product's availability and its versions. Additionally, this service can integrate with Information Technology Infrastructure Library (ITIL) tools like ServiceNow and Jira to automate provisioning requests through service tickets.

11.3   Artificial Intelligence/Machine Learning-Powered SAP Value Chains

In several ways, you can empower your SAP business processes to automate tedious steps or extend the current business functions to enhance them with AI/machine learning-powered services. The following sections provide some examples to help you understand the pattern.

11.3.1   Overview

You can power your SAP value chains with AI/machine learning models in the following three ways:

- You can build your own model based on your business data and your requirements. You can host them and derive inferences from it. You can leverage Amazon Sage-Maker to run the complete MLOps pipeline.

- You can also host a pretrained model for a specific use case and make inferences by feeding your data into it. You can find these pretrained machine learning models in the AWS Marketplace. You can use Amazon SageMaker to deploy and manage these models. You can build your own solution using AWS Lambda, Amazon API Gateway, and Amazon SageMaker Endpoints to interact with the deployed model. Visit *http://s-prs.co/v5776126* for more information on pre-trained machine learning models.

- Alternatively, you can make use of a pre-built AI/machine learning solution without the overhead of hosting and management. You can directly call the solution's API and integrate them into your SAP value chains. AWS provides a rich set of pre-built AI/machine learning services that you can readily use in your business processes. This list is growing every day. Table 11.1 lists some of the popular examples.

AI/Machine Learning Service	Use Case	Example
Amazon Comprehend	Text analysis, sentiment analysis, language detection, and entity recognition	Analyze customer reviews to understand sentiment and extract key information from unstructured text data
Amazon Rekognition	Image and video analysis, object detection, facial recognition, and content moderation	Automatically tag and categorize images, detect inappropriate content, or identify faces in photos or videos

Table 11.1 AWS-Native Prebuilt AI/Machine Learning Solutions

AI/Machine Learning Service	Use Case	Example
Amazon Translate	Language translation and localization	Translate customer inquiries or product descriptions into multiple languages to reach a global audience
Amazon Polly	Text-to-speech conversion with natural-sounding voices	Generate voiceovers for multimedia content, such as videos or podcasts
Amazon Transcribe	Automatic speech recognition (ASR) for converting spoken language into written text	Transcribe customer support calls or interviews for further analysis
Amazon Textract	Optical character recognition (OCR) for extracting text and structured data from scanned documents and images	Automate data entry from invoices, forms, or receipts into your business applications
Amazon Personalize	Recommendation engine for delivering personalized product or content recommendations	Enhance your e-commerce platform with personalized product recommendations for individual customers
Amazon Comprehend Medical	Medical text analysis for extracting insights from clinical notes, medical records, and research documents	Analyze medical records to identify trends, diagnoses, or adverse events
Amazon Fraud Detector	Fraud detection and prevention using machine learning models	Automatically detect and prevent fraudulent transactions or activities within your business processes
Amazon Kendra	Enterprise search service with natural language understanding (NLU) capabilities	Improve knowledge management by enabling employees to find relevant information quickly

Table 11.1 AWS-Native Prebuilt AI/Machine Learning Solutions (Cont.)

11.3.2   SAP Process Automation

SAP process automation using AI/machine learning can significantly enhance efficiency and decision-making within an SAP environment. Though the possibilities are vast, some examples include the following:

- **Invoice processing**
  You can implement AI/machine learning models to automatically extract information from invoices received in various formats (PDFs, images, etc.) within your SAP system.

Extracting Text in Python

Listing 11.1 is a simple Python code snippet using libraries like Tesseract and PyTesseract to extract text from images.

```python
import pytesseract
from PIL import Image

# Load the image
image = Image.open('invoice.png')

# Use Tesseract to extract text
extracted_text = pytesseract.image_to_string(image)

print(extracted_text)
```

Listing 11.1 Extracting Text from Documents/Images Using Python AI/Machine Learning Libraries

- **Predictive maintenance**
  Utilize machine learning to predict when SAP-related systems or equipment might fail based on historical data, thus reducing downtime.

Predictive Maintenance in Python

Use libraries like scikit-learn to build predictive maintenance models. For instance, a decision tree classifier, as shown in Listing 11.2.

```python
from sklearn.tree import DecisionTreeClassifier
from sklearn.model_selection import train_test_split

# Load historical data
X, y = load_data()

# Split data into training and testing sets
X_train, X_test, y_train, y_test = train_test_split(X, y, test_size=0.2)

# Create and train a decision tree classifier
clf = DecisionTreeClassifier()
clf.fit(X_train, y_train)
```

```
# Make predictions
predictions = clf.predict(X_test)
```

Listing 11.2 Building Predictive Maintenance Models Using scikit-learn Libraries

- **Chatbots for customer support**
  Implement a chatbot powered by natural language processing (NLP) to handle common customer queries related to SAP services on AWS.

Building a Chatbot in Python

Use a framework like Rasa to build a chatbot. Listing 11.3 is a simplified Python code snippet for a basic chatbot.

```python
from rasa.core import training
from rasa.core.agent import Agent

# Define intents, stories, and NLU data

# Train the agent
agent = Agent('domain.yml')
training_data = agent.load_data('data')
agent.train(training_data)

# Run the chatbot
response = agent.handle_text("How do I deploy SAP on AWS?")
print(response)
```

Listing 11.3 Building Chatbot Using Rasa Libraries

11.3.3   Business Process Extensions for SAP

Business process extensions using AI/machine learning for SAP on AWS can bring added intelligence and efficiency to your operations. Let's consider a few examples:

- **Inventory management optimization**
  Use AI/machine learning to predict demand patterns for products in your SAP-based inventory. This capability can help you optimize stock levels, reduce excess inventory, and minimize stockouts.

Time Series Forecasting Model in Python

Listing 11.4 is a simplified Python code example using a time series forecasting model with ARIMA (autoregressive integrated moving average).

```python
import pandas as pd
from statsmodels.tsa.arima_model import ARIMA
```

```
# Load historical inventory data
data = pd.read_csv('inventory_data.csv')

# Fit an ARIMA model
model = ARIMA(data['demand'], order=(5,1,0))
model_fit = model.fit(disp=0)

# Make future demand predictions
forecast = model_fit.forecast(steps=10)
```

Listing 11.4 Envisioning Inventory Status Using Machine Learning Forecasting Models

- **Customer churn prediction**
  Employ AI/machine learning models to predict which SAP services or products are more likely to be discontinued by customers, thus enabling proactive customer retention strategies.

Customer Churn Prediction Model in Python

Listing 11.5 is a Python code snippet using scikit-learn to build a customer churn prediction model with a logistic regression classifier.

```
from sklearn.model_selection import train_test_split
from sklearn.linear_model import LogisticRegression

# Load customer data and features
X, y = load_data()

# Split data into training and testing sets
X_train, X_test, y_train, y_test = train_test_split(X, y, test_size=0.2)

# Create and train a logistic regression classifier
clf = LogisticRegression()
clf.fit(X_train, y_train)

# Make predictions
predictions = clf.predict(X_test)
```

Listing 11.5 Predicting Customer Churn Using scikit-learn Libraries

- **Intelligent invoice approval workflow**
  Implement an AI-powered invoice approval system that uses NLP to extract key information from invoices and automatically route this information for approval within your SAP system on AWS.

Text Extraction in Python

You can use Python libraries like spaCy for NLP and Amazon Textract for document text extraction within an AWS Lambda function, as shown in Listing 11.6.

```python
import spacy
import boto3

# Load spaCy NLP model
nlp = spacy.load('en_core_web_sm')

# Process invoice text using spaCy
doc = nlp(invoice_text)

# Extract relevant information (e.g., invoice amount, date, vendor)
amount = extract_invoice_amount(doc)
date = extract_invoice_date(doc)
vendor = extract_invoice_vendor(doc)

# Submit the extracted data to your SAP workflow for approval
```

Listing 11.6 AWS Lambda Function to Extract Text from a Business Document in a Workflow

These examples demonstrate how AI/machine learning can be applied to extend and enhance SAP processes on AWS. By infusing intelligence into your business processes, you can make data-driven decisions, optimize your workflows, and ultimately improve the efficiency and effectiveness of your SAP ecosystem.

11.3.4   Generative Artificial Intelligence

With the emergence of generative AI, there is a significant increase in opportunities to enhance your business processes. Generative AI has the potential to fundamentally transform your business operations and how you utilize SAP in the coming years. However, unlike traditional AI/machine learning models, managing large language models (LLMs) in generative AI is a nuanced endeavor. Building LLMs like GPT-3 on AWS is a complex and resource-intensive process that typically demands substantial computational resources and expertise in machine learning and deep learning. Moreover, developing such models goes beyond the typical AWS use cases and may entail substantial costs and infrastructure investments.

Nevertheless, you can incorporate generative AI into your applications by following these approaches:

- **Training foundational models**
  Train foundational models specific to your requirements and use them to generate inferences with your data. Utilize AWS's purpose-built GPU chips like AWS Trainium for training these models and AWS Inferentia for running inferences on them.

- **Building tools and APIs**
  Develop tools using LLMs and foundational models, while managing and orchestrating interactions through APIs. Leverage Amazon Bedrock, a fully managed, serverless generative AI orchestration service, to seamlessly integrate LLMs and foundational models into your business workflow. Amazon Bedrock supports integration with a wide range of foundational models through APIs. For more details, visit *https://aws.amazon.com/bedrock/*.

- **Utilizing prebuilt solutions**
  AWS offers native solutions powered by LLMs and foundational models that you can readily adopt within your organization. Some notable solutions include:

  - *Code Whispers*: Enables developers to receive accurate code suggestions in popular integrated development kits (IDEs) and for various programming languages using natural language prompts. It supports customization, allowing you to train it on your custom repositories securely, thereby improving code reusability and adherence to development standards and governance frameworks. Learn more at *http://s-prs.co/v5776127*.

  - *Amazon Q*: A generative AI-powered assistant designed to assist you in your work. It comes pre-trained on AWS's knowledge base and can provide precise answers and guidance on AWS capabilities, services, and solutions. It helps you select the best AWS service for your use case, initiates your project, and connects to your business data and systems for tailored conversations, problem-solving, content generation, and relevant actions. Explore more at *https://aws.amazon.com/q/*.

These approaches allow you to leverage the potential of generative AI in your organization to enhance your business processes and operations.

11.4   Autonomous Operations and Autonomous Security

Besides your business processes, you can also incorporate AI/machine learning models into your operations to let systems take remediation actions through automation runbooks. In the following sections, we'll cover some areas where you can add AI/machine learning intelligence into the automation of your business operations.

11.4.1   Automating Incident Response

Automating incident response on AWS involves setting up processes and tools to detect, analyze, and respond to incidents without manual intervention. The basic anatomy of this process includes the following steps:

- **Create automation playbooks**

  Develop predefined incident response playbooks that outline the steps to be taken for specific types of incidents. You can prioritize your efforts based on historical incident data. Ensure these playbooks incorporate robust contexts, constraints, and security measures to minimize disruptions or unintended consequences from the outlined actions in the automation documents.

- **Detection**

  Implement Amazon CloudWatch alarms, AWS Config rules, or AWS CloudTrail for real-time monitoring of AWS resources and services. Define specific thresholds and rules that trigger alerts when anomalies or security breaches occur.

- **Evaluation**

  Upon receiving an alert, conduct initial triage to assess the severity and impact of the incident. Gather relevant context- and incident-related data.

- **Automate remediation**

  Utilize AWS Lambda functions or AWS Systems Manager automation documents to automate responses. For critical remediation actions, consider using AWS Step Functions to create complex decision steps. Implement automated remediation actions, such as service restarts, IP address blocking, or notification triggers.

- **Machine learning model training**

  Incorporate historical incident response data into the training of the machine learning model using supervised learning techniques. Continuously refine and improve responses over time to leverage the model's inferences during the evaluation stage.

11.4.2   Self-Healing Architecture

Self-healing architectures on AWS are designed to automatically detect and address system failures or issues without the need for manual intervention. While existing mechanisms are in place that can automatically handle some failures, these mechanisms are not SAP-aware and do not guarantee the availability of SAP applications.

Let's consider some use cases and see how we can improve them:

- **Amazon EC2 host auto-recovery**

  - Amazon EC2 host auto-recovery ensures that impaired Amazon EC2 instances are replaced on new hardware.

  - However, this feature doesn't automatically restart the SAP application running inside the Amazon EC2 instance and thus leaves the application in a failed state.

- Remediation:
  - Upon an auto-recovery event, Amazon EventBridge receives an event, allowing you to automatically trigger a subsequent automation sequence to start the SAP application in the recovered Amazon EC2 instance.
  - Create a local run-level startup script at the operating system (OS) level to trigger SAP start, with the specific implementation depending on your OS distribution.

- **Augmenting Pacemaker HA**
  - Pacemaker HA safeguards against single points of failure by failing over to a redundant application host in the cluster.
  - This solution initiates a failover when cluster components are in a failed state but doesn't respond if the application component is unresponsive.
  - For example, if storage I/O is suspended on an SAP HANA instance, causing the entire system to stall, high availability (HA) failover won't be triggered.
  - Remediation:
    - Monitor the situation and manually force Pacemaker to initiate a failover. However, this decision involves a multi-dimensional evaluation, considering system resource metrics, user experience patterns, and application process response service level agreements (SLAs).
    - Training a machine learning model on these aspects can reduce false positives. AWS Fault Injection Service (FIS) can be used to test assumptions and gather input data for the machine learning model. Learn more at *https://aws.amazon.com/fis/*.

- **Multi-availability zone HA setup**
  - In an ideal multi-availability zone (AZ) HA setup, application servers are evenly distributed across AZs to withstand AZ failures.
  - However, during an AZ failure, the failover setup may have only 50% of the application layer, potentially causing insufficient capacity for user traffic.
  - Remediation:
    - Build automation to trigger upon AZ failure after a successful HA failover. This automation can immediately redeploy the remaining 50% of the application instances in the target AZ to address user traffic.

- **Proactive autoscaling for SAP with machine learning models**
  - Traditional autoscaling doesn't inherently support SAP application scaling.
  - You can develop automation to track application server resource utilization and enable autoscaling for SAP application instances reactively.
  - However, this approach may result in a higher mean time to respond (MTTR), leading to potential service disruption for end users.

– Remediation:
  - Incorporate machine learning models trained on SAP resource consumption data to proactively adjust resource scaling in anticipation of demands. These models can also detect anomalies in SAP performance and trigger automated responses, thus reducing the impact of incidents on end users.

In summary, these examples illustrate some challenges in making systems self-healing and how automation and machine learning can be employed to address these challenges. A self-healing architecture continuously monitors SAP components, automates remediation actions, maintains resource states, scales resources as needed, and leverages AWS services to ensure the resilience and availability of the SAP system, ultimately reducing downtimes and enhancing reliability.

11.4.3 AIOps

In previous sections, we discussed the integration of AI/machine learning into SAP business processes and systems administration. Similarly, we can extend these capabilities to cloud-based operations by training AI/machine learning models on monitoring data from all the cloud services you use. This approach enables the correlation of service events, the identification of patterns, the analysis of resource utilization, and the initiation of remediation actions.

Some opportunities exist where AI/machine learning can enhance your cloud operations. For instance, you can direct observability data into an Amazon S3 bucket and use it to train AI/machine learning models. If you already have setups for observability data lakes and security data lakes, you can integrate AI/machine learning services, such as the following, to derive insights from these data sources:

- Amazon CloudWatch metrics, logs, traces, and application insights
- SAP application logs and traces, including SAP HANA trace files, SAP work process logs, and traces as well as Pacemaker logs
- AWS CloudTrail events
- Amazon GuardDuty threat detection findings
- Amazon Inspector scans
- AWS Security Hub Findings
- AWS Config findings
- AWS Trusted Advisor findings
- AWS Systems Manager inventory, AWS State Manager data, and AWS Application Manager data
- Any custom metrics, logs, or traces relevant to your environment

AIOps represents a shift from traditional IT operations to a more proactive and predictive approach. By harnessing AWS services, your organization can implement AIOps to automatically detect and resolve issues, predict outages before they occur, and optimize operations for improved performance and cost efficiency. These features not only enhance operational resilience but also can liberate your IT teams from routine firefighting so they can concentrate on strategic initiatives and innovation.

11.5   Model-Driven Architectures

A model-driven architecture (MDA) is an approach to software design and development that places a strong emphasis on using models to define various aspects of a system's architecture and behavior. The key concept in MDA is the separation of a system's functionality and implementation from the intricate details of specific platforms.

MDA relies on the utilization of standardized modeling languages and transformations to automatically generate code, configuration files, and other artifacts tailored to particular target platforms. The Object Management Group (OMG) is a prominent organization associated with MDA, promoting standardized modeling languages like the Unified Modeling Language (UML) and the MDA framework.

Alternatively, SAP Signavio is another tool that can be leveraged to create models using Business Process Model and Notation (BPMN) and other modeling languages.

MDA serves as a catalyst for fostering collaboration and coordination among various teams within an organization, including development teams, application architects, enterprise architects, and business architects. It streamlines the development process by enabling developers to concentrate on creating high-level models, which, in turn, can be employed to automatically generate code and configurations suitable for diverse SAP architectures and platforms. This approach not only ensures consistency but also diminishes the need for manual coding efforts, thereby enhancing the overall maintainability of intricate SAP systems.

Table 11.2 offers more specific examples of how an MDA framework can be applied in SAP and AWS.

In SAP	In AWS
SAP Process Orchestration (SAP PO) Converts BPMN models into integration scenarios.	**AWS Step Functions** Enables visual design to orchestrate complex workflow steps.

Table 11.2 MDA Scenarios in SAP and AWS

In SAP	In AWS
SAP Fiori UI development A UI designer can transform a visual model into SAPUI5 code.	**AWS Glue DataBrew** Create data transformation recipes as models rather than writing code.
SAP HANA data modeling Models in SAP HANA Studio/SAP HANA cockpit can be converted into SQL scripts and database artifacts.	**Amazon Textract custom models** Using template designs, you can instruct Amazon Textract on how to interpret specific document layouts.
Eclipse IDE You can convert UML models into ABAP code.	**Amazon SageMaker machine learning pipelines** You can use high-level abstractions and APIs to describe your machine learning pipelines.

Table 11.2 MDA Scenarios in SAP and AWS (Cont.)

11.6   No-Code Development

In the previous section, we discussed MDAs for generating code using a visual environment with prebuilt components and templates. This approach can also be loosely referred to as the low-code development paradigm.

However, no-code development platforms take simplification a step further by enabling users to build applications and workflows entirely without writing any code. These platforms typically offer a highly visual, drag-and-drop interface, making them accessible to non-developers.

No-code solutions from SAP include the following:

- **SAP Intelligent Robotic Process Automation (SAP Intelligent RPA):** This solution empowers business users to create and manage bots without any coding. Users can automate repetitive tasks by designing workflows through a visual interface.

- **SAP Business Workflow:** SAP's no-code workflow management system allows business users to define and customize approval workflows, notifications, and processes without the need for coding.

No-code solutions within AWS include AWS PartyRock, a no-code generative AI app-building platform backed by Amazon Bedrock. Users can create apps in just a few clicks through a web interface and easily share apps with their communities. More information about AWS PartyRock can be found at *https://partyrock.aws/*.

11.7 Summary

In this chapter, we delved into various opportunities that SAP workloads can pursue as the next steps in innovation on AWS. We started our discussion in this chapter to establishing a foundation for innovation.

Later in this chapter, our focus shifted to a comprehensive list of potential innovation areas for SAP on AWS. These areas encompass SAP modernization as the next logical steps, empowering end users through self-service analytics and provisioning capabilities. We also delved into optimizing SAP business processes, system administration, and cloud operations through integration with AI/machine learning models. Moreover, we explored methods to empower SAP business processes further with generative AI models.

Finally, we concluded this chapter by examining the availability of low-code/no-code development tools and services within both the SAP and AWS ecosystems, offering valuable insights into innovative approaches for application development and automation.

The Authors

Ravi Kashyap has been in SAP ecosystem his entire career and has focused on the cloud (private and public) over the past decade. He currently works for AWS as an SAP Solutions Architect. He is also the author of SAP PRESS book *SAP on Microsoft Azure*, where he shared best practices for migrating and managing SAP workloads in Azure. Having learned multiple clouds for SAP, throughout the book, he was conscious of presenting the information from a reader's point of view and not bogged down with making it everything for everyone.

When not working, Ravi likes to spend time with his family, listen to podcasts, and read books—his latest favorite is *The Alignment Problem: Machine Learning and Human Values*, which explores our biases and blind spots with AI.

Rajendra Narikimelli currently serves as an SAP Solutions Architect at AWS, where he aids SAP customers in harnessing the power of AWS to transition into cloud-native architectures. Prior to this role, he was a Services Delivery Architect in the MaxAttention CCoE at SAP America, conducting advisory workshops for large and premium SAP customers. With 22 years of IT experience, he has provided thought leadership to several multinational organizations globally, focusing on the management of complex SAP landscapes.

In his leisure time, Rajendra enjoys walking through the aisles of books in public libraries. He humorously notes that he is yet to find an official term for this passion!

With 17 years of hands-on experience in the dynamic realm of SAP, **Rozal Singh** brings a wealth of expertise encompassing the full spectrum of SAP system implementation, management, and operations. His proficiency extends to both cloud-based and on-premise SAP environments, positioning him as a versatile and seasoned professional in the SAP space. Before joining the AWS, Rozal distinguished himself as an SAP Technical Lead Architect and a migration specialist at a consulting company. His career is marked by a steadfast dedication to technology and a keen focus on developing robust architectural foundations to bring his customers' strategic visions to fruition.

Outside of work, Rozal can often be found indulging in the delightful offerings of his 2-year-old's ice cream shop, where the exclusive menu choice, as dictated by his daughter, is always chocolate ice cream. Beyond family ice cream adventures, Rozal is also an enthusiastic soccer fan who dedicates his few free weekend hours playing the sport at a local meetup group.

Index